Emerging Trends in Environmental Biotechnology

Emerging Trends in Environmental Biotechnology

Edited by
Sukanta Mondal, Shivesh Pratap Singh and
Yogendra Kumar Lahir

CRC Press
Taylor & Francis Group
Boca Raton London

CRC Press is an imprint of the
Taylor & Francis Group, an **informa** business

First edition published 2022
by CRC Press
6000 Broken Sound Parkway NW, Suite 300, Boca Raton, FL 33487–2742

and by CRC Press
2 Park Square, Milton Park, Abingdon, Oxon, OX14 4RN

CRC Press is an imprint of Taylor & Francis Group, LLC

Library of Congress Cataloging-in-Publication Data
Names: Mondal, Sukanta, editor. | Singh, Shivesh Pratap, 1963- editor. | Lahir, Yogendra Kumar, editor.
Title: Emerging trends in environmental biotechnology / edited by Sukanta Mondal, Shivesh Pratap Singh and Yogendra Kumar Lahir.
Description: First edition. | Boca Raton : CRC Press, 2022. | Includes bibliographical references and index.
Identifiers: LCCN 2021048854 (print) | LCCN 2021048855 (ebook) | ISBN 9781032030258 (hardback) | ISBN 9781032030289 (paperback) | ISBN 9781003186304 (ebook)
Subjects: LCSH: Bioremediation.
Classification: LCC TD192.5 .E53 2022 (print) | LCC TD192.5 (ebook) | DDC 628.5—dc23/eng/20220105
LC record available at https://lccn.loc.gov/2021048854
LC ebook record available at https://lccn.loc.gov/2021048855
ISBN: 978-1-032-03025-8 (hbk)
ISBN: 978-1-032-03028-9 (pbk)
ISBN: 978-1-003-18630-4 (ebk)

DOI: 10.1201/9781003186304

Typeset in Times
by Apex CoVantage, LLC

Dedicated to
Our beloved Parents and Families

for making everything worthwhile and cheerfully tolerating many hours of absence involved in writing the book

Contents

Preface

As we move into the 21st century, the environment faces considerable pressures like forecasting, increase in the world population to more than 8.5 billion, limitations of available land and water resources, low increase in crop yields, demands for cheaper and safer food and the need to sustain and improve the surrounding environment. The environment is threatened every day by the activities of man. Drastic changes have been reported since last half century that has impacted biodiversity and global climate changes. Climatic change has become the concern of everybody, because it has brought about severe and possibly permanent derogative alterations to our planet's geological, biological, environmental and ecological systems. These changes have led to the emergence of large-scale environmental hazards to biota and abiota, including human health; these changes include extreme weather, ozone depletion, loss of biodiversity, stresses to food- and energy-producing systems and others. Biotechnology is being considered as an emerging technology in environmental protection by (1) controlling environmental pollution through biodegradation, biotransformation, bioaccumulation of toxic compounds like organics, metals, oil and hydrocarbons, dyes, detergents and the like; (2) production of nonconventional, nonpolluting energy sources, such as biodiesel methanol, bioethanol, biogas, biohydrogen and others; (3) agricultural applications of biofertilizers and biopesticides; (4) recovery of resources from toxic or nontoxic wastes through biotechnological approach and (5) pollution monitoring by using biosensors.

Environmental protection is an integral component of sustainable development. In the recent past environment has undergone threatening derogative changes due to disproportional anthropogenic activities, uneven expansion of global population, quantity and quality of waste generated, indiscriminate destruction of forest, vegetation, mangroves, wetlands and other water bodies, among others. Despite of ever-escalating efforts, eco-friendly regulations/legislations to check the waste production and accumulation, overuse of chemicals, promotion of the concept of recycle, reuse, replenish and so on over the consumption of natural resources, our environment is still facing threat and its protection has become a compulsion. Biotechnology has the potential to identify the fluctuations in environment and changes in the energy budget, assess normal environmental conditions and provide remedial solutions in order to enhance the degree of sustainability of the environment. There are several books dealing with these issues separately. However, environmental protection using biotechnology is the need for a new resource in this area because it contributes more effectively to gains in agricultural productivity, enhances food security and provides the problem-solving solutions to biomedical, genetic, physiological and environmental disorders, problems related to food technology, cosmetics and any type of industries. Hence, there is immense need, as well as a lot of interest has been shown, by the students and professionals for an advanced book that integrates basic and advanced concepts of environmental biotechnology, along with future developments.

This book consists of 20 chapters that focus on current approaches and strategies for sustainable development under climate change scenario. Chapter 1 demonstrates an overview of environmental biotechnology and its application for sustainable development. Chapter 2 involves the traditional and innovative methods of waste management and the use of biocatalysts for detoxification of harmful chemicals, metals and other wastes released by industry into the environment. Chapter 3 focuses on the absorption process, adsorption process, factors affecting adsorption process, adsorption isotherms, adsorption of heavy metals, fluorides, pesticides, herbicides and other organic pollutants. Chapter 4 illustrates the classification and construction of wetlands for wastewater management, methodology for wastewater management and optimization of pollutants concentrations in sewage treatment. Chapter 5 includes the potential of the multi-soil layering system for the treatment of domestic wastewater. Chapter 6 discusses nanomaterials involved in environmental applications; environmental remediation and sensing and detection of pollutants; interactions among nanomaterials and tissue, cells, cell organelles and macro- and micromolecules; and the intricacies of the interactions among nanomaterials, biosystems and environment. Chapter 7 illustrates the effect of climate change on livestock reproduction and production, as well as strategies for minimizing the impact of stress for fertility augmentation. Chapter 8 deals with the production of biosurfactants, environmental applications, enhanced oil recovery, petroleum hydrocarbons and biodegradation of nitroaromatics. Chapter 9 describes the approaches the development of process for second-generation biofuel and chemical production following a biorefinery scheme, the current status of biofuels and biofuels of first and second (from lignocellulosic biomass) generations, as well as the potential of the developing third- (from algal biomass) and fourth- (metabolic engineering of algae) generation biofuels. Chapter 10 focuses on application of genome engineering technique (CRISPR/Cas9) on environmental biotechnology in combatting various environmental challenges, including sustainable agriculture, bioremediation, biofuels, metabolic engineering, biosensing, greenhouse gas emission, pesticide reduction and water management. Chapter 11 discusses versatile use of natural fibers for development of a clean and green environment, biotechnological interventions for improving natural fiber–producing species, including transgenic development and the role of white biotechnology to address critical issues such as biomass recalcitrance. Chapter 12 deals with threats to environment and how can it improve sustainable development, the use and regulation of biological systems for remediation of contaminated environmental constituents, such as land, air and water, and adopt environmentally friendly processes, such as green manufacturing technologies, the use of clean energy and waste management for sustainable development. Chapter 13 involves opportunities and innovation in environmental biotechnology, policy frameworks, environmental laws and legislation, as well as initiatives taken by Government of India on environmental protection. Chapter 14 focuses on effluent released in the environment from various industries, conventional methods used for treatment of industrial effluents and enzymatic treatments of pollutants. Chapter 15 illustrates the sources, existing and likely of health concerns due to biotechnological applications and scope of biotechnology and their relation to health. Chapter 16 deals with various views on defining *socio-economic* status (SES) in developed and developing countries, parameters influencing socioeconomic status, the correlation between pollution and SES in relation to developing countries and futuristic approaches. Chapter 17 illustrates chronic kidney disease of unknown etiology (CKDU), histopathology and risk factors, organochlorine pesticides and disease, the association of organochlorine pesticides and CKDU, xenobiotic-metabolizing enzymes polymorphism and gene–environment interaction. Chapter 18 covers the biodegradation of hazardous contaminants and pesticides by microorganisms such as bacteria, microalgae and fungi and *ex situ* and *in situ* approaches for management of contaminated sediments. Chapter 19 highlights public perception of environmental biotechnology interventions and processes, as well as measures for improving public perception by dissemination of information, participation, funding and impetus from government and nongovernment agencies.

This book is for researchers, students, instructors and extension experts in agriculture, veterinary and fishery sciences. The contributors to the book are internationally recognized experts in their field across the globe. It is a pleasure to publicly acknowledge our enormous debt to contributors who assisted materially in the preparation of this book. We are grateful to our families who cheerfully tolerated and supported many hours of absence for finishing the book project. Thanks are also due to David Hemming, Emma McCann and the publishing team for their extra care and patience in publishing the book.

Sukanta Mondal
Shivesh Pratap Singh
Yogendra Kumar Lahir

Acknowledgment

As always it is pleasure to publicly acknowledge our enormous debt to contributors who assisted materially in the preparation of this book. We express our gratitude to all those who helped directly or indirectly in the accomplishment of this work with their support, valuable guidance and innumerable suggestions. We wish to express special appreciation to the editorial and production staffs of CRC Press for their excellent work. The team of CRC Press publishing group has played a great role throughout, always helpful and supportive. Special thanks are also due to Jyotsna Jangra, Renu Upadhyay and entire publishing team for their extra care and patience in publishing the book. Most important, we must express our ongoing appreciation for the support and tolerance of our families.

Sukanta Mondal
Shivesh Pratap Singh
Yogendra Kumar Lahir

Editors

Dr. Sukanta Mondal is currently working as Principal Scientist, Physiology Division, ICAR-National Institute of Animal Nutrition and Physiology, Bangalore, India. He completed his postgraduation from National Dairy Research Institute and received a PhD from the Indian Veterinary Research Institute, Bareilly. He has 24 years of research experience in environmental genomics. He has 134 research publications, written 13 book chapters, and edited three books published by Elsevier. He is editor in chief of *Current Research in Poultry Science*. He was awarded the G. P. Talwar Midcareer Scientist Award, the DBT Overseas Associateship Award and the ISSRF Young Scientist Award.

Dr. Shivesh Pratap Singh is Chairman, Central Board of Studies, M.P. Higher Education, Bhopal and Chairman, Board of Studies A.P.S. University, Rewa and Professor and Head, P.G. Dept. of Zoology, Govt. Autonomous P.G. College, Satna (M.P.). He has edited 17 textbooks and published more than 80 research papers in reputed journals. He received the Appreciation Award in 1997 from the Society for Science and Environment and received a Gold Medal in 1998, 2000, 2009 and 2012 from the Zoological Society of India. Dr. Singh was also awarded a Gold Medal by the ICCB & MSET and the Indian Academy of Environmental Sciences in 2009 and 2012, respectively. Recently, he was awarded the G.D. Bhalerao Gold Medal 2017 by the Zoological Society of India at Barrackpore.

Dr. Yogendra Kumar Lahir is Visiting Faculty of Department of Biophysics, University of Mumbai, Department of Zoology, Thakur College of Science and Commerce (Autonomous body), Kandivli (east), Mumbai. He is a life member of the Indian Biophysics Society, the Society for Radiation Research and the Environmental Mutagenic Society of India. He has published three books on toxicology, one book on interaction between nanomaterials and biomolecules and one chapter on fullerene. He has published 59 articles in national and international journals. He is a reviewer and serves on the editorial panels of six national and international journals.

Contributors

Environmental Biotechnology: An Overview
Kamlesh Choure, Shivesh Pratap Singh and Rashmi Singh
Professor and Head
Govt. Autonomous P.G. College, Satna, M.P.- 485001, India
Email: drshiveshsingh2004@yahoo.co.in

Biotechnology in Waste Management
Prakash D. Raut
Professor and Former Head,
Department of Environmental Science,
Shivaji university, Kolhapur-416 004 (India)
Email: drpdraut@yahoo.co.in

Role of Absorption and Adsorption in the Removal of Waste
Dimple V. Shah
Department of Applied physics
Sardar Vallabhbhai National Institute of Technology-SURAT, 395007, GUJARAT, INDIA
Email: dshah@phy.svnit.ac.in

Biotechnological Approach for Wastewater Management: Constructed Wetlands Is a Green Solution
Satish S. Patil and Vinayak P. Dhulap
Department of Environmental Science, Dr. B.A.M. University, Aurangabad

Multi Soil Layering System for the Treatment of Domestic Wastewater
Thorvat Akshay Rajan, Deshpande Vaishnavi Vyanktesh and S. Mondal
Department of Environmental Engineering,
Kolhapur Institute of Technology's College of Engineering (Autonomous)
Kolhapur 416234, Maharashtra, INDIA
Email: akshaythorvat@yahoo.co.in

Nanotechnology and Its Implications in the Environmental Biotechnology
Y K Lahir
Department of Biophysics, Mumbai University
Vidyanagari, Santa Cruz (E), Mumbai-400 098, India
Email: lahiryk@gmail.com

Climate Change Impact on Livestock Reproduction
S. Mondal, Ramanuj Banerjee, Mamoni Banerjee, S Nandi, IJ Reddy, PSP Gupta, A Mor and PK Agarwal
Principal Scientist
ICAR-National Institute of Animal Nutrition and Physiology
Adugodi, Bangalore – 560 030, Karnataka, India
Ph. No – 91–80–25711164/ Ext. 209
Mail: sukanta781@gmail.com

Biosurfactants and Their Applications in Environmental Cleanup
M Devasena, A. Thenmozhi and Shatirah Akib
P S G Institute of Technology and Applied Research
Coimbatore, Tamil Nadu -641062, India
Email: nandu_2000@hotmail.com

Shatirah Akib:
School of Architecture, Design and the Built Environment, Nottingham Trent University
Email: shatirah.akib@ntu.ac.uk.uk

Biobased Fuel and Chemicals from Lignocellulosic Biomass—Prospects and Challenges
R. Pontes, A. Romaní, M. Michelin, L. Domingues, J. Nunes and J. Teixeira
Universidade do Minho
Campus de Gualtar, 4710–057 Braga, Portugal
Email: jateixeira@deb.uminho.pt

Impact of CRISPR-Based Gene Editing in Environmental Biotechnology
Saurabh Srivastav, Mahino Fatima, Mir Hilal Ahmad and Amal Chandra Mondal
Laboratory of Cellular & Molecular Neurobiology
School of Life Sciences, Jawaharlal Nehru University
New Delhi-110 067, India
Email: amal_mondal@rediffmail.com

Biotechnological Approaches for Increasing the Production and Utilization of Natural Fibers for a Cleaner and Greener Environment
Laxmi Sharma, Suman Roy, Soham Ray, Amit Bera, Debabrata Sarkar, Surendra Kumar Pandey and Pratik Satya
Crop Improvement division
ICAR-Central Research Institute for Jute and Allied Fibres
Barrackpore, West Bengal, Kolkata-700120, India
Mail: pscrijaf@gmail.com

Sustainable Development in Environmental Biotechnology
Mozammel Hoque and S. Mondal
Principal Scientist
ICAR-Indian Veterinary Research Institute, Izatnagar-243122, INDIA.
Email: mhoq61@yahoo.com

Policies and Issues in Environmental Biotechnology Research
Ramanuj Banerjee, Mamoni Banerjee and S. Mondal
Department of Scientific & Industrial Research
Ministry of Science & Technology, Government of India, New Delhi, India

Enzymes in Environment Management
Tek Chand Bhalla, Neerja Thakur, Sheetal, Monika, Shashi Kant Bhatia and Jog Raj
Department of Biotechnology
Himachal Pradesh University
Summer hill, Shimla-171 005, India
Email: bhallatc@rediffmail.com

Health Issues and Environmental Biotechnology
Anish Singhal *, Prathamesh H. Kamble †, Prasunpriya Nayak
Department of Physiology
All India Institute of Medical Sciences
Basni, Phase II, Jodhpur, Rajasthan 34200
Email: nprasunpriya@gmail.com

Socioeconomic Status and Its Relation to Pollution
Jalaj Tripathi Devgan
Department of chemistry, MMH COLLEGE
Ghaziabad (NCR, National capital region of Delhi), Gaziabad-201001
Uttar Pradesh, India
Email: jalajtripathi@gmail.com

* Department of Physiology, All India Institute of Medical Sciences, Bibinagar.

† Department of Physiology, All India Institute of Medical Sciences, Nagpur.

Gene–Environment Interaction and the Risk of Chronic Kidney Disease of Uncertain Etiology
Manushi Siddarth, Basu Dev Banerjee, Rishila Ghosh, Neeru Singh and Om Prakash Kalra
University College of Medical Sciences & GTB Hospital
University of Delhi, Dilshad Garden, Delhi-110095, India
Email: banerjeebd@hotmail.com

Environmental Biotechnology: Public Perception
Harit Jha, Vivekananda Mandal, Kamal Kumar Sen, Roshni Tandey, Kavi Bhushan Singh Chouhan and Rajendra Mehta
Guru Ghasidas Vishwavidyalaya
School of Natural Resources, Bilaspur (C.G.) 495009
Email: drrmehta21@gmail.com
New Delhi- 110 067, India

Biodegradation of Hazardous Contaminants and Pesticides
Marion McClary (Jr)
School of Natural Sciences
Biology and Marine Biology, Metropolitan Campus
Fairleigh Dickinson University, 1000 River Road, H-DH4–03
Teaneck, New Jersey 07666, USA
Email: mcclary@fdu.edu

1

Environmental Biotechnology: An Overview

Kamlesh Choure, Shivesh Pratap Singh and Rashmi Singh

CONTENTS

1.1 Introduction

Environmental biotechnology is defined as the method of some eco-friendly products using microorganisms, as well as this branch deals with environmental problems. This field has emerged from other disciplines and deals with the environmental problems with the applications and technology of biotechnology. It focuses on cause of environmental problems and deals with the methods of biotechnology (Zylstraa and Kukor, 2005) to solve these environmental problems. As the population increases, municipal corporations and industries have had difficulty in dealing with liquid and solid wastes. With the application of environmental biotechnology, the proper use of selected strains of microbes can help degrade the waste material completely into simple organic compounds; thus, biological oxygen demand (BOD) decreases significantly. In agriculture, environmental biotechnology plays a very vital role by developing environment-safe technology for replacing the use of chemical-based pesticides and insecticides that are injurious to health. Various environmental problems, such as the release of harmful contaminants in the environment, various types of pollution, a decrease in fertility of soil, all kinds of waste production and the like, have emerged and been managed through many techniques by biotechnologists. These techniques are biodegradation (Starnecker and Menner, 1996), bioremediation (Adams et al., 2015), biosensing and biomarker (Bontidean et al., 2003; Lagadic et al., 1994), biopesticides and biofertilizers formulation (Rajasekaran et al., 2012; Mazid, 2011), phytoremediation (Bader, 1996), biofuel production (Service, 2007), biomining (Rawlings and Johnson, 2007), composting and vermicomposting (Domínguez et al., 1997) and biogas production (El Mashad and Zhang, 2010), among others.

Biotechnological methods are applied (Figure 1.1) in all areas of the environment where various types of pollutants are released and affect the area by decreasing the quality of that area, such as water pollutants (including herbicides, insecticides, inorganic waste, agriculture runoff, chlorinated solvents, chemical wastes, fertilizers, personal care products, sewage), air pollutants (including sulfur dioxide, greenhouse gases, carbon monoxide, nitrogen oxide, volatile organic compounds, aerosols, radioactive pollutants, fuels emissions) and land pollution (including discharge from organic and inorganic industries,

DOI: 10.1201/9781003186304-1

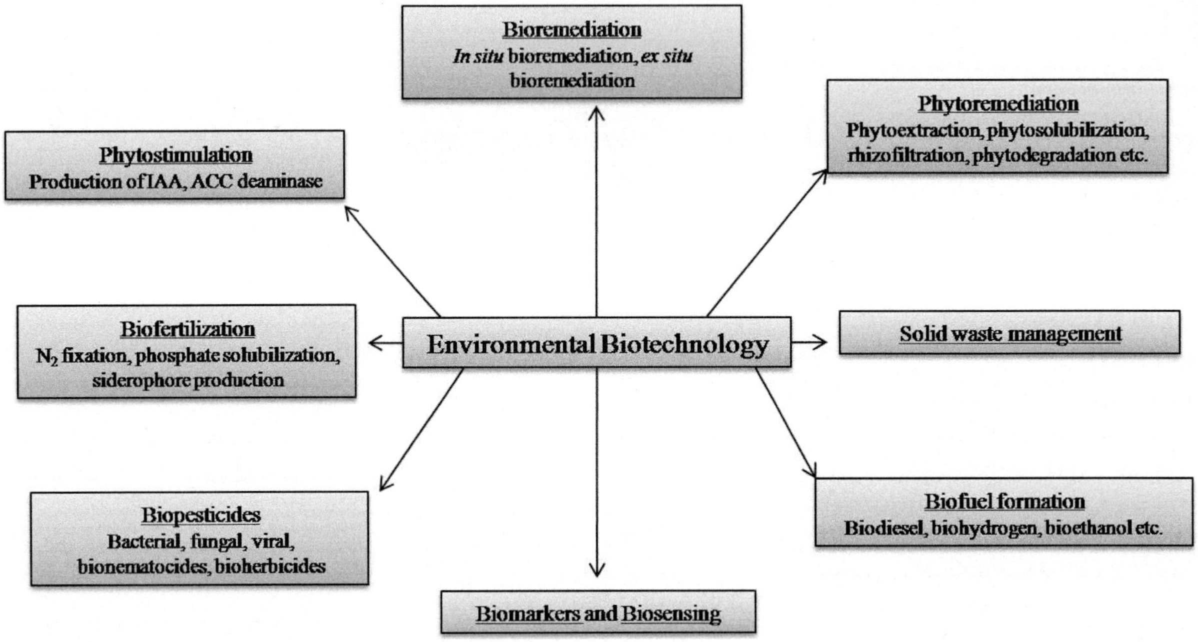

FIGURE 1.1 Applications of environmental biotechnology.

TABLE 1.1

Microorganisms Involved in Bioremediation

Microorganisms	Type	Degrading and Removal Capability
Pseudomonas putida	Aerobic bacteria	Toluene, naphthalene
Dechloromonas aeromatica	Aerobic bacteria	Perchlorate and aromatic compounds
Nitrosomonas europaea	Aerobic bacteria	These microorganisms are able to remediate ammonium and nitrate from industrial wastewater to control the eutrophication
Nitrobacter hamburgensis	Aerobic bacteria	
Paracoccus denitrificans	Aerobic bacteria	
Methylibium petroleiphilum	Aerobic bacteria	Methyl tetra-butyl ether (MTBE)
Deinococcus radiodurans	Aerobic bacteria	Ionic mercury and toluene in radioactive environment
Methylotrophs	Aerobic bacteria	Chlorinated aliphatic trichloro-ethylene, 1,2-dichloroethane
Bacillus subtilis, Bacillus amyloliquifaciens	Aerobic bacteria	Removal of heavy metals such as Cu, Ni, Pb, Hg
Phanerochaete chrysosporium	Aerobic fungi	Lignin of plant biomass, pesticides, polyaromatic hydrocarbons, PCB, dioxins, dyes, TNT, cyanide, azide, carbon tetrachloride, pentachloro-ethane.
Aspergillus niger, Penicillium spinulosum	Aerobic fungi	Removal of Cu, Cd, Zn
Rhizopus arrhizus, Saccharomyces cerevisiae	Aerobic fungi	Uranium

Source: Dubey (2014).

heavy metals, plastic, glass silt, oil spills). Major sources of pollutants are farming and gardening, industries, farming, acid rain, spray, vehicles, industries, construction sites and others. These pollutants enter the environment through anthropogenic activities and from natural sources. The presence of contaminants in the environment exerts bad effects on human health and disturbs ecosystems. This shows humans' irresponsibility toward the environment.

Environmental biotechnology is applied to restore the contaminated sources and helps decrease the release of organic and inorganic contaminants. This branch of biotechnology is beneficial in the following aspects:

1. Treatment of the solid waste and wastewater by microorganisms
2. Removal and cleanup of organic and inorganic pollutants at the site or in the lab by bioremediation
3. Energy production from biomass
4. Transformation of chemical compounds to control and protect the environment by native and genetically engineered microorganisms
5. Biomonitoring of pollutants in soil, water and so on for environmental control

1.2 Biotechniques Used for Environmental Problems

1.2.1 Bioremediation

Bioremediation is the process used to detoxify contaminants in the soil and environment through microorganisms (Shanker et al., 1998; Menn et al., 2008; Vidali, 2001; Beaudette et al., 2002). Plants and microorganisms break down the inorganic and organic wastes and restore the sites. Bioremediation is done by naturally occurring microorganisms and genetically engineered microorganisms (Table 1.1). This technology applies on site; the condition of the site is determined by indigenous microorganisms, type, quantity, toxic contaminants and so on. The degradation of waste is facilitated by the supply of nutrients, the enhancement of microbial activities, controlled environmental conditions and the like.

Types of bioremediation
There are two types of bioremediation:

1. ***In situ* bioremediation**: In this type, treatment of the waste is done at the site where waste is present.
2. ***Ex situ* bioremediation**: Ex-situ treatment involves the transport of contaminants or wastes from the origin to the treatment area.

B. Process of bioremediation

1. **Bioventing:** This involves the stimulation of degrading contaminants in soil by providing controlled oxygen airflow to microorganisms.
2. **Bioaugmentation:** This process involves adding specific or genetically engineered strains of microorganisms to the site (Adams et al., 2015).
3. **Biostimulation:** In this process, microorganisms are stimulated by giving them a supply of oxygen and inorganic nutrients, such as phosphorus, nitrogen, carbon and electron donors/acceptors, among others (Adams et al., 2015).
4. **Biosorption:** This physico-chemical process involves removing heavy metals and radionuclides by binding them to the microorganisms (Ansari and Malik, 2007), for example *Pseudomonas aeruginosa* and *Bacillus subtilis.*
5. **Biodegradation:** This refers to degrading organic wastes by microorganisms, such as bacteria and fungi (Norris, 1993; Adams et al., 2015), for example *Pseudomonas putida*, *P. cepacia* and *Fusarium oxysporium*, among others.
6. **Use of a Bioreactor:** A bioreactor or a large tank is used in *ex situ* bioremediation. It involves removing contaminants from the contamination site and adding contaminants in the bioreactor. In the bioreactor, waste is degraded in a controlled environment, and the monitoring of waste degradation is also done.
7. **Landfarming:** Landfarming is also an *ex situ* waste treatment method in which contaminants such as sludge, sediments or contaminated liquids are transported to the landfarming site, followed by periodic delivery of soil to mix air into the waste (Shanker et al., 1998; Menn et al., 2008).
8. **Biopiling:** Biopiling is a technology in which excavated soil is mixed with stimulated microbes in addition to oxygen, minerals, nutrients and moisture (Vidali, 2001; Beaudette et al., 2002). Biopiles are mainly implicated for petroleum constituents contaminated soil.

1.2.2 Phytoremediation

Phytoremediation implies the direct use of green plants for removing and degrading contaminants found in soil, water and so on. Plants are used to detoxify the contaminated site in many ways, such as degradation, immobilization or stabilization and volatilization, among others. Plants absorb the contaminants through the root system, resulting in removal of toxic wastes from soil. The root surfaces provide areas for the absorption and accumulation of contaminants (Figure 1.2). Plants indirectly stimulate microbial activities in the rhizosphere by exudating a secretion that causes biodegradation of contaminants near the rhizosphere (Barceló and Poschenrieder, 2003; Bellin and O'Connor, 1990; Bell, 1992; Burken and Schnoor, 1997; Dushenkov et al., 1997; Das et al., 2008; Lon e et al., 2008; Yoon et al., 2006; Singh, 2012; Dushenkov et al., 1995; Thangavel and Subbhuraam, 2004).

Types of phytoremediation:

1. **Phytosequestration/Phytostabilization:** This involves the precipitation or immobilization of con taminants in plant roots by the absorption of waste (Barceló and Poschenrieder, 2003).
2. **Phytoextraction:** In this method, plants accumulate contaminants in their tissues, and when plants are harvested, contaminated sites become free from wastes such as lead (Pb), cadmium (Cd), nickel (Ni), zinc (Zn), arsenic (Ar), copper (Cu) and others.

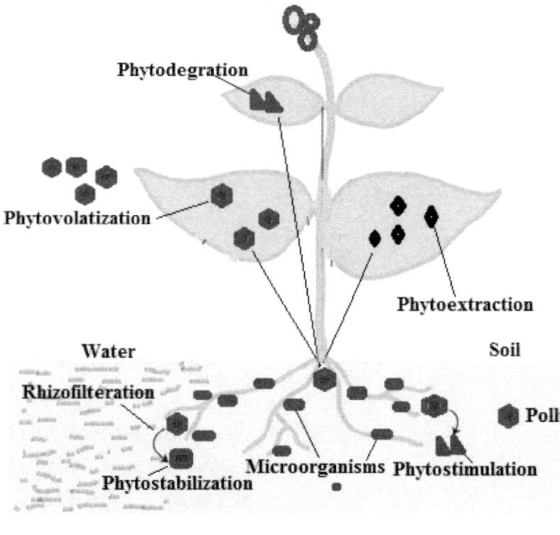

FIGURE 1.2 Phytoremediation of pollutants.
Source: Ojuederie and Babalola (2017)

3. **Phytovolatization:** In this process, plants absorb volatile compounds through their roots and release those compounds through leaves (Barceló and Poschenrieder, 2003).
4. **Phytodegradation:** Phytodegradation involves the biotransformation of contaminants in plant tissues.
5. **Rhizofiltration:** In this process, plants are used to clean up wastewater through their roots.
6. **Phytostimulation:** It is characterized with the use of plants having symbiotic association with microorganisms, for example, *Brassica juncea*, sunflower, *Salix* species, poplar trees and others.

1.2.3 Biofuels

Biofuels are fuels derived from biomass, such as plants, algae or animal wastes. These are classified under four categories (Figure 1.3). Biofuels are considered nonfossil fuels or renewable energy. Fossil fuels, such as petroleum, coal, natural gas and others, contribute to global warming, air pollution and the like. Biofuels are made by fermentation, chemical reactions and other processes and are eco-friendly. In the whole world, many countries are using various kinds of biofuels (Service, 2007; Sergeeva et al., 2008; Strobel et al., 2008; Spakowicz et al., 2015; Li et al., 2010; Kaltner et al., 2005; Hamelinck et al., 2006; Chisti, 2007, 2008; Cohen, 2001; Das and Mukherjee, 2007; Das et al., 2008; El-Sheekh et al., 2009). Ethanol is used in Brazil, biodiesel is available in Europe and other biofuels are used in the United States, Canada, Australia, China, India, Thailand and other countries (Table 1.2). Biomass, such as maize grain, sugarcane, palm oil, soybean, manure and the like, is used as a substrate for biofuel production.

1.2.3.1 Advantages of Biofuels

1. Biofuels are made from renewable resources.
2. Biofuels produce much lower level of pollutants after combustion. Biofuels are used in existing automobiles. There are no significant changes are required to the engine.
3. Biofuels are made from a variety of substrates, so many countries will be self-reliant in the production of fuel according to the demand.
4. Biofuel production can employ thousands of people by creating new jobs in biofuel production plants.
5. Biofuels can be used as additives to fossil fuels.
6. Biofuels are biodegradable; hence, there are fewer chances of soil contamination.
7. They improve engine efficiency and require low maintenance.

1.2.3.2 Types of Biofuels

Biofuels are found in solid, liquid and gaseous forms. Liquid forms of fuel are found more convenient for the automobile.

1. **Bioethanol**: Ethanol is a common biofuel produced by fermenting sugar or starch. This is also used as a blending agent with gasoline and helps decrease emissions.

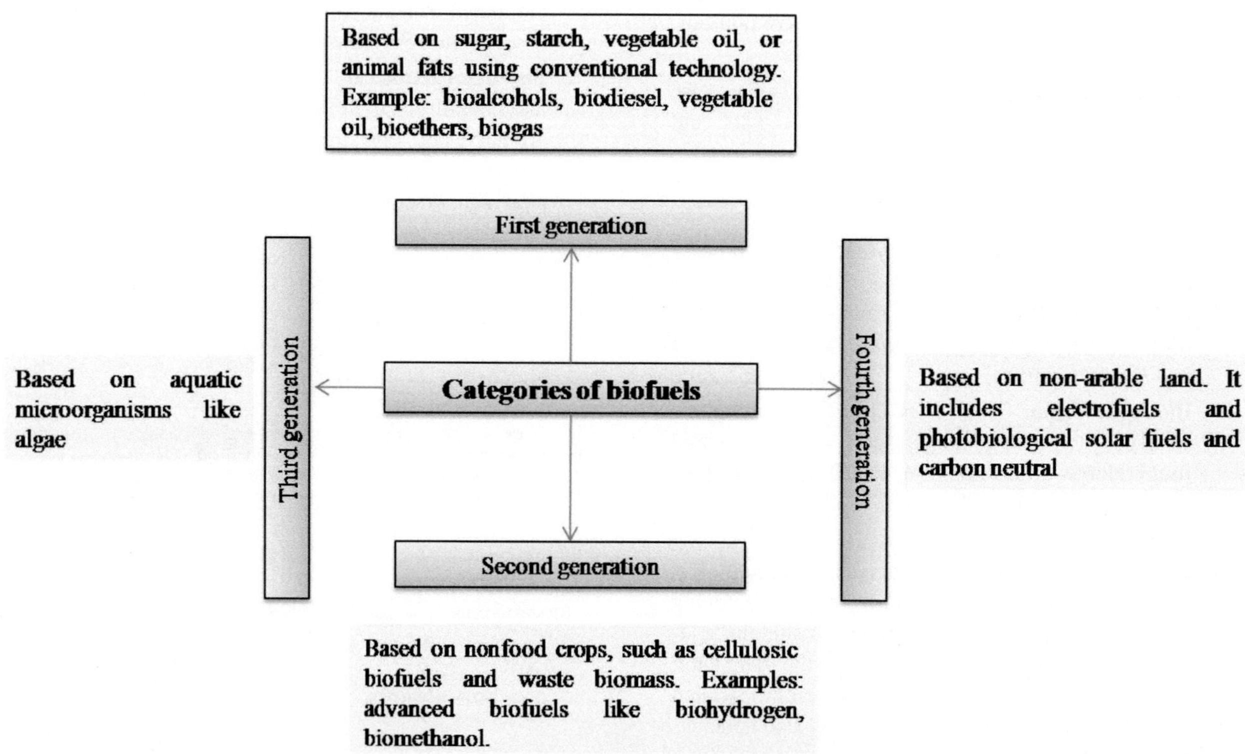

FIGURE 1.3 Categories of biofuels.

TABLE 1.2

Total Production of Biofuels

Country	Thousands of Barrels per Day
US	940
Brazil	449
Germany	68
China	59
Argentina	52
France	50
Indonesia	38
Canada	36
Thailand	24
Colombia	17

Source: US Energy Information Administration.

2. **Biodiesel**: This is produced as a result of a chemical reaction between alcohol and fatty acids, alkyl esters derived from vegetable oil, animal fats or waste cooking oil. Production of biodiesel requires the following three steps:
    a. Base-catalyzed transesterification of the oil
    b. Direct acid-catalyzed transesterification of the oil
    c. Conversion of the oil to its fatty acids and then to biodiesel
3. **Biobutanol**: This is also produced by microbial fermentation of biomass.
4. **Biohydrogen:** It is produced by pyrolysis, gasification or fermentation; it involves light, dark, or combined reaction for the production of hydrogen.
5. **Biogas:** Biogas is a mixture of different gases produced by anaerobic digestion of organic matter produced by agriculture waste, municipal waste, manure, plant waste materials or food waste.
6. **Syngas:** Syngas is a mixture of synthetic gases, such as coal emissions, waste emissions and others, that consists of hydrogen, carbon dioxide and carbon monoxide. It is converted into fuel by the gasification of carbon.

1.2.4 Biofertilization

Biofertilizer is a well-known living microorganism formulation that colonizes the rhizosphere or interior of the plant and promotes growth by increasing the supply or availability of major nutrients to the host plant when applied to seeds, plant surfaces or soil. Biofertilizers render nutrients available through natural processes, such as nitrogen fixation, phosphate solubilization and the production of plant growth hormones. They will minimize the utilization of chemical fertilizers and pesticides, as well as their harmful effects, in order to restore the soil's natural nutrient cycle and increase soil organic matter (Mishra et al., 2012; Saikia et al., 2011; Vance et al., 2000; Wani and Lee, 1995; Katyal et al., 1994; Venkatashwarlu, 2008).

1.2.4.1 Microbes Used as Biofertilizers

The soil consists of organic matter, minerals, air and water, which act as a medium for microorganisms. The diversity of microbes in soil depends on various factors, including soil texture, temperature, pH, oxygen, cation exchange capacity and others. Moreover, microbial diversity also depends on the rhizosphere region. The rhizosphere region is broadly enriched with millions of microorganisms that feed on the exudates of the root in the soil. Some microorganisms in the area of the rhizosphere do not affect the plants, but some of these interact with the plant's root and exert beneficial effects on plant growth. More than 1 million microbes may occur in 1 gram of soil. Among these bacteria is comparatively more in number. There may be more than 10,000 different species. Bacteria make available organic matter and minerals for plants, and other microbes play different roles to enhance soil quality and plant growth. These microbes are employed as microbial inoculants to the plant growth. There are a variety of microflora involved in enhancing the rhizosphere effect (Table 1.3).

TABLE 1.3

List of Microorganisms Exploited as Biofertilizers

Group	Classification	Microorganisms
Nitrogen-fixing biofertilizers	Free living	*Azotobacter, Beijerinkia, Clostridium, Klebsiella, Ananbena, Nostoc*
	Symbiotic	*Rhizobium, Frankia, Anabena azollae*
	Associative symbiotic	*Azospirillum*
Phosphate-solubilizing biofertilizers	Bacteria	*Bacillus megaterium var. phosphaticum, Bacillus subtilis, Bacillus circulans, Pseudomonas, Azotobacter*
	Fungi	*Penicillium sp. Aspergillus awamori*
Phosphate-solubilizing biofertilizers	Arbuscular mycorrhiza	*Glomus sp. Gigaspora sp. Aeaulospora sp., Scutellospora sp., and Sclerocystis sp.*
	Endomycorrhiza	*Laccaria sp., Pisolithus sp., Boletus sp., Amonita sp.*
	Ericoid mycorrhizae	*Pezizella ericae*
	Orchid mycorrhizae	*Rhizoctonia solani*
Biofertilizers for micronutrients	Silicate and zinc solubilizers	*Bacillus sp.*
Plant growth–promoting rhizobacteria (PGPR)		*Pseudomanas spp., Bacillus spp., Rhizobium spp., Azotobacter spp.*

These are bacteria, actinomycetes, algae and fungi. Among these microflorae, bacteria have received more attention of biotechnologists. Current fertilizers used to manage agriculture, are mainly inorganic chemical-based fertilizers, which increases unavoidable health problems, as well as creates environmental pollution. Currently, beneficial microbes are exploited in the form of biofertilizer to maintain a balance in the soil, promote the plant growth and reduce pollution. These beneficial microbes are known as 'biofertilizer'. The microbial inoculants are carrier-based preparations of beneficial microbes. These microorganisms fix atmospheric nitrogen, mineralize organically bound forms of nutrients, exchange reactions and perform more functions that make surety of available nutrients to the plants.

1.2.4.2 Reasons for Using Biofertilizers

Using biofertilizers in crops is a conservative agricultural method that is cheap and used with natural inputs. Biofertilizers are in demand due to their unavoidable qualities:

1. They help enhance nutrient fixation in the rhizosphere.
2. Bioinoculants produce growth stimulants or phytohormones that motivate the plant to produce its own beneficial hormones naturally.
3. They improve soil stability and mediate biological control by suppressing plant pathogens.
4. They play important role in nutrient recycling and improvement of soil quality.
5. They also promote mycorrhiza symbiosis.
5. They supply biologically available nitrogen to crops through biological nitrogen fixation.
6. They solubilize phosphorus from insoluble phosphate compounds such as hydroxyapatite.
7. They secrete siderophores that solubilize sequester iron from the soil and make it available to plants.
8. They make soil rich in all kinds of micro- and macronutrients and maintain the level for the long term.
9. Biofertilizers are nonpolluting and improve soil texture.
10. Chemical fertilizers have been increased crop production and yield according to the requirements of the population. But they produce serious hazards for the soil, its texture and quality; for the air and water; for microflora of the soil and others. They cause unavoidable hazards to the ecosystem, the environment and human health.

1.2.5 Biopesticides

Biopesticides refer to pesticides that come from living organisms or are made from living organisms (Table 1.4). They are very effective in very small quantities and cause less or no pollution (Mazid, 2011; Boyetcho et al., 1999).

1.2.5.1 Advantages of Biopesticides

1. Biopesticides are nontoxic.
2. Microbial pesticides are nonpathogenic to nontarget organisms.
3. They have a specific action.
4. They potentially control vectors.
5. Biopesticides are often effective in very small quantities.
6. They decompose in the environment.

Biopesticides include various types of potential products based on bacteria, fungi, virus and botanical biopesticides and plant incorporated protectants:

1. **Bacterial biopesticides:** Bacteria, fungi and viruses have shown significant action against pests and weeds. The first and most widely used bacterial pesticide is spores of *Bacillus thuringiensis.* Bacterial biopesticides are divided into four categories:
    a. Crystalliferous spore formers—*Bacillus thuringiensis*
    b. Obligate pathogens—*B. papilliae*
    c. Facultative pathogens—*Pseudomonas aeruginosa*
    d. Potential pathogens—*Serratia marcesens*
2. **Viral biopesticides:** There are more than 700 viruses have been known. Mainly, viruses from the Baculoviridae group have been found to be potentially active against pests. Subgroups of baculoviruses, nuclear polyhedrosis viruses (NPVs) and granulosis viruses (GVs), are most common viral biopesticides. They enter insects by ingestion of plant material, initiate infection in the midgut of insects followed by replication and inhibit the growth of insects.
3. **Fungal biopesticides:** More than 1,000 species of fungi that infect and kill or disable insects. The most commonly used fungus is *Beauveria bassiana* that infects a wide range of insects. *Trichoderma* is a mold that releases enzymes that break down insects and other pathogens.

TABLE 1.4
Example of Microbial Pesticides

Microorganisms	Pathogen/Infection
Bacillus thuringiensis	*Lepidopterous* pest
Agrobacterium radiobacter	Crown gall disease
Pseudomonas fluorescens	*Phythum, Rhizoctoniasolani, Gaemannomyces graminis*
Trichoderma	*Nacteia galligena*
Metarizium anisopliae	Spittlegbugs, rhinoceros beetles
Beauveria bassiana	Colorado potato beetle
Verticillium lecanii	Aphids, whiteflies
Nomuraea riley	Soybean caterpillars
Bacilovirues	*Lepidopterous* and *hymenopterous* pests
Nuclear polyhedrosis virus	Vegetables and field crop pests
Granulosis virus	Vegetables and field crop pests
Nosema	Lepidopteran insects
Vairimorpha	Orthopteran pest

4. **Botanical biopesticides:** Plant materials such as neem oil, ash, corn gluten, garlic oil, black pepper and others have been found to control pests' behavior and activity. In addition, they repel pests and decrease their survival rate.
5. **Plant-incorporated protectants:** Plant-incorporated protectants are genes and proteins that are genetically engineered into plants, resulting in enhanced immunity and the ability of plants to defend themselves against pests and insects.

1.3 Waste Management

Waste management refers to the collection and treatment of waste materials. Improper waste discarding leads to various pollution and unsensitized area. Hence, they must be properly treated and disposed of. For waste management, first, wastes are separated into solid waste and liquid waste (Mata Alvarez et al., 2000; Benítez et al., 2005; Mitchell, 1997; Tognetti et al., 2007).

1. Solid waste management: For solid waste management, first, waste materials must be separated. Waste materials such as papers, paperboard, organic waste, container, packaging, household and office waste, human excreta and the like are managed by different methods such as landfills, biogas production, composting, vermicomposting and phytoremediation. Landfill refers to disposing or dumping waste materials into pits or dumps. Composting and vermicomposting are two natural processes used for degrading solid waste materials. Composting refers to degrading organic matter by microorganisms under controlled conditions that produce organic fertilizers. Vermicomposting involves a joint action between earthworms and microorganisms to degrade waste. Composting and vermicomposting both are anaerobic biodegradation processes done by bacteria, actinomycetes and fungi, including *Alcaligenes faecalis*, *Arthrobacter*, *Bacillus brevis*, *B. licheniformis*, *B. subtilis*, *Clostridium* sp., *E. coli*, *Serretia*, *Pseudomonas*, *Frankia*, *Streptomyces*, *Aspergillus fumigates*, *Penicillium* sp., *Trichoderma* sp. and others. Biogas production involves degradation of organic matter in oxygen-lacking environments, resulting in the production of a mixture of gases, such as methane and carbon dioxide. Methane production has importance due to its wide range of uses (Isci and Demirer, 2007; Westerholm et al., 2012; Zhang et al., 2013a, 2013b). Biogas production occurs in an anaerobic digester (tank), where the anaerobic digestion process occurs with the help of bacteria in anaerobic conditions.
2. Liquid waste management: Liquid waste and wastewater management includes sewage and wastewater treatment. This is the oldest method of solid waste management.

1.4 Biomining

Biomining refers to extraction of metals by using microorganisms from rock ores or mine wastes. This technique is helpful for restoring metals from polluted sites. Metals are found in bound form in solid minerals. Microorganisms metabolize or oxidize them and dissolve them in water to get energy. In this way, metals are easily recovered in their dissolved form than as solid minerals. Various metals, such as copper, iron, uranium, nickel and gold, are mainly found in sulfidic form. Microorganisms oxidize sulfidic ores and convert them into metals. The following microorganisms are applied in the biomining process: *Thiobacillus ferrooxidans, T. prosperus, Leptospirillum ferrooxidans, Sulfobacillus acidophilus, S. thermosulfidooxidans, L. thermoferrooxidans, Acidianus brierleyi, A. infernus, A. ambivalens, Sulfurococcus yellowstonii, T. thiooxidans, T. acidophilus, T. caldus, Sulfolobus solfataricus, S. rivotincti* and *S. yellowstonii* (Brandl and Faramarzi, 2006; Mc Donald and Whittington, 2008; Rawlings and Johnson, 2007; Brierley, 1982; Rawlings et al., 2002).

1.4.1 Methods of Biomining

1. **Bioleaching:** In this method, low-grade ore and mineral concentrates are dumped into a heap and then soaked in a weak sulfuric acid wash, resulting in a reaction between acid and sulfidic ores. Bacterial growth extracts metals in fluid form. This method is implicated by the use of *Thiobacillus ferroxidans* and *T. thiooxidans*. Tank bioleaching have also been practiced.
2. **Bio-oxidation:** This is an oxidation process caused by microorganisms in which valuable metal is enriched in the solid phase and the solution can be discarded. This method is mainly used for the extraction of gold from ores.

Microorganisms have adopted two mechanisms in biomining: oxidation and reduction.

1. **Oxidation**: It is based on the activity of *Thiobacillus ferroxidans* that oxidizes ferrous ions to ferric ions.

$$4Fe^{++} + O_2 + 4H^+ \rightarrow Fe^{+++} + 2H_2O$$

Bacteria bind to the surface of ore and cause oxidation by direct and indirect methods. In the direct method, the oxidation of ore occurs by direct contact with the compound.

$$2FeS_2 + 7O_2 + 2H_2O \rightarrow 2FeSO_4 + 2H_2SO_4$$

In the indirect method, bacteria produce strong oxidizing agents during oxidation and produce sulfur from metal sulfides. Hence, the production of ferric ion causes oxidation of metal sulfide and separation of heavy metal and sulfur.

$$CuS + Fe^{+++} \rightarrow Cu^+ + S + Fe^{++}$$

2. **Reduction**: It involves the reduction of sulfates and the formation of hydrogen sulfides by *Desulfovibrio desulfaricans.*

$$4H_2 + H_2SO_4 \rightarrow H_2S + 2H_2O$$

1.5 Biosensors and Biomarkers

Biotechnology is now very much concerned with the development of effective biosensors and biomarkers for the real-time detection and monitoring of poisonous substances, including dyes, gases, smoke, chemicals and industrial effluents, which are major causes of pollution (Mauriz et al., 2006; Rodríguez-Mozaz et al., 2004, 2006). Early detection and monitoring of these pollutants can facilitate taking appropriate actions before spreading the pollution (Maletesta and Guascito, 2005). Some of the important biosensors used in environmental pollution monitoring are gas biosensors, immunosensors, BOD sensors and others. Gas biosensors detect gases such as sulfur dioxide (SO_2), methane, carbon dioxide and others; microbial biosensors have also been developed. For example, Thiobacillus-based biosensors can detect the pollutant SO_2, whereas methane (CH_4) can be detected by immobilized *Methalomonas.*

1.5.1 Immunoassay Biosensors

These biosensors are able to detect low levels of pollutants such as triazines, malathion and carbamates with the help of immunoassays that apply antibodies (Marty et al., 1993, 1998)., For the detection of BOD levels to determine the level of organic pollution, BOD biosensors have been developed with the help of the yeast *Trichosporon cutaneum* and an oxygen probe (Dubey, 2014).

The term *biomarker* has been defined as a xenobiotically induced variation in cellular or biochemical components or processes, structures or functions that is measurable in a biological system or sample. Biomarkers can be generally, broadly categorized as markers of exposure and effect (Hyun et al., 1993; Lagadic et al., 1994; Lam and Gray, 2003). These early warning biomarkers can be used in a predictive way, allowing to initiation of bioremediation strategies before irreversible environmental damage of ecological consequence occurs. These biomarkers are then defined as short-term indicators of long-term biological effects (Hyun et al., 1993; Lagadic et al., 1994; Lam and Gray, 2003).

1.6 Concluding Remarks: Environmental Biotechnology Challenges and Future Perspectives

The environment continues to face challenges due to pollution and mismanage use of the earth's resources. Hence, to deal with these problems, environmental biotechnology is playing a very crucial role as a key player to evolve new biotechnologies for the protection and restoration of the environment. These new biotechnologies are exploiting the potential of microorganisms and plants to make eco-friendly and strong agents for cleaning and protecting the environment in a variety of situations. These biotechnologies include development of bioengineering for improved biodegradation, bioremediation, solid and liquid waste management and the development of effective biosensors and biomarkers to monitor and detect pollutants. To enhance the process of biodegrading pollutants, evolutionary concepts and genetic engineering have been adopted to design effective microbial strains. To reuse the wastewater, biomembrane reactor technology and decentralization of sanitation are being used. To develop the alternative of nonrenewable energy resources, the anaerobic digestion process of biowaste material has been implemented. Bioremediation technology is extensively used for cleaning up contaminated soil and water. For fulfilling the future perspectives, environmental biotechnology should continue to engage in development of new technologies in the field of renewable energy resources, pollution prevention, biodegradation and bioremediation, biomining and real-time detection and monitoring of pollutants for addressing the challenges faced by the environment and restoring and sustaining the environment.

ACKNOWLEDGMENTS

Authors are grateful to AKS University, Satna (India) and Govt. Autonomous P.G. College, Satna (India) for all kind of support to complete this chapter.

REFERENCES

Adams, Godleads, Prekeyi Tawari-Fufeyin, Samson Okoro, and Igelenyah Ehinomen. 2015. Bioremediation, Biostimulation and Bioaugmention: A Review. *International Journal of Environmental Bioremediation and Biodegradation* 3: 28–39.

Ansari, M I and A Malik. 2007. Biosorption of Nickel and Cadmium by Metal Resistant Bacterial Isolates From Agricultural Soil Irrigated With Industrial Wastewater. *Bioresource Technology* 98(16): 3149–53.

Bader, D F. 1996. Phytoremediation of Explosives Contaminated Groundwater in Constructed Wetlands. *IBC Conference*, May 1996: 8–10.

Barceló, J, and C Poschenrieder. 2003. Phytoremediation: Principles and Perspectives. *Contributions to Science* 2: 333–44.

Beaudette, Lee, M B Cassidy, L England, J L Kirk, M Habash, H Lee, and J T Trevors. 2002. *Bioremediation of Soils*. John Wiley, 722–37.

Bell, R M. 1992. *Higher Plant Accumulation of Organic Pollutants from Soils*. Cincinnati, OH: Risk Reduction Engineering Laboratory.

Bellin, Cheryl A., and George A. O'Connor. 1990. Plant Uptake of Pentachlorophenol from Sludge-Amended Soils. *Journal of Environmental Quality* 19(3): 598–602.

Benítez, E, H Sainz, and R Nogales. 2005. Hydrolytic Enzyme Activities of Extracted Humic Substances During the Vermicomposting of a Lignocellulosic Olive Waste. *Bioresource Technology* 96: 785–90.

Bontidean, I, J Ahlqvist, A Mulchandani, W Chen, W Bae, R Mehra, A Mortari, and E Csoregi. 2003. Novel Synthetic Phytochelatin Based Capacitive Biosensor for Heavy Metal in Detection. *Biosensors and Bioelectronics* 18: 447–553.

Boyetchko, S, E Pedersen, Z Punja, and M Reddy. 1999. Formulation of Biopesticides. In *Biopesticides: Use and Delivery Methods in Biotechnology*, edited by Hall F R and Menn J J, 487–508.

Brandl, H, and M A Faramarzi. 2006. Microbe-Metal-Interactions for the Biotechnological Treatment of Metal-Containing Solid Waste. *China Particuolog* 4: 93–7.

Brierley, C L. 1982. Microbiological Mining. *Scientific American* 247(2): 42–51.

Burken, J G, and J L Schnoor. 1997. Uptake and Metabolism of Atrazine by Poplar Trees. *Journal of Environmental Sciences* 31: 294–306.

Chisti, Y. 2007. Biodiesel from Microalgae. *Biotechnology Advances* 25(3): 294–306.

Chisti, Y. 2008. Biodiesel from Microalgae Beats Bioethanol. *Trends in Biotechnology* 26(3): 126–31.

Cohen, Yariv. 2001. Biofiltration – the Treatment of Fluids by Microorganisms Immobilized Into the Filter Bedding Material: A Review. *Bioresource Technology* 77(3): 257–74.

Das, K, and A K Mukherjee. 2007. Crude Petroleum-Oil Biodegradation Efficiency of Bacillus Subtilis and Pseudomonas Aeruginosa Strains Isolated from a Petroleum-Oil Contaminated Soil from North-East India. *Bioresource Technology* 98: 1339–45.

Das, K, S Das, and S Dhundasi. 2008. Nickel, Its Adverse Health Effects and Oxidative Stress. *Indian Journal of Medical Research* 128: 412–25.

Domínguez, J, C A EdwArds, and S Subler. 1997. A Comparison of Composting and Vermicomposting. *Biocycle* 4: 57–59.

Dubey, R C. 2014. *Advanced Biotechnology*. New Delhi, India: S. Chand Publications.

Dushenkov, S, D Vasudev, Y Kapulnik, D Gleba, D Fleischer, Ting K C, and B Ensley. 1997. Removal of Uranium from Water Using Terrestrial Plants. *Environmental Science* 31: 3468–74.

Dushenkov, V, P B A Nanda Kumar, H Motto, and I Raskin. 1995. Rhizofiltration: The Use of Plants to Remove Heavy Metals from Aqueous Streams. *Journal of Environmental Sciences* 29: 1239–45.

El Mashad, H M, and R Zhang. 2010. Biogas Production from Codigestion of Dairy Manure and Food Waste. *Bioresource Technology* 101(11): 4021–8.

Hamelinck, C N, Van Hooijdonk, and A Faaij. 2006. Ethanol from Lignocellulosic Biomass: Techno-Economic Performance in Short, Middle and Long Term. *Biomass and Bioenergy* 28: 384–410.

Hossain, Ferdaus, Benjamin M Onyango, Brian J. Schilling, and William K. Hallman. 2003. Public Perceptions of Biotechnology and Acceptance of Genetically Modified Food. *Journal of Food Distribution Research, Food Distribution Research Society* 34(3): 1–15.

Hyun, C K, E Tamiya, T Takeuchi, and I Karube. 1993. A Novel BOD Sensor Based on Bacterial Luminescence. *Biotechnology and Bioengineering* 41: 750–7.

Isci, A, and G N Demirer. 2007. Biogas Production Potential From Cotton Wastes. *Renewable Energy* 32(5): 750–7.

Kaltner, F, G Azevedo, I Campos, and A Mundim. 2005. *Liquid Biofuels for Transportation in Brazil: Potential and Implications for Sustainable Agriculture and Energy in the 21st Century*. Fundação Brasileira Para o Desenvolvimento Sustentável.

Katyal, J C, B Venkateshwaralu, and S K Das. 1994. Biofertilizer for Nutrient Supplementation in Dry Land Agriculture. *Fertilizer News* 39: 27–32.

Lagadic, L, T Caquet, and F Ramade. 1994. The Role of Biomarkers in Environmental Assessment (5). *Invertebrate Populations and Communities* 3: 193–208.

Lam, P K S, and J S Gray. 2003. The Use of Biomarkers in Environmental Monitoring. *Marine Pollution Bulletin* 46: 182–6.

Li, H, A F Cann, and J C Liao. 2010. Biofuels: Biomolecular Engineering Fundamentals and Advances. *Annual Review of Chemical and Biomolecular Engineering* 1: 19–36.

Lone M I, Z He, P J Stoffella, and X Yang. 2008. Phytoremediation of Heavy Metal Polluted Soils and Water: Progresses and Perspectives. *Journal of Zhejiang University Science A* 9: 210–20.

Maletesta, C, and M R Guascito. 2005. Heavy Metal Determination by Biosensors Based on Enzyme Immobilized by Electro-Polymerization. *Biosensors and Bioelectronics* 20: 1643–47.

Marty, J L, B Lecea, and T Naguer. 1998. Biosensors for the Detection of Pesticides. *Analysis* 26: 144–48.

Marty, J L, N Miovetto, T Noguer, F Ortega, and C Roux. 1993. Enzyme Sensor for the Detection of Pesticides. *Biosensors and Bioelectronics* 8: 237–80.

Mata Alvarez, J, S Macé, and P Llabrés. 2000. Anaerobic Digestion of Organic Solid Wastes. *An Overview of Research Achievements and Perspectives* 74: 3–16.

Mauriz, E, A Calle, A Montoya, and L M Lechuga. 2006. Determination of Environmental Organic Pollutants with a Portable Optical Immunosensor. *Talanta* 69: 359–64.

Mazid, S. 2011. A Review on the Use of Biopesticides in Insect Pest Management: Biopesticides – A Safe Alternative to Chemical Control of Pests. *International Journal of Science and Advanced Technology* 1(7): 169–78.

Mc Donald, R G, and B I Whittington. 2008. Atmospheric Acid Leaching of Nickel Laterites Review. Part II. Chloride and Bio-Technologies. *Hydrometallurgy* 91: 56–69.

McHughen, A. 2007. Public Perceptions of Biotechnology. *Biotechnology Journal* September 2(9): 1105–11.

Menn, F-M, J P Easter, and G S Sayler. 2008. Genetically Engineered Microorganisms and Bioremediation. *Biotechnology* 441–63.

Mishra D J, Singh Rajvir, and Mishra U K. n.d. Shahi Sudhir Kumar. 2012. Role of Bio-Fertilizer in Organic Agriculture: A Review. *J. Recent Sciences* 2: 39–41.

Mitchell, A. 1997. Production of *Eisenia fetida* and Vermicompost from Feed-Lot Cattle Manure. *Soil Biology and Biochemistry* 29(3–4): 763–6.

Norris, Robert A. 1993. *Handbook of Bioremediation*. London: CRC Press, 272 pages.

Ojuederie, O B, and O O Babalola. 2017. Microbial and Plant-Assisted Bioremediation of Heavy Metal Polluted Environments: A Review. *International Journal of Environmental Research and Public Health* 14(12): 1504.

Rajasekaran, S, K Ganesh Shankar, K Jayakumar, M Rajesh, C Bhaaskaran, and P Sundaramoorthy. 2012. Biofertilizers Current Status of Indian Agriculture. *International Journal of Environment and Bioenergy* 4(3): 176–81.

Rawlings, D E, and D B Johnson. 2007. The Microbiology of Biomining: Development and Optimization of Mineral-Oxidizing Microbial Consortia. *Microbiology (Reading)* 153(2): 315–24.

Rodríguez-Mozaz, S, M J López de Alda, and D Barceló. 2006. Biosensors as Useful Tools for Environmental Analysis and Monitoring. *Analytical and Bioanalytical Chemistry* 386: 1025–41.

Rodríguez-Mozaz, S, M P Marco, M J López de Alda, and D Barceló. 2004. Biosensors for Environmental Applications: Future Development Trends. *Pure and Applied Chemistry* 76: 723–52.

Rogers, K. 1995. Biosensors for Environmental Applications. *Biosensors Bioelectron* 10: 533–41.

Saikia, D N, J Sarma, P K Dutta, and D K Baruah. 2011. Sustainable Productivity of Tea Through Conservation of Bio-Mass, Addition of Bio-Fertilizers and Reduction of Inorganic Fertilizer. *Two and a Bud* 58, 109–17.

Sergeeva, Y E, L A GAlanina, D A Andrianova, and E P Feofilova. 2008. Lipids of Filamentous Fungi as a Material for Producing Biodiesel Fuel. *Applied Biochemistry and Microbiology* 44: 523–27.

Service R F. 2007. Cellulosic Ethanol: Biofuel Researchers Prepare to Reap a New Harvest. *Science* 315: 1488–91.

Shanker, R, H J Purohit, and P Khanna. 1998. Bioremediation for Hazardous Waste Management: The Indian Scenario. In edited by Irvine R L and Sikdar S K, 81–96.

Singh, S. 2012. Phytoremediation: A Sustainable Alternative for Environmental Challenges. *The International Journal of Earth Sciences* 1: 133–9.

Spakowicz, D J, and S A Strobel. (2015). Biosynthesis of Hydrocarbons and Volatile Organic Compounds by Fungi: Bioengineering Potential. *Applied Microbiology and Biotechnology* 99(12): 4943–51.

Starnecker, A, and M Menner. 1996. Assessment of Biodegradability of Plastics Under Simulated Composting Conditions in a Laboratory Test System. *International Biodeterioration and Biodegradation* 37: 85–92.

Strobel, G, B Knighton, K Kluck, Y Ren, T Livinghouse, M Griffin, D Spakowicz, and J Sears. 2008. The Production of Myco-Diesel Hydrocarbons and Their Derivatives by the Endophytic Fungus Gliocladium Roseum (NRRL 50072). *Microbiology* 154: 3319–28.

Thangavel, P, and C Subbhuraam. 2004. Phytoextraction: Role of Hyperaccumulators in Metal Contaminated Soils. *Science Academy* 70: 109–30.

Tognetti, C, M J Mazzarino, and F Laos. 2007. Cocomposting Biosolids and Municipal Organic Waste: Effects of Process Management on Stabilization and Quality. *Biology and Fertility of Soils* 43: 387–97.

Vance, C P, P H GraHam, and D L Allan. 2000. Biological Nitrogen Fixation: Phosphorus In *Nitrogen Fixation from Molecules to Crop Productivity*, edited by Hungria M, Yates M, and Newton W E. Dordrecht: Kluwer Academic Publishers, 509–18.

Venkatashwarlu, B. 2008. Role of Bio-Fertilizers in Organic Farming: Organic Farming in Rain Fed Agriculture. Hyderabad: Central Institute for Dry Land Agriculture, Kisan Forum Pvt. Ltd., 85–95.

Vidali, M. 2001. Bioremediation. An Overview. *Pure and Applied Chemistry* 73(7): 1163–72.

Wani, S P, and K K Lee. 1995. Microorganisms as Biological Inputs for Sustainable Agriculture. In *Organic Agriculture Theory and Practices*, edited by Thampan P K. Cochin: Peekay Tree Crops Development Foundation, 39–76.

Westerholm, M, M Hansson, and A Schnurer. 2012. Improved Biogas Production from Whole Stillage by Co-Digestion with Cattle Manure. *Bioresource Technology* 114: 314–19.

Yoon, J, X Cao, Q Zhou, and Ma L Q. 2006. Accumulation of Pb, Cu and Zn in Native Plants Growing on a Contaminated Florida Site. *Science of the Total Environment* 368: 456–64.

Zhang, C, G Xiao, L Peng, H Su, and T Tan. 2013a. The Anaerobic Co-Digestion of Food Waste and Cattle Manure. *Bioresource Technology* 129: 170–6.

Zhang, Cunsheng, Gang Xiao, Liyu Peng, Haijia Su, and Tianwei Tan. 2013b. The Anaerobic Co-Digestion of Food Waste and Cattle Manure. *Bioresource Technology* 129 (February): 170–6.

Zylstraa, G J, and J J Kukor. 2005. What is Environmental Biotechnology? *Current Opinion in Biotechnology* 16(3 Spec. Iss.): 243–5.

2

Pollution and Pollution Control

Vitthal T. Mohite

CONTENTS

2.1 Introduction

The word *pollution* has become important these days because pollution has had a very adverse effect on all of humanity and the environment. As we know, our environmental balance is needed to maintain a healthy existence for organisms and humanity on the earth. The earth is the only planet in the universe that has suitable environmental and climatic conditions for the origin and the survival of life. Billions of years ago, organisms originated and since have evolved, sustained and maintained on the earth. During the history or the geological time scale, continual evolution has brought into organisms from prokaryotes to eukaryotes to unicellular eukaryotes to multicellular eukaryotes. For billions of years on Earth, the natural environment has undergone continual and sequential changes in a cyclic manner. Even before human beings, changes were brought into the environment. Humans are the

DOI: 10.1201/9781003186304-2

dominant creatures among all the organisms present on Earth. They started changing and modifying their surroundings per their needs. During ancient times, when humans were hunter-gatherers, they started altering their environment, but the impact was much less because of their limitations in terms of technology, population, awareness and so on. The discovery of fire and wheels, the introduction of agriculture, human settlement, colonization and domestication of animals started putting more burdens on the surroundings. When humans used migratory agricultural activities, they started cutting forest areas for agriculture activity, having an adverse impact on the environment.

Humans have stacked to leave in the calories, and the waste produced from these calories has started to put a burden on the surrounding environment. The human impact on the environment has become larger when humans started agricultural activity more than 11,000 years ago. About 7,000 years ago, the development of agriculture and burning and land-clearing activities increased and resulted in an increase in carbon dioxide (CO_2) levels. Methane in the atmosphere has also had a role in small amounts for 5000 years. The anthropogenic began much earlier and are linked to the first greenhouse gases appearing about 7000 years ago. During evolution, it can be suggested the blueprint for anthropogenic global warming originating from around the discovery of fire. Humans started putting their control over the surrounding environment along with development. Humans' increasing interference in the environment often created conflicts between human goals and natural processes. In order to achieve larger fields, humans started filtering their surroundings. To achieve their greedy objectives, humans changed the flows of energy, bypassing natural processes in the food chain. Some of these changes are irreversible environmental degradations.

During the colonization, competition increased for the calories for humans' dominance and survival and resulted in warfare, which is as old as humanity, but primarily, it has shown very little environmental effect. Warfare was due to a lack of food, shelter, and water and to acquire land; sanitation resulted into a risk of infectious diseases and psychological trauma. Environment destruction occurred around the world before colonization. Colonization did have devastating consequences for the indigenous people and nature around the world.

2.2 Pollution and the Industrial Revolution

During the Industrial Revolution in England, large numbers of proactive cottage industries were started. During the Industrial Revolution, only the processing of the raw material into good-quality products was carried out, but no thought was given to the waste discharge. This resulted in increases in pollution in European countries during the Industrial Revolution. Tremendous amounts of gas, solid, and liquid pollutants were discharged into the atmosphere, lithosphere, and hydrosphere. In Europe, the pollution released by industries created a deadly cloud of smoke and fog, known as smog, causing the deaths of thousands due to respiratory diseases. A result of this was more than 4,000 people were killed due to acid rain, discovered in the 1850s as a resulting problem from the coal-powered plants. Water pollution intensified during the Industrial Revolution; rivers that passed through urban areas became racetracks for human waste products, and factories began discharging diverse pollutants directly into rivers and streams. By the 1850s, inhabitants began experiencing epidemics of cholera, which was very deadly, and typhoid-like diseases.

2.3 World War and Pollution

World War II was an enormous, devastating, destructive event in human history during which 50 to 70 million people lost their lives. This war left behind deep marks on the natural environment on war-affected areas, home fronts, and areas occupied by war industries. This war, which played out in Europe and Asia, left global warfare garbage, war left teas, half-sunken ships, empty bases and bombed allures. The most severe environment threat created by this war was the long impact of the development of nuclear weapons. It also resulted in the emerging global environmental problems, which includes the chemicalization of industrial production, the adoption of environmental terrains and nuclear fallout. In this war, different governments adopted the use of DDT and other terrains. World War II had a wide range of effects on all of humanity and animals through material destruction. The effects of World War II on the environment usually fall into the destruction of habitat, widespread defenestration, unchecked unties, soil erosion and contaminated lands. The production, testing, transport and the use of these advanced weapons are perhaps the single most destructive use of biological, chemical and nuclear weapons. The environmental impact of the weapons used in warfare, whether they were biological, chemical or nuclear, was catastrophic. There are also great environmental risks present in the manufacture and disposal of these weapons. The effects of biological weapons, once used, are irreversible. The reaction produced by the explosion of nuclear bombs and by weapons made from depleted uranium lasts for hundreds or thousands of years.

2.4 Modern Development and Pollution

Since World War II, the rate of industrial development has increased manifold all over the world. The world population has started increasing very quickly, and along with it, the industrialization rate of urbanization has been also increased. Pollution has become a persistent problem due to industrialization and urbanization. The introduction of the factory system and its aligned services put tremendous pollution stress on the surrounding environment. The development of small-scale industries in different areas resulted in an increase in air and water pollution and contamination of land. Large amounts of chemical fertilizer and pesticide from the petrochemical, oil refining, metal smelting, iron and steel, food-processing and paper industries played a major role. Producing large amounts of waste and pollutants, some companies have not succumbed to social, political and governmental pressures, and they deny responsibility for the pollution even as they face its strong, adverse effects. After World War II, most countries in the world started agricultural development. In countries such as India in the 1960s, a green revolution started. The use of fertilizers and pesticides, weed sides, and herbicides has increased manifold. All these pesticides and fertilizers are nonbiodegradable, and in their overuse, they have begun accumulating in the surrounding environment and desecrating its quality. Tremendous amounts of pesticides and fertilizers residing in the environment have been discharged into river systems all over the world. Humans enjoy having a water source near them. But the continued contamination of water with infectious agents makes it unsafe for any type of use.

In the modern world, most goods we use are manufactured, and in all manufacturing processes, chemicals are used at one point or another. These chemicals are thoracic or can pollute the environment. A variety of manufacturing companies can process raw materials and convert them into good-quality products. Manufacturing companies, such as refineries, the paint industry, the food-processing industry, fertilizer production, chemical smelters, the automobile industry, and chemical industries are discharging their toxic chemicals as a waste material, which can pollute the surrounding environment. Most of these waste products are synthetic chemicals, or some are man-made. These synthetic and artificial chemicals are not part of the natural environment, are not easily been accepted by nature and can cause environmental health hazards. Among the manufacturing industries with higher emissions of heavy metals are mining, metal, leather, distilleries, butter and thermal power plants. Toxic metals are added into the aquatic stream from industrial processes and domestic sewage discharge. After World War II, the rate of industrialization and urbanization has been increased in many places. Industrial belts have been widely developed as villages turn into towns, towns into cities, and cities into metro cities. This development has had a very high impact on the environment through sewage and domestic waste discharge and industrial effluents; agricultural discharge has increased manifold.

After World War II, several important and serious natural and man-made hazardous incidences have occurred, such as the following:

1. Minamata disease
2. Itai-itai disease
3. The Bhopal gas tragedy
4. The Chernobyl accident
5. The Fukushima power plant explosion

2.4.1 Minamata Disease

Minamata disease was caused due to discharge of methyl mercury in Minamata Bay, Japan, in 1956. Contaminated with methyl mercury, industrial wastewater from the Chisso Corporation was discharged into Minamata Bay. This highly toxic chemical, containing methyl mercury, bioaccumulated in the fish and invertebrates (shellfish) in Minamata Bay. Local people and cats in this area who caught the fish and shellfish died due to mercury poisoning. This incident led researchers to believe that theft break was caused due to mercury poisoning. Research conducted on this issue has checked that the disorder is due to heavy metal poisoning that probably enters the human body through fish and shellfish consumption.

2.4.2 Itai-Itai Disease

The name **itai-itai** disease was coined by locals for the severe pain in the spine and joints experienced by those with the condition. In 1940, cadmium poisoning occurred in Japan that affected only poor, elderly women living in one local. Their diet consisted primarily of rice grown in paddies contaminated with cadmium. They developed a disease called **'itai-itai', characterized** by kidney damage and weakened, painful bones. Other **eating** the same diet were not adversely affected. Cadmium poisoning leads to severe problems, including ailments of the lungs, and death results from pulmonary **Odem, and shortness** of breath. It has also been shown to affect the liver, kidneys, and pancreas.

2.4.3 The Bhopal Gas Tragedy

In the 1980s, a devastating incident occurred in Bhopal, India, that has changed our way of looking at the world. The Bhopal gas tragedy was a gas leak incident that occurred on 2 December 1984 in the pesticide manufacturing factory, American-owned Union Carbide India Limited, in the state of Madhya Pradesh, India.

It was considered as to world's worst industrial disaster. In this plant methyl isocyanate (MIC) reacts violently with water. At the height of 2 December 1984, water entered the tanks of MIC, resulted in an explosion and released 4 tons of MIC and other chemicals into the city. Approximately 2,500 residents of Bhopal were killed overnight, and about 8,000 died in the next 3 days. Approximately 1,20,000 to 1,50,000 people remain chronically ill.

The disaster happened because water was either intentionally or accidentally introduced into a tank containing MIC. This caused a chemical reaction that resulted in the buildup of much carbon dioxide among other things. This has resulted in an increase in temperature inside the tank, reaching more than 200°C, greater than the tank could withstand. The tank had valves to control the pressure, which were triggered in an emergency to reduce the pressure. As a result, toxic gas was released into the environment. The rusty iron pipes made the reaction faster. All the contents of the tanks were released within a period of about 2 hours. Whatever may be the reason for the sequence of events that led to water entering the tank, thousands of people have died and several thousand more people have lifelong injuries, giving a lesson to all of humanity regarding atmospheric pollution.

2.4.4 The Chernobyl Accident

The Chernobyl accident occurred on 25–26 April 1986 in the No 4 light-water graphite-moderated **reactor** at the Chernobyl nuclear power plant in northern Ukraine, a republic of the Soviet Union. The incident occurred during a late-night safety test, which simulated a station blackout due to a power failure; because of this, the safety system turned off, resulting in uncontrolled reaction conditions. Water flashed into steam, generating a destructive steam explosion and an open graphite fire. This fire produced considerable updrafts for about 9 days. These lifted plumes of fission products into the atmosphere. The estimated radioactive inventory released during this very hot fire phase was approximately equal in magnitude to the arbor fission products released in the initial destructive explosion. This radioactive material precipitated onto western USSR and Europe. During the accident, steam-blast effects caused two deaths because of lethal doses of radiation, and 134 servicemen were hospitalized with acute radiation sickness, of which 28 firemen and employees died within the month from the radiation effects. In addition, approximately 14 radiation-induced cancer deaths among this group of 134 hospitalized servicers were found within the next 10 years. The Chernobyl accident is considered the most disastrous nuclear powerplant accident in history both in terms of cost and causalities.

2.4.5 Fukushima Daiichi Nuclear Disaster

Coastal Pollutions

The world is divided into five different continents, and the countries of the world have 1,634,701 km of coastline. Canada, with 20,2080 km of coastline, has the longest. Coastal India, including the islands of Andaman and Nicobar, has 7514 km of coastal line. Estuarine and brackish water ecosystems have been heavily influenced by anthropogenic activities all over the world. More than 80% of coastal pollution originates from or is related to pollution sources such as industry, urban areas and agricultural activities, among others. Worldwide, most megacities are located along the coast, putting a very heavy burden on the coastal areas around the cities. In India, three of the four megacities are located in coastal areas. Most pollutants that enter the coastal areas flow from the surrounding land via rivers, carrying the highest concentrations of pollutants and contaminants. After entering the coastal areas, these pollutants mix with coastal water bodies and are diluted and widely distributed. All these pollutants readily accumulate in the basins of coastal areas, start entering the food chain and food web and start causing hazardous environmental effects on coastal areas. The environmental effects of these toxic pollutants depend on bioaccumulation and biomagnifications. Bioaccumulation is the ratio of a chemical in the organism and the concentration in the surrounding medium. This factor is a function of the stability of the chemical but also how it will accumulate in the fat of the body, that is, its lipophilicity. Accumulation presents a risk to consumer organisms, including humans.

Biomagnifications means a higher concentration in an organism than in its diet. It is related to the biological availability of contaminants to organisms and to their metabolism and excretion rates. Therefore, identical levels of a specific contaminant can have different effects due to the fact that they can be present in different forms with different availability for uptake.

2.5 Air Pollution and Climate

2.5.1 Properties of Air Pollution

1. **Particulate Matter:** In general, the term particulate relates to all atmospheric substances that are not gases. They can be suspended droplets or solid particles or mixtures of the two particulates. It can be composed of inert or extremely reactive materials ranging in size from 100 micrometers down to 0.1 micrometers smaller. This is a complex mixture of solid particles and aerosols (liquid particles) suspended in air. They can be inhaled, and when done so, the particles impair respiratory functions especially in individuals with chronic respiratory problems. The inert materials do not react readily with the environment, nor do they exhibit any morphological changes as a result of combustion or any other process, whereas the reactive materials could be further oxidized or may react chemically with the environment.
2. **Dust:** It contains particles at the size ranging from 1 to 200 mm. These are formed by natural disintegration of rock and soil or by the mechanical process of grinding and spraying. They have large settling velocities and are removed from the air by gravity and other inertial processes. Fine dust particles act as centers of catalysis for many chemical reactions taking place in the atmosphere. All over the world, pollution by the dust is a very big problem. Dust storms from deserts cover thousands of square kilometers, reduce visibility in the atmosphere, and create many hazardous to health and respiratory disorders.
3. **Smoke:** It contains fine particles of the size ranging from 0.01 to 1 mm, which can be liquid or solid, and are formed by an incomplete combustion of firewood, agricultural materials, and other chemicals. Smoke may have different colors depending on the nature of the material burned. Every year, thousands of forest land are burned in forest fires, producing tremendous amounts of smoke destroys, that deplete forest land and the thousands of creatures into it.
4. **Fumes:** These are solid particles approximately 0.1 mm in size and are normally released from chemical or metallurgical processes.
5. **Smog:** Fog is the mist in which the liquid is water and is sufficiently dense to obscure vision. Mist is made up of liquid droplets generally smaller than 10 mm that are formed by condensation in the atmosphere or are released from industrial operations. Particularly in the urban areas and megacities, the smoke from the automobile industry and fog combine to from the smog, a severe air pollution problem. Smog formation releases primary and secondary pollutants into the atmosphere. During the winter season, fossils fuels, agricultural wastes and firewood are used to heat homes and buildings. These burning activities increase the discharge of smog, and smog formation releases primary and secondary pollutants. Photochemical is main formation mechanism during summer and winter smog formation, which is very much common. Cities from the northern part of India face the problem of smog every winter.

2.5.2 Sources of Air Pollution and Their Effects

1. **Volatile organic compounds (VOCs):** These include materials such as gasoline, paint solvents, and organic cleaning solutions. VOCs are the prime agents of ozone formation.
2. **Carbon monoxide (CO):** This is an invisible, odorless gas that is highly poisonous to air-breathing animals because it has the ability to block the delivery of oxygen to the organs and tissue.
3. **Nitrogen oxides (NO_X):** These include several gaseous nitrogen–oxygen compounds. They are converted nitric acids in the atmosphere and are a major source of acid deposition. Nitrogen dioxide is a lung irritant that can lead to acute respiratory disease in children.
4. **Sulfur oxides (SO_X):** Sulfur dioxide is a gas that is highly poisonous to both plants and animals. Children and older adults are especially sensitive to SO_2. It is converted to sulfuric acid (H_2SO_4) in the atmosphere, and it's also a major source of acid deposition.
5. **Lead and other heavy metals:** Lead is very dangerous in low concentrations and can lead to brain damage and death. It accumulates in the body and impairs many tissue and organs.
6. **Hydrocarbons:** These are organic compounds composed of hydrogen and carbon. The evaporation of gasoline through carburetors of cars, industries, and the like leads to the emission of hydrocarbons.
7. **Methane:** It is a colorless, odorless, tasteless, flammable gaseous hydrocarbon. It is produced during the decomposition of aquatic vegetation, garbage, and so on. Some percentage of methane present in natural gas is burned in homes and factories and is released into the air through poor maintenance.
8. **Ozone and other photochemical oxidants:** Recall that ozone in the upper atmosphere must be preserved to shield us from ultraviolet radiation. However, ozone is also highly toxic to both plants and animals. It damages lung tissue and is responsible for many lung disorders. Therefore, ground-level ozone is a serious pollutant.
9. **Air toxins and radon:** Toxic chemicals in the air include carcinogenic chemicals, radioactive materials, and other chemicals, such as asbestos, vinyl, chloride, and benzene, that are emitted as pollutants but are not included in the preceding list of conventional pollutants. The Clean Air Act identifies 188 hazardous air pollutants (HAPs) in this category. Many of which are known human carcinogens. Radon is a radioactive gas produced by natural processes within the earth. All radioactive substances have the potential to be damaging to any living matter they contact.

2.5.3 Effects of Air Pollution

Substantial evidence has been accumulated to prove that air pollution affects atmospheric properties, damages vegetation and the health of humans and animals, deteriorates soil materials, and affects climate. It also reduces visibility and solar radiation and impairs production processes, which contribute to safety hazards and generally interferes with the enjoyment of life and property.

2.5.4 Effects of Air Pollution on Atmospheric Properties

Air pollutants affect atmospheric properties in the following ways:

- Visibility reduction
- Fog formation and precipitation
- Solar radiation reduction
- Temperature and wind distribution alteration

These effects are primarily associated with the urban atmosphere. In addition, there has been much awareness currently of the possible effects of air pollutants, mainly carbon dioxide and particles, on the atmosphere as a whole.

In addition to reducing visibility, air pollution affects urban climates with respect to increased fog formation and reduced solar radiation. The frequency of fog formation has been observed to be higher in cities than in the country. The explanation for this observation lies in the mechanism of fog formation. Scattering and absorption of both solar and infrared radiation, as well as the emission of radiation, occur within the polluted layer. The net effect of these radiative processes during the night is a marked cooling of the pointed layer.

Pollutants which are known as phytotoxicants (substances harmful to vegetation) are sulfur dioxide, peroxyacetyl nitrate (an oxidation product in photochemical smog), and ethylene; of a somewhat lesser severity are chlorine hydrogen, chloride, ammonia, and mercury. In general, the gaseous pollutants enter the plant with the air through the stomata in the course of the normal respiration of the plant. Once lodged in the leaf of the plant, the pollutants destroy chlorophyll and disrupt photosynthesis. The damage caused can range from a reduction in growth rate to complete death of the plant. Symptoms of damage usually manifest in the leaf, and the particular symptoms often provide evidence of the responsible pollutant.

2.5.5 Effects of Air Pollution on Human Health

Pollutants enter the body through the respiratory system which can be divided into the upper respiratory system consisting of the nasal cavity, the trachea, and the lower respiratory system, consisting of the bronchial tubes and the lungs. At the entrance to the lungs, the trachea divides into two bronchial trees that consist of a series of branches of successively smaller diameters. The entire bronchial tree consists of more than 20 generations of bifurcations ending in bronchioles with a diameter of about 0.05 cm. At the end of the bronchioles are large collections of tiny sacs called alveoli. It is across the alveolar membranes that oxygen diffuses in the opposite direction. Although an individual alveolus has a diameter of only about 0.02 cm, there are several hundred million alveoli in the entire lungs, providing a total surface area of roughly 50 m^2 for gas transport. The respiratory system has several levels of defense against invasion by foreign materials. Large particles are filtered from the air stream by hair in the nasal passage and are trapped by the mucus layer lining the nasal cavity and the trachea. These large particles are unable to negotiate the sharp bends in the nasal passage and because of their inertia on the walls of the cavity as the air rushes down towards the lung.

2.5.6 Global Warming or Climate Change

Increasing evidence from around the world suggests that the global climate is already changing as a result of human actions. Precipitation has increased by about 1% over the world's continents in the last century. High latitudes are tending to see more rainfall while precipitation has actually declined in some tropical areas. Warmer temperature apparently is causing more severe storms. Hurricanes, typhoons, tornadoes, severe rainstorms, and flooding seem to be more frequent and more damaging, although part of the increased damage results from more people living in dangerous locations along the coasts and floodplains. Satellite images and surface temperature show that growing seasons are now as much as 3 weeks longer in a band across northern Eurasia than they were 3 years ago. Arctic sea ice is 40% thinner now on the edge. At the Arctic Sea, the ice pack is, on average, about 500 km (312 mi) further north than a century ago. Arctic wildlife, including polar bears, walruses, beluga whales, caribou, and musk oxen, are declining or changing their migration and feeding patterns as a result of changing climatic conditions. In the south and west, ice sheets have shrunk 21% over the past 25 years, including several giant icebergs. The largest is about the size of Delaware.

Wildlife is suffering from a loss of habitat and diminishing food supplies. The population of Adele penguins, for instance, has declined precipitously across the Antarctic as the ice on which they rest and lay their eggs is disappearing. In 1972, Venezuela had six glaciers, and now it has only two. All the glaciers in Montana's Glacier Park will be gone by 2070 if the current trend continues.

Over the next 50 years, at least half of all alpine glaciers in the world could disappear. Without this resource, the agriculture industry, power generation, and drinking water supplies will suffer. About 1.7 billion now live in areas where water supplies are light. This number could increase to 5.4 billion by 2025. Sea levels have risen worldwide by approximately 15 to 20 cm (6–8 in.) in the past century. About one quarter of this increase is due to the melting of alpine glaciers; roughly half is caused by the thermal expansion of ocean water. Many wild plants and animal species are being forced out of their current ranges as the climate warms in Europe and North America.

Around the world, coral reefs are bleaching because of higher water temperatures, rivaling tropical rainforests in their biological diversity and net productivity. Coral reefs already are threatened by human activities, such as dynamite fishing and limestone excavation. If sea temperatures continue to rise, most coral reefs will be wiped out in the next 50 years. Similarly, the Cape, the floral kingdom of South Africa, that is, one of the most unique terrestrial ecosystems in the world, could disappear due to changing weather patterns.

2.5.7 Greenhouse Effect

The sun supplies the earth with an enormous amount of energy, but the energy is not evenly distributed over the globe. The solar energy that reaches the outer atmosphere about one quarter is reflected by clouds and atmospheric gases, another quarter is absorbed by carbon dioxide and water vapor, and another quarter is absorbed by ozone, methane, and a few gases. This energy absorption warms the atmosphere slightly. Absorbed energy heats the absorbing surface (such as an asphalt lot in summer), evaporates water, or provides the energy for photosynthesis in plants; following the second law of thermodynamics, absorbed energy is gradually reemitted as lower quality heat energy.

Most solar energy comes in the form of intense high-energy light or near-infrared wavelengths. Thus, short-wavelength energy passes relatively easily through the atmosphere to reach the earth's surface. Energy rereleased from the earth's warmed surface (terrestrial energy) is of a lower intensity and a longer wavelength energy in the form of the infrared part of the spectrum. Atmospheric gases, especially carbon dioxide and water vapor, are absorbed by the lower atmosphere and are leaked out to space slowly. This terrestrial energy provides most of the heat in the lower atmosphere. If the atmosphere were as transparent to infrared radiation as it is to visible light, then the earth's average temperature would be about 18°C (0°F) to –33°C (59°F) colder than it is now.

This phenomenon is called the "greenhouse effect" because the atmosphere is loosely comparable to the glass of a greenhouse. It transmits sunlight while trapping heat inside the greenhouse. This effect is a natural atmospheric process that is necessary for life as we know it. However, too much greenhouse effect is caused by burning fossil fuels and deforestation and may result in harmful environmental changes.

2.6 Water Pollution

2.6.1 Domestic Water Pollution

It is mainly caused by sewage. Sewage is defined as waterborne waste derived from home, animal or food-processing plants and includes human excreta, soaps, detergents, paper and cloth. They are the largest group of water pollutants. Water pollution is caused by the uncontrolled dumping of waste from villages, towns and cities into ponds, streams, lakes and rivers. A major ingredient of most detergents is phosphate. Phosphates support a luxurious growth of algae. Algae withdraw large quantities of oxygen from water, which becomes detrimental to other organisms. Domestic waters are the primary sources of water pollution.

2.6.2 Industrial Water Pollution

Pollution is mainly caused by the discharge of industrial effluents into the water body is known as industrial pollution. These effluents contain a wide variety of inorganic and organic substances, such as oils, greases, plastics, plasticizers, methyl wastes, suspended solids, phenols, pesticides, heavy metals and acids. Major industries of the country are located on or near the coastline or riversides.

The discharge of industrial wastes has serious consequences on our environment:

1. Organic substances deplete the oxygen content.
2. Inorganic substances render the water unfit for drinking and other purposes.
3. Acids and alkalis adversely affect the growth of fish and other aquatic organisms.
4. Dyes change the color of water and affect aquatic life.
5. Toxic substances cause serious damage to flora and fauna.
6. Oil and other greasy substances interfere with the self-purification mechanism of water.

2.6.3 Biological Pollution

Infectious Agents: Among the most important waterborne diseases are typhoid, cholera, bacterial and amoebic dysentery enteritis, polio, infectious hepatitis and schistosomiasis, malaria, yellow fever, and filariasis, which are transmitted by insects that have aquatic larvae. Altogether, at least 25 million deaths each year are blamed on these waterborne diseases.

The main source of these pathogens is untreated or improperly treated human wastes. Animal wastes from feedlots or fields near waterways and food-processing factories with inadequate waste treatment facilities also are sources of disease-causing organisms.

Fecal coliform bacteria and oxygen demand: Detecting specific pathogens in water is difficult, time-consuming, and costly, so water quality is usually described in terms of concentrations of coliform bacteria. Of the many types that live in the colon or intestine of humans and other animals, the most common of those is Escherichia coli (E. coli), which lives symbiotically in many animals, but other bacteria such *Shigella*, *Salmonella*, or *Listeria* can cause fetal diseases. If only coliform bacteria are present in a water sample, infectious pathogens are usually assumed to be present also.

The amount of oxygen dissolved in water is a good indicator of water quality and of the kinds of life it will support. An oxygen content above 6 parts per million (ppm) will support game fish and other desirable forms of aquatic life. At oxygen levels below 2ppm, water will support mainly worms, bacteria, fungi, and other detritus feeders and decomposers. When turbulence, mixing rates, and photosynthesis of green plants, algae, and cyanobacteria are high, therefore, turbulent, rapidly flowing water is constantly aerated, so it often recovers quickly from oxygen-depleting processes. Oxygen is removed from water by respiration and chemical processes that consume oxygen, plant nutrients, and cultural eutrophication.

Water clarity (transparency) is affected by sediments, chemicals, and the abundance of plankton organisms and is a useful measure of water quality and water pollution. Rivers and lakes that have clear water and low biological productivity are said to be oligotrophic (oligo = little + trophic = nutrition); by contrast, eutrophic (eu + trophic = well nourished) waters are rich in organisms and organic materials.

Human activities can greatly accelerate eutrophication, an effect called "cultural eutrophication". Cultural eutrophication is mainly caused by increased nutrients input into a water body. Increased productivity in an adequate system sometimes can be beneficial. Fish and other desirable species may grow faster, providing a welcome food source. Often, however, eutrophication produces blooms of algae or thick growths of aquatic plants stimulated by elevated phosphorus or nitrogen levels.

Inorganic Pollutants: Some toxic inorganic chemicals are naturally released into water from rocks by weathering processes. Humans accelerate the transfer rates in these cycles thousands of times above natural background levels through mining, processing, using and discarding minerals.

Among the chemicals of greater concern are heavy metals such as mercury, lead, tin, and cadmium. Super-toxic elements, such as selenium and arsenic, also have reached hazardous levels in some waters. Other inorganic materials, such as acids, salts, nitrates, and chlorine, that are not toxic at low concentrations may become concentrated enough to lower water quality and adversely affect biological communities.

Metals: Many metals, such as mercury, lead, cadmium, zinc and nickel, are highly toxic in minute concentrations because these metals are highly persistent. They accumulate in food chains and have a cumulative effect in humans.

A mercury poisoning disaster appears to be in process now. Lead poisoning has been known since Roman times to be dangerous to human health. Lead pipes are a serious source of drinking water pollution, especially in older homes or in areas where water is acidic and therefore leaches more lead from pipes. Even lead solder in pipe joints and metal containers can be hazardous. Mine drainage and leeching of mining wastes are serious sources of metal pollution in water. A survey of water quality in eastern Tennessee found that 43% of all surface streams and lakes and more than half of all groundwater used for drinking supplies was contaminated by acids and metals from mine drainage; in some cases, metal levels were 200 times higher than what is considered safe for drinking water.

Nonmetallic Salts: Desert soils often contain high concentrations of soluble salts, including toxic selenium and arsenic. Irrigation and drainage of desert soils mobilize these materials on a larger scale and can result in serious pollution problems. Salts such as sodium chloride that are not toxic at low concentrations also can be mobilized by irrigation and concentrated by evaporation, reaching levels that are toxic for plants and animals.

Acids and Bases: Acids are released as by-products of industrial processes, such as leather tanning, metal smelting and plating, petroleum distillation, and organic chemical synthesis. Coal mining is an especially important source of acid water pollution. Sulfur compounds in coal react with oxygen and water to make sulfuric acid.

Groundwater Pollution: Most of the population of the world depends on underground water aquifers for their drinking water. Industrial, agricultural, and domestic contaminants threaten this vital resource.

The most toxic contaminants come from waste disposal sites. The greatest volume and area of contamination come from agricultural chemicals and wastes because deep underground aquifers often have times of thousands of years.

Nitrates from fertilizers often exceed safety standards in rural drinking water. These high nitrate levels are dangerous for infants (nitrates combine with hemoglobin in the blood and result in blue-baby syndrome)

DDT, other agricultural fertilizers, and industrial waste are the main sources of groundwater pollution.

2.6.4 Ocean Pollution (Marine)

Although we don't use seawater directly, ocean pollution is serious and one of the fastest growing water pollution problems. Coastal bays, estuaries, shoals, and reefs are often overwhelmed by pollution. Dead zones and poisonous algae blooms are becoming increasingly widespread. Toxic chemicals, heavy metals, oils, sediments and plastic affect some of the most attractive and productive ocean regions. The potential losses caused by this pollution amount to billions of dollars each year.

Because plastic is lightweight and nonbiodegradable, it is carried thousands of miles on ocean currents and lasts for years. Even the most remote beaches of distant islands are likely to have bits of polystyrene from containers or polyethylene packaging material that were discarded half a world away. It has been estimated that some 6 million metric tons of plastic bottles, packaging material, and other litter are tossed from ships every year into the ocean where they choke seabirds, mammals, and even fish.

Oil pollution affects beaches and open seas around the world. Oceanographers estimate that somewhere between 3 million and 6 million metric tons of oil are discharged into the world's oceans each year from oil tankers, fuels leaks, intentional discharges of fuel oil and coastal industries. Rivers carry a lot of chemicals in the oceans because most chemical and paper industries are situated riverside; also, cities' waste, as well as thermal pollution of power plants, affect marine water.

2.6.5 Freshwater Pollution

Freshwater pollution is mainly caused by chemical industries, acid rain, agricultural pesticides, fertilizers, paper industries, open dumping and ground dumping. The freshwater sources include rivers, lakes and groundwater because the ocean water has lots of bacteria and salt and it cannot be used for drinking purposes.

Freshwater is polluted because we use pesticides and fertilizers. During the rainy season, chemicals, with the help of water flow, go into the rivers, lakes and groundwater. Chemical industries are mostly situated on the side of rivers, and city sewers open into the rivers; because of that, river water and fresh groundwater are polluted.

Radioactive elements also pollute water because most electrical power is made with the help of water and thermal power plants, which are also set up along rivers; because of that, all radioactive waste goes into water, and the water gets polluted. Because of industries, air pollution increases, acid rain comes, polluting fresh water, land water and groundwater. Open dumping and ground dumping create problems because many materials, such as paints, oils, metals and toxic substances, are in the garbage and pollute the water.

2.6.6 Soil Pollution

Soil pollution adds toxicants, pollutants, and infectious agents to the thin upper layer of the earth's crust. The earth's crust provides support for the biota in terms of food and shelter. Healthy and fertile soil provides media for plant growth, developing the food chain and food web, to maintain the balance of ecosystem. Soil also contains air, water, minerals, sand, mud, clay and slit, and different types of soil provide suitable supplements for growing a variety of plants and animals. The growth of forests in agricultural production is helping maintain the balance of the ecosystem.

2.7 Pollution Control

2.7.1 Air Pollution Control

Dilution is the solution to pollution. It was one of the early approaches to air pollution control. Tall smokestacks were built to send emissions from the source where they become unidentifiable and largely untraceable. But dispersed and diluted pollutants are new sources of our most serious pollution problems.

2.7.2 Reducing Production

- The most effective strategy for controlling pollution is to minimize pollution activities. Because most air pollution in the developed world is associated with transportation and energy production, the most effective strategy would be conservation, reducing electricity and consumption, insulating homes and offices, and developing public transportation, all of which could greatly reduce air pollution.
- Alternative energy sources, such as wind and solar power, produce energy with little or no pollution, and these and other technologies are becoming economically competitive. In addition to conservation, pollution can be controlled by technological innovations.
- Particulate removal involves filtering air emissions. Filters trap particulates in a mesh of cotton cloth, spun glass fibers, or asbestos cellulose.
- Industrial air filters are generally giant bags 10 to 15 m long and 2 to 3 m wide. Effluent gas is blown through the bag much like the bag on a vacuum cleaner. Every few days or once a week, the bags are opened to remove the dust cake.
- Electrostatic precipitators are the most common particulate controls in power plants.
- Sulfur removal is important because sulfur oxides are among the most damaging of all air pollutants in terms of human health and ecosystems.
- Viability switching from soft coal with high-sulfur content to low-sulfur coal is the surest way to reduce sulfur emission.
- Switching to cleaner oil or gas would eliminate.
- Coal can be crushed washed and gasified to remove sulfur and metals before combustion.
- This improves heat content and firing properties but may replace air pollution with further solid waste and water pollution problems.

- Sulfur can also be removed to yield a usable product instead of simply a waste disposal problem.
- Elemental sulfur, sulfuric acid or ammonium sulfate can all be produced using catalytic converters to oxidize or reduce sulfur.
- NO_x can be reduced in both internal combustion engines and industrial boilers by as much as 50% by carefully controlling the flow of air and fuel. Staged burners can control burning temperature and oxygen flow to prevent the formation of NO_x. The catalytic converter in a car uses platinum–palladium and rhodium catalysts to remove up to 90% of NO_x, hydrocarbons, and CO at the same time.
- Hydrocarbons control mainly involves complete combustion or controlling evaporation. Hydrocarbons and volatile organic compounds are produced by incomplete combustion of fuels by solvent evaporation from chemical factories or plants dry cleaning plastic manufacturing, printing, and other industrial processes. Closed systems that prevent escape of fugitive gases reduce many of these emissions. In automobiles, for instance, positive crankcase ventilation (PCV) systems collect oil that escapes from around the pistons and unburned fuel and channel them back to the engine for combustion. Controls on fugitive losses from industrial valves, pipes, and storage tankers can have a significant impact on air quality; burning is often the best method for destroying volatile organic chemicals in industrial stacks.

2.7.3 Water Pollution Control

The cheapest and most effective way to reduce pollution is to avoid producing it or releasing it in the first place. Careful handling of oil and petroleum products can greatly reduce the amount of water pollution caused by these materials.

Industry can reduce pollution by recycling or reclaiming materials that otherwise might be discarded in the waste stream. These approaches usually have economic as well as environmental benefits. Companies can extract valuable metals and chemicals and sell them instead of releasing them as toxic contaminants into the water system.

Nonpoint <u>sources</u>: Farmers have long contributed a huge share of water pollution, especially in the developed world; increasingly, however, farmers are finding ways to save money and water quality while maintaining soil fertility as well as protecting water quality. Precise application of fertilizer, irrigation water, and pesticides saves money and reduces water contamination, preserving wetlands that act as natural processing facilities for removing sediments and contaminants and help protect surface and groundwater.

<u>Sewage Treatment</u>: Human and animal wastes usually create the most serious health-related water pollution problems. More than five types of disease-causing (pathogenic) bacteria, viruses, and parasites can travel from human to animal through water.

2.7.4 Municipal Sewage Treatment

<u>Primary Treatment</u>: Physically separates large solid waste from the waste stream with screens and settling tanks. Settling tanks allow _ and some dissolved (suspended) organic solids to fall out as sludge. Water drained from the top of settling tanks still carries up to 75% of the organic matter, including many pathogens. These are treated by secondary treatment.

<u>Secondary Treatment</u>: In this step, aerobic bacteria break down dissolved organic compounds. Effluent is aerated often with sprayers or in an aeration tank in which air is pumped through the microorganism-rich slurry. Fluids can also be stored in a sewage lagoon, where sunlight, algae, and air process waste more cheaply but more slowly. Effluent from secondary treatment processes is usually disinfected with chlorine, ultraviolet light, or ozone to kill harmful bacteria before it is released to a nearby waterway.

<u>Tertiary Treatment</u>: This process removes dissolved metals and nutrients, especially nitrate and phosphates, from the secondary effluent, although wastewater is usually free from pathogens and organic material after a secondary treatment. It still contains high levels of these inorganic nutrients. If discharged into surface waters, these nutrients stimulate algae blooms and eutrophication. Allowing effluents to flow through a wetland or lagoon can remove nitrates and phosphates; alternatively, chemicals often are used to bind and precipitate nutrients.

Industrial sites generally contain a variety of fertilizers, pesticides, oil, rubber, tars, lead (from gasoline), and other undesirable chemicals. Unfortunately, heavy storms often overload the systems, especially when systems are old and already overtaxed. As a result, large volumes of raw sewage and toxic surface runoff are dumped directly into receiving waters. To prevent this overflow, cities are spending hundreds of millions of dollars to separate storm and sanitary sewers.

Open Dumps: Often, the way people dispose of waste is to simply dump it someplace. Open unregulated dumps are still the predominant method of waste disposal in most developing countries. The many third-world megacities have enormous garbage problems. Most of this torrent of waste is left in giant piles exposed to the wind and rain, as well as rats, flies, and other vectors. Increasingly, toxic chemicals are showing up in the groundwater supplies on which nearly most of the population of the world depends. An alarming small amount of oil or other solvents can pollute large quantities of drinking or irrigation water.

<u>Ocean Dumping</u>: The oceans are vast but not so large that we can continue to treat them as carelessly as has been our habit. Every year, some 25,000 metric tons of packaging, including half a million bottles, cans and plastic containers, are dumped into the sea. Beaches, even in the remote regions, are littered with nondegradable thermocol and plastic from industrial society; in the garbage, many materials, chemicals and metal substances are involved that create a lot of damage to marine life.

Landfills: A particular area is chosen. The _ is spread out uniformly in a thin layer and then covered by earth, which is compacted by rollers. By this method, public health problems are reduced. It leads to water pollution if the landfill site is not properly chosen. In addition, there are aerobic reactions taking place, depending on availability of oxygen. Once oxygen is removed completely, anaerobic reactions take place, giving rise to methane, H_2S, CO_2 and water.

Solutions

1. **Reuse:** In many cities, glass and plastic bottles are routinely returned to beverage producers for washing and refilling. The reusable, refillable bottle is the most efficient beverage container. It is better for the environment than remelting and more profitable for local communities. A reusable glass container covers, on average, 15 round trips between the factory and the customer before it becomes so scratched and chipped that it has to be recycled. Reusable containers also favor local bottling companies and help preserve regional differences. Since the advent of cheap, lightweight disposable food and beverage containers, many small local breweries, canneries, and bottling companies have been forced out of business by huge national conglomerates. These big companies can afford to ship food and beverages great distances as long as it is a one-way trip. If they had to collect their containers and reuse them, canning and bottling factories serving large regions would be uneconomical.
2. **Recycling:** – The term recycling has two meanings in common usage. Sometimes, recycling really means reusing something such as refillable beverage containers. In terms of solid waste management, however, recycling is the reprocessing of discarded materials into new useful products. Some recycling processes reuse materials for some purposes for instance old aluminum cans and glass bottles are usually melted and recast into new cans and bottles. Other recycling processes turn old materials into entirely new products. Old tires for instance are shredded and turned into rubberized road surfacing. Newspapers become cellulose insulation. Kitchen waste becomes a valuable solid amendment, and steel cans become automobiles and construction materials.
3. **Composting:** Pressed for landfill space, many cities have banned yard waste from municipal garbage. Rather than bury this valuable organic material, they are turning it into a useful product through composting. Biological degradation or breakdown of organic matter under aerobic (oxygen-rich) conditions the organic compost resulting from this process makes a nutrient-rich soil amendment that aids water retention, slows soil erosion and improves crop yields. A home compost pile is an easy and inexpensive way to dispose of organic waste in an interesting and environmentally friendly way. Vegetable waste, fallen leaves, wood chips or other organic matter are kept in an out-of-the-way place, kept moist and turned over every week. Within a few months, naturally occurring microorganisms will decompose the organic material into rich, pleasant-smelling compost that can be used as a soil amendment.
4. **Hazardous and Toxic Waste:** Most hazardous waste is recycled and converted to nonhazardous forms and are stored or otherwise disposed of on-site by generators, chemical companies, petroleum refineries, and other large industrial facilities. So that it does not become a public problem still, the hazardous waste should not enter the waste stream or the environment, which would represent a serious environmental problem. Orphan waste left behind by abandoned industries still remains a serious threat to both environmental quality and human health.
5. **Options for Hazardous Waste Management:**
   - **Produce less waste:** As with other wastes, the safest and least expensive way to avoid hazardous waste problems is to avoid creating the waste in the first place. Manufacturing processes can be modified to reduce or eliminate waste production. Recycling and reusing materials also eliminate hazardous wastes and pollution. Many waste products of one process or industry are valuable commodities in another.
   - **Convert to less hazardous substances:** Several processes are available to make hazardous materials less toxic. Physical treatment ties up or isolates substances. Charcoal or resin filters absorb toxins. Distillation separates hazardous components from aqueous solution, precipitation, and immobilization in ceramics, glass, or cement isolates to from the environment so that they become essentially nonhazardous. One of the few ways to dispose of metals and radioactive substances is to fuse them in silica at high temperatures to make a stable, impermeable glass suitable for long-term storage plants. Bacteria and fungi can also concentrate or detoxify contaminants.
   - **Store permanently:** There will be some materials that we can't destroy or make into something else or cause to vanish. We will have to store them out of harm's way.
   - **Retrievable storage:** Dumping wastes in the oceans or burying them in the ground generally means that we have lost control of them. Permanent retrievable storage means placing waste storage containers in a secure building, salt mine, or bed-rock cavern where they can be inspected periodically and retrieved, if necessary, for repacking or transfer if a better means of disposal is developed. This technique is more expensive than burial in a landfill because the storage area must be guarded and monitored continuously to prevent leakage, vandalism or other dispersal of toxic materials. Remedial measures are much cheaper with this technique, however, and it may be the best system in the long run. Secure landfills, or landfilling, are one of the most popular solutions for hazardous waste disposal. The first line of defense in a secure landfill is a thick bottom cushion of compacted clay that surrounds the pit like a _. Moist clay is flexible & resists cracking if the ground

shifts. It is impermeable to groundwater and will safely contain wastes. A layer of gravel is spread over the clay layer, and perforated drainpipes are laid in a grid to collect seepage that escapes from the stored material. A thick polyethylene liner protected from punctures by soft padding materials covers the gravel bed. A layer of soil or absorbent sand cushions the inner liner, and the waste is packed in drums, which then are placed into the pit separated into small units by thick _ of soil or packaging material.

When the landfill has reached it maximum capacity, a cover much like a sandwich bottom of clay, plastic, and soil, in that order, caps the site vegetation, stabilizes the surface, and improves the appearance. Sump pumps collect any liquids that filter through the land fill either from rainwater or leaking drums. This is treated & purified before being released.

REFERENCES

1. Sethi M.S. and Iqbal S.A. (1998) *Encyclopedia of Nature & Environment*, Discovery Publishing House, New Delhi.
2. Fewthrell L., Kaufmann R.B. and Preuss, A. (2003) *Assessing the Environmental Burden of Disease at the National and Local Level: Lead*. Environmental Burden of Disease Series 2, World Health Organization, Geneva.
3. Bhatia H.S. (1998) *A Textbook of Environmental Pollution and Control*, Galgotia Publications Limited, New Delhi.
4. Abbasi S.A. and Vinthan S. (2000) *Environmental Impact of Industries on Environment*, Discovery Publishing House.
5. Chatterjee A.K. (2003) *Introduction to Environmental Biotechnology*. New Age International Publisher, New Delhi, pp. 10–12.
6. Sharma B.K. (2005) *Environmental Chemistry*, Goel Publishing House, Meerut.
7. Tripathi A.K. and Bhatt V.B. (1999) *Changing Environmental Ideologies*, Ashish Publishing House, New Delhi.
8. Aunan K., Patzay G., Aaheim H.A., Seip H.M. (1998) Health and Environmental Benefits from Air Pollution Reductions in Hungary. *Science of the Total Environment*, 212, 245–268.
9. Dhara V.R. and Dhara R. (2002) The Union Carbide Disaster in Bhopal: A Review of Health Effects. *Archives of Environmental Health*, 57(5), 391–404.
10. Intergovernmental Panel on Climate Change. (2001) *Climate Change 2001*, World Meteorological Organization, Intergovernmental Panel on Climate Change, Geneva.
11. Ofiara D. and Seneca J. (2006) Biological Effects and Subsequent Economic Effects and Losses from Marine Pollution and Degradations in Marine Environments: Implications from the Literature. *Marine Pollution Bulletin* 52, 844–864.
12. Takada H. and Yamashita R. (2016) Chapter 7.2: Pollution Status of Persistent Organic Pollutants. In: *Large Marine Ecosystems: Status and Trends*. UNEP, Nairobi, pp. 165–176.
13. Grant A. and Briggs A.D. (2002) Toxicity of Sediments from Around a North Sea Oil Platform: Are Metals or Hydrocarbons Responsible for Ecological Impacts? *Marine Environmental Research*, 53, 95–116.
14. Laane R.W.P.M., Sonneveldt H.L.A., Van der Weyden A.J., Loch J.P.G. and Groeneveld G. (1999) Trends in the Spatial and Temporal Distribution of Metals (Cd, Cu, Zn and Pb) and Organic Compounds (PCBs and PAHs) in Dutch Coastal Zone Sediments from 1981 to 1996: A Model Case Study for Cd and PCBs. *Journal of Sea Research*, 41, 1–17.
15. Mateo-Sagasta J., Zadeh S.M. and Turral H. (2017) *Water Pollution from Agriculture: A Global Review*. Executive Summary, FAO (UN) and International Water Management Institute.
16. Boyden C.R. (1974) Trace Element Content and Body Size in Mollusks. *Nature London*, 251, 311–314.
17. Chhapgar B.F. (1991) *Sea Shore Life of India*, Oxford University Press, India Published by World Wide Fund for Nature, India.
18. Paul A.K. (2009) *Living Green Pollution*, World Book Publication.
19. Bishop A.B. (1973) *Public Participation in Environmental Impact Assessment*. Foundation Conf. on Preparation of Environmental Impact Statements, New England College, Henniker, New Hampshire, USA, July 29–August 3.
20. Khitoliya R.K. (2007) *Environmental Pollution: Management and Control for Sustainable Development*, S. Chand Publication, New Delhi.
21. Dutta A.K. (2000) *Introduction to Environmental Science and Engineering*, Oxford & IBH Publishing Co. Pvt. Ltd., New Delhi, pp. 126–129.

3

Biotechnology in Waste Management

Prakash D. Raut

CONTENTS

3.1 Introduction

Waste is a product, by-product or residue that cannot be used profitably. A pollutant is a waste when it damages the environment. Waste may be biological, chemical or physical in nature and may originate from processes such as manufacturing, agriculture, energy production, transport, house building and domestic activities. The hazards produced by wastes and pollutants will vary depending upon their nature. The biological agents present in the waste spread diseases, contaminate food or water or may be mediated by a vector. The chemical

DOI: 10.1201/9781003186304-3

wastes are in various forms, such as solid, liquid or gas, and may be toxic or nontoxic. They may be biodegradable, poorly biodegradable, or nonbiodegradable. There are many physical wastes, such as dust, storms, radiations and the like, which may possibly be reduced by using cleaner technologies.

3.2 Importance of Biotechnology in Waste Management

Environmental pollution is today the most imperative issue in every nation. Biotechnology offers numerous treatment strategies to defeat this contamination issue. Biotechnology in solid administration attempts to discover new creative applications in the treatment of wastewater by biological technique and disposal of solid waste by composting method. The strategies in view of biotechnology in wastewater treatment are sludge treatment, trickling filters, oxidation lakes, biofilters and anaerobic treatment. The strategies such as solid waste fertilizing the soil, biotrickling channels and biosorption are the biotechnological applications in waste administration. In all these techniques, it is fundamental to find appropriate microorganisms that will degrade organic substances and finish the treatment process in favorable conditions.

Conventional procedures can be considered to treat the waste materials for recycling purposes. However, the general methodology costs, both operational and capital, will be impacted by a few criteria. The cost of the process relies on (1) the concentration of metal in a solution, (2) the operational method of equipment, (3) the need of secondary treatment and (4) the disposal of secondary waste, for example, sludge. Industries are required to put huge amounts of funds in these sectors, and consequently, any new type of technology significantly overcomes such hindrances in both cost-effective and eco-friendly ways. In this way, biological techniques can be taken as suitable procedures over the comprehensive array of technologies for the treatment of metals present in waste. As microorganisms have the capacity to accommodate a variety of pollutants, both organic and inorganic, it is necessary to understand from the beginning that microorganisms cannot destroy metals. By changing their chemical and/or physical characteristics, they can influence metal mobility in the environment.

The activity of microorganisms to act with metal ions relies on the nature of organisms whether they are prokaryotic or eukaryotic. The mechanism of microbes and metal interaction is shown in Figure 3.1. Microorganisms always bind metal ions present in the environment at the cell surface or transport them into the cell for different intracellular functions. The method of biosorption and bioaccumulation has received considerable attention for removing metals from various aqueous mediums. The metal ion is exchanged for a counter-ion attached to biomass which involves a typical ion-exchange process.

Bioleaching is a process in which microbes dissolve metals present in the solid matrix of waste into a soluble form. Many acidophilic microorganisms play a key role in the process of bioleaching. Acidic mediums enhance the bioleaching process whereby the metal ions get easily mobilized in an aqueous system. These microorganisms are referred as biomining or bioleaching microbes as they play a very important role in the bioleaching process. This method of industrial wastewater treatment is very inexpensive and quite flexible. These microbes can easily adapt to variations in the conditions and metabolize or co-metabolize the substrates present in the concerned medium. This methodology is perceived as a green

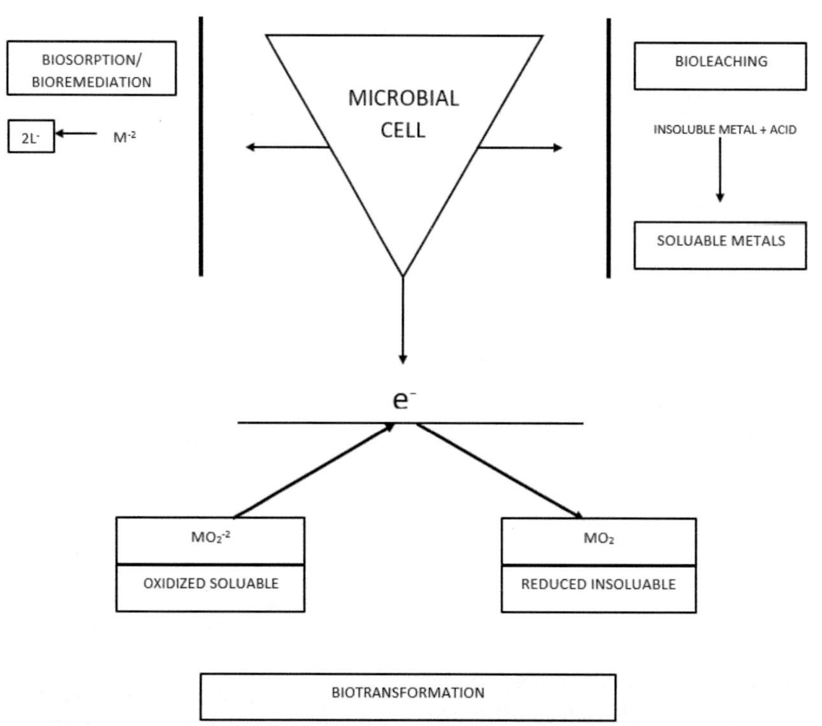

FIGURE 3.1 Microbe–metal interaction mechanism.

technology, and therefore, microbiological processes are gaining interest over the conventional technologies for the treatment of industrial wastes in order to recover valuable metals.

Modern biotechnology had given a new tool in which naturally occurring biocatalysts can be used to detoxify harmful chemicals, metals and other wastes that are released by industry into the environment. At a commercial level, these novel bacteria are exposed to waste compounds in a bioreactor, where bacteria consume the noxious waste material and convert it into water, carbon dioxide and salts to completely degrade the waste material. For the removal of uranium from drainage waters in mining operations and from contaminated groundwater, a bacterial species, *Geobacter metallireducens*, is used.

Biotechnology contributes to the isolation and characterization of various important genes. A wide range of pollutants can be degraded by developing new strains. Molecular manipulations in the genetic constitution of microorganisms can modify bacteria to use them for removing specific toxicants.

3.3 Biotechnical Methods

3.3.1 Traditional Methods

3.3.1.1 Landfilling

In vast nonurbanized systems, a typical technique for discarding biodegradable waste is the low-cost anaerobic landfill technology. The solid wastes are stored in low-lying and low-value sites. Solid waste can include a variety of materials found in household units close to commercial and industrial areas in this methodology. Solid waste is usually separated into the accompanying sorts, such as plastic waste, paper/card waste, tins and metals, ceramics and glass waste, among others. The waste deposit is compacted and secured by a layer of soil every day. These landfill regions house a wide variety of microscopic organisms; few of them are equipped for degrading particular types of waste. The main limitation in this procedure is that these microscopic organisms set aside an extensively long time to degrade the waste. However, modern biotechnology has empowered researchers to consider the accessible microscopic organisms, which are associated with the degradation of the waste—including hazardous substances. The most effective strains of these microorganisms can be cloned and duplicated in huge amounts and, in the long run, used for the specific locales. This ensures quick degradation of the waste material.

3.2.1.2 Composting

Composting is an anaerobic microbial process that converts organic waste into a stable sterile humus-like material. This material has the capacity to be securely returned to the natural surroundings. This technique is in low-moisture, solid-substrate fermentation. In large-scale operations, household solid waste is generally utilized for soil improvement. In some operations, raw substrates, such as straw, fecal matter and the like, are used to transform the waste into the compost, which is used as a substrate to produce mushrooms.

The prime aim of a composting task is to get fertilizer with a desirable compost quality in a constrained time. Oxidation is the basic biological reaction of the composting procedure of the mixed organic substrates to develop carbon dioxide, water and other organic by-products. It is essential to guarantee that treating the composting plant functions under environmentally safe conditions. Composting has for some time been perceived not just as a method for securely treating organic solid waste but additionally as a system for reusing organic matter. This method will progressively assume a noteworthy part in future waste management schemes as it empowers the recycling of organic material got from household, farming and food industry waste.

3.2.2 Modern Methods

3.2.2.1 Bioleaching of Solid Waste

Bioleaching is getting attention as an application for metal recovery from ores/concentrates, such as chalcopyrite, calcite, pyrite and arsenopyrite. Presently this process has demonstrated its promising impact in treatment of contaminated wastes for the metal recovery. The main bacterial groups involved in bioleaching are chemolithotrophic acidophiles namely, *Acidithiobacillus ferrooxidans*, *Acidithiobacillus thiooxidans* and *Leptospirillum ferrooxidans* and heterotrophs like *Sulfolobus*. The eukaryotic fungal species such as *Penicillium* and *Aspergillus niger* are bioleaching microorganisms that are used in metal recovery from industrial wastes.

For many years, bioleaching major wastes has been considered key technology for fly ash treatment, electronic waste, spent batteries and spent oil catalysts. The main microorganisms used are *Acidithiobacillus thiooxidans*, *Acidithiobacillus ferrooxidans* and fungal species like *Aspergillus niger* to recover metals like aluminum (Al), nickel (Ni), zinc (Zn), copper (Cu), cadmium (Cd), and chromium (Cr) from coal fly ash. Iron (Fe) and sulfur (S) are the main energy sources, and therefore, ferrous sulfate and elemental sulfur are supplied externally for bacterial activity [Kapoor and Viraraghavan 1998]. The alkaline nature of fly ash reduces bacterial growth as well as the leaching of metals. The metal dissolution and bioleaching process enhancement requires a proper adaptation of bacteria with fly ash. The bacterial genus *Acidithiobacillus* is used to leach metals like Cu, Ni, Zn and Cr. The contaminated sites are improved by recovering metals. After bioleaching process, this waste sludge can be used for land applications.

In the coming years, electronic waste or e-waste will be the major industrial waste, and researchers are using techniques like bioleaching for its degradation. Mesophilic and thermophilic acidophiles are getting consideration for bioleaching of metals. A moderate thermophile, *Sulfobacillus thermosulfidooxidans*, leach electronic waste and recovered around 80% of Ni, Cu, Zn and Al. Heavy metals, such as Cu, Ni and Zn, have been dissolved by a microbial consortium, including acidophilic bacteria and fungi, namely, *Aspergillus niger* and *Penicillium simplicissimum*. A one-step or two-step bioleaching process is being proposed for more effective and substantial dissolution of metals. A two-step process is more appropriate to increase leaching efficiency. The bacteria-generated acidic medium in the two-step process is dealt with as a lixiviant for metal dissolution. The independent lixiviant generation of the two-stage process removes the link between the bioprocess and the chemical process and optimizes each process independently to maximize productivity. Compared to the one-step process, the two-step process treats a higher waste concentration, leading to increased metal yields.

The mechanism of bioleaching varies with the different metal species and energy source used which is supported by lithium (Li)–ion battery or spent Ni-Cd battery source specially used in digital cameras, cellular phones or laptops by using pure culture of *Acidithiobacillus ferrooxidans*. The major solid waste of petroleum industries is released spent petroleum catalyst. Al, vanadium (V), molybdenum (Mo), Fe, tin (Sn), antimony (Sb), cobalt (Co) and Ni metals are found in the catalysts which facilitate different hydrocarbon transformations. The discarded spent catalysts usually contain 7% to 20% V + Ni, 15% to 25% coke, 7% to 15% S and 5% to 10% residual oil, together with active metals (Mo and Co or Ni) and Al_2O_3. The spent catalyst has to be pretreated before bioleaching as it contains a substantial amount of oil that is harmful to organisms and bacteria. The metal recovery from spent refinery catalyst is by the oxalic acid discharged by the fungus, which could disintegrate the profitable metals into watery media.

3.3.2.2 Major Bioleaching Microorganisms

The autotrophic acidophilic microorganisms participate in the disintegration of metals from wastes. These microorganisms can grow at low pH in inorganic medium and at high metal ions. The oxidation of Fe (II) to Fe (III) and S to H_2SO_4 is the two primary elements of these microorganisms essentially control the response energy.

The microorganism species involved in oxidizing Fe (II) to Fe (III) and S to H_2SO_4 are *Acidithiobacillus*, *Sulfolobus*, *Acidianus*, *Leptospirillium* and *Sulfobacillus*. On the basis of tolerance to temperature, the acidophils are classified in to three groups: mesophiles, moderate thermoacidophiles and extreme thermoacidophiles.

1. **Mesophiles:** These are microorganisms that grow at room temperature, that is, 28 to 37°C, for example, *Acidithiobacillus ferrooxidans*. The growth conditions shown are pH 1.5 to 2.5 and a temperature range of 28 to 37°C. *Acidithiobacillus ferrooxidans*, being a lithotroph, gets energy for its growth by oxidizing Fe (II) to Fe (III) and sulfur, sulfide and different oxyanions of sulfur to sulfate. The assimilation of carbon dioxide (CO_2) by the microorganism is through the Calvin cycle catalyzed by the ribulose-biphosphate carboxylase enzyme. The ammonium compounds present in the medium are the fulfilling nitrogen demand of the microorganisms.
2. **Moderate thermophiles:** These are growing at 50°C and isolated from various geothermal environments; for example, *Sulfobacillus thermosulfidooxidans* has the capability to oxidize S and Fe. They are found in the core of the waste dumps as the temperature is higher than the outside temperature.
3. **Extreme thermophiles:** These are microorganisms growing even above 80°C, for example, the genus *Sulfolobus*. The characters shown are (a) anaerobic growth coupled with decrease of elemental sulfur, (b) aerobic growth coupled with oxidation of sulfur, (c) optimum growth temperature of 65°C to 70°C and (d) the capacity to oxidize both Fe(II) to Fe(III) and sulfur to sulfate. These microorganisms do not require heat exchangers as the temperature increases during the reaction.

3.3.2.3 Mechanism of Bioleaching

1. **Biochemical Reaction Mechanism**

Metal leaching is a chemical process that uses ferric ions and protons in a leaching reaction. Microorganisms synthesize the leaching chemicals in the space where the leaching reaction takes place. Microorganisms from exopolysaccharides (EPSs) layer where they adhere and provide a reaction space where the reaction takes place rapidly and efficiently. The metal dissolution reaction is not always identical for all metal sulfides, but it is different for metal sulfites with different intermediates. In short, a thiosulfate mechanism includes the oxidation of acid insoluble metal sulfides, such as pyrite (FeS_2) and molybdenite (MoS_2), and a polysulfide mechanism for acid-soluble metal sulfides, such as sphalerite (ZnS), chalcopyrite ($CuFeS_2$) or galena (PbS). The thiosulfate sulfate mechanism, whereby solubilization takes place via a ferric ion attack on acid-soluble metal sulfate, where the main intermediate is thiosulfate and the end product is sulfate.

The reactions may possibly be represented using pyrite as an example of a mineral:

$$FeS_2 + 6\ Fe^{3+} + 3\ H_2O \rightarrow S_2O_3^{\ 2-} + 7\ Fe^{2+} + 6\ H^+ \quad (3.1)$$
$$S_2O_3^{2-} + 8\ Fe^{3+} + 5\ H_2O \rightarrow 2\ SO_4^{2-} + 8\ Fe^{2+} + 10\ H^+ \quad (3.2)$$

The elemental sulfur acts as the main intermediate in the polysulfide mechanism. Solubilization of the acid-soluble metal sulfide is through a combined attack by ferric iron and protons. This elemental sulfur is comparatively stable but may be oxidized to sulfate by sulfur-oxidizing microbes such as *Acidithiobacillus thiooxidans* and *Acidithiobacillus caldus* (Equation 3.5).

$$MS + Fe^{3+} + H^+ \rightarrow M^{2+} + 0.5\ H_2Sn + Fe^{2+}\ (n \geq 2) \quad (3.3)$$
$$0.5\ H_2Sn + Fe^{3+} \rightarrow 0.125\ S_8 + Fe^{2+} + H^+ \quad (3.4)$$
$$0.125\ S_8 + 1.5\ O_2 + H_2O \rightarrow SO_4^{\ 2-} + 2\ H^+ \quad (3.5)$$

The ferrous iron produced in Equations 3.1 to 3.4 may be re-oxidized to ferric iron by iron-oxidizing microorganisms such as *Acidithiobacillus ferrooxidans* or bacteria of the genera *Leptospirillum* or *Sulfobacillus*.

$$2\ Fe^{2+} + 0.5\ O_2 + 2\ H^+ \rightarrow 2\ Fe^{3+} + H_2O \quad (3.6)$$

Therefore, the role of microorganisms in the solubilization of metal sulfides is to provide sulfuric acid (Equation 3.5) for a proton attack and to keep the iron in the oxidized ferric state (Equation 3.6) for an oxidative attack on the mineral.

2. Cellular mechanism:

In the cellular mechanism, acidophiles can grow at low pH and tolerate high concentrations of acids. This enables recycling and permits reverse electron transport to take place. In this process, a low pH is required for the iron cycle, whereby ferrous iron serves as an electron donor under aerobic conditions and ferric iron as an alternative electron acceptor if the oxygen concentration falls. Ferric iron is nearly insoluble at a neutral pH, whereas in acid solutions, its solubility is enhanced. As ferric iron is totally insoluble in neutral and aerobic environments, it is an alternate electron acceptor and readily available to the acidophiles but less available to aerobic neutrophiles.

The internal pH of the acidophiles is neutral, whereas the external pH of the environment in which the acidophiles group is low, for example, pH 1 to 2. When a weak reductant such as ferrous iron is used as an electron donor, it is useful to develop a pH gradient across the cell membrane, which is important for nutritional purposes. Nicotinamide adenine dinucleotide phosphate (NAD(P)H) requirement is high in autotrophic organisms to reduce their carbon source for the production of sugars, nucleotides, amino acids and other molecules required for the development of cell mass. Heterotrophs do not require much amount of NAD(P)H as a carbon source is more reduced to carbon dioxide and hydrogen removed from the nutritional source is enough to satisfy energy requirement. A reverse electron transport is found in chemolithotrophic autotrophs that require a large transmembrane proton gradient to generate the required motive force to synthesize NAD(P)H.

3.3.2.4 Advantages

1. Useful for single or mixed minerals
2. More economical and simpler to operate and maintain
3. Can be used for single to complex bioreactors
4. Does not require high temperature and pressure
5. More eco-friendly process as it is possible to neutralize the liquid effluent generated
6. Bacteria grown are growing naturally and there is less landscape damage
7. Less harmful impact on the environment, uses less energy and does not produce SO_2 emissions

3.3.2.5 Disadvantages

1. It is a very slow process as compared to smelting.
2. Some toxic chemicals are produced in these processes.
3. Environmental damage is caused by the leaking of sulfuric acids, and the H^+ ions formed can leak into the ground- and surface water, making it acidic.
4. Heavy metals such as Fe, Zn, and arsenic can be leaked during acid mine drainage.
5. The 'Yellow Boy' pollution is caused by a rise in pH of this solution; ions precipitate as a result of dilution by freshwater.

3.3.3 Concept of Biosorption

Biosorption is a methodology that is a very innovative new technology for the removal of metals from wastewater by the metal-binding capacity of various biological materials. Biosorption is the ability of biological materials to accumulate heavy metals from wastewater through metabolically mediated or physico-chemical pathways of uptake. The potential biosorbents of metal are algae, bacteria and fungi and yeasts [Volesky and Holan 1995]. Low-cost, high-efficiency minimization of chemical and or biological sludge; no additional nutrient requirement; the regeneration of the biosorbent; and the possibility of metal recovery are the main advantages of biosorption over traditional treatment methods [Kuyucak and Volesky 1988; Keskinkan *et al.* 2003].

The biosorption task has a solid phase (sorbent or biosorbent, biological material) and a liquid phase (solvent, normally water) containing a dissolved species to be sorbed (sorbate, metal ions). Because of the greater affinity of the sorbent to the sorbate species, the latter is attracted and bound there by different mechanisms. The process continues till equilibrium is established between the amount of solid-bound sorbate species and its portion remaining in the solution. The distribution between the solid and liquid phases determines the measure of sorbent affinity for the sorbate.

3.3.3.1 Biosorbent Material

Biosorbents are with a broad range, binding and collecting a majority of the heavy metals with no specific activity, but some are very specific and can bind only certain metals. A function of the chemical makeup of the microbial cells is a strong biosorbent nature of certain microorganisms for metallic ions. Mostly, these biosorbants are dead and metabolically inactive cells. Recent studies on biosorption showed the use of waste material, which is a by-product of large-scale industries, for example, the waste mycelia available from fermentation processes, olive mill solid residues, activated sludge from sewage treatment plants [Holan *et al.* 1993], biosolids [Loaec *et al.* 1997], aquatic macrophytes and others [Loaec *et al.* 1997; Ahalya *et al.* 2003], used dewatered waste-activated sludge from a sewage treatment plant for the biosorption of zinc from aqueous solutions. In comparison, with the seaweed *Durvillea potatorum,* the adsorption capacity of biosolids is more.

A preliminary study on the 'Use of olive mill' residues as heavy metal sorbent material was carried out showed that maximum absorption of copper was found in the range of 5.0 to 13.5 mg/g under different operating conditions. The simultaneous biosorption capacity of Cu, Cd and Zn on dried activated sludge [Holan *et al.* 1993; Macaskie *et al.* 1987; Peng and Koon 1993] were 0.32 mmol/g for metal systems, such as Cu and Cd; 0.29 mmol/g for Cu-Zn; and 0.32 mmol/g for Cd-Zn. The results showed that the biomass had a net preference for copper followed by cadmium and zinc [Friis *et al.* 1998, as cited in Ahalya *et al.* 2003].

Many authors recommended abundant natural materials of cellulosic nature that are used as bioadsorbants. The processes of biosorption are complex such as ion exchange, chelation, adsorption by physical forces, entrapment in inter- and intrafibrillar capillaries and spaces of the structural polysaccharide network which is a result of the concentration gradient and diffusion across cell walls and membranes. Many chemical groups that may attract and sequester the metals in biomass are acetamido groups of chitins; structural polysaccharides of fungi; amino and phosphate groups in nucleic acids; amido, amino, sulfhydryl and carboxyl groups in proteins; hydroxyls in polysaccharides; and chiefly carboxyls and sulfates in polysaccharides of marine algae that belong to the divisions Phaeophyta, Rhodophyta and Chlorophyta.

On the other hand, it does not essentially mean that the presence of some functional groups guarantees biosorption, possibly be due to steric, conformational or other barriers.

3.3.3.2 Choice of Metal for Biosorption Process

The choice of metal depends on the interest and can be categorized into (1) toxic heavy metals, (2) strategic metals, (3) precious metals and (4) radionuclides. From an environmental point of view, toxic heavy metals and radionuclides are more important. Some metals are precious, and they are important from a recovery point of view.

3.3.3.3 Mechanism

1. Biosorption Mechanisms

The biosorption is dependent on the metabolism of the individual microbial cell and therefore, biosorption mechanism is classified as

1. metabolism-dependent and
2. non-metabolism-dependent.

On the basis of location where the biosorption takes place, it can be classified as

1. extracellular accumulation/precipitation,
2. cell surface sorption/precipitation and
3. intracellular accumulation.

Metabolism-dependent biosorption of the metal across the cell membrane yields intracellular accumulation in viable cells. This process is dependent on cellular metabolism, which is dependent on the cell's metabolism. This is for the defense of the cell in the presence of toxic metal.

A non-metabolism-dependent biosorption process of metal uptake is by the interaction between functional groups present on the cell membrane and its physico-chemical reaction. It involves processes that are not reliant on cell metabolism, such as physical adsorption, ion exchange and chemical sorption.

Precipitation is the metal uptake both in the solution and on the cell surface [Francesca Pagnanelli *et al.* 2002]. It may be or may not be reliant on cell metabolism. Precipitation is the metal uptake both in the solution and on the cell surface [Francesca Pagnanelli *et al.* 2002]. It may be or may not be reliant on cell metabolism.

2. Transport across cell membrane

This mechanism is like the mechanism of transport of metabolically significant ions, such as potassium, magnesium and sodium. The transport system on the cell membrane gets confused and absorbs heavy metal ions, such as essential ions [Ercole *et al.*, 1994; Gadd, 1990; Jeffers, 1991].

3. Physical adsorption

This mechanism is like the mechanism of transport of metabolically significant ions such as potassium, magnesium and sodium. The transport system on the cell membrane gets confused and absorbs heavy metal ions like essential ions [Ercole *et al.* 1994; Gadd 1990; Jeffers, 1991].

4. Ion Exchange

Microorganisms cell walls have polysaccharides and bivalent metal ions that exchange with the counter-ions of the polysaccharides; for example, the alginates of marine algae occur as salts of K+, Na+, Ca^2+ and Mg^2+. The exchange of these ions with counter-ions, such as CO^2+, Cu^2+, Cd^2+ and Zn^2+, result in the biosorptive uptake of heavy metals [Lisa Norton *et al.* 2003; Chandra Sekhar *et al.* 1998]. The ion-exchange mechanism is involved in the biosorption of copper by fungi, *Ganod*, was also taken up by ion-exchange mechanism.

5. Complexation

The metals and active groups on the cell surface form a complex that removes metals from the solution [Aksu *et al.* 1992]. The only mechanism responsible for calcium, magnesium, cadmium, zinc, copper, and mercury accumulation by *Pseudomonas syringae* is complexation [Arica *et al.* 2003; Kratochivl and Volesky 1998a; Scott and Karanjkar 1992; Wang *et al.* 1996]. Organic acids, for example, citric, oxalic, gluonic, fumaric, lactic and malic acids, formed by microorganisms may chelate toxic metals to produce metallo-organic molecules. They assist in the solubilization of metal compounds and their leaching from their surfaces.

6. Precipitation

Precipitation is always a consequence of the chemical interaction between the cell surface and the metal. It may be dependent on or independent of cellular metabolism. In the first case, it is a defense system of the microorganism working, whereas in the independent system, it is a consequence of the chemical interaction between the metal and the cell surface. Both these mechanisms run simultaneously.

3.3.3.4 Use of Recombinant Bacteria for Metal Removal

Recombinant bacteria are used to remove metals from the contaminated water; for example, a genetically engineered *E. coli* was able to selectively accumulate 8 mmole Hg^{2+}/g cell dry weight, which expresses Hg^{2+} transport system and metallothionin (a metal-binding protein).

3.3.3.5 Factors Affecting Biosorption

The following factors affect the biosorption process:

1. Temperature seems not to influence the biosorption in the temperature range of 20°C and 35°C [Aksu *et al.* 1992].
2. pH affects the solution chemistry of the metals, the activity of the functional groups in the biomass and the competition of metallic ions [Gadd *et al.* 1988; Gulay Bayramoglu *et al.* 2002].
3. The concentration of biomass in the solution seems to influence the specific uptake. For lower values of biomass concentration, there is an increase in the specific uptake, and an increase in biomass concentration leads to interference between the binding sites [Gadd 1990].
4. To treat wastewater in which more than one type of metal ions would be present, the removal of one metal ion may be influenced by the presence of other metal ions.

3.3.3.6 Biosorption by Immobilized Cells

Immobilization of cells from the biomass consists of small particles with low density, poor mechanical strength and much less rigidity, and therefore, for metal accumulation, solid structures require material with the right size, mechanical strength and rigidity with porosity. The benefit of this method is that the beads and granules can be easily used for stripping metals. Many methods of immobilization are available for the application of biosorption on inert supports, entrapment in polymetric matrix, covalent bonds in vector compounds or cell cross-linking [Makaskie *et al.* 1987; Michel *et al.* 1986; Wong and Kwok 1992; Yan Guangyu and Viraraghavan 2000].

3.3.3.7 Adsorption on Inert Supports

The adsorption on inert support materials involves introducing support materials prior to sterilization and inoculation with starter culture. They are kept inside the culture for a period that leads to the development of a film of microorganisms on the support surfaces. The immobilization of *Rhizopus arrhizus* fungal biomass in reticulated foam biomass support particles is used successfully when activated carbon is used as a support [Mahan and Holcombe 1992; Torma *et al.* 1993; Zhou and Kiff 1991; Bai and Emilia 2003; Volesky 1986].

3.3.3.8 Entrapment in Polymeric Matrices

For entrapment in polymetric matrices, the polymers used are calcium alginate [Babu *et al.* 1993 as cited in Ahalya *et al.* 2003; Ercole *et al.* 1994; Sakaguchi and A. Nakajima 1991; Hammaini *et al.* 2003], polyacrylamide, polyacrylamide [Mahan and Holcombe 1992; Mittleman and Geesey 1985; Schneider and Rubio 1999 as cited in Ahalya *et al.* 2003; Whistler and Daniel 1985], polysulfone [Kapoor and Viraraghavan 1995; Brierley *et al.* 1993 as cited in Ahalya *et al.* 2003].

3.4 Bioremediation

Bioremediation uses naturally existing microorganisms to speed up the breaking down of biological substances and degradation of various materials. The process is also called as biorestoration or biotreatment.

Modern technology is posing a great threat to the environment by releasing many wastes in the environment, for example, chemicals, sewage and so on. The natural mechanisms of maintaining an ecological balance are breaking these chemicals by natural processes. Some of these chemicals released in the environment are complex in nature, and it is very difficult to break them down. These chemicals are getting accumulated in the environment and bioremediation is the process that helps the environment to get rid of these hazardous chemicals. This process works on the principle of breaking down of organic material into simple compounds, such as carbon dioxide, water, salts and other harmless products. It takes place in two main ways: aerobic and anaerobic processes.

3.4.1 Aerobic Bioremediation

Aerobic bioremediation is a process in which oxygen is provided as an electron acceptor for degradation of petroleum, phenols, PAHs and the like. Oxygen is required for initiating the oxidation process by activating some enzymes involved in the process. It is an electron acceptor that yields higher energy. It is possible to degrade a variety of hydrocarbons, including diesel, gasoline, kerosene and others. The molecular weight of the compound is the deciding factor of biodegradation. The rate of biodegradation decreases with an increase in the molecular weight of compound.

3.4.2 Anaerobic Bioremediation

Anaerobic degradation takes place in the absence of oxygen. It is always used to treat a wide range of contaminants, for example, chlorinated ethanes, choloromethanes, chlorinated cyclic hydrocarbons, perchlorates and nitrates. Anaerobic bioremediation involves adding an electron donor to deplete the background electron acceptor, including oxygen, nitrate, oxidized iron, manganese and sulfate. This stimulates the chemical or biological reduction of the oxidized pollutants.

3.4.3 Bioremediation to Clean Up the Environment

Bioremediation helps to clean up the environment in two ways. In the soil environment, nutrients can be added to promote microbial growth, which leads to acclimatizing themselves to the toxic wastes and degrading these pollutants. Another way to degrade pollutants from the environment is to use genetically engineered microorganisms, which are more efficient and can specifically degrade pollutants, for example, 'Flavobacterium' species to remove pentachlorophenol from contaminated soil. GS-15 microbes developed by genetic engineering can degrade uranium and convert it into water-insoluble particles that precipitate and settle at the bottom of wastewater from nuclear plants [Tsezos and Volesky 1982]. These particles can be easily collected and disposed of scientifically.

3.5 Biodegradation of Xenobiotic Compounds

Xenobiotic compounds are foreign in nature in the biological system. They are man-made compounds released into the environment in amounts that are unnatural due to human activity. They include polystyrene, pesticides, surfactants, benzopyrenes, nitrotolunes, dyestuffs and others. Because of the unnatural nature of these compounds, there is no specific mechanism for degradation, and they persist in nature. The process of degradation depends on the size, volatility and stability of the compound; also, the physical nature of the environment, including pH, susceptibility to light and weather, is also important.

The microorganisms in the contaminated areas use organo-pollutants as the source of carbon and electron donors for generating energy, which includes many xenobiotic compounds. Many bacteria have the capacity to degrade organic compounds, but a single bacterium does not possess the capacity to degrade many of the organic pollutants. Specialized enzyme systems and metabolic pathways are found in microorganisms that are geographically separated. These isolated microorganisms from various areas with their consortium can be used to degrade organic pollutants. Therefore, there is a need to develop microorganisms by genetic recombination to degrade a variety of xenobiotic compounds.

3.6 Oil-Eating Bugs

Oil spills are a great threat to the marine environment. It has a direct impact on marine organisms. There are ample numbers of oil-eating bacteria and fungi that exist in the aquatic environment. Scientists have successfully developed bacteria that can degrade oil into individual hydrocarbons. A gene for oil degradation has been introduced in the bacteria *Pseudomonos aureginos*. These bacteria have been released in the marine environment where the oil spill has occurred. Once these bacteria finish the oil from the surface, these engineered oil-eating bugs eventually die as there is no food to survive. There are few bacteria, such as *Penicillium* species and *Alcanivorax*, that also have the capacity to degrade oil, but they require much more time than the genetically engineered bacterium.

3.7 Treating Industrial Wastes

3.7.1 Wastes from the Pulp Industry

Pulp and paper industry waste mainly contains very high contents of cellulose and lignocelluloses that lead to treatment problems. When cellulose is bound to lignin, it becomes extremely resistant to enzymatic breakdown. It is not an easy task to delignify the pulp, as lignin and carbohydrates are interlinked in wood. A new enzymatic pulp-bleaching technique has been developed by researchers by using the xylanase-producing organism *Bacillus stearthermophilus*, which is isolated from soil, that prevents bleach waste formation by removing or reducing chlorine consumption as well as the minimum use of water in pulping and bleaching.

Enzyme xylanase shows high temperature stability and high working in an alkaline medium, which helps in its firm binding with the substrate. Usually, the enzyme xylanse is synthesized by microorganisms with other polymers, such as cellulose and hemicelluloses. Therefore, recombinant DNA technology is usually expressing xylanase genes in noncellulolytic hosts. Various microorganisms are being used to express xylanase from which actinomycete *Chainia* from the deserts of Rajasthan.

The wood pulping industry also produces sulfite waste liquor containing a mixture of other organic compounds with ligno-sulfate (60%) and sugar (36%). Yeast and *Candida albicans* have the capacity to ferment sugar from the wood pulp.

3.7.2 Wastes from Dairy Industry

Whey is the main white-colored waste of the dairy industry. It is leftover after the separation of the curd, and for every 1 kilogram of cheese produced, 9 liters of whey are generated. However, it has some nutrients; the use is restricted to ice creams and animal feed. With the huge production of whey in the dairy industry, disposal has become a big problem. The release of whey into natural water bodies leads to an increase in biochemical oxygen demand (BOD). The following are methods for handling the whey disposal problem:

1. Treating whey with proper strains of microbes and nutrients
2. Direct fermentation of lactose to ethanol
3. Using yeast like *Kluyveromyces fragilis* and *Candida intermedi*
4. Hydrolysis of lactose to glucose and galactose

3.7.3 Wastes from Dye Industry

Until the discovery of synthetic dyes, the textile industry was using natural dyes extracted from natural sources, such as plants and shellfish. Presently, many dyes used by the textile industry are produced by synthetic procedures, and they are being released to the natural water bodies. Even though most of the dyes are not toxic or carcinogenic to fish and other aquatic organisms, many of them pose a serious hazard. There are both chemical and biological methods for treating these dyes. The concentration of the dye in the water bodies becomes a constraint. Therefore, many groups of scientists are busy finding microorganisms, such as bacteria and fungi, to degrade textile dyes. Some of the microorganisms, such as *Proteus sp., Enterococcus sp., Streptococcus sp., Bacillus subtillis, Cariolus vesicolor* and others, are used for degrading textile dyes.

3.8 Conclusion

Waste is a product, by-product or residue that cannot be used profitably and damages the environment. The nature of the waste is biological, chemical or physical and may originate from processes such as manufacturing, agriculture, energy production, transport, house building and domestic activities. The hazards produced by wastes and pollutants will vary depending on their nature. The biological agents present in the waste spread diseases, contaminate food or water or may be mediated by a vector. Modern biotechnology has provided a new tool in which naturally occurring biocatalysts are used to detoxify harmful chemicals, metals and other wastes released by industry into the environment. The processes such as composting and landfilling are the traditional methods of waste management. Bioleaching of major wastes is considered a key technology for treating fly ash, electronic waste, spent batteries and spent oil catalysts. Mesophilic and thermophilic acidophiles microorganisms are getting consideration for bioleaching of metals. Biosorption is a very innovative new technology for removing metals from wastewater. Nowadays, the immobilization of cells for bioabsorption is widely used. Bioremediation uses naturally existing microorganisms to speed up the breaking down of biological substances and degrading various materials. This technology is also used to degrade wastes from the dairy, dye and paper and pulp industries.

REFERENCES

Ahalya N., Ramchandra T. V. and Kanamadi R. D. 2003. Biosorption of heavy metals. *Res. J. Chem. Environ.*, 7(4): 71–79.

Aksu Z., Sag Y. and Kutsal T. 1992. The biosorption of copper (II) by *C. vulgaris* and *Zramigera. Environ Technol.*, 13: 579–586.

Arica M. Y., Arpa C., Kaya B., et al., 2003. Comparative biosorption of mercuric ions from aquatic systems by immobilized live and heat inactivated *Trarnetes versicolor* and *Pleurotus sajur-caju. Bioresour Technol.*, 2: 145–154.

Babu G. R. Y., et al. 1993. Degradation of inorganic cyanides by immobilized *Pseudomonas putida* cells. In A. E. Torma, M. L. Apel and C. L Brierley (Editors), *Biohydrometallurgical Technologies*, Vol. II, The Minerals, metals and materials Society.

Bai R. Sudha and Emilia Abraham. 2003. Studies on chromium (VI) adsorption-desorption using immobilized fungal biomass. *Bioresour. Technol.*, 87: 17–26.

Brierley C. L. 1990. Bioremediation of metal contaminated surface and Groundwater. *Geomicrobiol. J.*, 8: 201–223.

Brierley C. L., et al. 1993. Immobilization of Biomass for industrial application of Biosorption. In: A. E. Torma, M. L. Apel and C. L. Brierley (Editors), *Biohydrometallurgical Technologies*, Vol. II, The Minerals, Metal and Materials Society.

Chandra Sekhar, et al. 1998. Removal of metal ions using an industrial biomass with reference to environmental control. *Int. J. Miner. Proc.*, 53: 107–120.

Costa A. C. A. and Leite S. G. F. 1991. Metals Biosorption by sodium alginate immobilized *Chlarella homosphaera* cells. *Biotechnol Left.*, 13(8): 55–562.

Ercole C., Veglio F., Toro L., Ficara G. and Lepidi A. 1994. Immobilisation of microbial cells for metal adsorption and desorption. In: *Mineral Bioprocessing II*, Snowboard, UT.

Francesca Pagnanelli, Luigi Toro and Francesco Veglio. 2002. Olive mill solid residues as heavy metal sorbent material: A preliminary study. *Waste Manag.*, 22: 901–907.

Friis F. and Keith P. 1998. Biosorption of uranium and lead by *Streptomyces longwoodensis. Biotechnol. Bioeng.*, 28: 21–28.

Gadd G. M., et al. 1988. Heavy metal and radionuclide by fungi and yeasts. In: P. R. Norris and D. P. Kelly (Editors), *Biohydrometallurgy*. A. Rowe, Chippenham.

Gadd G. M. 1990. Fungi and Yeasts for metal accumulation. In: M. Ehrlich and C. L. Brierly (Editors), *Microbial Mineral Recovery*, New York, McGraw-Hill, pp. 249–276.

Galun M., Galun E., Siegel B. Z., et al. 1987. Removal of metal ions from aqueous solutions by Pencillium biomass: Kinetic and uptake parameters. *Water, Air and Soil Pollution.*, 33: 359–371.

Gulay Bayramoglu, Adil Denizli, Sema Bektas, et al. 2002. Entrapment of *Lentinus sajor-caju* into Ca-alginate gel beads for removal of Cd (II) ions from aqueous solution: preparation and biosorption kinetic analysis. *Microchem. J.*, 72: 63–76.

Hammaini A., Felisa Gonzalez, Antonio Ballester, et al. 2003. Simultaneous uptake of metals by activated sludge. *Minerals Eng.*, 16: 723–729.

Holan Z. R., Volesky B. and Prasetyo I. 1993. Biosorption of cadmium by biomass of marine algae. *Biotechnol. Bioeng.*, 41: 819–825.

Huang C. and Huang C. P. 1996. Application of *Aspergillus oryzae* and *Rhizopus oryzaefor* Cu (II) removal. *Water Res.*, 9: 1985–1990.

Jeffers T. H. 1991. Biosorption of metal contaminants from acidic mine waters. In: R. W. Smith and M. Misra (Editors), *Mineral Bioprocessing*, The Minerals, Metal and Materials Society.

Kapoor A. and Viraraghavan T. 1995. Fungal Biosorption: An alternative treatment option for heavy metal bearing wastewater: A review. *Bioresour. Technol.*, 53: 195–206.

Kapoor A. and Viraraghavan T. 1998. Biosorption of heavy metals on *Aspergillus niger:* Effect of pretreatment. *Bioresour. Technol.*, 63: 109–113.

Keskinkan O., Goksu M. Z. L., Yuceer A., Basibuyuk M. 2003. Heavy metal adsorption characteristics of a submerged aquatic plant (*Myriophyllum spicatum*). *Process Biochem.*, 1–5.

Kratochivl D. and Volesky B. 1998a. Biosorption of Cu from ferruginous wastewater by Algal Bio-mass. *Water Res.*, 32: 2760–2768.

Kuyucak N. and Volesky B. 1988. Biosorbents forrecovery of metals from industrial solutions. *Biotechnol Left.*, 10(2): 137–142.

Lisa Norton, Baskaran Kanagaratnam and McKenzie Trevor. 2003. Biosorption of zinc from aqueous solutions using biosolids. *Adv. Env. Res.*, 8(3–4): 629–635.

Loaec Martine, Rene Olier and Jean Guezennec. 1997. Uptake of lead, cadmium and Zinc by a novel bacterial exopolysaccharide. *Water Res.*, 31(5): 1171–1179.

Macaskie L. E., Wates J. M. and Dean A. C. R. 1987. Cadmium accumulation by a citrobacter sp. Immobilized on gel and solid supports: Applicability to the treatment of liquid wastes containing heavy metal cations. *Biotechnol. Bioeng.*, 30: 66–73.

Mahan C. A. and Holcombe J. A. 1992. Immobilisation of algae cells on silica gel and their characterization for trace metal preconcentration. *Anal. Chem.*, 64: 1933–1939.

McGahren W. J., Perkinson G. A., Growich J. A., et al. 1984. Chitosan by fermentation. *Process Biochem.*, 19: 88–90.

Michel L. J., Macaskie L. E. and Dean A. C. R. 1986. Cadmium accumulation by immobilized cells of a Citrobacter sp. Using various phosphate donors. *Biotechnol. Bioeng.*, 28: 1358–1365.

Mittleman M. W. and Geesey G. G. 1985. Copper binding characteristics of exopolymers from a freshwater sediment bacterium. *Appl. Environ. Microbiol.*, 49: 846–851.

Peng T. Y. and Koon K. W. 1993. Biosorption of cadmium and copper by *Sacchromyces cerevisiae. Microb. Util. Renewable Resour.*, 8: 494–504.

Sakaguchi T. and Nakajima A. 1991. Accumulation of heavy metals such as uranium and thorium by microorganisms. In: R. W. Smith and M. Misra (Editors), *Mineral Bioprocessing*, The Minerals, Metals and Materials Society.

Schneider I. A. H. and Rubio J. 1999. Sorption of heavy metal ions by the non-living biomass of freshwater macrophytes. *Environ Sci. Technol.*, 33: 2213–2217.

Scott J. A. and Karanjkar A. M. 1992. Repeated cadmium biosorption by regenerated *Enterobacter aerogenes* biofilm attached to activated carbon. *Biotechnol. Lett.*, 14(8): 737–740.

Torma A. E., Apel M. L. and Brierley C. L. (Editors). 1993. Biohydrometallurgical Technologies. *The Minerals, Metals and Materials Society.*, II: 799.

Tsezos M. and Volesky B. 1982. The mechanism of uranium biosorption by *Rhizopus arrhizus. Biotechnol. Bioeng.*, 24: 385–401.

Volesky B. 1986. Biosorbent Materials, Biotechnoi. *Bioeng Symp.*, 16: 121–126.

Volesky B. and Holan Z. R. 1995. Biosorption of heavy metals. *Biotechnol Prog.*, 11: 235–250.

Wang T. C., Weissman J. C. and Ramesh G. 1996. Parameters for removal of toxic heavy metals by water milfoil *Bull Environ. Contam. Toxicol.*, 57: 779–786.

Whistler R. and Daniel T. R. 1985. Carbohydrates. In: Owen R. Fennema (Editor), *Food Chemistry*, pp. 96–105.

Wong P. K. and Kwok S. C. 1992. Accumulation of nickel ion by immobilized cells of Enterobacter species. *Biotechnol. Left.*, 14(7): 629–634.

Yan Guangyu and Viraraghavan T. 2000. Effect of pretreatment on the bioadsorption of heavy metals on *Mucor rouxii. Water SA.*, 26: 1.119.

Zhou J. and Kiff R. J. 1991. The uptake of copper from aqueous solution by immobilized fungal biomass. *J. Chern Technol. Biotechnol.*, 52: 317–330.

4

Role of Absorption and Adsorption in the Removal of Waste

Dimple V. Shah

CONTENTS

4.1 Introduction

Industries such as textile, paper, plastics and leather use dye in large quantities. Among them, the textile industry is the prime user for dyeing multiple types of fiber. Wastewater from the textile industry not only contains dyes but other polluting matters such as toxic acids, bases, organic and inorganic residues as well. Dangerous chemicals producing dyes like benzidine or metals are carcinogenic and mutagenic. The dye effluence in wastewater not only affects the aesthetic nature but also obstructs the transmittance of sunlight and so reduces photosynthesis and disturbs ecosystems and water life's natural balance (Demirbas, Koboya, and Sulak 2008; Sun et al. 2008; Emanue, G. Kumar). Industrial wastewater must be treated to an environmentally acceptable limit. Industrial contamination has dissolved organic compounds, suspended solids, phenols, heavy metals, pesticides and so on. Generally, rivers are capable of self-purification, and natural decomposition takes place in conventional wastewater treatment plants. The current problems of wastewater treatment have increased due to the increasing pollution from organic compounds that are nonbiodegradable (Olafadehan and Aribike 2000). For the treatment of industrial wastewater, various methods include preliminary treatment, primary treatment, secondary treatment and tertiary treatment. The main aim of preliminary treatment and primary treatment is the removal of gross solids, such as large floating and suspended solid matter, grit, oil and grease if present in sizable quantities. Secondary treatment is a biological process that involves bacteria and other microorganisms that remove the dissolved and colloidal organic matter present in the wastewaters. Tertiary treatment (final treatment) is meant for "polishing" the effluent from the secondary treatment to improve the quality further. Wastewaters can be treated effectively and economically. These methods are evaporation, reverse osmosis, precipitation, ion exchange, electrolysis and adsorption.

In recent trends, adsorption is the most widely accepted method for environmental treatment all around the world. The

DOI: 10.1201/9781003186304-4

liquid–solid interface adsorption phenomenon is based on the capacity of solids to concentrate unwanted substances from solutions onto their surfaces. This principle can be applied for removing solid pollutants from wastewater.

Chemical waste is further subdivided into organic and inorganic pollutants. The most common form of water pollution is organic pollution. These pollutants can be classified into two groups: biodegradable (natural compounds) and nonbiodegradable (synthetic compounds). Aquatic organisms that multiply excessively or are undesirable, harmful or injurious are classified as biological pollutants. These pollutants arise as an effect of pollution by sewage or industrial wastes. Activated carbon makes adsorption a fiscally feasible process. Nowadays, nonconventional adsorbents have come into applications. Nonconventional adsorbents include inorganic adsorbents, organic adsorbents and biosorbents capable of removing heavy metal ions and organic pollutants from wastewater. Nonconventional adsorbents materials, such as chitin, chitosan, modified cotton, keratin, fly ash, oil cakes, fruit peels, barks and others, are used for removing heavy metals.

Treatment of industrial wastewater

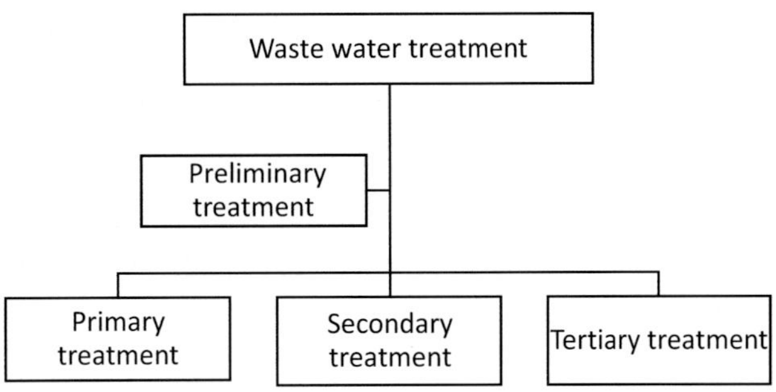

Treatment modalities

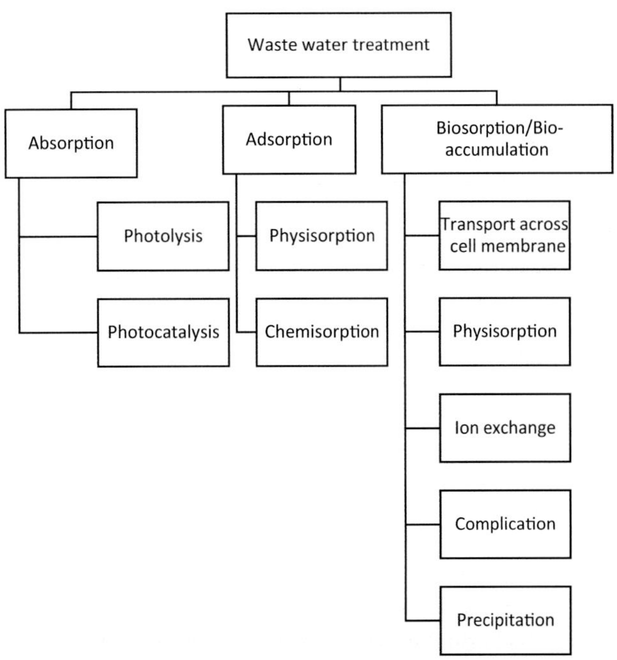

4.2 Absorption Process

Absorption is used to purify water (iron removal, oxidation, disinfection, biological purification) or a polluted gas. Gases to be dissolved are usually not very soluble. In general, in wastewater treatment applications, absorption is accompanied by a chemical reaction that is usually a fast oxidation reaction. When a gas reacts strongly with certain constituents found in water, the mass transfer coefficient is greater than in pure water. Not only are gas solubility and diffusion capacity accounted for, but the chemical kinetics will also limit the concentration in water of the O_2, O_3 and CO_2 absorbed.

Physico-chemical methods just transfer pollutants from one phase to other, complete degradation is impossible. The stable dyes, such as azo and azin compounds, are not degradable by biological methods. Semiconductor-mediated photocatalytic oxidation is a reliable technique in degrading organic effluents into innocuous products. In nature, purification is under sunlight, initiating the breaking of organic molecules into simpler molecules and ultimately to carbon dioxide and other mineral products. The already-existing methods such as adsorption involve only a phase transfer of the pollutants without degradation, and biological treatment cannot be applied to contaminants that are toxic to the bacteria. The photolysis process is one of the recent modes that can be used for detoxification. During photolysis, a direct photochemical transformation takes place when energy from light attacks the bond within a molecular compound, thereby degrading the compound. Photoexcitation is the first step in a photochemical process in which the reactant is elevated to a higher energy of excited state and then degraded toward the base constituents. This process occurs when the compound efficiently absorbs light and is highly active photochemically. The C–H, C–C, and C–O bonds with bond strength of 80 to 100 kcal/mole are commonly found in organic molecule structures. Correspondingly, ultraviolet (UV) light in the range of 100- to 400-nm wavelength and of specific concentration (71–286 kcal/mole) can be used to break these bonds, but this technique is less suitable for all the compounds.

Photocatalysis is a reaction in which a substance is activated by absorbing the light to enhance the rate of a chemical reaction without involving itself. The photocatalyst modifies the rate of chemical reaction by utilizing light. The photocatalyst creates a strong oxidation agent and electronic holes to break down the organic matter into water and carbon dioxide in presence of light. Several UV/semiconductor systems are used for the photocatalytic degradation of organic pollutants. Mostly, these semiconductor particles are metal oxides, such as TiO_2, WO_3, ZnO, and Fe_2O_3, and metal chalcogenides, such as CdS and MoS_2. Also, ozone, H_2O_2, hypocholorous acid and chlorine are used for water treatment.

4.3 Adsorption Process

Adsorption is the partitioning process in which a contaminant gets transferred from dissolved state in the aquatic phase to the surface of the solid phase. In adsorption, a solid surface comes into contact with a solution that has the tendency to accumulate and form a layer of solute molecules. The rate of adsorption is influenced by many factors. The general effects include more important factors, such as the nature of adsorbent and adsorbate, the extent of agitation, concentration, pH, temperature, contact time, shape and size of adsorbent particles, solubility of the solute, salinity and concentration of medium and others. The rate of adsorption increases as the pH is increased, and this observation is explained through surface modifications of the adsorbent and the associations of the adsorbate in the medium. The concentration of the salts in the medium decreases the degree of adsorption and varies according to the cationic and anionic nature of the salt. The observations are explained in terms of the nature of the electrical double layer that may be formed near the surface of the adsorbent. The order, rate and rate constants are found dependent on different adsorbent dose, pH and temperature

Inside the liquid, a molecule is surrounded by other liquid molecules on all sides exerting an equal amount of intermolecular forces of attraction in all directions. Whereas a molecule on the surface of liquid is surrounded by a larger number of molecules located inside liquid than the fewer molecules on outer surface that are in a vapor phase. An imbalanced force of intermolecular interaction is experienced by the molecules at a surface of a liquid, which contributes to the surface energy and causes surface tension. This causes a preferential accumulation of substrate molecules at the surface is called adsorption, which is a surface phenomenon. A similar imbalanced inward force of attraction exists in the solid at the surface. In transition metals, there are unutilized free vacancies at the surface. It is because of these unbalanced inward forces of attraction or free valences that when the solids and liquids come into contact, they have the property to attract and retain the molecules of a gas or a dissolved substance on their surfaces.

Adsorption is apparently an electrical effect due to unbalanced electronic forces. Adsorption is associated with the evolution of heat. The energy liberated when 1 mg of mole gas is adsorbed on the surface of the solid is called heat of adsorption.

The adsorption process is associated with the terms such as the following:

1. Adsorbate: The substance that gets adsorbed on any surface is called an adsorbate.
2. Adsorbent: The substance on the surface of which adsorption takes place is called an adsorbent.
3. Desorption: The removal of the adsorbed substance from a surface is called a desorption.

In a few instances, adsorption and absorption occur simultaneously which could not be differentiated so for this reason, noncommittal term sorption is frequently used.

For example, suppose in a solution of dye color, a powder of activated charcoal is added; there will be a decrease in the intensity of the color. The explanation for this is that the molecules of the dye color are attracted toward the solid surface change of charcoal. Suppose that these molecules remain on the surface of the solid and do not enter into the solid; then the concentration of the solution on the surface will be more than the concentration of the solution in the inner bulk. Thus, a change in concentration on the surface is observed. This process, which involves the molecules being maintained on the solid or liquid surface and the concentration of molecules on the surface becoming more than the concentration of the molecules in the bulk, is called adsorption.

4.3.1 Factors Affecting Adsorption Process

Nearly all solids absorb gases to a larger extent. The actual amount of a gas adsorbed depends on the following factors:

1. **Surface area and type of the adsorbent:** At the same temperature, the same amount of gas adsorbed by different adsorbents is different. Furthermore, the greater the surface area of the adsorbent, the greater is the volume of the gas adsorbed. Substances such as charcoal and silica gel are excellent adsorbents because they have highly porous structures and hence large surface areas.
2. **Nature of the gas being adsorbed:** At the same temperature, different gases are adsorbed to different extents by the same adsorbent. It has been observed that if the critical temperature of a gas is higher, a greater is the amount of that gas is adsorbed.

   The critical temperature of a gas is defined as the minimum temperature above which the gas cannot be liquefied; however, high pressure may be applied.
3. **Temperature:** The adsorption increases with a decrease in temperature and vice versa when the adsorption of any particular gas is by some particular adsorbent. The decrease of adsorption with an increase of temperature can be interpreted as any other equilibrium process; adsorption involves a true equilibrium. Two opposite processes are involved. Condensation means desorption of gas molecules on the surface of the solid and evaporation means adsorption of the gas molecules from the surface of the solid into the gaseous phase. As the process of condensation is exothermic, the equilibrium may be represented as:

$$\text{Gas (Adsorbate)} + \text{Solid (Adsorbent)} \rightleftharpoons \text{Gas adsorbed on solid} + \text{Heat}$$

4. **Pressure:** At a constant temperature, the adsorption increases with an increase in a gas, that is, an increase in pressure. It is observed that at a low temperature, initially, the adsorption of a gas increases very rapidly as the pressure is increased from small values.
5. **Activation of the solid adsorbent:** Methods to increase the adsorbing power of an adsorbent. Increasing the surface area will increase the adsorbent. Following is the process to achieve it:
   a. By process to make the surface of the adsorbent rough, for example, electroplating by depositing finely dispersed metals on the surface of the

adsorbent, by mechanical rubbing or by some action by chemicals

b. Subdividing the adsorbent into smaller pieces or grains increases surface area. This method has practical limitations, like breaking the adsorbent into infinitesimal particles, making almost powder, does not allow the penetration of the gas and ultimately obstruct adsorption.
c. By removing the gases already adsorbed, that is, by desorption, for example, desorption of charcoal takes place at a temperature between 623 to 1273 K (when it is heating, in superheated steam or in a vacuum). This helps in the adsorbed gas to be desorbed from the charcoal.

4.3.2 Types of Adsorption

There are two types of adsorption:

1. Physical Adsorption or Physisorption: In physical adsorption, the molecules are adsorbed through van der Waals forces. Physical adsorption is a physical process. The value of enthalpy is about 20 to 40 k $Jmol^{-1}$. Physical adsorption generally results at low temperatures. As the temperature increases, physisorption decreases. It is not specific but is instantaneous. In physical adsorption, multiple layers can be formed on the adsorbent. It is a reversible process. Less energy is required for this type of process. Physical adsorption usually depends on the nature of the gas.
2. Chemical Adsorption or Chemisorption: In chemical adsorption, the attraction forces are of chemical bond type. The value of adsorption enthalpy is about 80 to 240 k $Jmol^{-1}$. Chemisorption results at high temperatures. There is no peculiar effect of change in temperature. But it is specific and not instantaneous; it can be slow or fast. In chemical adsorption, a unimolecular layer is formed on the adsorbent. It is irreversible in nature. More energy is required for this type of process. Chemical adsorption also depends on the nature of the gas.

4.3.3 Adsorption Isotherms

The extent of adsorption on a given surface increases with an increase in pressure (for gases) and concentration (for solutions) at a constant temperature. At low temperatures, the adsorption of a gas increases rapidly with the pressure. When the temperature is high, the increase in adsorption is relatively less. When the adsorbent and the adsorbate are in a closed vessel, initially, the pressure of the gas decreases, and later, it reaches constant value also the amount of gas adsorbed reaches a constant value because at equilibrium, the rate of adsorption becomes equal to the rate of desorption.

The amount of gas adsorbed depends on the surface area of the adsorbent or on its mass if the adsorbent is taken in the form of powder.

When equilibrium is reached, the extent of adsorption is usually expressed as x/m, where m is the mass of the adsorbent and x is the mass of the adsorbate.

The specific surface area of a solid (in the form of a powder or porous mass) is the surface area in square meters per gram of the adsorbent. Highly active solids with large surface area (several hundred square meters per gram) are used as adsorbents. **The relationship at a given temperature between the amounts of the substance adsorbed the concentration of the adsorbate at equilibrium and the amount of the adsorbent (pressure in case of gaseous adsorbents) is known as adsorption isotherm.**

4.3.3.1 Freundlich Adsorption Isotherm

A relationship between the extent of adsorption (x/m) and the equilibrium pressure (p) can be derived as follows:

At low values of p, the graph is nearly straight and sloping, the equation is

$$\frac{x}{m}\alpha p^1 \ or \ \frac{x}{m} = constant \times p^1$$

At high pressure, *x/m* becomes independent of the values of *p*:

$$\frac{x}{m}\alpha p^0 \ or \ \frac{x}{m} = constant \times p^0$$

In the intermediate range of pressure, *x/m* will depend on *p* raised to powers between 1 and 0, that is, fractions. For a small range of pressure values, we can write

$$\frac{x}{m}\alpha p^{\frac{1}{n}} \ or \ \frac{x}{m} = constant \times p^{\frac{1}{n}},$$

where *n* is a positive integer and n and K are constants depending on the nature of the adsorbate and adsorbent.

This relationship was originally put forward by Freundlich and is known as Freundlich adsorption isotherm.

The simplest type of adsorption isotherm is shown in Figure 4.1.

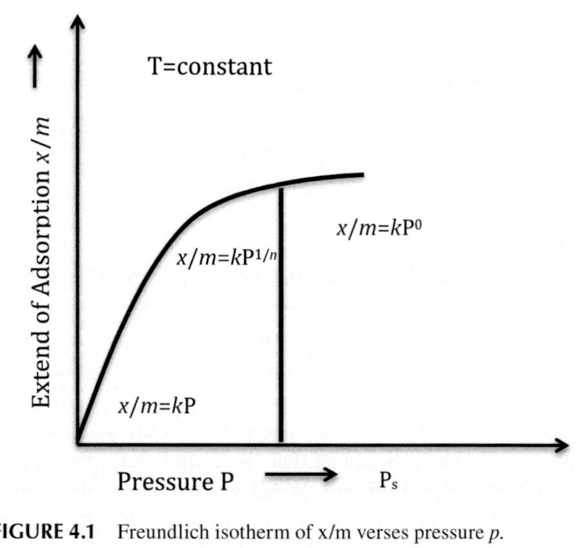

FIGURE 4.1 Freundlich isotherm of x/m verses pressure *p*.

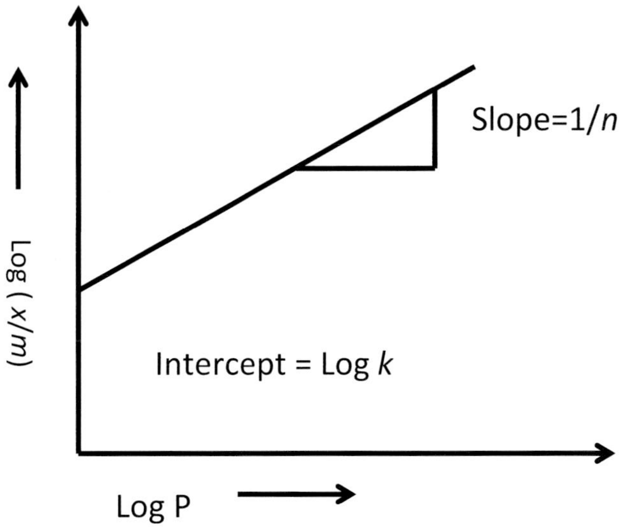

FIGURE 4.2 Freundlich isotherm of log (x/m) verses log *p*.

At equilibrium pressure P_s, or saturation pressure, *x/m* reaches its maximum value, and then it remains constant even though the pressure p is increased. This type of adsorption isotherm is observed when the adsorbate forms a uniform molecular layer on the surface of the adsorbent.

To test the validity of the equation $\frac{x}{m} = kp^{\frac{1}{n}}$, taking logarithms on both sides,

$$log\frac{x}{m} = \frac{1}{n}\log p + logK .$$

If log x/m is plotted against log p, a straight line is obtained.

Hence, the equation is comparable to the general equation of a straight line (y = mx + n; see Figure 4.2).

The slope of the line (tan θ) gives the value of $\frac{1}{n}$ while the value of log K is given by the intercept of the line.

The experimental values, when plotted, however, show some deviation from linearity, especially at high pressures. Hence, the relation is considered as an approximate one and is suitable only at low pressures.

Adsorption from Solutions: Solid surfaces can also adsorb solutes from the solutions. Activated charcoal can adsorb coloring impurities from the solutions of organic compounds. Adsorption from solutions can also involve colorless solutions. This type of adsorption is also affected by temperature and concentration. The extent of adsorption decreases with an increase in temperature and increases with an increase in concentration.

The isotherm for the adsorption of solutes from solutions (by the solid adsorbents) is found to be similar to that shown in Figure 4.1.

Hence, the relationship between x/m (mass of the solute adsorption per gram of the adsorbent) and the equilibrium concentration, C, of the solute in the solution is also similar, that is,

$$\frac{x}{m} = KC^{\frac{1}{n}} .$$

Taking logarithms on both sides of the equation,

$$log\frac{x}{m} = \frac{1}{n}\log C + logK .$$

This equation implies that a plot of log x/m against log C should be a straight line with slope $\frac{1}{n}$ and intercept log K. This is for small ranges of concentration (see Figure 4.2).

The equation for adsorption from solutions gives improved results than for adsorption of gases by solids.

4.3.3.2 Langmuir Adsorption Isotherm

Langmuir deduced a relationship on the basis of adsorption theory, which he put forward as follows:

> Suppose at any instant, the fraction of the surface covered by the adsorbed molecules = θ.
> The fraction of the surface bare and available for adsorption = $1 - \theta$.
> Suppose the pressure of the gas at that instant = P
> Then according to Langmuir's theory of adsorption, now the rate of condensation is given by $\alpha(1 - \theta) \times P = k_1 (1 - \theta) P$,
> where, k_1 is a constant of proportionality.
> Now the rate of evaporation is given by $\alpha\,\theta = k_2\theta$,
> where k_2 is another constant of proportionality.
> When the adsorption equilibrium is attained, $k_1 (1 - \theta) P = k_2\theta$

or

$$\theta\,(k_2 + k_1 P) = k_1 P$$

$$\theta = (1+x)^n = \frac{k_1P}{k_2 + k_1P} = \frac{\left(k_1/k_2\right)P}{1+\left(k_1/k_2\right)P} = \frac{bP}{1+bP} ,$$

where $\frac{k1}{k2}$ = b is another constant.

As the gas molecules are assumed to form a layer one molecule thick on the surface, the gas adsorbed per unit area or per gram of the adsorbent must obviously be proportional to the surface covered, that is,

$$\frac{x}{m}\alpha\theta or \frac{x}{m} = k_3\theta ,$$

where x is the mass of the gas adsorbed on m gram of the adsorbent and k_3 is a constant of proportionality:

$$\therefore \frac{x}{m} = k_3\frac{bP}{1+bP}$$

or

$$\frac{x}{m} = \frac{aP}{1+bP} ,$$

where $k_3\, b = a$ is another constant.

This is called the Langmuir adsorption isotherm. The constants a and b depend on the type of the gas adsorbate, the nature of the solid adsorbent and the temperature. Their values can be found from the experimental data.

To verify the validity of the Langmuir adsorption equation, both sides of the equation may first be divided by P so that we get

$$\frac{x/m}{P} = \frac{a}{1+bP}.$$

Taking the reciprocal of both sides, we get

$$\frac{P}{x/m} = \frac{1}{a} + \frac{b}{a}P$$

or

$$\frac{P}{x/m} = \frac{b}{a}P + \frac{1}{a}$$

As a and b are constants, a plot of $\frac{P}{x/m}$ versus P should be a straight line with slope equal to b/a and an intercept on the y-axis equal to $1/a$. This is actually found to be as shown in Figure 4.3 for the adsorption of N_2 on mica at 90 K.

Alternatively, taking reciprocal of both sides of Langmuir equation, we have

$$\frac{m}{x} = \frac{1+bP}{aP}$$

or

$$\frac{m}{x} = \frac{1}{aP} + \frac{b}{a}$$

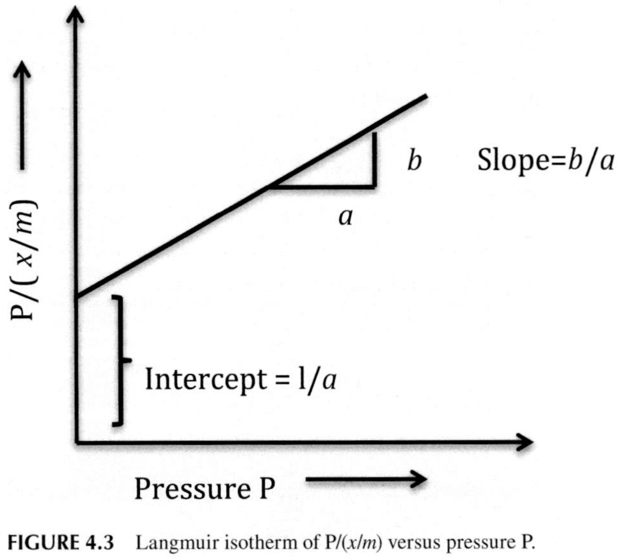

FIGURE 4.3 Langmuir isotherm of P/(x/m) versus pressure P.

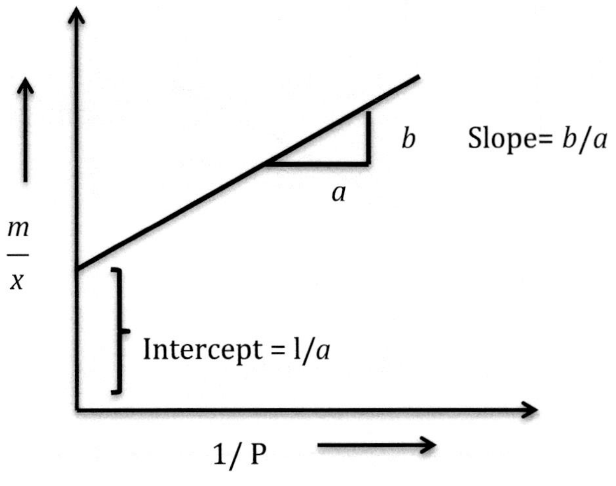

FIGURE 4.4 Langmuir isotherm of m/x verses 1/P.

or

$$\frac{m}{x} = \frac{b}{a} + \frac{1}{a} \times \frac{1}{P}.$$

Thus, a plot of m/x versus $1/P$ will be a straight line with slope equal to $1/a$ and an intercept equal to b/a (see Figure 4.4).

For adsorption from solution, the Langmuir equation is written in the form

$$\frac{x}{m} = \frac{aC}{1+bC},$$

where C is the equilibrium concentration in the solution of the solute; it can be tested in a similar way to that of adsorption of gases, that is, by plotting $\frac{C}{x/m}$ versus C or m/x versus $1/C$

$$\frac{C}{x/m} = \frac{b}{a}C + \frac{1}{a} \text{ and}$$

$$\frac{m}{x} = \frac{b}{a} + \frac{1}{a} \times \frac{1}{C}.$$

It can be tested in a similar way to that of adsorption of gases, that is, by plotting $\frac{C}{x/m}$ vs C and $\frac{m}{x}$ vs $\frac{1}{C}$

The Langmuir equation can be written in two situations:

1. At low pressure: In equation $\frac{x}{m} = \frac{ap}{1+bp}$, if the value p is taken to be low; then the value of bp can be neglected in comparison to 1. Hence, $\frac{x}{m} = ap$, that is, at low pressure, the adsorption of gas is directly proportional to the pressure.

Types of Adsorbents

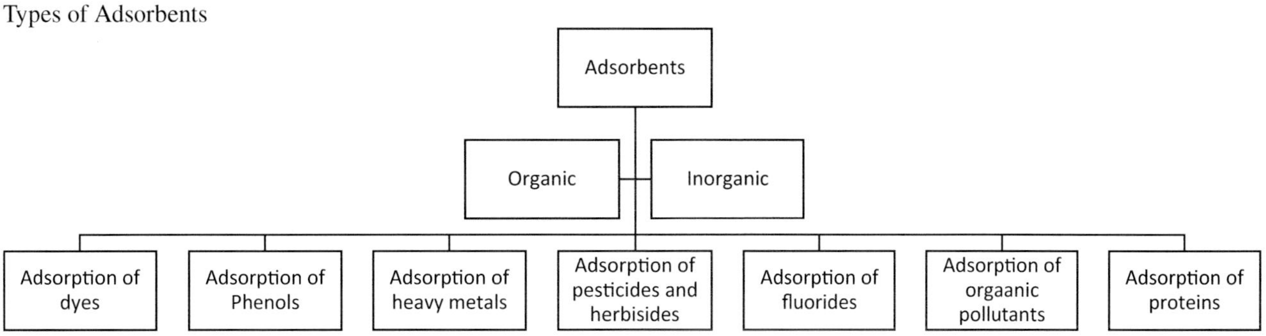

2. At high pressure: In the equation $\frac{x}{m}=\frac{ap}{1+bp}$, if the value of p is taken high then, 1 can be neglected in comparison to the value of bp.

Hence, $\frac{x}{m}=\frac{a}{b}=consant.$ Thus, at high pressure, adsorption will be almost constant. To determine the constants a and b, the equation $\frac{x}{m}=\frac{ap}{1+bp}$ can be written by inverting it, that is, it will be $\frac{m}{x}=\frac{1+bp}{ap}$. Now, $\frac{m}{x}=\frac{1+bp}{ap}=\frac{1}{bp}=\frac{1}{ap}+\frac{b}{a}$.

The equation is also the equation for straight line, and so if a graph of $\frac{m}{x}$ versus $\frac{1}{p}$ is plotted, then the value of the slope will be equal to $\frac{1}{a}$, and the value of intercept will be equal to $\frac{b}{a}$. From this relationship, the values of constants a and b can be obtained.

The Freundlich adsorption isotherm shows that in the beginning, the values of adsorption $\frac{x}{m}$ increases with the increase in pressure of gas, but when the value of pressure increases, the isotherm tends to become constant; that is, the adsorption attains saturation, and more adsorption will not happen, which supports the previous two situations.

4.3.4 Types of Adsorbents

Adsorbents are classified into two types: natural adsorbents and synthetic adsorbents. Charcoal, clays, clay minerals, zeolites and ores are natural adsorbents. These natural materials are cheap and abundant, have significant potential for modification and have enhancement capabilities. On the other side, synthetic adsorbents are made from agricultural products and their wastes, household wastes, factory wastes, sewage silt and polymers. Each adsorbent has different characteristics, such as porosity, pore structure and nature of its adsorbing surfaces, as well as the selection of adsorbate. The waste products generally used are fruit wastes, fertilizer wastes, coconut shell, scrap tires, bark and other tannin-rich materials, medicinal waste, rice husk, sawdust, petroleum wastes, fly ash, blast chitosan, furnace slag, seafood processing wastes, seaweed and algae, peat moss, clays, red mud, zeolites, sediment and soil, ore minerals and others. Adsorbents such as activated carbons from organic pollutants have complex adsorption processes and are still considered difficult in processing. The main cause of this difficulty includes electrostatic, dispersive and chemical interactions, intrinsic properties of the solute (e.g., solubility, ionization constant), intrinsic properties of the adsorbent (such pore size distribution), solution properties (in particular, pH) and the temperature of the system. Activated carbons (ACs, both granular activated carbon [GAC] and powdered activated carbon [PAC]) are used for the removal of unwanted odor, color, taste and other organic and inorganic impurities from domestic and industrial wastewater. ACs (both GAC and PAC) have large surface areas, micro-porous structures and nonpolar characteristics and economic availability and are mainly used. ACs have a major carbon constituent that is 95% of its mass weight, and other constituents are 5%. ACs contain hydrogen, nitrogen, sulfur and oxygen atoms in smaller percentages from different sources that have attached during activation.

4.3.5 Adsorption of Dyes

Commercial systems for decolorizing wastewaters use activated carbon as the sorbent because of its excellent absorption ability.

Synthetic and natural materials are used for the adsorptive removal of color. The use of solid natural organic materials is increasing these days (Liversidge et al. 1997; Robinson, Chandran, and Nigam 2002b, 28; Annadurai, Juang, and Lee 2002; Sivaraj, Namasivayam, and Kadirvelu 2001; Robinson, Chandran, and Nigam 2002a, 36; Gupta et al. 2000).

The natural materials and agricultural wastes are low-cost adsorbents and are environmentally friendly. There are many studies related to the use of cotton, corn cobs, palm fruit parts, eucalyptus bark, sunflower stalks and wheat straw as decolorizing adsorbents (El-Geundi 1991; Nassar, Hamoda, and Radwan 1995; Nassar and Magdy 1997; Nassar 1999; Morais et al. 1999; Bouzaida and Rammah 2002; Sun and Xu 1997).

GAC was also used for methylene blue dye adsorbent. AC produced from jackfruit peel was used to treat and remove malachite green from wastewater (Ibaraj and Sulochana 2002). GAC was also used for methylene blue dye adsorbent.

Activated carbon produced from Jack fruit peel was used to treat and remove malachite green from waste water (Ibaraj and Sulochana 2002).

Waternut carbon (WNMC) is used removing malachite green (triphenylmethane) and carbon red (azo) dye from an aqueous solution by batch and column process (Rais and Mondal, 2010) [19]. Treated *coconut hard shell* (TCHS) is used for removing methylene blue (Hasfalina, Akinbile, and Jun 2015) [19]. Treated *coconut hard shell* (TCHS) is used for removing methylene blue (Hasfalina, Akinbile, and Jun 2015).

Due to the nonpolar nature of non-AC, the AC is used. In most cases, dyes have a polar nature. Therefore, activation is performed intentionally to introduce polar groups for different applications of carbon adsorbents. So AC is neither economical nor cost-effective. The adsorption of dyes not only depends on the characteristics of the dyes and their structures but also on the surface chemistry of the adsorbent.

Several new techniques, such as immobilization of the ash with cement, acid neutralization, wet chemical treatment and thermal treatment, have been proposed for the removal of heavy metals. The main aim is to decontaminate toxic waste and render it inert so that it can be reused or deposited without risk. The major constituents in the ash, which includes cyclone ash and scrubber ash, are silica, aluminum and calcium. The waste melting process is the other way in which the most hazardous materials, such as heavy metals, are tightly fixed in a solid phase, and the slag generated by this process can be used as a construction material. Melting residues at temperatures exceeding 1300°C can generate glassy slag, after which the molten ash is water-quenched or air-cooled.

An adsorption process carried out using cheap adsorbents is simple, selective and cost-effective compared to conventional physical and chemical processes for the treatment of colored wastewater. The use of the natural waste adsorbent ash has been investigated for removing a textile vat dye Ostanthern blue GCD remaining after the dyeing of cotton textile. The ash was obtained as a waste material during the burning of brown coal in the heating station (MIODRAG et al. 2010; Bayat 2002, 251, 275; Kuo, Lin, and Tsai 2006; Johnson, Brandenberger, and Baccin 1995; Aricke, Van Gerven, and Vandecasteele 2006; Iori and Balg 1995; Sakai et al. 1990).

4.3.6 Adsorption of Phenols

Since 1860, phenol has been used in large amounts due to its basic nature as an antiseptic. During late the 19th century and onward, the use of phenol has been further extended to the synthesis of dyes, aspirin, plastics, pharmaceuticals and petrochemical and pesticide industries. Phenols are on the priority list of organic pollutants as they have deadly effects on plants, animals and humans, even in mild concentrations. The sources of phenolic are steel mills, petroleum refineries, pharmaceuticals, petrochemical, coke oven plants, paints, coal gas, synthetic resins, plywood industries and mine discharge. The wastewater with the highest concentration of phenol (>1000 mg/L) is typically generated from coke processing. Phenolic compounds are also obtained from resin plants. The Environmental Protection Agency has put a limit of 0.1 mg/L of phenol in wastewater. The phenol regulation is strictly observed by World Health Organization. It has set a limit of 0.001 mg/L of phenol concentration in potable drinking water. Adsorptions of phenolic compounds from aqueous solutions by ACs were studied. The phenol adsorption of mono-, di- and trichlorophenols from aqueous solutions on wood-based and lignite-based carbons were investigated (Zogorski, Faust, and Haas 1976).

For the removal of phenol from industrial wastewaters, sawdust after carbonization is used as a very low-cost adsorbent. Furthermore, *Bagasse ash* (BA) and *wood charcoal* (WC) were used for removing phenol from water. The adsorptive capability of some clays and natural zeolite materials is remarkable for phenol removal. Among the studied adsorbents, calcined hydrotalcite was the best. Also, as adsorbents, silica gel, activated alumina, AC, fitra-sorb 400 and Hisir 1000 adsorbent were examined for removing phenol from aqueous solution. They found that among the tested materials, Hisir 1000 was the best. Blast furnace flue dust (BFD) and slag were used to investigate phenol adsorption. Bromophenols (2-bromophenol, 4-bromophenol and 2, 4-dibromophenol) are considered one of the toxic organic phenols. Industrial wastes are used as low-cost adsorbents for the removal of these pollutants (Larous et al. 2008; Mukherjee Somnath et al. 2007; Yapar and Yilmar 2004; Roostari and Tezel 2004; Das and Patnaik 2005; Bhatnagar 2007). For the removal of phenol from industrial waste waters, sawdust after carbonization is used as very low cost adsorbent. Further *Bagasse ash* (BA) and *wood charcoal* (WC) were used for removal of phenol from water. The adsorptive capability of some clays and natural zeolite materials are remarkable for phenol removal. They found that among the studied adsorbents calcined hydrotalcite was the best. Also, silica gel, activated alumina, AC, fitra- sorb 400 and Hisir 1000 adsorbent were examined for the removal of phenol from aqueous solution as adsorbents. They found that among the tested materials Hisir 1000 was the best. Blast furnace flue dust (BFD) and slag was used to investigate phenol adsorption. Bromophenols (2-bromophenol, 4-bromophenol and 2, 4- dibromophenol) is considered as one of toxic organic phenol. Industrial wastes are used as low cost adsorbent for the removal of these pollutants (Larous et al. 2008; Mukherjee Somnath et al. 2007; Yapar and Yilmar 2004; Roostari and Tezel 2004; Das and Patnaik 2005; Bhatnagar 2007).

To remove highly odorous dissolved organic compounds from industrial wastewater, an adsorption test also showed that GAC removed color and phenol. The five phenolic compounds that were selected were phenol (Ph), 2-chlorophenol (2-CP), 4-chlorophenol (4-CP), 2, 4-dichlorophenol (DCP) and 2, 4, 6-trichlorophenol (TCP). They were removed by AC by adsorption (Hamdaoui and Naffrechoux 2007).

The phenol adsorption trend was studied using local GAC produced from *palm date* pits. Porous clay, natural clay, leaf litter, activated phosphate rock, mesoporous carbon and others are used for the removal of phenols (Sofía et al. 2005; Enamul et al. 2010; Atef, Alzaydien, and Waleed 2009; Djebbar et al. 2012).

4.3.7 Adsorption of Heavy Metals

The heavy metals adsorption of Ajwain (*Trachysperum copticum*) from synthetic, as well as electroplating, wastewater has been studied. The sorption of Cu (II) onto waste residue left after extracting juice from the *Indian gooseberry* (Amla) *fruit (Emblica officinalis)* was studied. Also, iron-oxide-coated sand, pyrite, sawdust, tire rubber, *Pseudomonas stutzeri, perlite*, ACs, *kyanite, pecan shell, peanut hull pellets, Ulothrixzonata, magnetic chitosan,* agriculture by-products, *rice husk, green coconut shell,* and others are used for removing Cu(II) (Shaista 2012).

The adsorption of Cr (VI) on *Afsanteen (Artemisia absinthium)* was performed. For removing Cr (VI) from synthetic wastewater, *gular (Ficusglomerata)* is used. The adsorption of heavy metals *on pomegranate (Punicagranatum)* peel and the removal and recovery of Cr (IV) from a multimetal system were carried out. Coniferous leaves, hazelnut shell, rice straw, chitosan, used tires and sawdust, hazelnut-shell AC, Spirogyra species, fly ash of poultry litter, cone biomass of *Pinus sylvestris*, Dunaliella species, low-cost abundantly available adsorbent such as *soya cake*, distillery sludge, calcined Mg-Al-CO_3, indigenous low-cost material, ferrous saponite, *maple sawdust*, Fe-modified steam-exploded wheat, *Chinese reed (Miscanthus sinensis)* and others are used for Cr (II) removal. Pig-iron sludge is also explored as an adsorbent for the treatment of Cr (IV) contaminated wastewater (Shaista 2012).

Use of *olive mill residues* as heavy metal sorbent material was studied. *Gaozaban (Onosma bracteatum)*, a medicinal herb, is used for the sorption of Cd (II) from an aqueous phase. *Fennel biomass (Foeniculum vulgari)*, a medicinal herb, is used for the biosorption of Cd (II) from an aqueous phase. The use of *Polyalthia longeifolia* seeds (seeds of the Indian mast tree) as adsorbent for removing Cd (II) from an aqueous solution. Transcarpathian clinoptilolite, pyrolusite, bone char, white fungus *Trametes versicolor, C. vulgaris*, calcium hydroxyapatite, de-alginated seaweed, geothite-coated sand, perlite, chitin, chitosan-based crab shells, muloorina illite and related clay minerals, sugarcane Bagasse pith and others are used for removing Cd(II) (Shaista 2012).

Fly ash, *Citrus reticulate* (fruit peel of orange), chlorella species, *C. vulgaris* and *C. Miniata*, pretreated biomass of marine macroalga *Durvillaea potatorum*, steel converter slag, AC prepared from coir pith, sludge ash, bentonite, alumina particles, clay-based beds, baker's yeast and others are used for removing Ni (II) (Shaista 2012).

Removing Pb(II) ions from an aqueous solution onto lignite has been studied. Recycled iron material, coir pith carbon, dried animal bones, *Aspergillus niger* biomass, bentonite, peat, nonliving biomass of marine algae, natural materials, sea nodule, tree fern, clays, AC, zeolite and sepiolite, palm kernel fiber, cellulose, coir and dye-loaded coir fibres and others are used for removing Pb(II). A polystyrene-alumina-AC composite as synthetic adsorbent was investigated for removing Pb(II) ions from aqueous solutions. Cr^{+3}, Cd^{+2} and Co^{+2} heavy metal ions from textile industry wastewater were adsorbed using AC, orange peel, bone charcoal, cornstalks, agro-waste material, palm shell, tamarind hull, agricultural waste, fabric cloth, cow dung, black locus leaves and others were reported (Lagergren 1898; Ho and McKay 1998; Horsfall and Spiff 2004; Pearson 1997; Chong and Volesky 1995; Langmuir 1918; Ahalya, Kanamadi and Ramachandra 2005; Mckay, Blair, and Gardener 1982; Freundlich and Zeitschriftfr 1907; Kardivalu and Namasivayan 2000; Juang, Wu, and Tseng 1999; Hsisheng 1998; Suárez-García, Martínez-Alonso, and Tascón, 2001; Molina-Sabio et al. 1995; Baquero et al. 2003; Moreno-Piraján, Carlos, and Giraldo 2010).

4.3.8 Adsorption of Fluorides

The two most suitable natural bioadsorbents *citrus limetta peels* and *Ficus religiosa* leaves are screened on the basis of the high removal capacity of fluoride, and these bioadsorbents are used for the fluoride removal from synthetic wastewater.

Biological treatments with bacterial whole cells are another important route for fluoride removal from water in which whole cells are used either in a bulk phase or are immobilized on a solid support. Immobilized whole cells produce biolayer on a bioadsorbent surface, and when treated with a solution having fluoride, the bacterial cells bioaccumulate it within the cell, and it gets adsorbed on the bioadsorbent surface.

Also, activated alumina, calcium, iron-based oxides/hydrocides/oxohydroxides/metal oxides and others are used for the adsorption of fluorides (Bhatnagar, Eva, and, Sillanpää 2011).

4.3.9 Adsorption of Pesticides and Herbicides

Pesticides and herbicides are ubiquitous in aquatic systems and easily occur in the form of complex mixtures. The main reason for organic pollution in a water stream is the leaching of chemical fertilizers and pesticides applied to agricultural and forest land. Pesticides and herbicides are considered deadly for the living world due to their toxicity, carcinogenicity and mutagenicity. The toxicity of pesticides and their degradation products makes these chemical substances potential hazards by contaminating the environment. Serious concerns have been raised about the aquatic ecosystem and human health because of the adsorption technique for removing organic pollutants from water and wastewater in the long term accumulation of toxicological effects. The contamination of groundwater, surface water and soils by pesticides and herbicides is currently of significant concern, and this is because of the increasing use of pesticides in agriculture and for domestic purposes (Shaarani and Hameed 2010; Gilliom, 2007; Liu et al. 2011; IARC 1987; IARC 1987; Cui et al. 2011). Phosphonothioates are used as pesticides, which can be adsorbed by cyclodextrin polymers (CDPs) (Memon et al. 2008). Porous polymeric adsorbents were used for the adsorption of herbicides namely alachlor, trifluralin, amitrole and prometryn from the liquid form of a solution.

Agricultural residues such as olive kernel, corncobs, rapeseed stalks and soya stalks are used to produce AC via physical steam activation. The prepared AC was tested for removing bromopropylate (BP) pesticide from water (Salman, Njokua, and Hameeda 2011).

The adsorption of pesticides named as pesticides ametryn [2- (ethyla-mino)-4-isopropylamino-6-methyl-thio-s-triazine], aldicarb (2-methyl-2-(methylthio) propionaldehyde o-methylcarbamoyloxime],diuron[N-(3,4-dichlorophenyl)-N,N-dimethyl urea] and dinoseb [2-(sec-butyl)- 4,6-dinitrophenol], from aqueous solution onto high specific-area AC-cloth adsorbent (ACC) were studied (Ayranci and Hoda 2005).

Lignocellulosic wastes of vegetable origin such as coffee grounds, melon seeds and orange peels are used to develop new AC. The adsorption efficiency of these new adsorbents was checked with organic pollutants: o-nitrophenol and p-nitrotoluene (Djilani et al. 2012).

MgAl-layered double hydroxides (LDHs) and their calcined products abilities to adsorb pesticides contaminants, 2, 4-dinitrophenol (DNP) and 2-methyl-4; 6-dinitrophenol (DNOC) from water was assessed (Chaara et al. 2010). MgAl layered double hydroxides (LDHs) and their calcined products to adsorb pesticides contaminants, 2, 4-dinitrophenol (DNP) and 2-methyl-4; 6-dinitrophenol (DNOC) from water was assessed (Chaara et al. 2010).

AC prepared from banana stalk by potassium hydroxide (KOH) and carbon dioxide (CO2) activation (BSAC) was explored for its capacity to remove the pesticides 2,4- dichlorophenoxyacetic acid (2,4-D) and bentazon (Salman, Njokua, and Hameeda 2011).

Magnetic and graphitic carbon nanostructures are also explored for the removal of a pesticide (2, 4-dichlorophenoxyacetic acid) from an aqueous solution (Maryam and Azizian 2012). Magnetic and graphitic carbon nanostructures are also explored for the removal of a pesticide (2, 4-dichlorophenoxyacetic acid) from aqueous solution (Maryam and Azizian 2012).

4.3.10 Adsorption of Other Organic Pollutants

Organic pollutants arise from household sewage, urban runoff, industrial effluents, agriculture wastewater, pulp and paper manufacturing and aquaculture, among others. The water is consumed in large quantities during the decomposition of organic pollutants, causing oxygen depletion and having severe consequences for aquatic life. Large quantities of suspended solids in wastewater cuts down the light availability to photosynthetic organisms. Again when these particles will settle on the riverbed, it will change riverbed's characteristics, creating an unfit environment for many invertebrates. The major organic pollutants include pesticides, fertilizers, hydrocarbons, phenols, plasticizers, biphenyls, detergents, oils, greases, pharmaceuticals, proteins and carbohydrates. Activated coconut shell has a better adsorption capacity of organic pollutants. Cassava is a major production in Colombia; only small amounts of cassava are consumed by traditional food industries, and the rest are used in starch industries (Ali, Mohd. Asim, and Khan 2012; Ali, Aboul-Enein, and Chiral 2004; Damià 2005). Cassava peel, which is a waste from the starch industry, can cause environmental problems, so AC made from the KOH activation of orange peel provides high surface areas and pore volumes for adsorption of cassava (Juan Carlos and Giraldo 2011).

The removal of amoxicillin (antibiotic) from pharmaceutical effluence using bentonite and AC as adsorbents was investigated. The adsorption of methane on GAC was also studied. It was found that with decreasing temperature or increasing methane uptake the efficiency of GAC decreases (Essa, Al-Zahrani, and Nesaratnam 2004; Putraa et al. 2009). The removal of amoxicillin (antibiotic) from pharmaceutical effluence using bentonite and activated carbon as adsorbents was investigated. Adsorption of methane on granular activated carbon was also studied. It was found that with decreasing temperature or increasing methane uptake the efficiency of GAC decreases (Essa, Al-Zahrani, and Nesaratnam 2004; Putraa et al. 2009).

4.3.11 Adsorption of Protein

Protein adsorption is a process that has a basic role in biomaterials. Indeed, biomaterials surfaces in contact with biological media, like blood or serum, are immediately coated by proteins. Therefore, living cells do not interact directly with the biomaterial surfaces, but with the adsorbed protein layer. This protein layer mediates the interaction between biomaterials and cells, translating biomaterial physical and chemical properties into "biological language". In fact, cell membrane receptors bind to protein-layer bioactive sites, and these receptors–protein-binding events are transduced through the cell membrane in a manner that stimulates specific intracellular processes that then determine cell adhesion, shape, growth and differentiation. Protein adsorption is influenced by many surface properties, such as surface wettability, surface chemical composition and surface nanometer-scale morphology. Surfactant is a similar phenomenon but uses surfactant molecules in the place of proteins.

4.4 Biosorption and Bioaccumulation

Biosorption is a physiochemical process that occurs naturally in biomass. It does not interact with residues but passively concentrate and bind contaminants onto its cellular structure. **Bioaccumulation** is an active metabolic process driven by energy from a living organism and requires respiration. Both bioaccumulation and biosorption occur naturally in all living organisms.

Biosorption is defined as the potential of biological materials to remove heavy metals from wastewater through metabolically mediated or physico-chemical pathways. Nowadays, biomass is used in environmental cleanup; scientists and engineers are hoping this phenomenon will provide an economical alternative for removing toxic heavy metals from industrial wastewater and aid in environmental remediation. Pollutants react naturally with bio-systems which is uncontrolled. Even small concentrations of heavy metals and other organic compounds are harmful to wildlife and humans. The most discarded waste includes eggshells, bones, peat, fungi, seaweed, yeast and carrot peels, which are effectively used for the removal of heavy metals due to their diversity, adaptability and versatility. Several bacterial species have an inherited metal detoxification property. Harmful compounds, such as methylmercury, are formed when mercury reacts in the environment, which is toxic to humans. Other harmful metals such as arsenic, lead, cadmium, cobalt, chromium and uranium can also be removed by using adsorbing biomass or biosorbents. Biosorption is one of the environmentally friendly filtering techniques. Many living bacteria were capable of recovering nitrogen and phosphorus from the bio-sewage when mixed in an aeration tank. Noticing the sequestering characteristic in dead biomass, research shifted from bioaccumulation to biosorption. Bioaaumulation is an activates sludge process widely used for wastewater treatment in plants. Chitosan is a biological adsorbent used for removing heavy metals without any damaging effects on environment. In search of new technologies for removing toxic metals from wastewaters has intended attention to biosorption, based on metal binding capacities of different biological materials. The most probable metal biosorbents are algae, bacteria and fungi and yeasts. The main advantages of biosorption over conventional treatment methods are their low cost, high efficiency, a minimization of chemical and biological sludge, no additional nutrient requirement, the regeneration of the biosorbent and the possibility of metal recovery. The biosorption process involves a solid phase, which may be bioserbent or biological material, and liquid phase, that is, water containing a dissolved species to be sorbed, say, by metal ions. Because of higher affinity of the sorbent for the sorbate species, the latter is attracted and bound there by different mechanisms. The process continues until equilibrium is reached between the amount of solid-bound sorbate species and its portion remaining in the solution. The sorbent affinity for the sorbate is calculated from its distribution between the solid and liquid phases.

4.4.1 Biosorption Mechanisms

The complex structure of microorganisms demonstrates that by many ways microbial cell deal with the metal. The biosorption mechanisms are plenty and not completely understood. They may be classified by various criteria. According to the dependence on the cell's metabolism, biosorption mechanisms can be divided into

1. metabolism-dependent and
2. non-metabolism-dependent.

During nonmetabolism biosorption, there is physio-chemical reaction between metal and the functional group present on the microbial cell surface. The cell wall of microbial biomass consists of polysaccharides, proteins and lipids. So functional groups, such as carboxyl, sulfate, phosphate and amino groups, are present, which interact with metal (Kuyucak and Volesky 1988).

Transport across the cell membrane: Heavy metal transport across microbial cell membranes may be mediated by the same mechanism used to convey metabolically important ions, such as potassium, magnesium and sodium. The metal transport systems may become confused by the presence of heavy metal ions of the same charge and ionic radius associated with essential ions without any metabolic activity. The biosorption phenomenon includes two steps: First, the metals are bound to the living organism cell walls, which is a metabolism-independent binding, and the second is metabolism-dependent in which metal ions are transported across the cell membrane, that is, intracellular uptake (Gadd 1990; Huang, C. and C P. Huang 1996).

Physical Adsorption: Van der Waals forces play a major role in physical adsorption. Electrostatic interactions between the metal ions in solutions and cell walls of microbial cells are responsible for the biosorption of uranium, cadmium, zinc, copper and cobalt by dead biomasses of algae, fungi and yeasts, where electrostatic interactions take place. The bacterium *Zoogloea ramigera* and the alga *Chiarella vulgaris* have been demonstrated to be responsible for copper biosorption in which fungi *Ganoderma lucidum* and *Aspergillus niger* are responsible for chromium biosorption (Kuyucak and Volesky 1988).

Ion Exchange: Cell walls of microorganisms have polysaccharides and bivalent metal ions exchange with the counter ions of the polysaccharides. For example, the alginates of marine algae occur as salts of K^+, Na^+, Ca^{2+} and Mg^{2+}. These ions can exchange with counter-ions such as Co^{2+}, Cu^{2+}, Cd^{2+} and Zn^{2+}, resulting in the biosorptive uptake of heavy metals. The biosorption of copper by fungi *Ganoderma lucidium* and *Aspergillus niger* was also up taken by the ion-exchange mechanism (Muraleedharan and Venkobachar 1990).

Complexation: Complex formation on the cell surface takes place after the interaction between the metal and the active groups for the metal removal from the aqueous solution. The biosorption of copper by *C. vulgaris* and *Z. ramigera* involves both phenomenon adsorption and the formation of coordination bonds between metals and amino and carboxyl groups of the cell wall polysaccharides. It was found that complexation is the only mechanism responsible for calcium, magnesium, cadmium, zinc, copper and mercury accumulation by Pseudomonas syringae. Microorganisms may also produce organic acids (e.g., citric, oxalic, gluonic, fumaric, lactic and malic acids), which may constitute toxic metals that result in the formation of metallo-organic molecules. These organic acids help in the solubilization of metal compounds and their leaching from their surfaces. Metals may be biosorbed or complexed by carboxyl groups found in microbial polysaccharides and other polymers.

Precipitation: Precipitation may be either dependent on the cellular metabolism or independent of it. Active defense system of the microorganisms is associated metal removal from solution. These microorganisms interact with toxic metal residue and produce compounds that enable the precipitation process. The process, which is independent of cellular metabolism, follows the chemical interaction between the cell surface and the toxic metal. So, by their interaction, a compound is formed that favors precipitation. All biosorption mechanisms mentioned earlier can occur simultaneously.

Metal Removal by Recombinant Bacteria: Organic and inorganic compounds have strong influence of physicochemical parameters, such as ionic strength, pH and the concentration on the metal removal by adsorbents from water and wastewater. Recombinant bacteria are being experimented with for removing specific metals from contaminated water. For example, a genetically engineered E. coli, which expresses Hg^{2+} transport system and metal lothionin (a metal-binding protein), was able to selectively accumulate 8/1mole Hg^{2+}/g cell dry weight. Bioaccumulation is not affected by the presence of chelating agents Na^+, Mg^{2+} and Ca^{2+}.

The design of the equipment depends on the investigation of the efficacy of the metal uptake by the microbial biomass, essential for the industrial application of biosorption. This investigation gives information about the equilibrium of the process.

4.4.2 Biosorbent Material

It has been observed that the strong biosorbent behavior of a few microorganisms against metallic ions is a function of the chemical makeup of the microbial cells. Generally, this biosorbent is made up of dead and inactive cells. Many types of biosorbents have a broad range, binding and collecting the majority of heavy metals with no specific activity; that is, they will adsorb mostly all types of heavy metals, while others are very specific for certain metals. Biosorption properties can be strengthened by processing raw biomass. Biosorption experiments have focused attention on waste materials, which are by-products or the waste materials from large-scale industrial operations, for example, the waste mycelia available from fermentation processes, olive mill solid residues activated sludge from sewage treatment plants, biosolids and the like. The biosorption of zinc from aqueous solutions was treated by dewatered waste-activated sludge from a sewage treatment plant. Another inexpensive source of biomass in abundance is in oceans as seaweeds, representing many different types of marine macro-algae. Nowadays, researchers are experimenting the uptake of toxic metals by live marine and freshwater algae focused on the toxicological aspects, metal accumulation, and pollution indicators by live, metabolically active biomass. The biosorption mechanism is complex, mainly ion exchange, chelation, adsorption by physical forces, entrapment in inter- and intrafibrillar capillaries and spaces of the structural polysaccharide network as a result of the concentration gradient and diffusion through cell walls and membranes are the processes that take place. There are several chemical groups that would attract the metals in biomass: acetamide groups of chitin; structural polysaccharides of fungi; amino and phosphate groups in nucleic acids; amido, amino, and carboxyl groups in proteins; hydroxyls in polysaccharides; and mainly carboxyls and sulfates in polysaccharides of marine algae that belong to the divisions Phaeophyta, Rhodophyta and Chlorophyta.

Municipal and industrial wastewater contains high amounts of antibiotics or other persistent molecule conjugate and metabolites in ng/L or μg/L concentration that may bear a high threat to the environment and to the living organisms. The conventional wastewater treatment technologies could not eliminate these micropollutants with enough efficiency. Although several compounds and immobilized enzymes (decarboxylases and/or phenol oxidases, e.g., lactase, tyrosinase, peroxidases, etc.), or microorganisms can degrade biologically active contaminants even in ng/L concentration using advanced oxidation processes. These biochemical processes realize the oxidation process at atmospheric pressure and low temperature with the formation of hydroxyl radicals/*OH/ by a multistep reaction. Horseradish peroxidase, lignin peroxidase and manganese peroxidase can mineralize some stable compounds. The immobilization and/or binding of enzymes or microbes to membranes as enzymatic membrane bioreactors can effectively precipitate or transform even extremely stable compounds or micro-pollutants.

Biotechniques used for cleaner, fermentation, bioremediation technologies based on the use of fungi, plants and bacteria to ensure phytostimulation, biofertilization and biocontrol of the ecosystem to retain ecological homeostasis. Mycoremediation involves the role of the specific fungus in the ecosystem to degrade the organic substances into simpler subunits. This breakdown is accomplished by the extracellular enzymes and acids to digest lignin and cellulose; these are the components of plant fibers. Phytoremediation includes the role of plants to contain and eliminate the toxic and chemical compounds, such as pesticides, solvents, explosives, crude oil and others, from the air, water and soil. Microbial remediation is brought about by the degradation of the organic contaminants and binding the use of metals in less bioavailable-using aerobic and anaerobic microbes. Genetically engineered microbial techniques are involved.

Due to the disposal problem of adsorbent substances, the biodegradation process has remained the most efficient method, although it takes a long time for treatment. Therefore, the biological process has emerged as the best process for removal of toxic dye.

Aerobic granules were investigated as a visible biomaterial for biodegradation and adsorption of malachite green.

Anaerobic digestion has been successfully employed for more than one organic molecule, like dyes. The main aim of this anaerobic digestion is the degradation and destruction of dye molecules with consequent sludge stabilization and pathogen reduction.

Removing axo dyes in biological wastewater treatment systems is based on an anaerobic treatment in combination with an aerobic oxidation treatment. A mixed culture of bacteria was used to degrade azo dyes acid yellow-36 (AY-36) by exposing anaerobic granular sludge and aerobic aromatic amine-degrading bacterial enrichment cultures under anaerobic conditions. The removal of pentachlorophenol was done using different strains of fungal biomass.

4.5 Conclusion

Dye adsorption as a purification method has advantages because other processes, which include alternation or the destruction of chromphore dye to decolorize the solution, do not remove residual from the wastewater that can be harmful to the environment. Adsorption removes the complete molecule without any remnants. Adsorption is important because metals, such as Cr, Co and Cu, bound to dye are completely removed from the wastewater. It is very important for humans to understand various aspects of the environment and their interactive dependence because we are one of the components of domain—"biota" of nature. One has to identify related problems, find the appropriate solutions and apply them as remedial modes causing minimum or no disturbance to the environment. One must keep in mind that the environment has its own ways to sustain; we need to assist and promote such natural remedial modes. Some low-cost adsorbents are agricultural wastes, natural materials and industrial wastes.

Biosorption is an emerging technology in which different types of microbial biomass are used as biosorbents for the treatment of heavy metal contaminated water bodies. Biosorption of heavy metals using actinobacterial biomass has proved their potential as promising agents in the formation of "green technology" for the control of pollution by heavy metals.

REFERENCES

Ahalya, N., R. D. Kanamadi and T. V. Ramachandra (2005), Biosorption of chromium (VI) from aqueous solutions by the husk of Bengal gram (Cicerarientinum), *Electronic J. Biotechn.*, 8(3), 258.

Ali, I. and H. Y. Aboul-Enein (2004), *Chiral Pollutants: Distribution, Toxicity and Analysis by Chromatography and Capillary Electrophoresis*, John Wiley & Sons, Chichester, https://doi.ogr/10.1002/0470867825.ch.7.

Ali, I., Mohd. Asim and Tabrez A. Khan (2012), Low cost adsorbents for the removal of organic pollutants from wastewater, *J. Environ. Manag.*, 113, 170–183.186 Organic Pollutants – Monitoring, Risk and Treatment.

Annadurai, G., R. S. Juang and D. J. Lee (2002), Use of cellulose-based wastes for adsorption of dyes from aqueous solutions, *J. Hazard. Mater. B.*, 92, 263.

Aricke, S., T. Van Gerven and C. Vandecasteele (2006), Accelerated carbonation for treatment of MSWI bottom ash, *J. Hazard. Mater.*, 137, 235.

Atef, S. Alzaydien and Waleed Manasreh (2009), Equilibrium, kinetic and thermodynamic studies on the adsorption of phenol onto activated phosphate rock, *Int. J. Phys. Sci.*, 4(4), 172–181.

Ayranci, Erol and Numan Hoda (2005), Adsorption kinetics and isotherms of pesticides onto activated carbon-cloth. *Chemosphere.*, 60, 1600–1607.

Baquero, M. C., L. Giraldo, J. C. Moreno, F. Suárez-García, A. MArtínez-Alonso and J. M. D. Tascón (2003), Adsorption of Cu(II) ions in aqueous solutions by HCL activated carbon of oil palm, *J. Therm. Anal. Appl. Pyrol.*, 70, 779.

Bayat, B. (2002), Comparative study of adsorption properties of Turkish fly ashes in the case of nickel (II), Copper (II) and zinc (II), *J. Hazard. Mater.*, 3, 251.

Bhatnagar, Amit (2007), Removal of bromophenols from water using industrial wastes as low cost adsorbents, *J. Hazard. Mater., B.*, 139, 93–102.

Bhatnagar, Amit, Eva Kumar and Mika Sillanpää (2011), Fluoride removal from water by adsorption – A review, *Chem. Eng. J.*, 15, 811–840.

Bouzaida, I. and M. B. Rammah (2002), Adsorption of acid dyes o treated cotton in a continuous system, *Mater. Sci. Eng. C.*, 21, 151.

Chaara, D., I. Pavlovic, F. Bruna, M. A. Ulibarri, K. Draoui and C. Barrig (2010), Removal of ni-trophenol pesticides from aqueous solutions by layered double hydroxides and their calcined products, *Appl. Clay Sci.*, 50, 292–298.

Chong, K. H. and H. Volesky (1995), Description of two-metal biosorption equilibria by Langmuir-type models, *Biotechnol. Bioeng.*, 47, 451.

Cui, N., X. Zhang, Q. Xie, S. Wang, J. Chen, L. Huang, X. Qiao, X. Li and X. Cai (2011), Toxicity profile of labile preservative bronopol in water: the role of more persistent and toxic transformation products, *Environ. Pollut.*, 159(2), 609–615.

Damià, B. (2005), *Emerging Organic Pollutants in Waste Waters and Sludge*, Springer, Berlin.

Das, C. P. and L. N. Patnaik (2005), Removal of phenol by industrial solid waste, *Practice Period. Hazard. Toxic Radioactive Waste Manage. ASCE.*, 9(2), 135–140.

Demirbas, E., M. Koboya and M. T. Sulak (2008), Decolorization of a textile vat dye by adsorption on waste ash, *Bioresource Technol.*, 99, 5368.

Djebbar, M., F. Djafri1, M. Bouchekara, and A. Djafri (2012), Adsorption of phenol on natural clay, *Afr. J. Pure Appl. Chem.*, 6(2), 15–25, 15.

Djilani, Chahrazed, Rachida Zaghdoudib, Ali Modarressid, Marek Rogalskid, Faycal Djazia, and Abdelaziz Lallame (2012), Elimination of organic micropollutants by adsorption on activated carbon prepared from agricultural waste, *Chem. Eng. J.*, 189–190, 203–212.

El-Geundi, M. S. (1991), Colour removal from textile effluents by adsorption techniques, *Water Res.*, 25, 271.

Enamul, Haque, Nazmul Abedin Khan, Siddulu Naidu Talapaneni, Ajayan Vinu, Jonggeon Jegal and Sung Hwa Jhung (2010), Adsorption of phenol on mesoporous carbon CMK-3: Effect of textural properties, *Bull. Korean Chem. Soc.*, 31(6).

Essa, M. H., M. A. Al-Zahrani and S. A. Nesaratnam (2004), Step towards national reliance using locally produced activated carbon from dates. Third Saudi Technical Conference and Exhibition (STCEX-3), Riyadh, Saudi Arabia, pp. 291–297.

Freundlich, H. and Zeitschriftfr (1907), Uber die adsorption I Losunge, *PhysikalischeChemie*, 57, 385.

Gadd, G. M. (1990), Fungi and yeasts for metal accumulation. M. Ehrlich and C. L. Brierly (Eds.), *Microbial Mineral Recovery*, McGraw-Hill, New York, pp. 249–276.

Gilliom, R. J. (2007), Pesticides in U.S. streams and groundwater, *Environ. Sci. Technol.*, 41(10), 3408–3414.

Gupta, V. K., D. Mohan, S. Sharma and M. Sharma (2000), Removal of basic dyes (Rodamine B and methylene Blue) from aqueous solutions using bagasse fly ash Sep. *Sci. Technol.*, 35, 2097.

Hamdaoui, O. and E. Naffrechoux (2007), Modeling of adsorption isotherms of phenol and chlorophenols onto granular activated carbon Part I. Two-parameter models and equations allowing determination of thermodynamic parameters, *J. Hazard. Mater.*, 147, 381–394.

Hasfalina, C. B. M., C. O. Akinbile and C. X. Jun (2015), Coconut husk adsorbent for the removal of methylene blue dye from wastewater, *BioRes.*, 10(2), 2859–2872.

Ho, Y. S. and G. McKay (1998), Sorption of dye from aqueous solution by peat, *Chem. Eng. J.*, 70(2), 115.

Horsfall, M. and A. L. Spiff (2004), Studies on the effect of pH on the sorption of Pb2+ and Cd2+ ions from aqueous solutions by caladium biocolor (Wild cocoyam) biomass, *Electron J. Biotech.*, 7(3), 1.

Hsisheng, C. T. H. (1998), Heavy metals ions adsorption from wastewater using activated carbon from orange peel, *Ind. Eng. Chem. Res.*, 39, 3618.

Huang, C. and C. P. Huang (1996), Application of *Aspergillus oryzae*and *Rhizopus oryzae*for Cu (II) removal. *Water Res.*, 9, 1985–1990.

Ibaraj, S. and N. Sulochana (2002), Effects of agitation time and adsorbent dosage on the adsorption of dyes, *Indian J. Chem. Technol.*, 9, 201–208.

International Agency for Research on Cancer (IARC) (1987), Overall evaluations of carcinogenicity: *An updating of IARC monographs*, 1–42, Supplement 7, WHO, Lyon, France.

Iori, J. and J. Balg (1995), Decolorization of a textile vat dye by adsorption on waste ash, *Residue Treat.*, 6, 9–10.

Johnson, C. A., S. Brandenberger and P. Baccin (1995), Decolorization of a textile vat dye by adsorption on waste ash, *Environ. Sci. Technol.*, 29, 142.

Juan Carlos, Moreno-Piraján and Liliana Giraldo (2011), Heavy metal ions, adsorption from wastewater using activated carbon from orange peel, *E-J. Chem.*, 9(2), 26–937, http://dx.doi.org/10.1155/2012/383742.

Juang R. S., F. C. Wu and R. L. Tseng (1999), Adsorption removal of copper (II) using chitosan from simulated rise solutions containing chelating agents, *Water Res.*, 33, 2403.

Kardivalu, K. and C. Namasivayan (2000), Heavy metals ions adsorption from wastewater using activated carbon from Orange Peel, *Environ Technol.*, 21, 1091.

Kuo, Y. M., T. C. Lin and P. J. Tsai (2006), Immobilization and encapsulation during vitrification of incineration in ashes in a coke bed furnace, *J. Hazard. Mater.*, 133, 75.

Kuyucak, N. and B. Volesky (1988), Biosorbents for recovery of metals from industrial solutions. *Biotechnol Left.*, 10(2), 137–142.

Lagergren, S. (1898), About the theory of so-called adsorption of soluble substances. Kungliga Svenska Vetenskapsakademiens. *Handlingar.*, 24(4).

Langmuir, I. (1918), The adsorption of gases on plane surfaces of glass, mica and platinum, *J. Am. Chem. Soc.*, 40, 1361.

Larous, S., A-H. Meniai, M. F. F. Sze, V. K. C. Lee and G. McKay (2008), The use of sawdust as by product adsorbent of organic pollutant from wastewater: Adsorption of phenol, *De-salination.*, 218, 323–333.

Liu, Huihui, Xiyun Cai, Yu Wanga and C. Jingwen (2011), Adsorption mechanism-based screening of cyclodextrin polymers for adsorption and separation of pesticides from water, *Water Res.*, 45, 3499–3511.

Liversidge, R. M., G. J. Lloyd, D. A. J. Wase and C. F. Forster (1997), Removal of basic blue 41 dye from aqueous solution by linseed cake, *Process Biochem.*, 32, 473.

Maryam, Khoshnood and Saeid Azizian (2012), Adsorption of 2,4-dichlorophenoxyacetic acid pesticide by graphitic carbon nanostructures prepared from biomasses, *J. Ind. Eng. Chem.*, 18(5), 1796–1800.

McKay, G., H. S. BlaiR and J. R. Gardener (1982), Te adsorption of dyes on chitin, *J. Appl. Polym. Sci.*, 27, 3043.

Memon, G. Zuhra, M. I. Bhanger, Mubeena Akhtar, Farah N. Talpur and Jamil R. Memon (2008), Adsorption of methyl parathion pesticide from water using watermelon peels as a low cost adsorbent, *Chem. Eng. J.*, 138, 616–62.

MIODRAG ŠMELCEROVIĆ, DRAGAN ĐORĐEVIĆ*, MILE NOVAKOVIĆ and MIRJANA MIZDRAKOVIC (2010), Decolonization of textile vat dye by adsorption on waste ash, *J. Serb. Chem.* Soc., 75(6), 855–872.

Molina-Sabio, M., F. Rodríguez-Reinoso, F. Caturla and M. J. Sellés (1995), Porosity in granular carbons activated with phosphoric acid, *J M, Carbon.*, 33, 1105.

Morais, L. C., O. M. Freitas, E. P. Goncalves, L. T. Vasconcelos and C. G. Gonzalez (1999), Reactive dyes removal from wastewaters by adsorption on eucalyptus bark: Variables that define the process, *Water Res.*, 33, 979.

Moreno-Piraján, Juan Carlos and Giraldo Liliana (2010), Study of activated carbons by pyrolysis of cassava peel in the presence of cholride zinc, *J. Anal. Appl. Pyrol.*, 87(2), 288.

Mukherjee, Somnath, Sunil Kumarb, Amal K. Misra and Maohong Fan (2007), Removal of phe-nols from water environment by activated carbon, bagasse ash and wood charcoal, *Chem. Eng. J.*, 129, 133–142.

Muraleedharan, T. R. and C. Venkobachar (1990), Mechanism of biosorption of copper (II) by Ganoderma lucidum. *Biotechnol Bioeng.*, 35, 320–325.

Nassar, M. M. (1999), Intraparticle diffusion of basic red ad basic yellow dyes on palm fruit bunch, *Water Sci. Technol.*, 40, 133.

Nassar, M. M., M. F. Hamoda and G. H. Radwan (1995), Adsorption equilibria of basic dyestuff onto palm-fruit bunch particle, *Sci. Technol.*, 32, 27.

Nassar, M. M. and Y. H. Magdy (1997), Removal of different basic dyes from aqueous solutions by adsorption on palm-fruit bunch particles, *Chem. Eng. J.*, 66, 223.

Olafadehan, O. A. and D. S. Aribike (2000), Treatment of industrial wastewater effluent, *J. Niger. Soc. Chem. Eng.*, 19, 50–53.

Pearson, R. G. (1997), Applications of density functional theory to chemical reactivity, *Chemical Hardness*, Wiley-VCH, Weinheim, Germany.

Putraa, E. K., R. Pranowo, J. Sunarsob, N. Indraswati and S. Ismadji (2009), Performance of activated carbon and bentonite for adsorption of amoxicillin from wastewater: Mechanisms, isotherms and kinetics, *Water Res.*, 43, 2419–2430.

Rais, A. and P. K. Mondal (2010), Application of modified water nut carbon as a sorbent in Congo red ad malachite green dye contaminated wastewater remediation, *Sep. Sci. Technol.*, 45(3), 394–403.

Robinson, T., B. Chandran and P. Nigam (2002a), removal of dyes from a synthetic textile dye effluent by biosorption on apple pomace and wheat straw, *Water Res.*, 36, 2824.

Robinson, T., B. Chandran and P. Nigam (2002b), Removal of dyes from an artificial textile dye effluent by two agricultural waste residues, corncob ad barely husk, *Environ. Int.*, 28, 29.

Roostari, N. and F. H. Tezel (2004), Removal of phenol from aqueous solution by adsorption, *J. Environ. Manage.*, 70, 157–164.

Sakai, S., M. Hiraoka, N. Takeda and T. Tsunemi (1990), Overview of MSWI residues recycling by thermal processes, *Water Sci. Technol.*, 22, 329.

Salman, J. M., V. O. Njokua and B. H. Hameeda (2011), Adsorption of pesticides from aqueous solution onto banana stalk activated carbon, *Chem. Eng. J.*, 174, 41–48.

Shaarani, F. W. and B. H. Hameed (2010), Batch adsorption of 2,4-dichlorophenol onto activated carbon derived from agricultural waste, *Desalination.*, 255(1–3), 159–164.

Shaista, Ikram (2012), Thesis Entitled "Sorption of Heavy Metals on Non–Conventional Adsorbents" Aligarh Muslim University.

Sivaraj, R., C. Namasivayam and K. Kadirvelu (2001), Orange peel as an adsorbent the removal of acid violet 17 (acid dye) from aqueous solution, *Waste Manage.* (Oxford), 21, 105.

Sofía, Arellano-Cárdenas, Tzayhrí Gallardo-Velázquez, Guillermo Osorio-Revilla, Ma. del Socorro López-Cortéz1 and Brenda Gómez-Perea (2005), Adsorption of phenol and di- 190 organic pollutants – monitoring, risk and treatment chlorophenols from aqueous solutions by Porous Clay Heterostructure (PCH) *J. Mex. Chem. Soc.*, 49(3), 287–291.

Suárez-García, F., A. Martínez-Alonso and J. M. D. Tascón (2001), Production of activated carbon from pine cone ad evaluation, *J M D, Carbon.*, 39, 1111.

Sun, G. and X. Xu (1997), Sunflower stalks as adsorbents for color removal from textile wastewater Ind, *Eng. Chem. Res.*, 36, 808.

Sun, X. F., S. G. Wang, X. W. Liu, W. X. Gong, B. Y. Gao and H. Y. Zhang (2008), Biosorption of malachite green from aqueous solutions onto aerobic granules: Kinetic and equilibrium studies, *Bioresource Technol.*, 99, 3475.

Yapar, S. and M. Yilmar (2004), Removal of phenol by using montmorillonite, clinoptilolite and hydrotalcite, *Adsorption.*, 10, 287–298.

Zogorski, J. S., S. D. Faust, and J. H. Haas Jr. (1976). The kinetics of adsorption of phenols by granular activated carbon, *J. Colloid Interface Sci.*, 55, 329–341.

5

Biotechnological Approach for Wastewater Management: Constructed Wetlands Is a Green Solution

Satish S. Patil and Vinayak P. Dhulap

CONTENTS

5.1 Introduction

Most water resources, surface as well as groundwater, are contaminated by manmade activities majorly due to the discharge of industrial effluent, agricultural runoff and most domestic sewage. Domestic wastewater that contains a large amount of inorganic and organic impurities and cyst of pathogens, bacteria and viruses causes waterborne diseases. Organic pollution is due to presence of high-molecular-weight compounds, such as sugars, fats, oils, proteins obtained from food processing industries such as sugar, dairy, starch and others. Domestic sewage causes bad odors, colors and taste, and this forms algal growth. Today, every community generates both liquid and solid waste. The liquid portion wastewater is essentially to the water supply of the community after it has been fouled by a variety of uses from the standpoint of sources of generation.

Generally, wastewater is defined as a combination of the liquid- or water-carried wastes removed from residences, institutions and commercial and industrial establishments. Due to discharges, the pollution of surface waters occurs too much from undesirable or harmful substances that flows into a body of water, exceeding the natural ability of that water body to remove the undesirable material and dilute it into harmless forms.

DOI: 10.1201/9781003186304-5

Water pollutants are categorized as released from point or nonpoint sources. All the segments of our society (urban, rural, industrial, agricultural and military) may contribute to the problem of water pollution. Today almost all the industries discharge water containing effluent or wastewater at some stage of their manufacturing process, but industrial wastes are not the same in every case. Every industry produces its own characteristic process wastes because of many operations, such as washing operations, washing of raw materials, the formation of intermediates and final products, washing of product containers and others. The discharged wash waters usually contain a small quantity of the process materials, which impart their characteristics to the process wastes due to these wastes and contaminants water resources become polluted. The contamination of the industrial wastes that comes into the water is mostly dangerous. Similarly, the sewage from big cities is often drained into streams, rivers and other water bodies. The sewage promotes the growth of phytoplankton. This excessive growth depletes the oxygen in the water. This reduction of oxygen and the presence of poisonous wastes affect the aquatic faunal population. This contaminates the water with germs of various diseases, such as cholera, dysentery, hepatitis, typhoid, gastroenteritis, snail fever, enteric fever, malaria and others.

Municipal sewage systems used in developed countries are often too expensive to build and operate in the developing world, where low-cost, low-tech alternatives for treating wastes are needed. One option is effluent sewerage, a hybrid between traditional septic tank and full sewer system. Another alternative is to use natural or artificial wetlands to dispose of wastes. Wetland waste treatment systems are now operating in many developing countries.

5.2 Wetland

Wetland is the nature's gift for control of environmental pollution particularly water pollution and it sustains livelihood. Wetlands are the transitional areas between land and water. These encircle a broad range of wet environments, including marshes, bogs, swamps, meadows, tidal wetlands, floodplain, and ribbon (riparian) wetlands along stream channels that behave like a kidney of the earth ecosystem. These wetlands are located at the ecotones between dry terrestrial systems and permanently flooded deepwater aquatic systems, such as rivers, lakes, estuaries or oceans. As such, they have an intermediate hydrology, a biogeochemical role as a source, sink or transformer of the chemicals, and have generally high productivity if they are open to hydrologic and chemical fluxes. The wetlands also perform vital functions in the water cycle and are important for flood management, climate and biodiversity. They sustain the livelihoods of hundreds of millions of people. For this reason, it is important to maintain the wetlands in an excellent condition while using them for water treatment. Wetlands (both natural and constructed) are able to purify wastewater because of their ability to filter suspended solids and to take up nutrients from the water or wastewater.

5.3 Constructed Wetlands for Wastewater Management

The system of planting aquatic plants such as reeds or bulrushes in a wet (often gravel) substrate medium for gray water recycling is called a "constructed wetland" (CW) or "artificial wetland" or "human-engineered wetland". The first artificial wetlands were built in the 1970s. By the early 1990s, there were more than 150 constructed wetlands treating municipal and industrial wastewater in the United States. According to the U.S. Environmental Protection Agency in 1999, constructed wetland treatment systems can be established almost everywhere, including on land with limited alternative uses. This can be done relatively simply where wastewater treatment is the only function required. They can build in natural settings, or they may entail extensive earthmoving, the construction of impermeable barriers or building of containments such as tanks or trenches. Wetland vegetation has been established and maintained on substrates ranging from gravel or mine spoils to clay or peat. Some systems are set up to recharge at least a portion of the treated wastewater to underlying groundwater. Others act as flow-through systems, discharging the final effluent to surface waters. CWs had many applications and is found across the country and around the world. They can often be an environmentally acceptable, cost-effective treatment option, particularly for small communities (Dyck, 2006).

CWs are designed especially for the water pollution control and exist in locations where natural wetlands are not present. Generally, two types of CWs are in common use today. One type exposes the water's surface, and the other maintains the water surface below the level of the gravel. Some designs combine elements of both. Subsurface flow constructed wetlands are also referred to as "vegetated submerged bed", "root zone method", "rock reed filter", "microbial rock filter", "hydrobotanical method", "soil filter trench", "biological-macrophytic marsh bed" and "reed bed treatment". Subsurface-flow wetland is considered advantageous compared to the open-surface wetlands and are more commonly used for individual households. By keeping the water below the surface of the gravel medium, there is less chance of odors escaping, less human contact, less chance of mosquito breeding and faster treatment of the water (due to more of the water being exposed to the microbially populated gravel surfaces and plant roots). The subsurface water is also less inclined to freeze during cold weather. Constructed wetlands generally consist of one or more lined beds, or cells. The gravel media in the cells should be as uniform in size as possible and should consist of small to medium-sized gravel or stone, from 1 to 3 feet in depth. A layer of sand may be used at either the top or the bottom of the gravel medium or a layer of mulch and topsoil may be applied over the top of the gravel. In some cases, gravel alone will be used with no sand, mulch or topsoil. The sides of the wetlands are bare to prevent rainwater from flowing into them, and the bottom may be slightly sloped to aid in the flow of gray water through the system. A CW for a household, once established, requires some maintenance, mainly the annual harvesting of the plants. In any case, the roots of aquatic plants will spread through the gravel as the plants grow. The most common species of plants used in the wetlands are the cattails, bulrushes, sedges and reeds. Gray water is filtered through the gravel, thereby keeping the growing environment wet and bits of organic material from the gray water become trapped in the filtering medium. Typical retention times for gray water in a subsurface-flow wetland system range from 2 to 6 days. During this period, the organic material is broken down and utilized by microorganisms living in the medium and on the roots of the plants. A wide range of organic materials can also be taken up directly by the plants themselves. Bacteria, both aerobic and anaerobic, are among the most plentiful microorganisms in wetlands and are thought to provide the majority of the wastewater treatment. Microorganisms and plants seem to work together symbiotically in constructed wetlands, as the population of microorganisms is much higher in the root areas of the plants than in the gravel alone. Dissolved organic materials are taken up by the roots of the plants, while oxygen and food are supplied to the underwater microorganisms through the same root system. Aquatic microorganisms have been reported to metabolize a wide range of organic contaminants in wastewater, including benzene, naphthalene, toluene, chlorinated aromatics, petroleum hydrocarbons and pesticides. Aquatic plants can take up and sometimes metabolize water contaminants, such as insecticides and benzene. The water hyacinth can removes phenols, fecal coliforms, suspended particles and heavy metals including lead, mercury, silver, nickel, cobalt, and cadmium from contaminated water. In the absence of heavy metals or toxins, water hyacinths can be harvested as a high-protein livestock feed. It can also be harvested as a feedstock for methane production. Reed-based wetlands can remove a wide range of toxic organic pollutants. Duckweed also removes organic and inorganic contaminants from water, especially nitrogen and phosphorous. When the outdoor air temperature drops below a certain point (during the winter months in cold climates), wetland plants will die and microbial activity will drop off. Therefore, CWs will not provide the same level of water treatment year-round. These artificial wetland systems constitute a relatively new approach to water purification and the effects of variables such as temperature fluctuations are not completely understood. Nevertheless, wetlands are reported to perform many treatment functions efficiently in winter. It is estimated that one cubic foot of artificial wetland is required for every gallon per day of gray water produced. For an average single bedroom house, this amounts to about a 120 square feet system, 1 foot deep. However, it is better to overbuild a system than to underbuild. Some CW situations may not have enough drainage water from a residence to keep the system wet enough. In this case, extra water may be added from rainwater collection or other sources.

5.4 Working of CWs in Wastewater Management

CWs are manmade, engineered systems that utilize natural treatment processes to reduce the pollution levels in wastewater. The combination of soil, plants and microorganisms efficiently remove organic pollutants, nutrients and toxic contaminants in wastewater using a variety of physical, biological and chemical processes. Mainly, the sun provides the energy that sustains all these processes. Investigations on the capacity of marsh plants in the reduction of organic pollutants and nutrients in water started in the 1950s in Germany. Since then, various designs of CW systems have been developed and thousands of facilities are currently in use in Europe, Australia and the United States (Cooper et al., 1996; Shrestha, 1999; Kadlec and Knight, 2004). Recent years have seen the proliferation of constructed wetland systems in Africa and Asia. There are two basic types of constructed wetland, such as surface flow and subsurface flow (SSF). Surface-flow wetlands are essentially shallow ponds planted with floating and emergent species, the water flow occurring almost entirely above the soil. The flow in SSF wetlands occurs under the surface and within the soil planted with emergent species such as cattails (*Typha*), rushes (*Juncus*) and reeds (*Phragmites*). SSF is either horizontal with the water moving parallel to the surface or vertical where the influent is distributed across the surface, percolates down the media and collected in a bottom layer. Subsurface wetlands are generally more efficient and require less land area.

Wastewater treatment in constructed wetlands occurs by several mechanisms, such as dilution with rainfall, chemical reactions and biological activity that transform and filter the wastewater. Wastewater strength is generally measured by how much oxygen is required to reduce wastewater contents to chemical compounds that are stable or nearly stable in the environment. Oxygen is necessary for most of the chemical and biological processes that treat or reduce animal wastewater biochemical oxygen demand (BOD) and nutrient content to more desirable levels.

Wetland plants transport oxygen from leaves through stems to roots and thus into the wastewater solution. This routing of oxygen helps to satisfy the high BOD of the wastewater introduced into the pond. Therefore, the selection of appropriate plants for constructed wetlands is critical. The most commonly used emergent plants in constructed wetlands for animal waste treatment are the reeds, bulrushes, and cattails (*Scirprs, Juncus,* and *Typha*, respectively). The saturation culture of soybean and flooded rice has also successfully treated swine wastewater. The roots of the plants play an important role in the treatment process, providing surfaces for bacterial growth, the filtration of solids, nutrient uptake and oxygen to promote the processes of nitrification and denitrification.

During different seasons, weather conditions and times of day the oxygen concentration in wetlands varies. These daily and seasonal changes affect the biological processes, which removes carbon, nitrogen and phosphorus along with reducing the high BOD in wastewater.

Streams, pond, marsh and wetland ecosystems all contain many different physical and biological processes, which help to purify water flowing through them. CWs are engineered wetland systems designed to harness these useful natural processes and optimize them for treating polluted wastewater, including sewage, agricultural slurries, industrial wastes, and even road runoff. Reed beds are a particular type of CW usually consisting of a gravel-filled container or bed planted with reeds. Wastewater flows through the gravel and reed roots and is purified by the actions of millions of bacteria, fungi and algae (microorganisms) that digest the sewage. They can be used in several ways: sewage treatment companies commonly use reed beds as a 'polishing' stage at rural works following conventional treatment to give a cleaner effluent. Reed beds also effectively provide complete sewage treatment for households and small communities not served by main sewerage. There are two basic types of reed bed, vertical flow and horizontal flow. Before running sewage to the reed beds, you need to separate the solids from the liquids in a settlement tank or conventional septic tank.

Performance data suggest that a well-designed reed bed system can remove more than 98% of the organic matter in sewage, 60% to 80% of the nitrogen and up to 60% of the phosphates. In the United Kingdom, reed beds continue to function during the winter but less effectively than in summer. This is because bacteria work more quickly at higher temperatures. Sewage is usually relatively warm, so a frozen reed bed system is unusual in U.K. climates. Maintenance is required during the first year, a certain amount of weeding may be necessary so the reeds are not choked out by other plants. Once they are established, reed beds require very little maintenance. In a vertical-flow reed bed, the flow should be switched between parallel beds each week (which takes about 3 minutes) and the system should be checked over for blockages (a further 3 minutes). Surplus or dead plants should be removed, and there is a minimal amount of weeding to be done each spring. In comparison with conventional treatment systems, CWs have lower energy and chemical requirements. A wetland is fundamentally powered by the sun, and if topography allows, water flow occurs by gravity. Plants obtain their nutrients from the wastewater, so fertilizer application is generally not needed. Another attraction of constructed wetlands is simplicity in operation, with less need for highly skilled personnel in day-to-day operations. CWs do not appear like conventional wastewater treatment systems that have an industrial look. They can be incorporated into the rural or urban landscape quite effectively, even in difficult terrain. In tropical countries such as the Philippines, where the annual variation of temperature is not very wide, the plants and their associated microorganisms do not encounter seasonal changes common in temperate countries. Given a relatively warm and stable climate, microbial degradation kinetics are higher and constant throughout the year, thus making wetlands a reliable system for removing biochemical oxygen demand and nutrients all year round.

The treatment mechanism in CWs appears to perform all the biochemical transformations of wastewater constituents that take place in conventional energy intensive, environmental engineering–based systems including activated sludge process, septic tanks, drain fields and other form of land treatments. These systems have also been found to be very effective in removal of BOD, Suspended Solids (SS), and nitrogen, phosphorous, metals, trace organics and pathogens more effectively than conventional means. However, phosphorous removal capacity varies from system to system and depends largely on site-specific factors. This reduction is accomplished by diverse treatment mechanisms, namely, sedimentation, filtration, chemical precipitation and adsorption, microbial interactions and uptake by vegetation (Lipkow, et al., 2010).

The Clean Water Act provision makes it imperative for communities to undertake sewage collection and treatment in order to protect surface and groundwater resources. However, conventional wastewater treatment facilities are usually associated with expensive capital outlays, high operating and maintenance costs and requiring highly skilled personnel to run such systems. These translate into high recurrent costs that will be ultimately borne by the waste generator for the public. Constructed treatment wetlands offer a more sustainable option for wastewater treatment. The capital expenditure for constructing a wetland treatment facility may be equivalent or lesser than that of a conventional treatment plant, but its operating cost is minimal, depending on how much energy is required.

A CW cell is typically built by digging a shallow excavation, lining it with an impermeable material such as plastic liner or concrete, filling up the space with filter media, such as soil, sand or gravel, and installing an influent storage/feeding and effluent collection/disposal system. Locally available construction materials are used, except for the use of plastic liner. After the civil works are completed, planting of the wetland will commence. In order to facilitate the speed of colony establishment, cuttings of the wetland plants may be gathered and rooted in an on-site nursery during the construction stage. The plants are usually acquired from local sources. Because constructed treatment wetlands are natural systems, little environmental impact is expected. In fact, the benefits far outweigh the negative impacts if there are any. The smaller energy requirement means fewer greenhouse gases; in fact, the plants sequester carbon dioxide into their biomass. However, special attention must be given to the choice of vegetation. Some wetland plants such as reeds (*Phragmites*) are regarded as noxious species, which can colonize irrigation ditches and rice fields if not managed properly; factors related with their spread must be controlled. Although their seeds are usually not viable, cuttings of rhizomes or stems disposed improperly in a suitable spot can establish a reed stand in no time, manual removal is very difficult. An established wetland will attract various animals, such as amphibians, insects, birds and even reptiles. It must be noted that natural existing wetlands, despite their inherent pollution assimilative capacity, should not be intended as receiving bodies of untreated wastewater.

Constructed wetlands are passive systems and generally require large areas of land compared to conventional systems. As a rule of thumb, 1 to 5 square meters of land are required per population equivalent. Consequently, CW systems are not suitable in areas where the cost of land is high. Variable wastewater flow rates from domestic sources or tourism establishments can be a problem for SSF wetlands, where a flow short-circuit may occur. This is usually solved by designing a flow-equalization basin. CWs generally require a long start-up period. The plants take several months to establish and colonize a wetland cell, although there may be cases in which wastewater feed can commence as early as 3 months after planting. Sewage and similar types of wastewater may need pretreatment prior to feeding into a wetland. Solids removal is essential to ensure that the wetland media will not clog, resulting in lower pollutant removal efficiency. Well-designed septic tanks or sedimentation chambers are common pretreatment options for removing suspended solids in sewage. There may be instances in which the use of water for sanitation purposes may be impractical, such as in dry mountainous areas with very little water supply (Gurung and Oh, 2012).

Aquatic plants possess an outstanding ability for assimilating nutrients and creating favorable conditions for microbial decomposition of organic matter. This ability can be exploited in the restoration process of natural streams, lakes and wetlands and in wastewater treatment systems.

CW technology can solve various respective environmental problems, namely, denitrification, whereby nitrate is denitrified under anaerobic conditions in a wetland and organic matter accumulated in the wetland and provides a carbon source for microorganisms converting nitrate to gaseous nitrogen–oxygen conditions that can be regulated by water flow rates. Adsorption of ammonium and metal ions by clay minerals, the adsorption process can be regulated by addition of various minerals during the filter design. Adsorption of metal ions, pesticides, phosphorus compounds by organic matter, complexing of metal ions by humic acids and other organic polymers, which can significantly, reduces the toxicity of this ions-stimulation of humus-forming processes. Decomposition of biodegradable organic matter, either aerobically or anaerobically, by microorganisms in the transition zone create proper microhabitats. The elimination of pathogens are those outcompeted by natural microorganisms within the transition zone. Ultraviolet radiation plays an important role. The assimilation of metallic contents and other toxic substances by macrophytes to varying degrees of efficiency, selection of plants and oxygen conditions using the water regime. Toxic organic compounds, through anaerobic processes in wetlands, depend on the biodegradability of the compounds and their retention time.

In the area of treating wastewater appropriately, there seems to be a number of technologies available. These technologies depending on their quantity and quality of wastewater and its desired cleaning objective of the requirements are completely different. For the treatment of wastewater, it may be wastewater from domestic waste, industrial waste or agriculture waste. Even waste from farming sector or from gardening, from smaller communities or from tourist resorts and hotels is used for this treatment.

In the ecological context, wetlands are intermediate between terrestrial and aquatic ecosystems. CWs are engineered systems designed and constructed to utilize the natural processes such as vegetation, soil/substrate and their associated microbial assemblages. Raw or pre-treated wastewater is piped into the plant and flows through the constructed wetland bed and the contaminants elimination processes take place during this passage. They are based upon various complex processes, such as physical, chemical and biological, within the association of substrate, macrophytes and microorganisms. These CWs have other functions in the ecosystems:

1. Settling of suspended particulate matter
2. Filtration and chemical precipitation through contact of the water with the substrate
3. Chemical transformation
4. Adsorption and ion exchange on the surfaces of plants, substrate, sediment and litter breakdown and transformation and uptake of pollutants and nutrients by microorganisms and plants
5. Predation and natural die-off of pathogens

CWs are of cost-effective and technically feasible approach to the treatment of wastewater for several reasons:

1. Operation and maintenance expenses (energy and supplies) are low.
2. Operation and maintenance require only periodic, rather than continuous, on-site labor.
3. They are often less expensive to build than other treatment options.
4. They are able to tolerate high fluctuations in flow.
5. They are able to treat wastewaters with very different constituents and concentration facilitate water reuse and recycling.

CWs are becoming increasingly popular for the treatment of domestic wastewaters at the world level. Their treatment capacity provides a sustainable alternative to mechanical-based wastewater treatments for small-scale waste producers. CWs are self-contained, artificially engineered wetland ecosystems. They are designed to optimize the microbiological, chemical and physical processes naturally occurring in a wetland. Removal mechanisms in the constructed wetland technology is based on the treatment power of three main mechanisms, such as microorganisms colonizing the system, the physical and chemical properties of the bed media (sand or gravel) and, finally, the plants themselves. Out of these, the microorganisms are the most important constituent.

5.5 Classification of CWs

CWs are classified according to the water flow regime into either free-water surface flow (FWS) or SSF CWs and according to the type of macrophytes plant, as well as flow direction. CWs used macrophytes plants which are aquatic plants that grow in or near water bodies. SSF CWs are designed to keep the water level totally below the surface of the filter bed. They can even be walked on. This avoids the mosquito problems of FWS CWs. Different types of CWs may be combined with each other to form hybrid systems in order to exploit the specific advantages of the different systems (Figure 5.1). The coarse sand used in SSF CWs contributes to the treatment processes by providing a surface for microbial growth and by supporting adsorption and filtration processes. This results in a lower area demand and higher treatment performance per area for SSF CWs, compared to FWS CWs. SSF CWs are the predominant wetland type in Europe. There are two different types of filtering material: sand or gravel. Gravel bed systems are widely used in North Africa, South Africa, Asia, Australia and New Zealand. The sand bed systems have their origin in Europe and are now used all over the world.

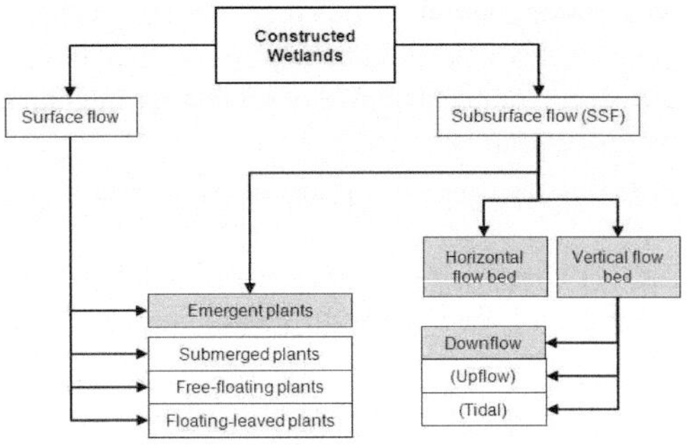

FIGURE 5.1 Types of constructed wetlands.

HFB and VFB are abbreviations for horizontal-flow bed and vertical-flow bed, respectively. Hybrid systems are also possible. "Emergent plants" are a type of macrophytes where the leaves are above the water level (Hoffmann, et al., 2011). The dashed ellipse signifies the focus of this document. HFB and VFB are abbreviations for Horizontal Flow Bed and Vertical Flow Bed, respectively. Hybrid systems are also possible. "Emergent plants" are a type of macrophytes where the leaves are above the water level (Hoffmann, et al., 2011).

CWs can also be classified according to their functions and mode of operation and the biological agents. The life-form of the macrophytes (plants) in the system such as floating macrophytes–based system, submerged macrophytes–based system and rooted emergent macrophytes–based system. In the following, only type 3 is described with more detail because it is the most common used in Europe and in the world. It can be categorized according to the flow pattern.

5.6 SSF

a. **Horizontal Subsurface Flow (HSF) Systems:** The sizing procedure can be performed using the scientifically approved methods, like the first order kinetic equations for the pollutants removal and the Darcy law for the hydraulic aspects. They are based on area coefficients like area per gram of COD [chemical oxygen demand]" (U.S. Environmental Protection Agency, 2000).

This type of root bed treatment system (RBTS) consists in a properly designed waterproof basin that contains a filter material, wetland plants (normally reeds) and microorganisms. The bed is fed with wastewater coming from a suitable primary treatment by a simple inlet device. The filling material (coarse gravel, fine gravel, coarse sand) has to offer not only an appropriate hydraulic conductivity but also a large surface for the biofilm growing. The water level remains always under the surface of the bed, the wastewater flows horizontally by a slope (about 1%) which is obtained by a sand layer under the membrane liner.

The subsurface flow prevents odors and mosquitoes and permits public access in the wetland area. This kind of CW is particularly efficient in suspended solids, carbon and pathogens removal, as well as for denitrification, while due to its prevalently anoxic conditions, nitrification is limited.

b. **Vertical Subsurface Flow (VF) Systems:** The sizing procedure for VF beds is mainly based on the nitrification process in fact, when the treatment goals required for ammonium concentration are fulfilled all the other parameters are satisfactory eliminated too. A common practice for dimensioning VF reed beds is to calculate an area coefficient per person equivalent like described in several national guidelines.

In this VF system, the wastewater is applied through a distribution system overall surface area and passes the filter in a more or less vertical path. The pretreated wastewater is passed on the bed in large batches (intermittent feeding), thus flooding the surface. During the time between the feedings, the pores within the filter media can fill up with air that is trapped by the next dose of liquid. Thus, oxygen requiring nitrifying bacteria is favored and full nitrification can be achieved, but only a small part of the formed nitrate is denitrified under aerobic conditions. The denitrification and thus total nitrogen elimination can be increased by a partial recirculation of the nitrified effluent into the first chamber of the septic tank. The treated water is collected in a bottom drainage system to be discharged. The water level can be maintained with a height of about 5 to 10 cm from the bottom of the bed, or otherwise the beds can be totally empty after each feeding pulse. This kind of CW is particularly efficient in nitrification, carbon and suspended solids removal. Due to its prevalently aerobic conditions, denitrification is poor (Luederitz, et al., 2001; Lantzke, et al., 1999).

5.7 FSW or Surface Flow

c. FWS Systems

Surface flow wetlands are densely vegetated basins optionally including open water areas. They need some sort of subsurface barrier to prevent seepage and soil or another suitable medium to support the emergent vegetation. The water flows through the unit at a relatively shallow depth. Particulates tend to settle and to be trapped in the system in such a way they enter the biogeochemical element cycles within the water column and surface soils of the wetland. At the same time, dissolved elements enter the overall mineral cycles of the wetland system. This kind of constructed wetlands is particularly efficient in the pathogens' removal due to the high exposure of the wastewater to the ultraviolet (UV) component of the sunlight. For this reason, and for a good denitrification power, these systems are often used as tertiary treatment.

5.8 Hybrid Systems—Combinations of a, b and c Systems

Using the different properties and abilities of the previously mentioned three types a wastewater treatment plant can be assembled out of the different types (ARTEC, 2003). This combined system is used to improve the performance of the

CW. A vertical subsurface flow wetland is followed by a horizontal flow subsurface wetland. The vertical stage will give a good nitrification whereas the horizontal stage will create the anaerobic conditions necessary for denitrification. So this system can enhance the removal of nitrogen (Deun and Dyck, 2006).

The treatment of domestic or municipal wastewater sludge is also possible using the principle of constructed wetlands. The sludge will be pumped into a similar designed wetland. Through evaporation and a drainage system for further treatment, the sludge will dry without any further addition of energy. The plants will grow along with the raising sludge level. Its roots are constantly penetrating the drying sludge. The volume reduction through this system is over 90% so that after 8 to 10 years, this process is finished, and the sludge is transformed into reusable earth.

Treating domestic or municipal wastewater is currently a conventional application. There is several thousand operating CWs worldwide, since more than 15 to 20 years, and the most used are SSF systems. There are numerous possibilities also for industrial wastewater like chemical industry, laboratory effluents, landfills, acid mines and agricultural or agro-food wastewaters, such as wineries, olive oil mills, dairy and, in general, all the high-organic-content-characterized waters. The ability of CWs is to clean wastewater, despite fluctuating contamination (red line); if properly planned, installed and maintained CWs can continuously match the different benchmarks, such as BOD, COD and NH_4-N. Within the last 20 to 30 years, various types of CWs have been developed in different countries. There is a wide acceptance and interest within the population because it has many advantages, such as being less expensive to build than other treatment options, having a simple construction, having an optimal aesthetic appearance, creating biotopes and others.

5.9 Operation Requirements in a CW System

a. **Pretreatment:** Wastewater generally contains high numbers of solids, that is, organic and inorganic. Some forms of waste treatment for solids removal are necessary prior to the water entering the wetlands. Mechanical solids separation, stack pads with drained-liquid collection for storage of wastewater or collection of the total wastewater, stream-in lagoons for solids settling, and treatment is necessary. Solids removal ensures that the wetlands are not so heavily loaded with carbon that they are totally anaerobic and incapable of supporting the interactive functioning of aerobic and anaerobic processes that are necessary for successful operation of the wetlands.
b. **Horizontal flow filter:** The operation and the maintenance routine for the HF system are easy to do and require no specialized personnel. The main periodic checks are wetland vegetation uniform diffusion, presence of weeds and presence of vegetal diseases or damages by insects or animals.

   The plant density must be kept over 10 plants per square meter; otherwise, it is necessary to plant new reeds. In the vegetation management, the first action is to cut the reeds mechanically or manually 3 years after the plant started. The reeds debris must be subsequently removed from the bed. Further vegetation cuttings will be performed every 2 years. The inlet pipe and the first meters of the filling medium should be checked every 6 months to verify if clogging by sedimentation of the suspended solids and biomass growing on medium surface has occurred.
c. **Vertical flow filter:** During the first year, attention is to be paid to the growing of the plants. Weeds should be removed until the reed is established. The following parts should be controlled in regular intervals such as functioning of the mechanic devices especially the intermittent feeding device, even distribution of wastewater on the filter area, water level in the filter bed and cutting of the plants is not necessary each year and depends on the climatic conditions.
d. **Free-water system:** The inlet settling zone has to be emptied whenever necessary. Bank vegetation maintenance is required with frequency depending on the used plants.
e. **System costs:** Costs are varying in the different countries. Since the technique is quite simple it is possible to reduce building expenses (cost of labor) considerably by involving the owner in the construction process.
f. **Horizontal Filter (HF):**
   1. *Superficial runoff, overload:* Enlarge the filter area or reduce hydraulic load.
   2. *Sludge drift from the pretreatment system:* Empty the pretreatment system or reconstruct pretreatment system.
   3. *Clogged inlet pipes:* Clean the pipe system consecutively, the inlet coarse rocky material has to be removed and washed at first 30 centimetre level.
   4. *Plant disease:* If vegetal diseases or damages by insect or animals have happened, an intervention of specialized personnel is required to achieve the right solutions.
   5. *Presence of weeds:* Remove them manually or by controlled flooding of the bed.
g. **Vertical filter (VF):** Apart from to the previously mentioned points in VF systems, the following troubles might occur:
   1. *Displacement of distribution system upon the filter area:* Adjust the distribution system.
   2. *Uneven building of the sand layer:* Adjust the sand layer.
   3. *Clogged distribution system:* Clean the pipe system.

      In the case of the long term, these malfunctions might lead to soil clogging. After the repairing, a complete restoration of the filtration capacity is possible given a resting period of few weeks. If a resting period is not possible, the filter will also recover after the clogging layer is opened and the reasons for clogging are restored (ARTEC, 2003).

h. Planning for CWs: Following elements must be considered in constructed wetlands:

1. Type of wastewater, which is influenced by the number and type of animals, a lot of runoff, rainfall collection and stack-pad drained liquid
2. Wastewater content, including BOD, total suspended solids, nitrogen, and phosphorus
3. Hydraulic flow through the system, which affects BOD reduction, fluid transport rate and odor
4. Seepage, evaporation and transportation losses
5. Suitable posttreatment of the outflow
6. Total land area needed for all components of the entire treatment system

i. Posttreatment: Because direct discharge of wastewater into lentic or lotic water bodies is not allowed, wetland effluent must be reused or applied to land. Wastewaters are usually irrigated either directly to surrounding land or through vegetated filter strips to croplands and woodlands. Wetland effluent must be further treated in receiving ponds or lagoons before being discharged, reused as flush water or recycled to the wetland. The amount of land available for land application of wastewater will determine whether constructed wetlands are appropriate at a particular site. Problems with ammonium nitrogen concentrations and potential toxins to wetland plants can be reduced by proper management of rainwater and lot run-off, as well as by dilution with freshwater recycle effluent (United States Environmental Protection Agency, 1993).

5.10 Macrophytes in the CW

Aquatic plants used in constructed wetland systems can be divided into two general groups, such as microscopic and macroscopic. Most of the microscopic plants are algae, which can be either single cell (such as *Chlorella* or *Euglena*) or filamentous (such as *Spirulina* or *Spirogyra*). Macroscopic (larger) plants can grow underwater (submergent) or above water (emergent and free floating). Some grow partially submerged and some partially emerged. Some examples of macroscopic aquatic plants are reeds, bulrushes, water hyacinths and duckweeds. Submerged plants can remove nutrients from wastewaters but are best suited in water where there is plenty of oxygen (water with a high level of organic material tends to be low in oxygen due to extensive microbial activity). Examples of floating plants are duckweeds and water hyacinths. Duckweeds can absorb large quantities of nutrients. Small ponds that are overloaded with nutrients such as farm fertilizer run-off can often be seen choked with duckweed, appearing as a green carpet on the pond's surface. Algae work in partnership with bacteria in aquatic systems. Bacteria break down complex nitrogen compounds and thereby make nitrogen available to algae. Bacteria also produce carbon dioxide, which is utilized by the algae.

The plants used in constructed wetlands are helophytes and hydrophytes. Helophytes are freshwater plants with perennating tissue in soil beneath the water, aquatic plants rooting at the bottom. Hydrophytes are freshwater plants with perennating tissue below the surface of the water, that is, free-floating water plants. The roles of the plants in the CWs are as follow:

1. Root system increases the surface available to bacterial colonization
2. Transfer oxygen to provide an aerobic/oxidized environment
3. Oxygen leakage from the roots (limited)
4. Nutrient assimilation (N and P; limited)
5. Maintain hydraulic pathways in the substrate
6. Plant litter provides substrate to the microorganisms
7. Accumulated litter serves as thermal insulation and aesthetics of the constructed wetland.

The most popular species of macrophytes are emergent ones, such as reeds than free-floating and submerged type of macrophytes. They are valuable in building biochemical barriers because they not only accumulate nutrients, which can be easily removed after plant harvesting, but some of them are able to oxygenate sediments, for example, *Phragmites* and *Typha* (Salati, 1987; Brix, 1994; Cooper, et al., 1996).

5.11 Methodology for Wastewater Management

To characterize sewage and treat it in a CW by using selective aquatic macrophytes and finding its treatment efficiencies, the core study focused on the scientific and technical aspects of the construction of wetland. The suitability of aquatic plants or macrophytes, the effect of concentration, the effect of retention time, the effect of flow rate, the effect of seasons, the effect of design and other operational, as well as environmental, factors were studied. The study also covers the design parameters, such as the type of media used for the plant growth, the impact of angle of inclination, the effect of filter media, the role of microbial population in the bed, the flow channel configuration for the wastewater flow and other factors determining the extent of sewage treatment were studied. Finally cost benefits of the optimized system of HSF CWs were carried.

The scientific and technical design parameter optimization of HSF CW increases the treatment efficiency, and this reduces the cost of treatment and time of treatment and increases the success rate of entire wetland. By realizing the importance of this emerging technology based on use of aquatic macrophytes, mainly common weeds offer greater significance and promise. Considering the need to develop such method and study its extent of effectiveness, including cost–benefit analysis is the matter of present research interest.

The purpose of present research work was to study and quantify the domestic wastewater. To examine physicochemical properties or characteristics of untreated and treated wastewater using CW models. The study also aims to develop, design, fabricate or construct an effective and economically low-cost biological treatment method or constructed wetland model for the wastewater treatment. In this study, a functional pilot scale of CWs is used to treat domestic wastewater. The main objective of the study is to determine the effect of different hydraulic loading rates (HLRs) on the treatment efficiency, characterize the wastewater and develop experimental setup for the treatment of wastewater of constructed wetland. The study uses of biological treatment to reduce pollution of wastewater, find pollution reduction efficiency of the root zone technology and, finally, assess the cost–benefit analysis of the system tested.

The present studies are aimed to investigate the pollutant reduction efficiency using *Cana indica*, *Colocasia esculenta*, *Pennisetum puerperium*, *Phragmites karka*, *Typha latifolia and Panicum maximum* macrophytes in HFS CW. The investigations are carried out to determine the treatment efficiency of these plants at different period intervals for reducing the pollutants from sewage, impact of change in design of constructed wetland on sewage treatment, impact of polyculture bed on the sewage treatment, impact of flow rates (hydraulic retention time [HRT]) on the treatment efficiency, identification of microorganism present in the constructed wetland bed and its role in the sewage treatment, among others. Considering the abundant availability of these plants or macrophytes in study area, these plants has been selected for the present study. These plants were locally adaptive, abundant, fast growing and easily available, which made it easy to select for the present investigation.

The experiments were carried out and studied on various concentrations of sewage samples with known and unknown concentrations for study of efficiency of sewage pollution removal using various aquatic plants (free-floating and emergent), construction designs, sewage flow rates (inflow and outflow), media of reed bed, plant biomass, plant culture (monoculture and poly/mix culture) seasonal study, microbial study etc.

A vertical plastic bucket/tub/drum or similar vessel was used as storage (inlet), supply tank for wastewater or test solution and outlet tank or cans for collection of treated water. The reed bed of different plants was prepared separately by using various types of plants and various sizes of media. The bucket, pot, or similar type of plastic container was used for the collection of treated water at the outlet of treatment plant.

5.12 Designing of CWs

Experimental setup and designs: (a) Wetlands Flow Model–I and (b) Wetlands Flow Model–II (Figure 5.2a and b). In the experimental setup and designs, these basic models of CWs on the designed on the basis of their flows.

5.13 Fabrication and Construction of CW Beds

The present horizontal-type bed of CW was developed in the departmental laboratory. The bed was fabricated using plastic materials in the form of plastic sheets with or without channels on which suitable medium was deposited for the better root growth of plants. The treatment pilot plant was prepared with a suitable outlet (Figure 5.3). The beds were made up of different capacities and sizes for different retention times to see the efficiency by retaining the wastewater for different times, which varied from a few hours to few days. In addition, the basement angle of the beds for the flow of wastewater was provided. This angle will be optimized for better results evidenced by pollution reduction.

In this study, selected locally available aquatic plants/macrophytes were grown at a time in different beds to getting quick and different results separately. These plants studied based on their fast growth rate, adaptive and abundantly available and has found efficient by earlier workers followed by other ones in other climatic conditions. Materials were used in the experiment analysis the said work is as under:

First, a vertical bucket is placed at the top level (upper side) than the root zone bed, which will help adjust the flow rate of wastewater sample. Second, a tap is fitted above 5 cm from the bottom of the inlet bucket. Third, a rectangular tub (test plant bed) used for the testing of sewage through plants or weed and this bed having a size of 62 cm, a length of 35 cm and a width and depth of 30 cm. Fourth, media, such as pebbles, gravel and sand of different sizes, were collected and arranged in different layers: (1) big size smooth pebbles and sharp gravels of 20-kg weight, (2) 15-kg weight of small size and smooth sand and (3) garden black soil up to 5 kg. Fifth, all these media materials were collected, the beds prepared and the plants or

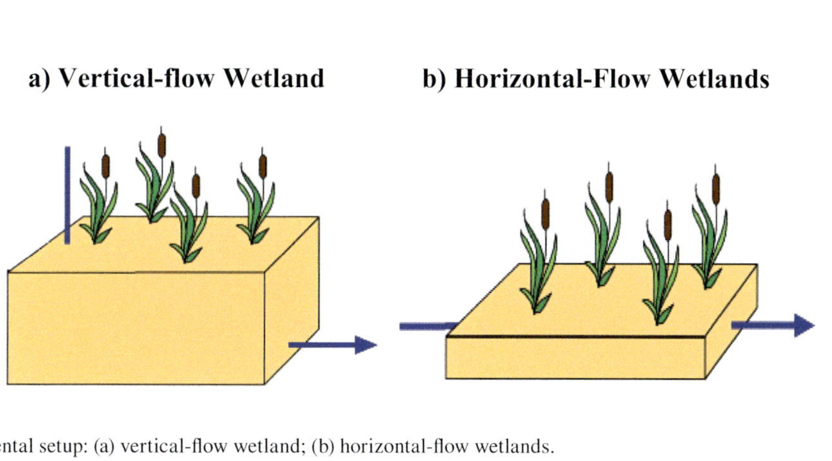

FIGURE 5.2 Experimental setup: (a) vertical-flow wetland; (b) horizontal-flow wetlands.

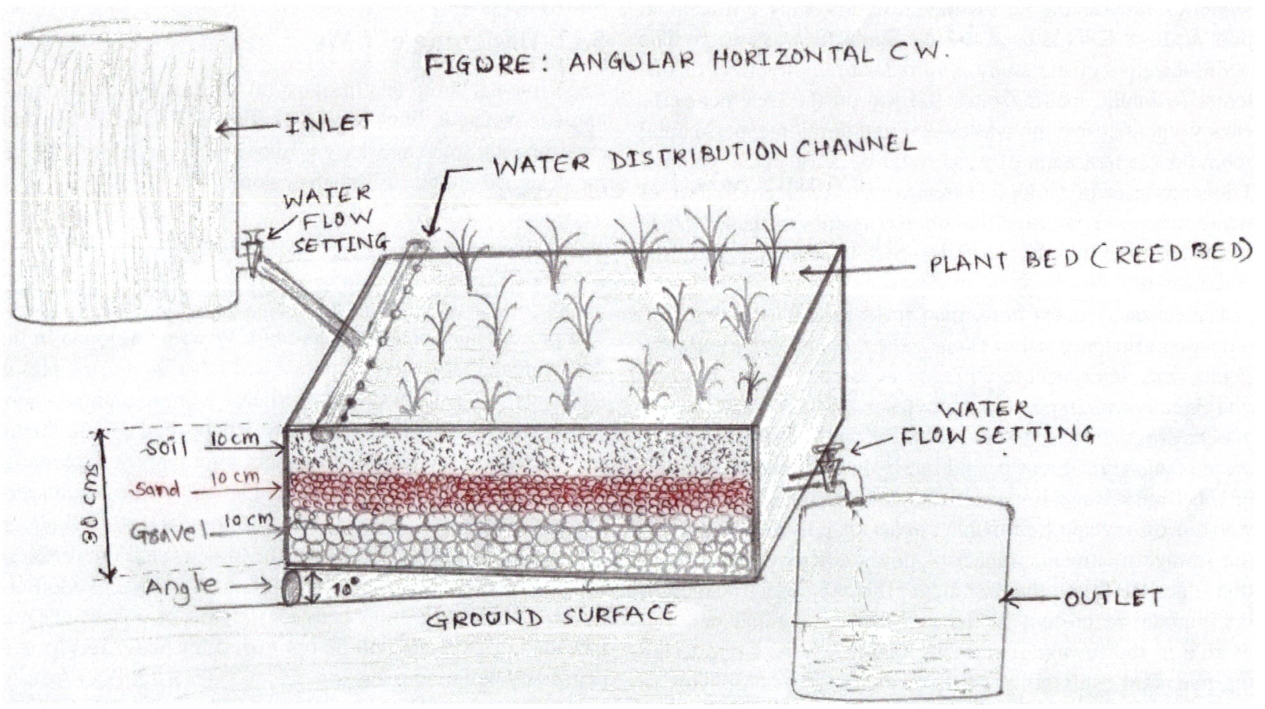

FIGURE 5.3 Angular horizontal SSF CW.

macrophytes planted in the beds, and these plants were acclimatized in a tub for 1 week for survival and to adjust these plants to the present climatic conditions. Finally, the sewage samples were passed through media in the plant bed, and the treated water was collected in cans at the outlet after given retention time. Two separate cans were used for collection of treated water. The analysis of treated and untreated sewage samples were carried out in the departmental laboratory.

All the experimental work carried at the Department of Environmental Science laboratory, using angular and angular with zigzag HSF CW method, which were used for treatment of sewage. *Typha latifolia, Cana indica, Phragmites karka, Colocasia esculenta, Pennisetum puerperium, Panicum maximum, Eichhornia crassipes and* control plant weeds were used in the present study. The plant weeds collection was followed by stabilization in the specific environment and then by acclimatization with a minimum concentration of sewage. The studies were continued for other higher concentrations of sewage thereafter.

Recycling and reuse of wastewater are not practiced in many urban centers. The study indicated that most treated wastewater is recommended for recycle and reuse in agriculture, horticulture, washing, cooling and cleaning purposes on the basis of reed and final characteristics of the treated water. Recycling of wastewater not only helps reduce pollution of land and water bodies but also has implications on water demand and additional investment requirements in water supply. Recycling and reuse of wastewater can reduce or postpone the need for developing new sources of water supply and helps improve the environment.

5.14 Analysis of Untreated and Treated Wastewater

Sewage was not directly flowed because the aim is to understand the variation in treated water using a percentage-wise dilution and to check the pollution removal efficiencies of each plant using various designs, flow rates, media, biomass and so on in the CW. An analysis of the sewage samples was carried out before and after treatment using different parameters (physico-chemical and biological) for understanding the pollution level (pollution removal efficiency) and the extent of treatment of water quality. These basic parameters were studied. These parameters mainly include the pH, Electrical Conductivity (EC), color, odor, solids, BOD, COD, chlorides, nitrates, phosphates and sulfates, as well as tested metals, from the plants and both sewage samples. Untreated and treated sewage samples of the constructed wetland were analyzed.

Wastewater can be formed by a variety of activities, including washing, bathing and using the toilet. Rainwater runoff is also considered wastewater. Wastewater treatment reduces the contaminants to acceptable levels to be safe for discharging into THE environment. Untreated wastewater generally contains high levels of organic material, numerous pathogenic microorganisms, as well as nutrients and toxic compounds. It thus entails environmental and health hazards and, consequently, must immediately be conveyed away from its generation sources and treated appropriately before final disposal. The ultimate goal of wastewater management is to remove as much of the solids (Total Suspended Solids (TSS) and total dissolved solids [TDS]) and nutrients as possible before discharging the water back to the environment. As solid material decays, it uses up O_2, which is needed by the plants and animals living in the water.

Wastewater quality consists of physical, chemical and biological characteristics. Physical parameters include color, odor, temperature and turbidity. Insoluble contents, such as solids, oil and grease, also fall into suspended and dissolved solids as well as organic (volatile) and inorganic (fixed) fractions. Chemical parameters associated with the organic contents of wastewater include BOD and COD. Inorganic chemical parameters include pH, chloride, sulfates, nitrates and phosphates. Bacteriological parameters include coliform, fecal coliforms, specific pathogens, fungi and viruses.

Physical, chemical and biological methods are used to remove contaminants from wastewater in industrial scale. The most important aerobic treatment system is the activated sludge process, based on the maintenance and recirculation of a complex biomass composed of microorganisms able to absorb and adsorb the organic matter carried in the wastewater. Anaerobic processes are widely applied in the treatment of industrial wastewater and biological sludge. Some wastewater may be highly treated and reused as reclaimed water. Modern systems include tertiary treatment by microfiltration or synthetic membranes. After membrane filtration, the treated wastewater is indistinguishable from water of natural origin of drinking quality. NO_3^- can be removed from wastewater by microbial denitrification, for which a small amount of methanol is typically added to provide the bacteria with a source of carbon. Some of wastewater with ecological approaches using reed bed systems such as CWs is an appropriate for treatment.

A properly designed municipal wastewater treatment system is facilitating water management and reuse practices. These sustainable practices will bring economical and environmental benefits for future development. Compared to the conventional engineering treatment systems, wastewater stabilization ponds (WSPs) and CWs provide various advantages. They are low-energy-consuming and biologically self-designing strategies and are of social and economic adherence. They also produce high-quality treated water suitable for almost any type of reuse.

In this study, sewage was treated in a CW by using selective aquatic macrophytes and was carried out for their treatment efficiencies. The core study focused on the scientific and technical aspects of the construction of wetland. The suitability of aquatic plants or macrophytes, the impact of concentration, the impact of retention time, the impact of flow rate, the effect of seasons, the impact of design and other operational, as well as environmental, factors of sewage treatment were studied. The study also included the design parameters such as the type of media used for the plant growth, the impact of angle of inclination, the impact of filter media, the role of microbial population in the bed, the flow channel configuration for the wastewater flow and other factors determining the extent of sewage treatment were studied. Finally, the cost and benefits of the optimized system of HSF CW were carried out.

5.15 Results and Discussion

Sewage was not directly flowed because the aim is to understand the variation in treated water using percentage wise dilution and to check the pollution removal efficiencies of each plant species using various designs, flow rates, media, biomass, seasonal impact, microbial role and others in the CW. An analysis of sewage samples was carried out before and after treatment using different parameters (physico-chemical and biological) for understanding the pollution level (pollution removal efficiency) and the extent of treatment of water quality. These basic pollution parameters were studied. These parameters mainly include the pH, EC, color, odor, solids, BOD, COD, chlorides, nitrates, phosphates and sulfates, as well as assessed metals from the plants and both sewage samples. Untreated and treated sewage samples of the constructed wetland were analyzed. Both types of samples were analyzed in the three categories, which were physical, chemical and biological parameters.

One free-floating and six emergent-type plant species, such as E. crassipes, Typha latifolia, Panicum maximum (forage grass), Colocasia esculenta Schott (alu), Cana indica, Pennisetum puerperium, Phragmites karka and the control bed (without using plant or macrophytes), were selected to treat the wastewater in laboratory conditions. The selection of plants is very important for wastewater treatment systems because of the following positive attributes: high growth rate and productivity, high nutritional value relative to other species and ease of harvesting and stocking. The study was designed to evaluate the nutrient removal capacity and determine the physical parameter of wastewater in monoculture and polyculture systems and one for control system (no aquatic plants). Our results show that planted system removed the high color, odor and pH compared with the control system. Significant removal was found in the polyculture system compared to the monoculture. Free-floating plants, such as E. crassipes, showed the highest removal of nutrients, mostly nitrogen in the monoculture. Because of the well-developed root system, Eichhornia crassipes provides a good habitat for bacteria to enhance the nitrification and denitrification, which resulted in higher nitrogen removal.

Experiment was also carried out to treat the wastewater in constructed or artificial wetland system by using six different emergent plant species, such as *Typha latifolia, Panicum maximum* (forage *grass), Colocasia esculenta Schott* (alu), *Cana indica, Pennisetum puerperium* and *Phragmites karka*, and a control *bed* (without *using plant or macrophytes)*. These plants were grown for 8 months and harvested every 20 days to measure the standing biomass. The harvested plants were separated into aboveground (leaf, stem) and belowground (root) tissue. Among all the plants, *Pennisetum puerperium, Panicum maximum* and *Cana indica* showed the maximum biomass in the order of stem > leaf > root. The total standing biomass was found higher in September/October than after, when the biomass was slightly decreased. It was also observed that biomass was increased from June/July to September / October; this is because the young plants absorbed more nutrients rather than did old plants. However, other selected species showed maximum biomass in the order of leaf > stem > root.

The removal of nutrient is very important for wastewater treatment. The concentration of phosphates was found to be lower than nitrates in all the plants. Plants, algae and microorganisms all utilize phosphates as essential nutrients. The maximum phosphates concentration was found to be higher in *Pennisetum puerperium* and *Panicum maximum* compared to other species. Most of the phosphates are taken by plant roots and absorption through leaves and are restricted to the shoots. In the present study, the overall performance of all plants were tested for the sewage treatment by using a CW. In this work, the various physico-chemical parameters were preferred for studying of impact of HRT, the seasons, design, roles and types of microorganisms, plant biomass, types of plants, types of flows and others on the sewage treatment or the purification of sewage.

5.15.1 Optimization of Pollutants Concentrations in Sewage Treatment Using SSF CW

a. ***Eichhornia crassipes* in SF CW:**
The study focused on the overall performances of *E. crassipes* for the treatment of domestic wastewater. Water quality parameters studied such as pH, Dissolved Oxygen (DO), BOD, COD, nitrate, chloride, TS, TDS, TSS, and heavy metals reflected different treatment efficiencies. The color and odor were removed, and hence, the treated samples were observed to be clear and odorless. In all sets of dilutions, the results obtained were near to the neutral form. In 20% dilution before treatment, the pH was 7.5, and after treatment, it was 7.2. In 100% dilution, the pH was 6.7 before treatment, but after treatment, it was 7.2. Solid contents, such as TDS, TSS and TS, estimated through this technique, reflected better treatment with phytoremediation. The maximum reduction of solids was found in 80% dilution. Better BOD reduction was observed from 20% to 80% and then observed less reduction. The COD reductions were found in increasing order up to 80%; then there was less reduction. In 80% and 100% sets of dilution, maximum COD reduction was observed. The values of nitrate showed variables nature at various dilutions. The presence of common heavy metals, namely, Cu, Ni, Co and Fe in domestic wastewater was determined before and after treatment. Metal concentrations were estimated in plant parts using atomic absorption spectroscopy (AAS). The metal reduction was noticed in all treatments. The Co was reduced by 78.78%, Cu was reduced by 28.90% and Fe by 23.42%.

b. **Emergent plants using SSF CW:** *Typha latifolia, Panicum maximum (*forage *grass), Colocasia esculenta Schott (*alu*), Cana indica, Pennisetum puerperium, Phragmites karka* and the control bed (without using plant or macrophytes)

1. ***Typha latifolia:*** In the present study, *Typha latifolia* is used for the treatment of sewage to test its pollutant absorption capacity. The HSF method was used for treating sewage for recycling and reuse. The treated and untreated samples of sewage with different dilutions, namely, 20%, 40%, 60%, 80% and 100%, were tested. The results reveal that the maximum reduction efficiency in EC was 31.73%, TSS by 72.53%, TDS by 27.77%, TS by 35.40%, BOD by 58.33%, COD by 93.96%, nitrates by 60.28%, phosphates by 72.47% and sulfates by 14.58%.
2. ***Panicum maximum:*** Study focuses on the overall performances of *Panicum maximum* Jacq (forage grass) for treating domestic wastewater. Water quality parameters studied, such as pH, EC, TSS, TDS, TS, COD, BOD, NO_3, PO_4 and SO_4, reflected different treatment efficiencies. The color and odor were removed, and hence, the treated samples were observed to be clear and odorless. The pH before treatment in the set of 10% dilution was 7.91, and after treatment, it was 7.6. In 100% concentration, it was 6.79 before treatment and 7.21 after treatment. The maximum reduction in EC was observed in an 80% concentration was 25.14%. The maximum reduction of solids was found in the 80% concentration. The TDS concentration 603 mg/L in a 10% dilution before treatment, which was reduced after treatment to 465 mg/L. A similar reduction was noticed in the 100% dilution. Before treatment, the TDS concentration was 1360, and after treatment, it was reduced to 664 mg /L. Better BOD_5 reduction was observed from 10% to 80%, and then reduction efficiency was decreased. The maximum BOD reduction found 59.25% in the 80% concentration. The COD reductions were found in increasing order up to 80%; then, there was decrease in COD reduction. The maximum COD reduction up to 61.05% was found in the 80% concentration. The maximum nitrate reduction up to 69.32% was found in the 100 % concentration of sewage. The maximum phosphates reduction up to 48.10% was found in the 80% sewage dilution. The values of sulfate before treatment and after treatment were varied in all sets of dilutions. The highest sulfate removal efficiency was 41.48%, found in the 80% sewage concentration.
3. ***Colocasia esculenta:*** The results in the set of *Colocasia esculenta* reveal that the maximum pollution reduction efficiency was observed in the 70% sewage concentration. The average wastewater pH values obtained in the plant bed for the inlet and outlet of angular HSF CW were 6.79 to 7.26, respectively. This increase of pH value may be due to the formation of some acidic components in the bioremediation process. The maximum TDS were reduced by 50.08%, BOD was reduced by 54.30% and COD was reduced by 58.69%. The maximum NO_3 was reduced by 59.48%, PO_4 was reduced by 46.99% and SO_4 was reduced by 39.32%. These pollutants were reduced due to the reed bed of *Colocasia esculenta*.
4. ***Cana indica:*** The color and odor were removed, and hence, the treated samples were observed to be clear and odorless. The pH before treatment in the set of 10% concentration was 8.10, and after treatment, it was 7.6. The maximum reduction, up to 54.48% in TSS, was observed in 80% concentration. The maximum reduction in TDS was observed in the 90% concentration. The highest BOD_5 reduction was observed up to

80% concentration, and then a reduction efficiency was decreased. The maximum BOD reduction found was to be 73.77% in the 80% concentration. The maximum COD reduction up to 75.19% was found in the 80% concentration. The maximum nitrate reduction up to 73.13% was found in the 100% concentration of sewage. Phosphate reduction increased up to 80% of sewage dilution; then there was decrease in it. The maximum reduction, up to 56.02%, was found in the 80% sewage concentration. The highest sulfate removal efficiency was 49.48%, found in the 90% sewage concentration. The overall results obtained from analysis of treated wastewater samples in present investigation indicated that the phytoremediation is more effective up to the sewage concentration of up to 80%.

5. ***Pennisetum puerperium:*** The color and odor were removed, and hence, treated samples were observed to be clear and odorless. The average wastewater pH values obtained in the plant bed for the inlet and outlet of angular HSF CW were 6.72 to 7.22, respectively. This increase of pH value may be due to formation of some acidic components in the bioremediation process. The maximum TDS was reduced by 56.18%, BOD was reduced by 76.65%, and COD was reduced by 77.51%. The maximum NO_3 was reduced by 74.62%, PO_4 was reduced by 57.81% and SO_4 was reduced by 51.06% in present study. These pollutants were reduced in the root zone bed due to plant uptake, the trickling process and a biofilm of aerobic and anaerobic microorganisms. As a result, the treatment efficiency is higher at 80% concentration of sewage, in experimental test set of *Pennisetum puerperium*, which has also been seen in plants through huge growth and flowering in the plant (Dhulap and Patil, 2014a).
6. ***Phragmites karka:*** The analysis of final effluent indicated that constructed wetland is capable to treat the wastewater and reduced TS by 61.64%, TDS by 60.37%, TSS by 63.19%, hardness by 57.15%, nitrate by 94.69%, phosphate by 92.95%, BOD by 61.47% and COD reduced by 64.74% respectively, by BOD up to 21.95%, COD up to 33.61%, removed hardness up to 57.15% and nitrates and phosphates up to 94.69 and 92.95%, respectively, in 4 days, or 96 hours, HRT.
7. ***Control (without plant):*** The treated and untreated samples of sewage with different dilutions viz. 10%, 20%, 30%, 40%, 50%, 60%, 70%, 80%, 90% and 100% were tested with a reference set as control (without plant). The results in the set of control bed reveal that the maximum pollution reduction efficiency was observed at 40 and 50% sewage concentrations. The color and odor were removed and hence treated samples were observed clear and odorless. The average wastewater pH values obtained in the control bed for the inlet and outlet of angular HSSF CW were 6.79 to 7.22. The maximum TDS were reduced by 32.66%, TS were reduced by 29.94%, BOD was reduced by 31.26% and COD was reduced by 39.81%. The maximum NO_3 was reduced by 24.21%, PO_4 was reduced by 20.89% and SO_4 was reduced by 20.75%. These pollutants were reduced in the reference set due to trickling process and biofilm of aerobic and anaerobic microorganism.

5.15.2 Impact of HRT on Sewage Treatment in the Angular HSF CW

In the impact or effect of HRT on the treatment of sewage study was carried using *Cana indica, Panicum maximum, Colocasia esculenta, Typha latifolia, Pennisetum puerperium, Schumach* and *E. crassipes*. These were studied at 50% and 100% sewage concentrations in the CW. In this work, the impact of wastewater flow or change in HRT on pollution reduction of 0 day to day 5 were studied.

In this work, the impacts of wastewater flow or change in HRT on pollution reduction were studied at 0 to 5 days. The pH level at 50% dilutions indicates that the maximum pH was changed from acidic level to alkaline in the test plant of *Typha latifolia*. In this plant, at day 0, the pH was 6.21 and at day 5 had changed by 7.61. The *Pennisetum puerperium* pH range was changed by 6.89 to 7.49, and *Panicum maximum*, by 6.92 to 7.39. The same study was carried at 100% sewage concentration for same plants, and results show the maximum pH was changed in *Typha latifolia*. In this plant, the pH range was changed by 6.62 to 7.82; in *Pennisetum puerperium*, by 6.72 to 7.38; and in *Panicum maximum*, by 6.79 to 7.29. All these plants are from the emergent-type category; only *E. crassipes* was selected from free-floating type. In both emergent and free floating categories of macrophytes based CW designed and developed through different types of CW models. In the *Eichhornia crassipes* plant bed at 50% dilution pH range was 6.7 to 7.3, and in the 100% dilution range, it was changed by 6.81 at day 0 to 7.34 at day 5.

TDS was changed in both the 50% and 100% sewage concentrations. The maximum TDS (mg/L) removal was observed in *Typha latifolia*. In angular HSF CW using *E. crassipes*, TDS was removed by 990 mg/L at day 0 to 273 on day 4, and on day 5, it was removed by 276 mg/L. In 100% sewage concentration treatment shows maximum TDS was removed by plant *Pennisetum puerperium*. In the surface flow type of CW using *E. crassipes*, TDS was removed by 1240 mg/L at 0 day to 545 mg/L at 4th and 547 mg/L at 5th day HRT.

In 50% sewage dilution or concentration, the maximum BOD (mg/L) removal was observed in *Typha latifolia*. In angular HSF CW using *E. crassipes*, removed BOD by 110 mg /L at day 0 to 55 at day 3 and 59 mg/L at day 4, and at day 5, it was removed by 67 mg/L. In 100% sewage concentration shows maximum BOD was removed by plant *Pennisetum puerperium*. In the surface flow type of CW using *E. crassipes*, BOD was removed by 230 mg/L at 0 day to 118 mg/L at 4th and 120 mg/L at day 5 HRT. In the 50% sewage dilution or concentration, the maximum COD (mg/L) removal was observed in *Pennisetum puerperium*. In angular HSF CW using *E. crassipes*, COD was removed by 159 mg/L at day 0 to 62 at day 3 and 71 mg/L at day 4 and at day 5, it was removed by 79 mg /L. The 100% sewage concentration shows the maximum COD removed by plant *Pennisetum puerperium*. In the surface flow type of CW using

E. crassipes, COD was removed by 315 mg/L at day 0 to 156 mg/L at day 4 and 162 mg/L at day 5 HRT.

In the 50% sewage dilution or concentration, the maximum NO_3 (mg/L) removal was observed in *Pennisetum purpureum*. The 100% sewage concentration shows maximum NO_3 was removed by *Pennisetum puerperium*. In the 50% sewage dilution or concentration, the maximum PO_4 (mg/L) removal was observed in *Pennisetum puerperium*. In this plant, at day 0, PO_4 was 12.46 mg/L; on day 4, it was removed by 3.12 mg/L; and on day 5, by 3.21 mg/L. The 100% sewage concentration shows maximum PO_4 was removed by *Pennisetum puerperium*. This plant PO_4 removed by 21.60 mg/L at day 0 to 2.32 mg/L at day 4, and at day 5, PO_4 was increased by 2.40 mg/L. In the 50% sewage dilution or concentration, the maximum SO_4 (mg/L) removal was observed in *Pennisetum puerperium*. In this plant, at day 0, SO_4 was 78.12 mg/L; in day 4, it was removed by 22.80 mg/L; and in day 5, 24.14 mg/L. The 100% sewage concentration shows maximum SO_4 was removed by plant *Pennisetum puerperium*. This plant SO_4 removed by 107 mg/L at day 0 to 18.20 mg/L at day 4. and at day 5, SO_4 was increased by 18.42 mg/L.

These plants also seem to be healthy greenery with flowering. Finally, each plant's pollution removal efficiencies at 50% and 100% were calculated. On the basis of the calculation, it shows that *Pennisetum puerperium* and *Panicum maximum* signify their suitability as the best plants for pollution removal at different HRTs of sewage treatment.

5.15.3 Impact of Plant Density on Sewage Treatment Using Angular HSF and Surface Flow CWs

In this study, the impact of plant biomass or plant growth on sewage treatment using 100% sewage concentration, which showed the highest pollution reduction efficiency. The plants studied include *Cana indica, Panicum maximum, Colocasia esculenta, Typha latifolia, Pennisetum puerperium, Schumach and E. crassipes.*

In this overall study, the *Pennisetum puerperium* of the emergent-type plants and *E. crassipes* of the free-floating plants removed maximum organic and inorganic pollutants from the sewage. The CW bed of all plants showed greenery, leafier and were seen flowering, and most of the faunal species, such as insects, caterpillars and avifauna's, were attracted to the experimental bed.

5.15.4 Impact of Design on the Sewage Treatment in Angular with Circular SSF CW

In the present work, study carried sewage treatment in Angular Horizontal Subsurface and circular or zigzag flow of CW by using the plants *Cana indica, Colocasia esculenta, Pennisetum puerperium, Panicum maximum* and the control (without using plants). These plants were selected on the basis of their high growth against sewage in the earlier experimental methods and on the basis of their pollutant absorption capacity or removal of pollutants from the sewage using modifying design of constructed wetland.

1. ***Cana indica*:** The study focuses on the overall performances and impact on treatment by a change in design (angular or circular horizontal flow) of the CW using *Cana indica* for treating domestic wastewater. The color and odor were removed, and hence, treated samples were observed to be clear and odorless. The pH values in the before- and after-treatment sets were changed at different concentration of treatment. The maximum reduction of TDS was found in 90% concentration, which was 55.17%. The BOD_5 were estimated in the samples before and after treatment. The maximum BOD reduction found 74.03% in the 80% concentration. BOD_5 removal efficiency is a function of HRT. The COD reductions were found in increasing order up to 80%; then there was decrease in COD reduction. The maximum COD reduction up to 75.52% was found in the 80% concentration.

   The maximum nitrates reduction up to 57.14% was found in 80% sewage dilution. The values of sulfate before treatment and after treatment were varied in all sets of dilutions. The highest sulfate removal efficiency was 50% found in the 80% sewage concentration.
2. ***Colocasia esculenta:*** The color and odor were removed, and hence, treated samples were observed to be clear and odorless. The pH before treatment in the set of 10% dilution was 7.89, and after treatment, it was 7.08. In the 100% concentration, it was 6.75 before treatment and 7.27 after treatment. The maximum reduction of solids was found in 70% concentration, which was 46.66%. The maximum BOD reduction found 59.63% in the 80% concentration. The COD reductions were found in increasing order up to 80%; then there was decrease in COD reduction. The maximum COD reduction up to 60.42% was found in the 70% concentration.

   The maximum reduction of nitrate up to 60.60% was found in 70% concentration of sewage. Phosphate maximum reduction up to 51.53% was found in 70% sewage dilution. The highest sulfate removal efficiency was 40.62%, found in the 80% sewage concentration.
3. ***Pennisetum puerperium:*** The pH before treatment in the set of 10% dilution was 7.82, and after treatment, it was 7.63. In the 100% concentration, it was 6.74 before treatment and 7.22 after treatment. The maximum reduction of solids was found in the 80% concentration, which was 52.86%. The maximum reduction of TDS was found in the 90% concentration, which was 55.79%. Better BOD_5 reduction was observed from 10% to 80% and then reduction efficiency was decreased. The maximum BOD reduction found 78.16% in the 80% concentration. The maximum COD reduction up to 78.61% was found in the 80% concentration. Nitrates in treated effluent at outlet were observed in the range of 3.1 mg/L to 6.6 mg/L, and the percentage reductions were between 46.55% to 75.73%. Maximum reduction up

to 75.73% was found in 100% concentration of sewage. Phosphate reduction increases up to 80% of the treated sewage then there were decreases in phosphate reduction. A maximum reduction up to 61.03% was found in 70% sewage dilution. The highest sulfate removal efficiency was 57.61%, found in the 80% sewage concentration.

4. ***Panicum maximum***: The pH values in before- and after-treatment sets were changed at different concentration of treatment. The pH before treatment in the set of 10% dilution was 7.93, and after treatment, it was 7.8. In the 100% concentration, it was 6.79 before treatment and 7.22 after treatment. The maximum reduction of TDS was found in 80% concentration, which was 52.93%. The BOD_5 before treatment in set of 10% concentration was 4.64 mg/L, and it was 3.48 mg/L after treatment. The maximum BOD reduction was 61.95% in the 80% concentration. The COD reduction was 29.41% to 59.57%. The maximum COD reduction, up to 62.30%, was found in the 80% concentration. The maximum nitrate reduction, up to 69.41%, was found in 100% concentration of sewage. The phosphate reduction increases up to 80% of the treated sewage then there were decreases in phosphate reduction. Maximum reduction, up to 48.27%, was found in the 80% sewage dilution. The highest sulfate removal efficiency was 42.10% found in the 80% sewage concentration.
5. **Control bed (without plants used):** The color and odor were removed, and hence, treated samples were observed to be clear and odorless. The pH before treatment in the set of 10% dilution was 7.94, and after treatment, it was 7.64. In the 100% concentration, it was 6.79 before treatment and 7.22 after treatment. The maximum reduction of TDS was found in the 40% concentration, which was 33.17%. The maximum BOD reduction found to be 32.54% in the 40% concentration. The maximum COD reduction, up to 41.14%, was found in the 40% concentration. Nitrates in the treated effluent at outlet were observed in the range of 4.7 mg/L to 23.8 mg/L. The maximum reduction, up to 31.93%, was found in 40% concentration of sewage. The maximum phosphate reduction, up to 28.88%, was found in the 40% sewage concentration, and the highest sulfate removal efficiency was 25.07%, found in the 40% sewage concentration.

5.15.5 Media-Based Soil Analysis in the Angular HSF CW

The soil was used as the medium, acting as role like of trickling filter and supports to the standing macrophytes or plants. On the same way the soil applied in the plant used bed and without plant bed (control), and the soil was analyzed before use in the treatment and after use in treatment as a medium. Soil was characterized and assessed through physico-chemical analytical methods. The analysis was carried out through selected parameters are nitrate (kg/ha), phosphorous (kg/ha), potassium (kg/ha), calcium (Meq), magnesium (Meq), sodium (Meq), pH range, salinity (TSS), lime (free) percentage, density of soil and organic carbon percentage.

The results reveal that nitrate (kg/ha) is an important component and element for plant growth. The maximum nitrate was observed in the *Colocasia esculenta*, and the minimum was observed in the control bed without plants. The phosphorous (kg/ha) in the soil samples observed by 7 (kg/ha) in nursery soil, 7 (kg/ha) in the *Cana indica* bed soil, 12 (kg/ha) in the *Colocasia esculenta* bed soil, 15 (kg/ha) in the *Panicum maximum* bed soil *and* 11 (kg/ha) in the control bed of soil. Less phosphorous was observed in the nursery soil sample, and the maximum was observed in the *Panicum maximum* soil bed.

The maximum potassium was observed in the *Panicum maximum* soil bed. The maximum calcium was observed in *Colocasia esculenta* soil bed. The maximum magnesium was observed in *Cana indica* soil bed. The maximum sodium was observed in *Cana indica* soil bed. The pH range in the soil samples varied from 7.90 in the nursery soil, 7.90 in the *Cana indica* bed soil, 7.70 in the *Colocasia esculenta,* 7.50 in the *Panicum maximum and* 7.80 in the control bed soil. The control bed soil was shown to be more alkaline than the other soil samples, and the *Panicum maximum* soil sample was shown to be less alkaline or near a neutral pH range. The maximum salinity was observed in the *Panicum maximum* soil bed. The maximum lime free (%) observed in *Cana indica* soil bed. The maximum density observed in *Colocasia* soil bed.

The maximum organic carbon (%) was observed in the *Colocasia esculenta* soil bed. In the entire soil assessment study, *Panicum maximum, Cana and Colocasia* all were highly pollutant absorbers from the sewage in presence of soil as a bed media, and the plants were grown very rapidly in this soil bed media. This media helped support standing biomass, helped natural filtration and helped as a support to microbial growth.

5.15.6 Impact of Polyculture Macrophytes on Sewage Treatment Using Angular HSF CW

In the study of the impact of polyculture emergent macrophytes plantation (mix plantation) on sewage treatment using angular SHF CW study, the parameters studied, such as pH, EC, TSS, TDS, TS, BOD, COD, NO_3, PO_4 and SO_4, were reflected in different treatment efficiencies. The results reveal that the color and odor were removed, and hence, the treated samples were observed to be clear and odorless. The maximum p^H value in the 10% concentration before treatment was 7.5 and minimum in after treatment was 7.0. In the 100% concentration, the maximum p^H before treatment was 8.4, and the minimum after treatment was 8.0. In the 100% concentration, the maximum TDS before treatment was 540 mg/l, and the minimum after treatment is 440 mg/L. In the 10% concentration, the maximum BOD before treatment was 3.68 mg/L and the minimum after treatment was 3.30 mg/L. In the 100% concentration, the maximum BOD before treatment was 14.7 mg/L, and the minimum after treatment was 14.3 mg/L. In the 10% concentration, the maximum COD before treatment was 180 mg/L, and after treatment, the COD was found to be 150 mg/L. In the 100% concentration, the maximum COD before treatment was 260 mg/L, and after treatment, the minimum COD was found to be 230 mg/L. In the 10% concentration, the maximum NO_3 before treatment was 5.4 mg/l, and after treatment, the NO_3 was found to be 5.0 mg/L. In the 100% concentration, the maximum NO_3 before treatment was 24.40 mg/l, and after treatment, the NO_3 was found to be 24.18mg/L.

In this entire study of angular HSF CW using polyculture emergent types of macrophytes and a comparative study carried with control bed (without plants) in a CW. The treatment plant bed included emergent aquatic macrophytes plants that showed flowering and a leafier appearance and greenery. The treated water is apparently cleared for common water uses in both household and industries.

5.15.7 Microbial Studies in the Soil Medium, on the Root (Rhizosphere), Inside the Root System (Endorhizosphere) and in Sewage (Treated and Untreated) Using Angular HSF CW

Microorganisms have a major role in the sewage treatment. In the root zone bed, the rhizosphere (root surface), the endorhizosphere (inside root), there were massive numbers of microorganisms present. These microorganisms degrade or convert complex organic matter into simpler or the nutrient form of organic matter. After this, these nutrient forms were obtained by plants for their growth. In this HSF and HSF circular flow CWs, following plants or macrophytes were used for sewage treatment: (1) *Typha latifolia*, (2) *Cana indica*, (3) *Panicum maximum*, (4) *Colocasia esculenta*, (5) *Pennisetum puerperium* and (6) *control* (without plants). There were different types of bacteria, and types of fungi were examined, observed and identified from the following samples:

1. *Soil sample (used as media in constructed wetland)*
2. *Root (soil + root)/rhizosphere*
3. *Root (inside) sample*
4. *Untreated wastewater samples with plants*
5. *Treated wastewater samples with plants*
6. *Untreated water samples in the control (without plant used in the bed)*
7. *Treated water samples in the control (without plant used in the bed)*

In this study of microbial examination, the results showed in the form of (+++) high frequency (lawn growth), (--) absence, (++) decrease frequency and (+) low frequency. *E. coli* was observed in the *Cana indica* plant bed, and it was found in decrease in frequency (++); in the *Panicum maximum, Pennisetum puerperium* and *Colocasia esculenta* beds, it was observed in low frequency (+), and in the sewage without treatment (before/inlet sample), it was shown as high frequency (+++) or lawn growth. *Staphylococcus aureus* was observed in the *Cana indica* plant bed and found in low frequency; in the *Panicum maximum, Pennisetum puerperium* and *Colocasia esculenta* beds, it was observed in absence (--); and the in sewage without treatment (before/inlet sample), it was shown as high frequency (+++) or lawn growth. *Streptococci* was observed in all the samples of *Cana indica, Panicum maximum, Pennisetum puerperium* and *Colocasia esculenta*, and in the sewage without treatment (before/inlet sample), it was shown as low frequency (+) or lawn growth. *Azotobacter* in the plant beds of *Cana indica, Panicum maximum* and *Pennisetum puerperium* found in decreasing frequency; in *Colocasia esculenta*, it was observed in low frequency and an absence of frequency (+/-), and in the sewage without treatment (before/inlet sample), it was shown as decreasing in frequency (++) or lawn growth.

Bacilli observed in the *Cana indica* plant bed was found in low frequency; in the *Panicum maximum* bed, it was *found* as decreasing *in frequency*; *in* the *Pennisetum puerperium* bed, it was observed in low frequency; in the *Colocasia esculenta* bed, it was observed in absence frequency (--).; and in the sewage without treatment (before/inlet sample), it was shown in decreasing frequency (++) or lawn growth. *Pseudomonas sp.* observed in the *Cana indica, Panicum maximum* and *Pennisetum puerperium* beds were shown as decreasing in frequency; in the *Colocasia esculenta* bed, it was observed in absence frequency (--); and in the sewage without treatment (before/inlet sample), it was shown as high frequency (+++) or lawn growth. *Aspergillus niger* was found to be low in all the types of plants bed, such as *Cana indica, Panicum maximum, Pennisetum puerperium* and *Colocasia esculenta*, but it was found in high frequency or lawn growth in the in sewage without treatment (before/inlet sample).

A. **Microbial testing in the soil sample:** A soil's bacteria are associated with plant roots that can exert beneficial effects on their hosts are designated as plant-growth-promoting rhizobacteria (PGPR). These soil bacteria were isolated by pure culture method, microorganisms were isolated and enumerated from soil by the serial dilution-agar plating method (or viable plate count method), fungi were isolated by the Warcup method, *Azotobacter* species were isolated from soil and protozoa were isolated from soil.
B. **Microbial testing in the soil and plant (rhizosphere) samples:** In this method, sample was investigated by isolating microorganisms from the rhizosphere and isolating and characterizing PGPR methods.
C. **Microbial testing in the endorhizosphere of plant samples:** In this examination study, microbe isolation and identification methods were carried out, such as (1) the isolation of root surface bacteria, (2) the isolation of endorhizosphere bacteria (inside root), (3) the semisolid enrichment cultures of root pieces for isolating microaerophilic diazotrophic bacteria, (4) nonselective media for isolation of rhizosphere bacteria in general and (5) the isolation of nodule endophytes.
D. **Identification of unknown microorganisms:** In this study, unknown and unidentified microorganisms in soil samples, which were isolated, were observed.

E. **Microbial testing of untreated and treated water samples:** In this study, treated sewage water samples were tested for microbes and microflora were observed. This microflora was determined by total bacterial population by standard plate count method and the isolation of fungi (aquatic) in the water sample.

5.15.8 Seasonal Impact on the Sewage Treatment in Angular HSF and Subsurface Flow CW

The list of plants studied under seasonal impacts on sewage treatment using angular HSF constructed wetland include (*1*) *Typha latifolia,* (2) *Colocasia esculenta,* (*3*) *E. crassipes,* (*4*) *Cana indica,* (5) *Panicum maximum,* (*6*) *Pennisetum puerperium* and (*7*) a control (without *plant*).

All these test plant species showed positive growth in the sewage water without obvious symptoms of toxicity or nutrient deficiency. After 365 days (1 year), the development of plant species were increased up to 6 to 8 feet in height. *E. crassipes, Cana indica, Pennisetum puerperium, Panicum maximum and Typha latifolia* showed maximum growth and increased their total biomass. Observations during plant establishment showed that new growth initially emerged mainly from rhizomes or rootstocks, with old shoots generally dying back during the first few weeks of growth.

There is a significantly seasonal factor to this wetland for BOD_5, COD, nitrate and phosphates when measured on a percentage reduction basis. BOD_5, COD, nitrate, phosphates, and sulfates removal efficiencies displayed seasonal variations compared with the control and removals being more efficient in winter and rainy season compared to those in summer. In the overall observational study, the *Pennisetum puerperium* and *E. crassipes* plant beds showed maximum pollution removal efficiency compared than the other plant beds and the control bed in all the three seasons (Dhulap et al., 2014b).

5.15.9 Cost–Benefits Analysis (CBA) of Angular HSF CW

In the cost–benefit analysis, before making any decision related to wastewater reuse or recycling, we should take into account the aspects related to both costs and benefits. The benefits include the agricultural output produced by recycled water, aquifer recharge, saving wastewater disposal cost and avoided costs by preventing health risks. The costs include wastewater treatment, storage and conveyance costs, agricultural production costs (fertilization and irrigation) and costs associated with environmental degradation, aquifer pollution, salinity, health risk and soil structure deterioration (Haruvy et al., 1999).

Researcher designed and fabricated pilot plant of angular horizontal surface and subsurface flow constructed wetland wastewater treatment, species of plant or macrophytes was tested, such as (1) *E. crassipes,* (*2*) *Phragmites karka,* (*3*) *Cana indica,* (*4*) *Colocasia esculenta,* (*5*) *Panicum maximum,* (*6*) *Pennisetum puerperium*, (*7*) *Typha latifolia and* (*8*) a control (without plant).

The total INR Rs. 1,52,865/- only expenditure cost for total 15 experimental pilot plants, which includes angular surface and angular HSF CW. The Rs. 10,191/- is an expenditure cost per pilot plant. Operation and maintenance costs include influent and effluent testing; water-level adjustment; maintenance of inflow, outflow and distribution devices; weed control; odor control; and control of nuisance pests and insects. The largest single cost is generally the cost of testing influent and effluent, which depends on the frequency of sampling, the number of water-quality parameters and the number of samples. Weeds should be controlled around the edges, and large weeds should be removed. Odors can be reduced by lowering water levels below the substrate or reducing solid loading. A significant potential maintenance activity for surface flow systems is the control of nuisance pests, including beavers, mosquitoes and other insects (ITRC, 2003). The difference in cost between the two types of systems may be because surface flow systems are typically larger for a given flow rate than subsurface flow systems.

These costs of data can be used to develop a rough idea of the costs for constructing and operating a surface and SSF CW.

5.16 Conclusion

In this angular HSS CW technology has following important aspects:

a. **Reuse:** The design of the SSF CW should be in line with the desired effluent quality for disposal or reuse. The most common type of reuse is irrigation, such as drip irrigation or subsurface irrigation, for lawns, green spaces or crop production. In this case, utilizing nutrients contained in wastewater rather than nutrient removal is desirable.
b. **Hygiene (pathogens):** Humans excrete many different types of disease-causing pathogens if they are infected or carriers of a disease. The human intestinal tract contains coliform bacteria that are discharged with the feces. These coliform bacteria are not pathogens but, if present in environmental samples, indicate that intestinal pathogens may also be present. Sewage that has been treated in SSF CWs usually meets the standards for pathogen levels for safe discharge of surface water without further treatment.
c. **Quantity:** Horizontal flow beds in warm climates can lose all the wastewater due to evapotranspiration. This needs to be considered in the water balance. The larger the surface area, the more significant are the effects of precipitation (rain) and evapotranspiration, especially in warm and dry climates.
d. **Color aspects:** The effluent from any biological wastewater treatment process, such as a CW, can have a yellowish or brownish color. This is caused by organic matter, such as humic substances, including humic acids or humins. The coloration may reduce the social acceptance of wastewater reuse. Humic acids

originate from the biological degradation of organic matter, being the unbiodegradable fraction of organic matter. Humic acids are a natural compound of soil, lake and river water. They are not harmful to the environment, but they have a negative impact on disinfection processes with chlorine or ultraviolet radiation. From all the results of the present study, sewage after treatment in a CW tends to be colorless and odorless.

This research work has provided several useful outcomes that can assist in future guidelines for wastewater management:

1. Pretreatment of wastewater before plant acclimatization could be potentially effective.
2. Plants are observed to be more effective in improving water quality compared to a control system (without plant).
3. Maximum removal of nitrate was observed, so it appears that the optimum condition for nitrate removal is in a synthetic medium.
4. Although the proportion of nutrient removal contained in plant biomass was low, the results indicate that wetland harvesting in the early growing period would be more effective.
5. For the total standing biomass, plants have more effective in September/October than after standing biomass was decreased in all the cases, so this is a suitable period for harvesting emergent plants for biomass calculation.
6. After the treatment, treated water can be used for agricultural system, lawn water and to make fountains, which can be environmentally friendly. These systems require low capital, have competitive operating costs and are easy to operate and maintain.

REFERENCES

ARTEC. 2003. AmbienteRicercaTecnologia AMBIENTE. via Costituzione, 25 42015 Correggio (RE) Italy.

Brix, H. 1994. Use of constructed wetlands in water pollution control historical development, present status and future perspectives. *Water Sci. Technol*. Vol. 30, Issue 8, 209–223.

Chavan, B.L. and Dhulap, V.P. 2012d. Optimization of pollutant concentration in sewage treatment using constructed wetland through phytoremediation. *International Journal of Advanced Research in Engineering and Applied Sciences* (IJAREAS), *Green Field Advanced Research Publishing House* (GARPH). Vol. 1, Issue 6, 1–16.

Cooper, P.F., Job, G.D., Green, M.B., and Shutes, R.B.E. 1996. Reed beds and constructed wetlands for wastewater treatment. Medmenham, Marlow, UK: WRC Publications, 184 pp.

Deun, R.V. and Dyck, M.V. 2006. Department agro and biotechnology, katholiekehogeschoolkempen, Geel Belgium. "Tisza – Tisa" – project: Capacity building in extensive wastewater treatment – Tools for improving water quality in the Middle and lower Tisza Area.

Dhulap, V.P., Ghorade, I.B. and Patil, S.S. 2014b. Seasonal study and its impact on sewage treatment in the angular horizontal subsurface flow constructed wetland using aquatic macrophytes. *Int. J. Res. Eng. Technol*. (IMPACT: IJRET) ISSN (E): 2321–8843; ISSN(P): 2347–4599. Vol. 2, Issue 213–224.

Dhulap, V.P. and Patil, S. S. 2014a. Removal of pollutants from sewage through constructed wetland using pennisetum purpureium. *Int. Eur. Acad. Res*. Vol. II, Issue 1.

Dyck, M.V. 2006. *Constructed Wetland Systems in Flanders*; KatholiekeHogeschoolKempen: Geel, Flanders Belgium.

Gurung, Anup and Oh, Sang-Eun. 2012. An overview of water pollution and constructed wetlands for sustainable wastewater treatment in Kathmandu Valley: A review. *Sci. Res. Essays*. Vol. 7, Issue 11, 1185–1194.

Haruvy, N., Hadas, A., Ravina, I. and Shalhevet, S. 1999. Cost assessment of averting groundwater pollution, the 7th International conference of the Israel society for ecology and environmental quality sciences on 'Environmental Challenges for the Next Millennium', Jerusalem, Israel.

Hoffmann, Heike, Platzer, Christoph, Winker, Martina and Muench, Elisabeth Von. 2011. Technology review of constructed wetlands subsurface flow constructed wetlands for greywater and domestic wastewater treatment. Deutsche Gesellschaft fur InternationaleZusammenarbeit (GIZ) GmbH Sustainable sanitation – ecosan program Postfach 5180, 65726 Eschborn, Germany.

Interstate Technology & Regulatory Council (ITRC). 2003. *Technical and Regulatory Guidance Document for Constructed Treatments Wetlands*. The Environmental Council of the States, Washington, DC.

Kadlec, Robert H. and Knight, Robert. L. 2004. *Treatment Wetlands*; Lewis Publishers: Boca Raton, FL.

Lantzke, I.R., Mitchell, D.S., Heritage, A.D. and Sharma, K.P. 1999. A model controlling orthophosphate removal in planted vertical flow wetlands. *Ecol. Eng*. Vol. 12, 93–105.

Lipkow, U., Munch, E.V. and Israel, C. 2010. Case study of SuSanA projects – Constructed wetland for a peri-urban housing area, Bayawan City, Philippines.

Luederitz, V., Eckert, E., Lange-Weber, M., Lange, A. and Gersberg, R. 2001. Nutrient removal efficiency and resource economics of vertical flow and horizontal flow constructed wetlands. *Ecol. Eng*. Vol. 18, 157–171.

Salati, E. 1987. Edaphic phytodepuration: A new approach to wastewater treatment. In *Aquatic Plants for Water Treatment and Resource Recovery*; Reddy, K.R., Smith, W.H., Eds.; Magnolia Publishing: Orlando, FL, pp. 199–208.

Shrestha, R.R. 1999. Application of constructed wetlands for wastewater treatment in Nepal. Ph.D. Dissertation, Department of Sanitary Engineering and Water Pollution Control, University of Agricultural Sciences Vienna, Austria.

U.S. Environmental Protection Agency. 2000. *Manual for Constructed Wetlands Treatment of Municipal Wastewaters*, EPA-625/R-99–010; U.S. Environmental Protection Agency: Cincinnati, OH.

United States Environmental Protection Agency. 1993. Constructed wetlands handbooks: A guide to creating wetlands for agricultural wastewater, domestic wastewater, coal mine drainage and storm water in the Mid-Atlantic region. (Volumes 1–5).

6

Multi-Soil Layering System for the Treatment of Domestic Wastewater

Thorvat Akshay Rajan, Deshpande Vaishnavi Vyanktesh and S. Mondal

CONTENTS

6.1 Introduction

Domestic wastewater is the wastewater that is derived from daily human activities in households such as bath, toilets, laundry, toilets and the like; thus, domestic wastewater contains pollutants which can adversely impact the surrounding environment, namely, groundwater and surface water. Water is said to be contaminated when it contains an adequate number of pollutants and makes it unfit for consumption for any human use. So there is a need for the treatment of domestic wastewater so as to avoid any negative impact on humans and the environment.

Wastewater treatment, which is also called sewage treatment, is the method of removal of pollutants/impurities from the wastewater before it reaches the natural waterways. Domestic wastewater treatment plays a significant role these days. Domestic wastewater treatment ensures that all domestic sewage is properly treated according to the standards to make it safe and appropriate for releasing it back into the environment.

Fresh water is one of the fundamental human necessities that need to be obtained from various sources. Today we are facing the problem of water pollution due to the discharge of wastewater directly into the water bodies. Wastewater treatment is the only way to overcome these ecological problems. Several natural and low-cost technologies can be used for the treatment of wastewater like lagoons, constructed wetlands, trickling filters, sand filters and others. Due to the nonavailability of full-fledged treatment facilities, a large quantity of domestic wastewater is released into the environment without proper treatment and/or no treatment. Even though the characteristics of wastewater from different societies differ considerably in quality and quantity, wastewater is generalized as having a high content of particulate matter, dissolved organic matter, nutrients and microbes. The receiving water bodies may become heavily polluted, resulting in serious environmental and ecological problems; this issue is mainly severe in developing countries. Therefore, it is a challenge to provide reliable and affordable wastewater treatment facilities to avoid serious environmental and ecological problems in such areas. The environmental impact of untreated wastewater is a considerable problem in society today. Wastewater contains a variety of pollutants, such as nitrogenous wastes, phosphorus- and sulfur-containing compounds, fecal coliform, other organic chemicals and heavy metals.

Unintentional discharge of wastewater in water bodies can lead to public health epidemics, such as infectious diarrhea,

DOI: 10.1201/9781003186304-6

hepatitis, heavy metal poisoning and so on. These contaminants may come from various sources, such as sewage plants, industries, mines, farms and dairies. Many studies have been mainly focused on the removal of priority pollutants. Recently, there have been increasing concerns for emerging contaminants. Therefore, it is of vital importance to investigate the technological solution, which will have various advantages such as low cost, easy operation and maintenance and use of local materials that are easily available for domestic wastewater treatment.

The multi-soil layering (MSL) system also has the advantage of simultaneous decrease in organic pollutants including biological oxygen demand (BOD) and chemical oxygen demand (COD), nitrogen, phosphorus and fecal coliform from wastewater. Another advantage is that this system can be constructed locally from easily available resources in almost any location. Furthermore, the system can treat extremely polluted wastewater, resulting in treated water that may be used for agricultural purposes.

Traditional and modern technologies have been used for decentralized sewage treatment; for example, many individual residences in rural areas rely on septic tanks to dispose of their wastewater. Although septic tanks are an easy and safe technique, the concentration of suspended solids (SS), BOD, total Kjeldahl nitrogen (TKN) and total phosphorus (TP) in effluents from septic tanks can hardly meet the discharge requirements. It is also noted that septic tanks are not correctly designed, built and operated in many situations. Urbanization increases the population density of residential areas, thus increasing the production of domestic wastewaters in these areas, which without proper treatment may severely affect the natural environment (Ho and Wang, 2015).

6.2 Background of the Study

The cost-effective and easy-to-construct natural treatment system (NTS) has achieved significant consideration in recent years. NTSs utilize natural elements, such as oxygen, soil, microbes and plants to decontaminate wastewater to effluent standards before discharging it into water bodies, thus reducing water pollution caused by the discharge of domestic wastewaters. The most common NTSs include constructed wetlands, cobble contact beds, and sand filters (Kao et al., 2001; Chen et al., 2009). NTSs also have several limitations. For example, domestic wastewater treatment by using wetlands is particularly time-consuming, and it also requires a large area of land to achieve the desired treatment efficiency. Thus, this type of NTS becomes too costly in areas where land cost is higher. Also, wetlands are usually areas with running water sources, are ideal habitats for mosquitoes and weeds and are therefore not favored by residents. In the cobble contact beds and sand filters, fine solids particles often form sediments within systems, thus causing clogging. This reduces the treatment efficiency and increases maintenance difficulty (Ho and Wang, 2015).

Wastewater collection and disposal is still one of the most common community health and environmental problems in many areas. Most of these areas do not have wastewater collection systems and rarely have wastewater treatment plants. In remote areas, a huge quantity of domestic wastewater is released into the water bodies without effective treatment and/or no treatment. Although the characteristics of wastewater from different communities vary significantly in quality and quantity, wastewater is able to be generalized as having a high content of particulate matter, dissolved organic matter, nutrients and microbes. This issue of environmental and ecological pollution is particularly distressing in villages and remote regions of developing countries. It is a challenge to provide reliable and affordable wastewater treatment in such areas around the world.

The treatment of wastewater usually involves one or more of mechanical, chemical and biological processes; each has various advantages as well as limitations. Mechanical processes often involve techniques such as sedimentation, flocculation, filtration, reverse osmosis, adsorption and others. Generally mechanical treatment systems are effective at capturing large or suspended particles but are less efficient at eliminating small or dissolved particles, organic materials and infectious biological agents. Furthermore, such systems often have high energy demands, need more space and require significant maintenance.

Chemical processes include reduction, oxidation, change in pH, and other processes, including the use of catalysts, and may cause precipitation, coagulation and modification of toxic chemicals to less harmful compounds. Although chemical processes are not effective for removing larger particles by using certain chemicals, they are advantageous in isolating out smaller, more soluble wastewater components that mechanical processes cannot.

Biological processes usually destroy or consume various wastewater components, by the use of aerobic and/or anaerobic microbes. Biological processes are thus very useful in eliminating nitrogenous or phosphorus-containing wastes, which is the main reason for the extreme growth of algae in water. However, one problem associated with such biological processes is that the microbes can overmultiply and cause undesirable biofilms (biomass) that clog wastewater purification system components.

Nowadays, it is desirable in the field of wastewater treatment to have a simple, compact, efficient, cost-effective and low-maintenance system that will significantly purify contaminated Wastewater. Such a method will optimally use a combination of chemical, biological and mechanical processes while minimizing possible disadvantages associated with the traditional processes. Such a system will also possibly generate beneficial renewable resource components from the wastewater such as fertilizers or treated water that can be used for agricultural purposes.

Among the available technologies, wastewater treatment by soil involves physical, chemical and biological processes (An et al., 2016) that have been known to be effective and that have been used in natural land treatment and soil trench systems. Such soil wastewater treatment systems cost less but need a wider land area than modern mechanical treatment systems. Where the land area is not limited and land cost is low, soil wastewater treatment systems have often been used for onsite wastewater disposal. However, there is a restriction in the use of soils for wastewater treatment systems because of the low permeability of soils, which reduces the discharge rate of wastewater into a treatment system. In soil wastewater treatment systems, the hydraulic loading rate (HLR) normally ranges from 10 to 40 $l/m^2/d$ for traditional soil trench systems and from 1 to 8 $l/m^2/d$ for slow-rate land treatment systems to avoid the accumulation of waste materials and the metabolic by-products that cause clogging. Variations in the chemical and biological properties of the soil types, which can influence the treatment ability of the soils, are also a possible limitation. Hence, it is required to consider these factors in order to use soils appropriately for wastewater treatment systems. It is important to select the most favorable soils depending on the treatment conditions required. However, the availability of soil types is, in fact, limited in the respective regions.

The MSL system has been developed and studied to use soils effectively for wastewater treatment through the alleviation of the above limitations of soils. The MSL system is composed of soil mixture layers and permeable layers that are arranged in a brick-layer-like pattern, and this structure keeps the water permeability high and reduces the risk of clogging the system. The environmental cleanup ability of soil is maximized in the MSL system while avoiding many problems encountered in the previous soil-based sewage treatment systems.

Few studies have been conducted to increase the treatment efficiency of the MSL system by altering the material composition and aeration conditions of the system. If the MSL system is used continuously without some kind of aeration, anaerobic conditions become prevailing. Strong anaerobic conditions reduce the removal efficiency of the MSL system in removing organic substances, nitrogen and phosphorus (Luanmanee et al., 2001). Therefore, aeration is essential to enhance the effectiveness of the MSL system for wastewater treatment (Luanmanee et al., 2001). Based on these studies, it appears that changing the material composition and aeration conditions by matching specific wastewater conditions could control the treatment characteristics of the MSL system. The knowledge and experience gained from previous studies may enable the design of better MSL systems. However, there is still a need for further information on the features that influence the treatment efficiency to properly design the MSL system depending on treatment conditions, such as the concentrations of the contaminants in wastewater, quality target of treated water and treatment speed (i.e., HLR).

Keeping this objective, the present study is aimed to demonstrate the pollutant removal efficiency of the MSL system with reference to variations in HLR, organic loading rate and effect of aeration and non-aeration methods. Also the study is subjected to check to what extent there is any occurrence of clogging in the MSL system due to the buildup of organic matter. The objectives of the present study are (1) to evaluate the domestic wastewater quality at the selected station over a period, (2) to examine the potential of the MSL system that is composed of locally available materials in remediating pollution of domestic wastewater, (3) to study the effects of HLR and organic loading rate on the removal efficiency of the MSL system, (4) to analyze effects of aeration and non-aeration methods on the efficiency of the MSL system and (5) to ensure whether there is any effect of clogging on the removal efficiency of the MSL system and giving remedial measure to overcome it.

6.3 MSL System

6.3.1 Technology

Soil is the most vital natural resource for waste and wastewater treatments, and it has been used for wastewater treatment for a long time. Soil facilitates wastewater treatment through a mixture of physical, biological and biochemical processes. These include physical and chemical adsorption, desorption, precipitation, solubilization, cation and anion exchange and decomposition by microorganisms. Many types of microbes are present in the system, like a monoculture, methanogenic bacteria, acidogenic bacteria, a mixed population of organisms or the microorganisms which is present in the inlet wastewater.

The system operates at a temperature range of about 16°C. to 32°C. The most serious problem when using soil as a 'natural' filter for wastewater treatment is 'clogging', which results in low loading capacity. Clogging can result from excessive loading of inorganic and organic materials, biofilms formation around soil particles as well as clay dispersion due to a high sodium concentration of wastewater.

Soil is considered a cost-effective decentralized wastewater treatment in rural areas in developing countries. It is available naturally, and so it has been used in various natural treatment systems, such as septic tanks, soil trench system, soil filter beds and others. To overcome the problem of clogging, some changes were made in the existing treatment unit that reduced the clogging and increased the treatment efficiency, and so the MSL system was introduced. The technique of MSL is a new technology developed in Japan by Wakatsuki et al., (1993). An MSL system improves the natural ability of the soil to decontaminate wastewater. The MSL system is a technology that uses natural soil in a unit to facilitate wastewater treatment (Wakatsuki et al., 1993). This has been successfully developed in Japan and Thailand to treat domestic and restaurant wastewater as well as polluted river water (Wakatsuki et al., 1993; Luanmanee et al., 2001). The system reduces concentrations of inorganic contaminants, such as nitrate, ammonium and phosphate, as well as organic contaminants as measured by high COD and BOD. This is a biphasic layered system that uses locally available materials, such as soil, iron particles, jute or sawdust, charcoal and zeolite or alternative materials (Attanandana et al., 2000; Luanmanee et al., 2001; Pattnaik et al., 2007). Soil is considered as a cost effective decentralized wastewater treatment in rural areas in developing countries. It is available naturally and so it has been used in various natural treatment systems like septic tanks, soil trench system, soil filter beds etc. To overcome the problem of clogging some changes were made in the existing treatment unit which reduced the clogging and increased the treatment efficiency and so the MSL System was introduced. The technique of MSL is a new technology developed in Japan by Wakatsuki et al., (1993). A MSL System improves the natural ability of the soil to decontaminate wastewater. The MSL System is a technology that uses natural soil in a unit to facilitate wastewater treatment (Wakatsuki et al., 1993). This has been successfully developed in Japan and Thailand to treat domestic and restaurant wastewater as well as polluted river water (Wakatsuki et al., 1993; Luanmanee et al., 2001). The system reduces concentrations of inorganic contaminants such as nitrate, ammonium, and phosphate, as well as organic contaminants as measured by high COD and BOD. This is a biphasic layered system that uses locally available materials such as soil, iron particles, jute or sawdust, charcoal, and zeolite or alternative materials (Attanandana et al., 2000; Luanmanee et al., 2001; Pattnaik et al., 2007).

When compared with traditional soil treatment systems, this system could withstand a higher HLR and is less prone to clogging. The MSL system uses high quantities of natural materials to produce reusable water for agriculture. In addition, this system can be maintained and operated at a minimum cost, also it requires only a small land area, and it is ideal for urban areas in developing countries. The MSL system mainly comprises soil mixture layers and permeable layers, which are arranged in a brick-like pattern.

A few of the characteristics of the MSL system are (1) it gives high purification ability, (2) it has low costs for operation and maintenance, (3) the MSL-treated water can be reused, (4) the soil used as a media for the treatment of wastewater can be reused and can be used as a good fertilizer, (5) the MSL system assures high removal efficiency for nitrogen and phosphorus and (6) in spite of the fluctuations in the raw wastewater, the MSL system gives stable performance at output.

The composition of the SML is approximately 70% to 80% soil and 20% to 30% additional materials, such as carbon powder, organic matter and iron (Ho and Wang, 2015). From the various materials that are included in the soil mixture block, the soil serves as a zone for microbes; the activated carbon powder adsorbs a high quantity of organic matter from wastewater, thus improving the efficiency of organic matter decomposition. The organic matter works as nutrients for microbes and iron materials efficiently adsorb phosphates. The materials are mixed together in a given proportion and packed into jute bags. The bags are then piled to form the soil mixture layers, with each layer separated by a permeable layer. The permeable layer includes aggregates of gravels or zeolite of approximately 1 to 5 mm in diameter. The aggregate of permeable layer adsorbs the organic matter from wastewater and serves as habitations for nitrobacteria. Therefore, both layers actively remove contaminants from domestic wastewater. In addition to the soil mixture layers and permeable layers, the MSL system also has inflow and outflow pipes arranged for wastewater.

Removing pollutants from wastewater through the MSL involves various physical and biochemical processes (An et al., 2016). The MSL system is characterized by its ability to handle higher HLRs and pollutant loads compared with some conventional land treatment systems (Guan et al., 2014). The MSL system supports higher HLRs and pollutant loads than other soil treatment systems due to its structure. This technique can overcome many constraints in the operation of the conventional functioning sewage treatment with soil (Guan et al., 2014; Lamzouri et al., 2016). The MSL system has several advantages, among others, namely, reducing the value of BOD, COD, TSS and color, and has a high ability to receive and absorb water that flowed into the system (Zein et al., 2016). The MSL Removal of pollutant from wastewater through the MSL involves various physical and biochemical processes (An et al., 2016). MSL system is characterized by its ability to handle higher hydraulic loading rates and pollutant loads compared with some conventional land treatment systems (Guan et al., 2014). MSL system supports higher hydraulic loading rate (HLR) and pollutant loads than other soil treatment systems due to its structure. This technique can overcome many constraints in the operation of the conventional functioning sewage treatment with soil (Guan et al., 2014; Lamzouri et al., 2016). The MSL system has several advantages, among others, viz. reduces the value of BOD, COD, TSS and color, has a high ability to receive and absorb water that flowed into the system (Zein et al., 2016). The MSL system is effective for the prevention of clogging and shortcut water flow which are major constraints in conventional soil-based wastewater treatment systems (Sato et al., 2011). The MSL system can be operated and maintained at a low/ minimum cost, also it requires small land area, and it is ultimate treatment system to be implemented in urban areas in developing countries. Hence it is valuable to study the MSL system as a cost-effective and feasible treatment facility and method for the removal of contaminant from domestic wastewater system is effective for preventing clogging and shortcut water flow, which are major constraints in conventional soil-based wastewater treatment systems (Sato et al., 2011). The MSL system can be operated and maintained at a low/minimum cost; also, it requires a small land area, and it is the ultimate treatment system to be implemented in urban areas in developing countries. Hence, it is valuable to study the MSL system as a cost-effective and feasible treatment facility and method for removing contaminants from domestic wastewater.

In areas where there are scattered households, it is essential to provide a wastewater treatment facility at an effective and low cost and with easy operation and maintenance so that it is beneficial for the scattered individual households. Hence, the proposed research is aimed to carry out an experimental investigation of the MSL technology using the lab-scale experimental setup for domestic wastewater treatment. The purpose of this research is to obtain more quantitative data and analysis for the removal of contaminants from domestic wastewater by using the MSL system. Also, the research is aimed to test the feasibility and removal efficiency of pollutants in the lab-scale model under different operating conditions.

In the MSL system, aeration could be utilized to relieve the clogging by breaking the sticky film on the media surface. Moreover, aeration could enhance the removal of organic contaminants as well through the biological oxidation reaction, and increment of aeration intensity could raise the removal efficiency effectively. However, excessive aeration may negatively affect the normal metabolism of microorganisms and intermittent aeration may provide a possible solution to balance the removal of COD, TN, and TP in the MSL system (Guan et al., 2015).

The wastewater that can be treated by using the MSL system includes wastewater from dairy, agricultural effluent, pond, crop effluent, sewage, marsh, and waste from crop plants, waste from greenhouse, Industrial facilities, aquaculture waste, food production waste, overflow runoff and so on. The MSL system can effectively work in polluted river water treatment (Masunaga et al., 2003).

6.3.2 Materials of the MSL System

The locally available materials are used as the treatment media so as to save the cost of materials and transportation. The materials used are usually grouped as follows.

6.3.2.1. Soil Mixture Layers

The material properties used in the experiments are as follows:

a. ***Soil:*** Soil is made up of organic matter, natural resources, gases, liquids and organisms. It helps influence microbial activity in the treatment of wastewater and so is used.
b. ***Powdered Activated Carbon:*** It is also known as activated charcoal. It is a form of carbon that has small, low-volume pores that enlarge the surface area available for adsorption and chemical reactions. It is commonly used to adsorb natural organic compounds, taste and odor compounds, and synthetic organic chemicals in wastewater treatment. Normally, activated carbons are made in particulate form as powders which have large surface-to-volume ratio and particle sizes smaller than 0.075 mm.
c. ***Sawdust:*** Sawdust is an organic matter that provides a carbon source for the microorganisms. Sawdust is as an effective adsorbent as it contains a high concentration of cellulose, which irreversibly adsorbs the pollutants contained in wastewater. Furthermore, it also functions as a carbon source for the microorganisms.
d. ***Iron Scraps:*** When iron scrap is added to MSL, it slowly oxidizes into ferrous iron or ferric. As a phosphorus soil, adsorption capacity is completely related to its composition in active iron and aluminum. The iron addition significantly increases the efficiency of adsorption of phosphorus in MSL systems. Adding iron scraps into the soil mixture layers facilitates phosphorus adsorption noticeably.

These materials are mixed together in the ratio as soil:powdered activated carbon:sawdust:iron scraps 7:1:1:1 by volume, and the mixture is then packed into jute bags (Chen et al., 2007). The jute bags are then piled in the treatment system, with each layer separated by the permeable layer material.

6.3.2.2. Permeable Layers

It consists of the material which acts as the permeable media for the wastewater to be treated. This material is generally used as per the local availability. Various permeable layers that can be used are pebbles, zeolite, cobblers, gravels, used granular activated carbon and so on. The efficiency of treatment varies with the use of permeable layer material. Zeolite is substituted with comparatively cheaper and more environmentally friendly materials, such as expanded clay aggregates, oyster shells and already-used granular activated carbon collected from water purification plants (Ho and Wang, 2015). However, natural zeolite is not easily available in India, and as it is costly, there is a need to replace the permeable layer material.

6.4 Case Study on the MSL System for Treatment of Domestic Wastewater

6.4.1 Study Area

The case study is taken from Kolhapur city, which is located In Maharashtra, India. The Kolhapur Municipal Corporation (KMC) supplies about 120 to 130 million liters per day (MLD) of water to Kolhapur city, with an estimated average per capita supply of about 135 liters per capita per day (LPCD), including water for domestic use, commercial use, the floating population, firefighting, industrial and institutional usage and so on. It is estimated that Kolhapur city generates about 83 MLD of wastewater. Out of this, the domestic wastewater accounts for 92% (76 MLD), followed by commercial establishments and sources 8% (7 MLD), the industrial sources account for a negligible 0.1 MLD of wastewater. The domestic wastewater is collected through a drainage line of about 222 km across four zones, namely, Jayanti, Dudhali, Line Bazar and Bapat camp for sewage management (Environmental Status Report 2015–16).

Deshpande and Thorvat (2018) carried out a study on the experimental setup of MSL for the removal efficiency of pollutants from domestic wastewater. The study area selected for the case study is the nallah carrying domestic wastewater in Bhagat Singh colony, Kolhapur, Maharashtra, India, at a latitude of 16° 41.04773′ N and a longitude of 74° 14.43288′ E. The purpose behind selection of this nallah is that the nallah carries domestic wastewater from the nearby area and is the sub-source of Jayanti nallah, which is one of the major nallah in the city. The nallah carries wastewater only from residential areas, namely, Ashwini Nagar, Ambai Defense Colony, Bhagat Singh Colony and Pratibha Nagar area.

6.4.2 Treatment Methodology

Soil is used for the domestic wastewater treatment from long time ago. Nowadays, wastewater treatment and disposal are the major problems observed everywhere. In remote areas, where combined wastewater treatment facility is not available, there is a need to provide a separate treatment facility for individual houses. The only solution to this problem is to provide decentralized treatment facilities and obtain the treatment of wastewater that is low in cost and easy to operate and maintain. The MSL system is a decentralized treatment system that allows the treatment of wastewater in a cost-effective manner. The present study is aimed to demonstrate the pollutant removal efficiency of the MSL system with reference to variations in HLR and the effect of aeration and non-aeration conditions.

The present study helps identify the effectiveness of the MSL system for treating domestic wastewater and its remediation. The change in aeration and non-aeration conditions and the effects of varying the HLR on the treatment efficiency of the MSL system helps optimize the remediation process and operating conditions for the treatment of domestic wastewater.

The study is mainly grouped into four stages:

A. Fieldwork composed of identifying the study area for the collection of domestic wastewater samples in Kolhapur city and finalizing the locally available soil to be used in the experimental setup for the treatment of domestic wastewater.
B. The preliminary laboratory analysis consists of analysis the physico-chemical characteristics of the soil samples to be used in the experimental setup and experimental analysis of raw domestic wastewater for the parameters, such as pH, BOD, COD, TKN, TP and total suspended solids (TSS).
C. The experimental setup and secondary laboratory analysis are composed of the preparation of lab-scale model for the MSL system using locally available materials, an experimental analysis of the MSL-treated wastewater three times in a week for the parameters, such as pH, BOD, COD, TP and TSS, and one per set for TKN and checking whether there is any effect of clogging of the MSL system due to the accumulation of the organic matter over a period.
D. The data analysis and documentation of outcomes are composed of demonstrating the effects of hydraulic behavior and fluctuations in wastewater pollution levels on the removal efficiency of the MSL system and checking the performance of the MSL system for aeration and non-aeration conditions for the treatment of domestic wastewater.

The permeability of the soil plays a vital role in selecting the soil to be used in the treatment media. So, the soil samples are collected from various locations and tested for Permeability. The soil selected is laterite soil from Punal region, Panhala Taluka, Kolhapur. The ability of the soil to pass the water quickly is good as it will not clog the filter. Laterite soil is also known as Tambad in the local language. It is brownish red in color. The preliminary laboratory analysis is done to analyze the physico-chemical characteristics of soil sample that is used in the MSL system. The laboratory analysis is carried out for the soil sample for the parameters, such as pH, conductivity, organic carbon, phosphorus and potash. The grain size distribution of the soil is also studied.

The experimental model for the MSL system is set by using locally available materials like laterite soil, activated carbon, sawdust, iron scraps and gravels. The study is then carried out for the domestic wastewater treated by the MSL system on a thrice-a-week basis for the parameters, such as pH, BOD, COD, TP and TSS, and one per set for TKN. The effects of hydraulic loading rate and fluctuations in wastewater pollution levels on the removal efficiency of the MSL system are demonstrated, and the performance of the MSL system under aeration and non-aeration conditions is checked.

6.4.3 Materials Used in the Experimental setup

The materials which are used in the experimental setup are natural materials that are available locally. Figure 6.1 shows the materials used in the Experimental setup. The materials used are usually grouped as follows:

a. **Soil:** In this study, laterite soil is used. The soil passing through a 4.75-mm sieve is used in the soil mixture block.
b. **Powdered Activated Carbon:** In this study, activated carbon is used in particulate form as powders with large surface-to-volume ratio and particle sizes smaller than 0.075 mm.
c. **Sawdust:** The sawdust passing through a 4.75-mm sieve and retaining on a 0.18-mm sieve is used in the experimental setup.
d. **Iron Scraps:** The iron scrap passing through a 4.75-mm sieve and retaining on a 2-mm sieve is used in the soil mixture block.
e. **Permeable Layers:** In this study, gravel with a diameter of 1 to 3 cm is used as permeable layer material.

These materials are mixed in the ratio as 7:1:1:1 (i.e., soil:powdered activated carbon:sawdust:iron scraps) by volume, and the mixture is then packed into jute bags, each of a size 12 cm × 20 cm × 5 cm. The jute bags are then piled in the treatment system with each layer separated by the permeable layer material. Figure 6.2 shows soil mixture block which is prepared by using laterite soil, activated carbon, sawdust and iron scraps.

6.4.4 Experimental Setup

The experimental setup is made up of acrylic material. The acrylic box is prefabricated for increasing the strength of the box. The dimension of the experimental setup is 50 (L) × 20 (W) × 60 (D) cm. Laterite soil, activated carbon, sawdust and iron scraps are used in the soil mixture block. They are added in the ratio soil:activated carbon:sawdust:iron scrap as 7:1:1:1 by volume. This mixture is filled in the jute bags. The jute bags are prestitched and have dimensions of 12 × 20 × 5 cm. They are piled in an alternating pattern in the acrylic box separated by a permeable layer composed of gravel of length 1 to 3 cm. Figure 6.3 shows the structure of the MSL system.

The aerator used is of capacity 0.30 m^3/h. The aeration pipe is mounted between the second and third layer of soil mixture and in the permeable layer material. The inlet distribution tank is of a 50-liter capacity and is placed at the height above the experimental setup so as to enable the flow of wastewater to the experimental setup by gravity. The adjusting valve is fixed to change the flow of incoming wastewater to the experimental setup as desired. The inlet pipe arrangement is made so as to allow uniform wastewater distribution throughout the experimental setup. The valve is connected to the outlet to allow the treated wastewater out of the experimental setup. The experimental analysis is carried out in 6 sets with different operating conditions, they are Set 1: non-aeration condition with flow rate 0.2 $m^3/m^2/d$, Set 2: aeration

(a) Soil

(b) Sawdust

(c) Iron scrap

(d) Activated carbon

FIGURE 6.1 Materials used in the experimental setup.

FIGURE 6.2 Soil mixture block prepared using laterite soil, activated carbon, sawdust and iron scraps.

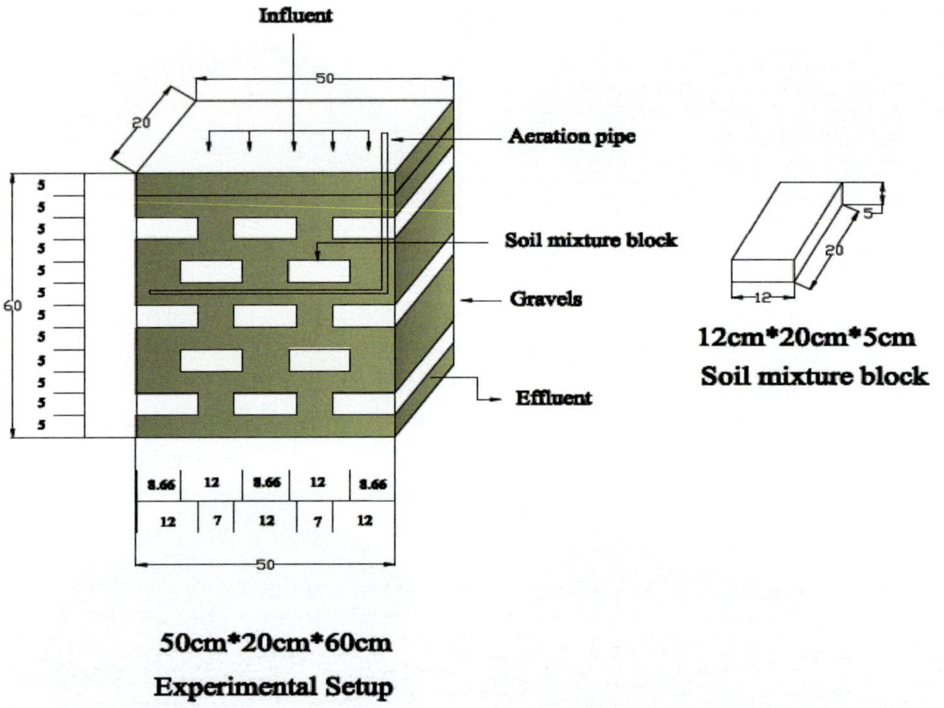

FIGURE 6.3 Structure of the MSL system.

TABLE 6.1

Parameters of the Soil Used in the Setup

Parameter	Unit	Value
pH	–	6.80
Conductivity of the soil	mS/cm	0.28
Organic carbon	%	0.62
phosphorus	kg/ha	72.44
Potash	kg/ha	495.43

condition with flow rate 0.2 $m^3/m^2/d$, Set 3: non-aeration condition with flow rate 0.3 $m^3/m^2/d$, Set 4: aeration condition with flow rate 0.3 $m^3/m^2/d$, Set 5: non-aeration condition with flow rate 0.4 $m^3/m^2/d$ and Set 6: aeration condition with flow rate 0.4 $m^3/m^2/d$.

6.4.5 Results and Discussion

a. **Soil Analysis:** The soil used in the setup is laterite soil. Basically, the soil treats wastewater in two most important ways: One is through soil's physical characteristics, and other is as a site for biological activity. Table 6.1 shows the parameters of the soil used in the study. It is observed that the soil is slightly acidic. When excess salts are present in soil, then electrical conductivity (EC) goes higher and increases the problem of salinization. The cation exchange capacity of the soil can be known from the conductivity of the soil. The soil organic carbon content is 0.62%. Higher soil organic carbon encourages soil structure means that there is greater physical stability. This improves soil aeration (oxygen content in the soil), water drainage and retention and reduces the risk of erosion and nutrient leaching. In addition, soil organic carbon is also essential to chemical composition and biological yield. The phosphorus and potash content in the soil is high. Phosphorus and potash are essential nutrients for sustaining all lifeforms. Table 6.2 shows the sieve analysis and grain size distribution of soil.

Grain size distribution is one of the vital characteristics of treatment media. Clogging generally occurs as a result of using fine soil. The most important role while using soil in treating wastewater is

TABLE 6.2

Sieve Analysis and Grain Size Distribution

Sr. No	Sieve Size (mm)	Mass Retained (gm)	Mass Retained Cumulative	% Retained Cumulative	% Finer/ Passing
1	2.00	185.20	185.20	18.52	81.48
2	1.70	112.00	297.20	29.72	70.28
3	0.6 0	485.30	785.20	78.52	21.75
4	0.42	93.40	875.90	87.59	12.41
5	0.30	63.80	939.70	93.97	6.03
6	0.15	30.70	970.40	97.04	2.96
7	0.09	16.20	986.60	98.66	1.34
8	Pan	13.40	1000.00	100.00	0.00

Table 6.3 Preliminary Domestic Wastewater Analysis

Parameter	Value
pH	6.82
TSS	680.00
TP	12.48
BOD	106.00
COD	260.00

Except for *pH, all values are in mg/l.*

not the soil grain but rather the pores the soil creates. So, it is a must to have soil grains consistent in nature. The wastewater treatment occurs on the soil surface where the SS gets trapped and the microbes grow. The percentage of various sizes of particles in the dry soil sample is found by a particle size analysis. Particle size analysis, or mechanical analysis, is the separation of soil into its different fractions. The results of the mechanical analysis are plotted to get a particle size distribution curve. This curve gives an idea about the type and gradation of soil. The curve of the soil shows that the soil is well-graded soil; that is, it includes particles of all sizes.

b. **Preliminary Wastewater Analysis and Results Obtained Using the MSL System:** The preliminary parameters of nallah domestic wastewater are listed in Table 6.3. The pH is slightly acidic. The TSS concentration is high. Solid waste is disposed of in the nallah by the inhabitants from the locality. This is one of the major reasons that the TSS content in the nallah wastewater is observed to be high. The TP concentration observed is 12.48 mg/l; the BOD value observed is 160 mg/l, while the COD value observed is 260 mg/l. From these parameters, it is concluded that it is medium- to high-strength domestic wastewater. Figure 6.4 shows the raw and treated wastewater through the MSL system. It shows the two bottles: The left one is raw domestic wastewater, and the right one is treated water from the MSL system.

FIGURE 6.4 Wastewater before and after treatment through the MSL system.

c. **Results of Set 1—Non-Aeration Condition with Flow Rate of 0.2m³/m²/d:** The average inlet and outlet values of pH observed during the sampling of Set 1 are 6.76 and 7.54, respectively. The average percentage of increase in pH observed is 11.5%. The average inlet and outlet values of BOD for Set 1 are 166.4 mg/l and 28.65 mg/l, respectively. The average percentage of removal in BOD observed is 82.44%. Due to the low flow rate, the efficiency observed is

TABLE 6.4

Average Values of Inlet and Outlet Parameters and Percentage of Removal Efficiency for Set 1

Sr. No.	Parameter	Inlet (mg/l)	Outlet (mg/l)	% Removal Efficiency
1	pH	6.76	7.54	11.5
2	BOD	166.40	28.65	82.44
3	COD	378.65	64.52	81.52
4	TSS	785.70	191.42	75.68
5	TP	16.62	1.02	94.26

TABLE 6.5

Average Values of Inlet and Outlet Parameters and Percentage of Removal Efficiency for Set 2

Sr. No.	Parameter	Inlet (mg/l)	Outlet (mg/l)	% Removal Efficiency
1	pH	6.75	7.59	12.45
2	BOD	203.35	24.62	87.63
3	COD	418.57	48.73	87.73
4	TSS	842.85	194.28	77.12
5	TP	14.14	0.71	95.45

high. The average inlet and outlet values of COD observed during the sampling of Set 1 are 378.65 mg/l and 64.52 mg/l, respectively. The average percentage of removal in COD observed is 81.52%. Due to the low flow rate, the efficiency observed is high. The average inlet and outlet values of TSS observed during the sampling of Set 1 are 785.71 mg/l and 191.42 mg/l, respectively. The average percentage of removal in TSS observed is 75.68%. Due to the low flow rate, the efficiency observed is high. The average inlet and outlet values of TP observed during the sampling of Set 1 are 16.62 mg/l and 1.02 mg/l, respectively. The average percentage of removal in TP observed is 94.26%. Due to the low flow rate the efficiency observed is high. Table 6.4 shows average values of inlet and outlet parameters and percentage of removal efficiency for Set 1.

Except for *pH, all values are in mg/l, and in* the *case of pH, it is percentage* of *increase in pH.*

d. **Results of Set 2—Aeration Condition with Flow Rate of 0.2m³/m²/d:** The average inlet and outlet values of pH observed during the sampling of Set 2 are 6.75 and 7.59, respectively. The average percent increase in pH observed is 12.45%. The average inlet and outlet values of BOD observed during the sampling of Set 2 are 203.35 mg/l and 24.62 mg/l, respectively. The average percentage of removal in BOD observed is 87.63%. Due to the aeration condition and low flow rate, the efficiency observed is higher than Set 1. The average inlet and outlet values of COD observed during the sampling of set 2 are 418.57 mg/l and 48.73 mg/l, respectively. The average percentage of removal in COD observed is 87.73%. Due to the low flow rate and aeration condition, the efficiency observed is high. The average inlet and outlet values of TSS observed during the sampling of Set 2 are 842.85 mg/l and 194.28 mg/l, respectively. The average percentage of removal in TSS observed is 77.12%. Due to the low flow rate, the efficiency observed is high. The average inlet and outlet values of TP observed during the sampling of Set 2 are 14.14 mg/l and 0.71 mg/l, respectively. The average percentage of removal in TP observed is 95.45%. Due to the low flow rate, the efficiency observed is high. Table 6.5 shows average values of inlet and outlet parameters and percentage of removal efficiency for Set 2.

Except for *pH, all values are in mg/l, and in* the *case of pH, it is percentage* of *increase in pH.*

e. **Results of Set 3—Non-Aeration Condition with Flow Rate of 0.3 m³/m²/d:** The average inlet and outlet values of pH observed during the sampling of Set 3 are 6.64 and 7.48, respectively. The average percent increase in pH observed is 12.45%. The average inlet and outlet values of BOD observed during the sampling of Set 3 are 166.4 mg/l and 28.65 mg/l, respectively. The average percentage of removal in BOD observed is 82.44%. Due to the low flow rate, the efficiency observed is high. The average inlet and outlet values of COD observed during the sampling of Set 3 are 320.85 mg/l and 74.28 mg/l, respectively. The average percentage of removal in COD observed is 76.91%. The average inlet and outlet values of TSS observed during the sampling of Set 3 are 734.28 mg/l and 205.71 mg/l, respectively. The average percent removal in TSS observed is 71.74%. The average

TABLE 6.6

Average Values of Inlet and Outlet Parameters and Percentage of Removal Efficiency for Set 3

Sr. No.	Parameter	Inlet (mg/l)	Outlet (mg/l)	% Removal Efficiency
1	pH	6.64	7.48	12.45
2	BOD	166.4	28.65	82.44
3	COD	320.85	74.28	76.91
4	TSS	734.28	205.7	71.74
5	TP	16.78	2.54	83.94

TABLE 6.7

Average Values of Inlet and Outlet Parameters and Percentage of Removal Efficiency for Set 4

Sr. No.	Parameter	Inlet (mg/l)	Outlet (mg/l)	% Removal Efficiency
1	pH	6.77	7.61	12.44
2	BOD	174.05	41.22	75.88
3	COD	336.00	67.71	79.01
4	TSS	720.00	197.14	72.79
5	TP	16.90	1.98	88.00

inlet and outlet values of TP for set 3 are 16.78 mg/l and 2.54 mg/l, respectively. The average percentage of removal in TP observed is 83.94%. Table 6.6 shows average values of inlet and outlet parameters and percentage of removal efficiency for Set 3.

Except for *pH, all values are in mg/l, and in* the *case of pH, it is percentage* of *increase in pH.*

f. **Results of Set 4—Aeration Condition with Flow Rate of 0.3 m³/m²/d:** The average inlet and outlet values of pH observed during the sampling of Set 4 are 6.77 and 7.61, respectively. The average percentage of increase in pH observed is 12.44%. The average inlet and outlet values of BOD observed during the sampling of Set 1 are 174.05 mg/l and 41.22 mg/l, respectively. The average percentage of removal in BOD observed is 75.88%. The average inlet and outlet values of COD observed during the sampling of Set 4 are 336 mg/l and 67.71 mg/l, respectively. The average percentage of removal in COD observed is 79.01%. As the flow rate is increased, the efficiency observed is low; however, the efficiency was increased with the same flow rate as Set 3 with the aeration condition. The average inlet and outlet values of TSS observed during the sampling of Set 4 are 720 mg/l and 197.14 mg/l, respectively. The average percentage of removal in TSS observed is 72.79%. The average inlet and outlet values of TP observed during the sampling of Set 4 are 16.90 mg/l and 1.98 mg/l, respectively. The average percentage of removal in TP observed is 88%. Table 6.7 shows average values of inlet and outlet parameters and percentage of removal efficiency for Set 4.

Except for *pH, all values are in mg/l, and in* the *case of pH, it is percentage* of *increase in pH.*

g. **Results of Set 5—Non-Aeration Condition with Flow Rate of 0.4 m³/m²/d:** The average inlet and outlet values of pH observed during the sampling of Set 5 are 6.64 and 7.46, respectively. The average percentage of increase in pH observed is 12.34%. The average inlet and outlet values of BOD observed during the sampling of Set 5 are 150.9 mg/l and 42.32 mg/l, respectively. The average percentage of removal in BOD observed is 68.67%. Due to the increase in flow rate, the efficiency observed is low. The average inlet and outlet values of COD observed during the sampling of Set 5 are 318.85 mg/l and 85.88 mg/l, respectively. The averag`e percentage of removal in COD observed is 71.42%. The average inlet and outlet values of TSS observed during the sampling of Set 5 are 717.14 mg/l and 220 mg/l, respectively. The average percentage of removal in TSS observed is 69.11%. Due to the low flow rate, the efficiency observed is high. The average inlet and outlet values of TP observed during the sampling of Set 5 are 15.04 mg/l and 3.05 mg/l, respectively. The average percentage of removal in TP observed is 79.49%. Table 6.8 shows average values of inlet and outlet parameters and percentage of removal efficiency for Set 5.

TABLE 6.8

Average Values of Inlet and Outlet Parameters and Percentage of Removal Efficiency for Set 5

Sr. No.	Parameter	Inlet (mg/l)	Outlet (mg/l)	% Removal Efficiency
1	pH	6.64	7.46	12.34
2	BOD	150.90	42.32	68.67
3	COD	318.85	85.88	71.42
4	TSS	717.14	220	69.11
5	TP	150.04	3.04	79.49

TABLE 6.9

Average Values of Inlet and Outlet Parameters and Percentage of Removal Efficiency for Set 6

Sr. No.	Parameter	Inlet (mg/l)	Outlet (mg/l)	% Removal Efficiency
1	pH	6.54	7.34	12.38
2	BOD	184.00	51.11	71.35
3	COD	355.88	91.55	72.52
4	TSS	725.71	217.14	70.08
5	TP	14.03	3.82	79.75

TABLE 6.10

Inlet and Outlet TKN and Percentage of Removal Efficiency of TKN in the 6 Sets

Sets	Inlet TKN (mg/l)	Outlet TKN (mg/l)	% Removal Efficiency
Set 1	64.40	14.00	78.26
Set 2	61.60	11.20	81.81
Set 3	70.00	19.60	72.00
Set 4	72.80	16.80	76.92
Set 5	78.40	25.20	67.85
Set 6	75.60	22.40	70.37

Except for *pH, all values are in mg/l, and in* the *case of pH, it is percentage increase in pH.*

h. **Results of Set 6—Aeration Condition with Flow Rate of 0.4 m^3/m^2/d:** The average inlet and outlet values of pH observed during the sampling of Set 6 are 6.54 and 7.34, respectively. The average percentage of increase in pH observed is 12.38%. The average inlet and outlet values of BOD observed during the sampling of Set 6 are 184 mg/l and 51.11 mg/l, respectively. The average percentage of removal in BOD observed is 71.35%. Due to the increase in flow rate, the efficiency observed is low. The aeration condition slightly improved the efficiency as compared to Set 5. The average inlet and outlet values of COD observed during the sampling of Set 6 are 355.88 mg/l and 91.55 mg/l, respectively. The average percentage of removal in COD observed is 72.52%. Due to the high flow rate, the efficiency observed is low, but with the same flow rate as Set 5 and aeration condition, the efficiency observed is slightly high. The average inlet and outlet values of TSS observed during the sampling of Set 6 are 725.71 mg/l and 217.14 mg/l, respectively. The average percentage of removal in TSS observed is 70.08%. Due to the high flow rate, the efficiency observed is low. The average inlet and outlet values of TP observed during the sampling of Set 6 are 14.03 mg/l and 3.82 mg/l, respectively. The average percentage of removal in TP observed is 79.75%. Table 6.9 shows average values of inlet and outlet parameters and percentage of removal efficiency for Set 6.

Except for *pH, all values are in mg/l and in* the *case of pH, it is percentage increase in pH.*

i. **Results of Inlet and Outlet TKN and Percentage of Removal Efficiency of TKN:** Table 6.10 shows the inlet and outlet TKN in the six sets. In the analysis of wastewater, TKN is the sum of organic nitrogen, ammonia (NH_3), and ammonium (NH_4^+). The TKN is analyzed once for each set. The removal efficiency observed is 70% to 80%.

The pH of the treated water was not majorly affected by flow rate change and aeration-non-aeration condition. Most microbes do well in a pH range of 6.5 to 8.5. The algal growth is observed in the model which is one of the reasons behind the increase in pH value of outlet. The soil is the main carrier of microbes and so it enables various microorganisms to grow in it. It is observed that the removal efficiency of BOD decreased as the flow rate is increased. The efficiency of removal increased as intermittent aeration is provided; intermittent aeration facilitates the decomposition of microbes at a higher rate. The COD removal efficiency of the MSL is observed good with a lower HLR and decreases as the HLR is increased. The intermittent aeration improves the removal efficiency of the COD. The intermittent aeration enhanced the removal efficiency of COD. The solids are trapped physically between the pores in the gravels and soil mixture blocks, so the TSS removal efficiency of the MSL is high. The TSS removal efficiency gradually decreases as the flow rate increases. Due to intermittent aeration, the solids are decomposed at a faster rate, which increases TSS removal efficiency. It is also observed that the MSL system is very effective for removing phosphorus from raw domestic wastewater. The phosphorus removal efficiency is observed higher at a lower flow rate and decreases gradually at a higher flow rate. The TKN removal efficiency observed is good. The aeration condition was more efficient than non-aeration condition. The system can be operated at varying flow rates and depends on various factors, such as the quality of the inlet wastewater, the materials present in the system, the size of the system, the temperature of the system, the number of layers present, the input pumping rate, the drainage pipe width and surrounding environmental conditions.

6.5 Conclusion

As per the available literature and the case study carried out, it is observed that the MSL system is efficient for treating domestic wastewater. Aeration and non-aeration conditions and variations in flow rates in the experimental setup revealed that the aeration system proved to be more efficient for treating domestic wastewater and its remediation process. The effects of varying hydraulic loading rates are studied on the basis of percentage of removal efficiency using the MSL system. It is observed that the flow rate of 0.2 $m^3/m^2/d$ with aeration condition is observed to be more efficient for treating domestic wastewater using the MSL system. Soil filters get clogged frequently due to the accumulation of void spaces by the solids. In the present study, the clogging problem is not observed during the running period of the experimental setup. This is due to the alternate assembly of soil mixture layer, permeable layer material and granular size of the permeable layer. This traps the solids and decomposes them biologically. The periodic aeration of the system allowed the solids to decompose at a faster rate and prevent the system to get clogged. The solids removal efficiency is around 60% to 80%. Further studies can be carried out for improving the efficiency of treatment by using different combinations in the soil mixture layers and permeable layers and the removal efficiency of the model by changing the aerator capacity.

ACKNOWLEDGMENTS

The authors are indebted to the CRC Press, Taylor and Francis Group, USA, the editor and the reviewers and wish to sincerely acknowledge their painstaking efforts in going through the chapter and providing philosophical and in-depth constructive comments. The authors are highly grateful to Kolhapur Institute of Technology's College of Engineering (Autonomous), Kolhapur, Maharashtra, India, for the support, cooperation and encouragement.

REFERENCES

An, C.J., McBean, E., Huang, G.H., Yao, Y., Zhang, P., Chen, X.J. and Li, Y.P. 2016. Multi-Soil-Layering Systems for Wastewater Treatment in Small and Remote Communities. *Journal of Environmental Informatics*, 27(2):131–144.

Attanandana, T., Saitthiti, B., Thongpae, S., Kritapirom, S., Luanmanee, S. and Wakatsuki, T., 2000. Multi-Media-Layering System for Food Service Wastewater Treatment. *Ecological Engineering*, 15(1–2):133–138. http://dx.doi.org/10.1016/S0925-8574(99) 00041-5.

Chen, C.F., Lin, J.Y., Huang, C.H., Chen, W.L. and Chueh, N.L. 2009. Performance Evaluation of a Full-Scale Natural Treatment System for Nonpoint Source and Point Source Pollution Removal. *Environmental Monitoring and Assessment*, 157(1):391–406.

Chen, X., Sato, K., Wakatsuki, T. and Masunaga, T. 2007. Effect of Structural Difference on Wastewater Treatment Efficiency in Multi-Soil-Layering Systems: Relationship Between Soil Mixture Block Size and Removal Efficiency of Selected Contaminants. *Soil Science and Plant Nutrition*, 53(2):206–214. http://dx.doi.org/10.1111/j.1747-0765.2007.00126.x

Deshpande, V.V. and Thorvat, A.R. 2018. Experimental Investigation of Treatment of Domestic Wastewater Using Multi Soil Layering (MSL) System. *Aquademia*, 2(2):5. https://doi.org/10.20897/awet/3963.

Guan, Y.D., Xu, D.F., Chen, X., Luo, A.C., Fang, H. and Song, Y.Z. 2014. Flow Patterns of Multi-Soil-Layering Systems. *Desalination and Water Treatment*, 52(22–24):4165–4169. http://dx.doi.org/10.1080/19443994.2013.801791.

Guan, Y.D., Zhang, Y., Zhong, C.N., Huang, X.F., Fu, J. and Zhao, D. 2015. Effect of Operating Factors on the Contaminants Removal of a Soil Filter: Multi-Soil-Layering System. *Environmental Earth Sciences*, 74(3):2679–2686. http://dx.doi.org/10.1007/s12665-015-4288-8.

Ho, C.C. and Wang, P.H. 2015. Efficiency of a Multi-Soil-Layering System on Wastewater Treatment Using Environment-Friendly Filter Materials. *International Journal of Environmental Research and Public Health*, 12(3):3362–3380. http://dx.doi.org/10.3390/ijerph120303362.

Kao, C.M., Wang, J.Y., Lee, H.Y. and Wen, C.K. 2001. Application of a constructed wetland for non-point source pollution control. Water Science and Technology, 44(11–12):585–590.

Kolhapur Municipal Corporation, Environmental Status Report 2015–16. Teri – The Energy and Resources Institute. https://www.teriin.org/sites/default/files/2018-02/2016MC02%20%20%20Kolahpur%20ESR%202015-16%20%28English%29_0.pdf.

Lamzouri, K., Mahi, M., Ouatar, S., Bartali, E., Masunaga, T., Latrach, L. and Mandi, L. 2016. Application of Multi-Soil-Layering

Technique for Wastewater Treatment in Moroccan Rural Areas: Study of the Operation Process for an Engineering Design. *Journal of Materials and Environmental Science*, 7(2):579–585.

Luanmanee, S., Attanandana, T., Masunaga, T. and Wakatsuki, T. 2001. The Efficiency of a Multi-Soil-Layering System on Domestic Wastewater Treatment During the Ninth and Tenth Years of Operation. *Ecological Engineering*, 18(2):185–199.

Masunaga, T., Sato, K., Zennami, T., Fujii, S. and Wakatsuki, T. 2003. Direct Treatment of Polluted River Water by the Multi-Soil-Layering Method. *Journal of Water and Environment Technology*, 1(1):97–104.

Pattnaik, R., Yost, R.S., Porter, G., Masunaga, T. and Attanandana, T. 2008. Improving Multi-Soil-Layer (MSL) System Remediation of Dairy Effluent. *Ecological Engineering*, 32(1):1–10. http://dx.doi.org/10.1016/j.ecoleng.2007.08.006.

Sato, K., Iwashima, N., Wakatsuki, T. and Masunaga, T. 2011. Quantitative evaluation of treatment processes and mechanisms of organic matter, phosphorus, and nitrogen removal in a multi-soil-layering system. *Soil Science and Plant Nutrition*, 57(3):475–486. http://dx.doi.org/10.1080/00380768.2011.590944.

Wakatsuki, T., Esumi, H. and Omura, S. 1993. High Performance and N & P-removable On-site Domestic Waste Water Treatment System by Multi-Soil-Layering Method. *Water Science and Technology*, 27(1):31–40.

Zein, R., Ningsih, Z.S., Novita, L., Swesty, N., Mukhlis and Novrian, H., 2016. Treatment of Waste Water Noodle Industry with a Multi-Soil-Layering (MSL) System. *Research Journal of Pharmaceutical, Biological and Chemical Sciences*, 7(6):88–94.

7

Nanotechnology and Its Implications in the Environmental Biotechnology

Y K Lahir

CONTENTS

7.1 Introduction

Nanotechnology is an offshoot of science that enables the modulation of matter at atomic, molecular and nanoscale levels. At these levels, the specific physicochemical features of the nanomaterials become very efficient and effective. These nanomaterials can comfortably move across most of the biological barriers, their size is more or less similar to the size of cellular components. These can be modified as biocompatible and bioavailable materials. Nanomaterials provide a large surface area and, hence, provide more capacity for interaction. These materials have a higher degree of dispersibility in the medium and can be suitably tailored with respect to the applications in different environmental conditions even for detection, sensing and remediation (Feng et al., 1997). These features make them very strong candidates as useful tools/means to understand the biochemical and molecular intricacies. Under the influence of laws of physics and chemistry, these nanomaterials become more relevant options in the field of biomedical, biotechnological, pharmaceutical, environmental, food technology and others (Buzea et al., 2007; Saini et al., 2010; Lahir, 2015, 2017; Lahir et al., 2016). In the recent past, nanotechnology has emerged as one of the multifaceted technologies that has the potential to deal with most of the current problems related to the environment. Nano-remediation involves the use of 'reactive nanomaterials' for either transformation or detoxification of various pollutants. This phenomenon is one of the most effective mode of pollution remediation, but still, in most cases, its theoretical approach is being considered.

All such advancements in nanotechnology are not possible without financial and economic support from governmental and private sectors. Nanotechnology is one of the fast-growing sciences, and it has global, economical and industrial support. The current expected global market for the products of nanotechnology has grown beyond US$2600 billion in comparison to US$500 million in 1999 and with about 10 million job avenues (UN-ESCAP, 2010). The global applications of nanoparticles can be represented as 31% in material science, 28% in electronics, 17% in medicines, 9% in chemistry, 6% in aviation and 9% in other fields (UN-ESCAP, 2010).

Biotechnology—a branch of an applied aspect of basic sciences—utilizes biosystems and biological processes for the development and formation of products. These are formulated to enhance the quality of life of humans, crops and domestic animals and the sustainability of the environment. Although this technology is pretty old and dedicated to agricultural, veterinary and beverage practices, in the current scenario, it provides practically the most suitable techniques for understanding biological and environmental intricacies. The prime

DOI: 10.1201/9781003186304-7

focus is on the improvement and development of disease-resistant crops and genetically improved varieties of domestic animals, such as poultry, sheep, goats, horses, camels, pigs and others. This aspect may include the formulations of pesticides, herbicides, wormicides, fungicides or fertilizers, preservatives or drugs, among others, as the set target.

Biotechnology has its diversions, such as red, white/gray, green and blue biotechnologies. Red biotechnology is dedicated to the medical field to produce new drugs by using stem cells for regenerating dysfunctional tissues or growth of a completely new organ, among others, using the biosystem as a model. White biotechnology—also referred to as grey biotechnology—is concerned with the development and production of new industrial chemicals, searching and developing new resources for fuel to be used in transportation. Green biotechnology is focused on the advancement of agricultural practice, such as developing pest-resistant grain, crops and animals of domestic and commercial utilities. The term *blue* biotechnology is not in common use. It encompasses investigations related to the advancement of marine and aquatic environments and control of the noxious aquatic organisms that cause heavy damage to the otherwise normal aquatic biota.

These useful technologies may be abused. There are legal authorities at local, national and international levels to control such functional discrepancies. The main object of these authorities is to study, investigate and the judicial use of biotechnologies, nanotechnology and their products. They also ensure safe use, transportation, storage, supply and disposal of unused, extra and damaged products and check the adverse effect on the environment. The advancement in the field of biotechnology, nanotechnology, applied physics, and chemistry has resulted in improvisations in tools and their applications. Most of these are being used in the varied domestic, agricultural, biomedical, electronic, pharmaceutical and varied industries. Although these developments have good intentions, their dark side is the production of wastes and their disposal. When these untreated wastes are released into the environment, they become the sources of contamination and pollution. Nanomaterials may be released directly by industries and inadvertently escape in the form of discarded cosmetics, food packages, products from pharmaceutical and other industrial waste or products whose shelf lives have expired or have been rejected due to nonadherence to the set standards (declared defective/substandard). Furthermore, because there is relatively less confirmed information regarding their toxic or harmful impacts on the abiota and biota of the environment; unhesitantly, these nanomaterials are released in the environment.

Natural and engineered nanoparticles can enter the environment during mining in the form of fine metals and metal oxide nanoparticles, as a result of combustion in the form of soot, carbon black, fly ash and carbon nanomaterials, while laundering Ag-nanoparticles from the antibacterial fibers and metal oxide nanoparticles from ultraviolet (UV)–protected fibers, the sources of energy generators such as quantum dots and metal oxide nanoparticles used in solar cells, discarded medicine as containing nanoparticles of silver and gold used as antibacterial and therapeutic agents and auto-traffic aerospace exhaust in the form of palladium and platinum nanoparticles, catalyst convertor, battery waste, fine carbon particles, wear and tear of tires and so on.

7.2 Nanomaterials Involved in Environmental Applications

Nanomaterials can be improvised with a view to solving the specific environmental remediation process. The ameliorated nanomaterials like nanosorbents, nanocatalysts, bioactive nanoparticles, nanostructured catalytic membranes, filtration devices facilitated by nanoparticles and other nanomaterials developed based on nanotechnology. Advancements in nanoscience and nanoengineering have resulted in cost-effective, eco-friendly nanomaterials that are feasible to treat wastewater, soil and air, enhancing the environmental sustainability (Jayavarthanan et al., 2017). The most common nanomaterials involved as remedial agents are nanosorbents, zeolites, redox and catalytically active nanoparticles, nanostructured membranes, bioactive nanoparticles; dendrimers enhanced ultrafiltration; and others in the sensing, detecting and remediation processes.

7.2.1 Nanosorbents

These are the ameliorated nanomaterials that have an extremely large surface area and the ability to get functionalized. These features elevate the degree of sorption of pollutants and contaminants. Multiwalled carbon nanotubes (MWCNTs) accomplish sorption of Pb (II), Cu (II), and Cd (II) present in wastewater with 97.08mg/g, 24.49 mg/g and 10.86mg/g, respectively, at room temperature and at pH 5.0. These carbon nanotubes have a 3 to 4 times higher degree of metal-ion-sorption in comparison to the powdered granular activated carbon (Savage and Diallo, 2005). Chitosan nanoparticles are functionalized by ionic gelation, and triphosphate has shown a higher degree of sorption up to $189m^2$ /g of Pb (II) (Peng et al., 2003). Carbon nanotubes supporting CeO_3 (cerium oxide) nanoparticles exhibiting a very large surface area ($189m^2$ /g) could adsorb As (V), (Peng et al., 2005). Unique nanocrystals made of 'akaganeit' [β-FeO (OH)] are very effective sorbent for As (V) and Cr (VI) (Deliyanmi et al., 2003; Lazaridis et al., 2005). Adsorbent derived from dried Withania frutescens plant could remove metallic ions, nitrates, phosphate ions from wastewater. This adsorbent material is low cost, has a size <500 nm and removed Pb around 99%, Cd around 98%, Cu around 92% and Zn around 93% from wastewater (Chiban et al., 2012). Biosorbents perform better related to the removal of Cu (II) from wastewater (Farouq and Yousef, 2015).

7.2.2 Zeolites

Zeolites are among the effective sorbents and suitable media for ion exchange. $Na_6Al_6Si_{10}O_{32}$ zeolite is a cost-effective nanostructure with having a very high density of Na-ion exchange sites. These can be synthesized involving hydrothermal activation of fly ash with low Si/Al ratio at 150° using 1.0 to 2.0 M NaOH'. This material can effectively remove heavy metals such as Cr (III), Ni (II), Zn (II), Cu (II), and Cd (II) from the acidic effluent discharge from acid mines and metal electroplating units. Zeolite, like nanosorbents, can be prepared as 'self-assembled monolayers' on mesoporous support. These are very helpful in removing toxic metal ions, anions and radio-nuclides because these have an enormous surface area (about 1000 m2/g) and enhanced density of sorption sites (Moreno et al., 2001; Alvarez-Ayuso et al., 2003; Yantasee et al., 2003; Kelly et al., 2001; Lin et al., 2005).

7.2.3 Carbon-Based Nanomaterials

The carbon element gets polymerized at the atomic level and produces a long carbon chain. Carbon is tetravalent and gets linked with other elements via a single, double or triple covalent bond. It exists in different molecular forms because of its specific electronic structure and smaller size. Most of the carbon nanomaterials may be grouped with respect to their geometrical structure, (Zaytseva and Neumann, 2016). Carbonaceous nanomaterials are engineered as good nanosorbents like activated carbon fibers (ACFs). These are synthesized with an average pore size of 1.16 nm and a surface area ranging from 171 to 483 m^2/g. These can effectively sorb organic molecules like benzene, toluene, p-xylene, ethylbenzene at 20°C (Magnum et al., 2001). Carbon nanotubes (CNTs) adsorb 1,2-dichlorobenzene with maximum capacity (30.8 mg/g) in just 40 min. MWCNTs are functionality better as sorbent specifically for volatile organic compounds (VOCs). These easily adsorbed water-soluble dyes, such as acridine orange, ethidium bromide, eosin bluish and orange-G, with a higher degree of sorption and capacity specifically when oriented in cage form (Fugetsu et al., 2004). A hybrid nanosorbent can be synthesized involving inorganic-organic compounds, such as dodecyl sulfate incorporated in magnesium–aluminum-layered double hydroxides. These resultant nanosorbents exhibit a higher degree of sorption for chlorinated alkenes, such as tetrachloroethylene and trichloroethylene, from an aqueous medium in comparison to organoclays (Zhao and Nagy, 2004). Fullerene is also an effective nanosorbent specifically for polycyclic aromatic compounds such as naphthalene while amphiphilic polyurethane nanoparticles exhibit bioavailability and a higher degree of sorption of a polycyclic aromatic compound from an aqueous medium (Cheng et al., 2004; Thugittiplakorn et al., 2004, 2005). Carbon-based nanomaterials are the basic material used in acetylcholinesterase sensors, and these sensors are very useful in the detection of organophosphate pesticide analysis (Ion et al., 2014).

The engineered nanosized graphene nanocomposites contain inorganic and organic molecules and exhibit better functionalities to extract pollutants from an aqueous medium. These are biocompatible and bioavailable in nature; hence, they do not pose any serious threat to the biota of the aquatic ecosystem. Graphene-based magnetic nanomaterials are very effective options to remove heavy metals from the aquatic system. These potent agents are convenient, cost-effective and capable to remove contaminants in comparison to most of the conventional methods. Because these overcome the problems related to solid-phase extraction involving adsorbent–adsorbate system and attain equilibrium in a relatively short time (Gautam and Chattopadhaya, 2016).

7.2.4 Nanocatalysts and Redox Active Nanoparticles

Some of the nanoparticles behave like nanocatalysts and redox-active nanoparticles. Nanoparticles, such as TiO_2, exhibit promising nanocatalytic and redox catalytic functional efficiencies. These nanomaterials are very effective to eliminate inorganic, organic pollutants from wastewater. This interaction of TiO_2 gets multifold increased in the presence of the UV range. These nanoparticles also degrade organic compounds, such as chlorinated alkanes, benzene, dioxins, furans and polychlorinated biphenyls, and are able to remove toxic metals, such as Cr (VI), Ag (II) and Pt (II), present in the wastewater, but under UV radiation (Amin et al., 2014). N-doped TiO_2 nanoparticles can readily photodegrade methylene blue under visible light range (Ashasi et al., 2001). Engineered TiO_2 nanoparticles modified by ruthenium complex sensitized with Pt deposits that act on trichloroacetate and carbon tetrachloride present in an aqueous medium and under visible light. This is accomplished by reductive dehalogenation (Bae and Cho, 2003).

Zero-valent iron nanoparticles (Fe°) and bimetallic Fe° nanoparticles exhibit a larger surface area and have a greater degree of redox reactivity in comparison to bulk zero-valent iron nanoparticles. Because of these features nano Fe° and bimetallic nanoparticles, such as Fe°/Pd°, Fe°/Pt°, Fe°/Ag°, Fe°/Ni° and Fe°/Co°, are exploited for the remedial purposes in environment pollution. These nanoparticles reduce various organic pollutants like chlorinated alkenes, chlorinated benzene, pesticides, organic dyes, nitroaromatics, polycyclic biphenyls and inorganic ions like nitrates present in wastewater. Toxic heavy metals such as Cr (VI) and other less toxic varieties are removed from an aqueous medium. Chlorinated organic compounds, perchloroethylene, trichloroethylene, carbon tetrachloride and others in wastewater are reduced catalytically as a result of active reductive dehalogenation (Dror et al., 2005).

7.3 Nanostructured and Reactive Membranes

Nanostructured and reactive membranes form another group of nanomaterials that are cost-effective and remove pollutants efficiently. One such type of nanostructured membrane is made of carbon nanotubes filters and hollow cylinders. The carbon nanotubes are arranged radially in these nanostructured membranes. These nanomembranes can be reused after cleaning by ultrasonication and autoclaving. These also remove microbial contaminants like *E. coli, Staphylococcus aureus, poliovirus sabin-1* and others, along with pollutants from wastewater (Shirvastava et al., 2004). The efficiency of MWCNTs is because of the matrix is made of uniform CNTs, which successfully separate virus/bacteria involving physicochemical depth filtration. The parameters that influence efficient filtration include the composition of water, ionic strength, monovalent and divalent ions/salts, pH, organic matter, alginate (anionic polysaccharide found in plants cell wall major component of viscous gum and biofilm produced by bacteria and pathogen), phosphate, bicarbonate and others. If the ionic strength is increased by adding NaCl and Ca^{++}, viral isolation gets enhanced, but when Mg^{++} are added, the rate of viral isolation declines (Brady-Estevez et al., 2010).

Another ultra-filter can be fabricated using A-alumoxanes with nanoparticles size 7 to 25 nm. The 'pore size' and 'molecular weight cut off' are related to the uniformity of A-alumoxanes (alumina) nanoparticles. In this case, the molecular weight cutoff ranges between 1000 and 10,000 Da, and the average pore size is around 4 nm. These dimensions do not allow the movement of synthetic dyes, such as 'Direct Red 81', 'Direct Blue-71' and 'Direct Yellow-71', to cross over these fabricated membranes. When the ultra-filtration membrane is doped with Fe, Mn or La (lanthanum), its selectivity and permeate flux increase (Defriend et al., 2003). To increase the water flux and divalent cations like Ca^{++}, Mg^{++} and Cl/SO4–, the nanofilter membranes (porous alumina) are modified by adding 4.5 to 5.0 layers with pairs of poly (styrene sulfonates/ poly (allylamine hydrochloride; Stanton et al., 2003). When these membranes are treated specifically by multilayered deposition of charged polypeptides [poly9L-glutamic acid or poly (L-lysine)], the pores present in these membranes and the degree of metal ion retention and permeate flux get elevated. These functionalized membranes are 10 nm thick and have a pore size of about 200 nm (Hallman and Bhattacharya, 2004). When zero-valent iron as bimetallic Fe°/Pt° nanoparticles are incorporated in cellulose acetate film, the resultant nanofilter membrane exhibits higher functional efficiency. Ultrafiltration membrane modified by embedding metal (average size 24 nm) reduces trichloroethylene sufficiently from the water samples (Meyer et al., 2004). Nanoreactive membrane functionalized by regulating their physical aspects and the surface chemistry of the pore present therein. If one can prepare uniform pores in these membranes, then efficient functionalized membranes are fabricated. Changes in their pore size, using plasma that increases 'chemical vapor deposition', and the addition of an atomic layer of silicon-di-oxide elevate the functionality of nanomembranes significantly (Srijanto et al., 2010).

When two types of hydroxyl groups with different binding energies are linked with nanostructured ZnO thin film, then their surface chemistry is improved. For example, hydroxyl groups like as $O\bar{H}$ group with binding energy of about 535.5eV and second C-OH group with different binding energy of about 268e V when bound to such a ZnO thin film, their surface chemistry gets improvised and changed. Furthermore, different C-concentration present at the surface of ZnO thin film is related to nonstoichiometry and it reflects on the surface morphology of the membrane, (Kwoka et al., 2018). The successful fabrication of nanostructured membranes should have an approximate pore size to thickness ratio as one. This feature facilitates attaining high permeability and an efficient potential for separation. Parameters such as a large area, self-organizing polymers, nanosphere lithography, anodization and solid-phase change the permeability and ability to separate the contaminants (Mireles and Gabroski, 2017). When a material is subjected to anodization (it takes place suddenly), it gets oxidized and cylindrical nanopores are formed parallel to each other and perpendicular to the metallic substrate place underneath (Mijangoes et al., 2016).

Gibbs free energy is a key factor during adsorption of the prototype anion at the interface. This energy of adsorption at the interface graphene/water can be determined using 'surface-sensitive spectroscopy'. The mechanism involved is interpreted following 'molecular dynamics simulation'. The mechanism of adsorption is qualitatively different from the one that operates in the case of an air/water interface. In most cases, the interface is represented by hydrophilic and hydrophobic materials (Survade et al., 2015). Graphene is effective and efficient among the carbon nanomaterials that play multiple roles in nanotechnology and nanoscience (Lahir, 2018). A graphene membrane is effective because the pore size can be manipulated using 'ionic bombardment' with respect to the set designed pore size of the membrane. These are readily fabricated by treating the graphene layer with oxygen plasma resulting in the monoporous membrane having a pore size of 1 nm dimension. The ratio between pore size and the membrane thickness is close to one for every pore size (1 nm) in the fabricated membrane. These nanomembranes are of great utility when very small molecules are to be separated (Survade et al., 2015).

7.4 Bioreactive Nanoparticles

Bioreactive nanoparticles induce interactions amongst biological species that may be harmful or beneficial in nature. Mostly strong oxidants (inorganic and organic) are used to disinfect aquatic, aerial and soil (terrestrial) media. For example, using chlorine to disinfect water produces by-products such as trihalomethanes, haloacetic acids, and aldehydes. The mechanism involved results in the impairment of cellular functions of a pathogen by destroying their cellular components, interfering with the cellular metabolic processes or inhibiting the growth process by blocking the synthesis of key cellular components, such as DNA, coenzymes, specific proteins and cellular proteins (Savage and Diallo, 2005). Nanomaterials that disinfect water and act as biocides are chlorine-free. MgO nanoparticles are one such effective agent. It is biocides specifically for gram-negative *E. coli* and gram-positive *Bacillus megaterium, Bacillus subillus* and bacterial spore. The prime target of these nanoparticles appears to be cell membranes, (Stoimenov et al., 2002). Stable Ag nanoparticles (narrow size range) prepared by the process of reduction of $AgNO_3$ with ascorbic acid in the presence of DAXA 19 as a stabilizer, act as an effective bactericide for *E. coli*, (Sondi and Salopek-Sondi, 2004).

Three antifouling compounds namely Irgaol-1051, Zn pyrithion, and E-conea were encapsulated in biodegradable poly (lactic acid) and used as nanoparticles. These encapsulated nanoparticles (311.9 nm and 1013.3 nm) exhibit very low or the least leachability and even distribution in the paints used for preventing biofouling (Kamtsikakis et al., 2017). The physicochemical properties of Ag nanoparticles embedded in organic/ inorganic matrices change with respect to the size and shape. Tetraethyl orthosilicate (TEOS) and phenyltriethyoxysilane (PhTESO) are two silica-based coatings that are used to enclose Ag nanoparticles. These fabricated nanoparticles exhibited a longer duration of biocidal activity. Out of the two matrices, PhTESO is a more suitable and preferred option for encapsulation of biocidal Ag nanoparticles, (Babapour et al., 2012). TiO_2 nanoparticles are considered to be environmentally friendly photocatalyst. When it is co-doped with silver, carbon and sulfur, the resultant products behave like a 'multifunctional generic biocide' and can be activated as photocatalyst in the presence of visible light. Nanometal oxides after being associated with polymer fibers act as very effective agents to remove heavy metals and As from wastewater (Hamal et al., 2010).

7.4.1 Dendrimers and Dendrimer-Enhanced Ultra-Filters

Dendrimers are polymeric forms and are referred to as dendrimers polymers. Architecturally, dendrimers include central core, intermediate inter-branched cell and peripheral terminal branch cell. Dendrimer polymers include a random hyperbranched polymer and dendrimer graft polymer. These soft nanoparticles exhibit a range in size (1–20 nm), a higher capacity and a recyclable hydrophilic ligand that traps toxic metal ions and radionuclides, including inorganic anions. These soft nanomaterials are recyclable unimolecular micelles used to recover organic solutes from an aquatic medium, as scaffolds and templates for fabricating redox nanoparticles. These redox nanoparticles are catalytically active and act as delivery vehicles for antimicrobial agents such as AgI (Chen and Cooper, 2002; Aorkas et al., 2003). These nanomaterials are very useful in molecular studies, isolating organic compounds present at the interphases. Thus, the dendrimers being ultrasensitive are a very practical option for environmental monitoring (Augustus et al., 2017). Polyamidoamine dendrimers having 1.8-naphthalin amide surface groups are very effective sensors specifically for photo-induced 'electron transfer fluorescence' and is capable to sense rare earth metals and metal cations (Rachid, 2011)

Membranes used in reverse osmosis (RO) have a pore size between 0.1 and 1.0 nm. These membranes are efficient in retaining dissolved inorganic and organic solutes whose 'molecular mass' is less than 1000 Da (Savage and Diallo, 2005). Dendrimers are extremely suitable to extract chemicals present between two phases. Cu++, Pb++ and Cd++ in ionic form can be separated from aqueous effluent by using functionalized SBA-15 mesoporous silica gel involving poly (amino-amine) dendrimer (Shahbasi et al., 2013; Augustus et al., 2017). Copper is adsorbed on chloro-propyl-silica gel surface that is functionalized with 3-poly (propyleneimine) hexa decamine dendrimer (DAB-Am-16; Serenko et al., 2017). Third- and sixth-generation poly (amidoamine) dendrimers and graphene oxide (GO) can remove naphthalene. These two nanomaterials, that is, third and sixth generations of poly (amidoamine) dendrimer and GO, separate polyaromatic hydrocarbon from aqueous effluent (Defever, et al., 2015). Dendrimers are a good option to remove heavy metals and arsenic 'As' for wastewater (Thingavimoghaddam et al., 2013).

7.4.2 Mesoporous Nanomaterials

Functionalized organic compounds as monolayer can be associated with silica. These compounds, say, mercaptopropylsilane, get bound covalently to the mesoporous silica. This arrangement is of cross-linked monolayer and covers the surface of mesoporous nanomaterial up to 76%. The resultant product efficiently separates heavy metals such as mercury from aqueous and nonaqueous waste effluent. It exhibits a distribution coefficient to the tune of 340,000 and is reusable. These mesoporous nanomaterials are further modified with rational design with respect to surface properties in order to comply with varied environmental and industrial applications (Feng et al., 1997; Shahbasi et al., 2013). Cobalt contents and the method of preparation of both factors regulate the morphology and dispersion of cobalt oxide (Co_3O_4) in fabricated composites. The Co_3O_4 is in the form of small patches and gets dispersed uniformly throughout the support medium, causing the least blockage of pores. These materials act as catalytic agents in the liquid phase, causing oxidation of cyclo-hexanol, benzyl-alcohol, 1-phenyl ethanol and norbornene. Such mesoporous nanomaterials are good options for nanoabsorbents (Thingavimoghaddam et al., 2013).

'Carbo-iron' and 'fine activated carbon' have similar sorption abilities; both can reduce metallic iron. These properties are utilized during the remediation of nondissolved hydrophobic pollutants. This is accomplished by generating a border reaction zone and enhancing its affinity for the targeted pollutants (Bleyl et al., 2013). Carbo-iron is a specific colloidal composite support having 'activated carbon colloid' (ACC) with particle size (0.8 μm) embedded in a cluster made of Fe°. Carbo-iron colloids are a suitable composite option for treating groundwater. This composite matter is relatively more efficient than nanosized zero-valent iron (n-ZVI). Activated carbon colloid being hydrophobic in nature shows increased affinity for carbo-iron-colloids over non-aqueous phase liquids. When water is undisturbed the reactive particles get collected at the interface between water and non-aqueous phase liquids, (Mackenzie et al., 2012). Comparative application of carbo-iron and nano-zero-valent-iron (n-ZVI) for groundwater remediation has shown that the carbo-iron supersedes the performance of n-ZVI. Carbo-iron support showed a relatively wider range of interaction in comparison with n-ZVI. A comparative study of the functionality of carbo-iron composite and n-ZVI proved that the carbo-iron composite interacted with pollutants and decomposed tetrachloroethane (also called perchloroethylene) and polycyclic ethylene at a higher rate than the rate shown by n-ZVI without forming intermediate product, such as vinyl chloride. Carbo-iron composite is a better performer in 'plume and source treatment' than an n-ZVI composite. Furthermore, when the carbo-iron is injected the microbial degradation of polychloroethylene (PCE) occurs but not in the case of n-ZVI (Mackenzie et al., 2016).

Nanoabsorbents are used to remove pollutants from aqueous media; carbo-iron removes pollutants from groundwater and soil. Nanoclays adsorb phosphorous and organic contaminants. Nanoagrogel separates uranium from groundwater. Nano-iron oxide adsorbs hormones and pharmaceuticals from wastewater effluent (Thingavimoghaddam et al., 2013). Polymer nano-composite material effectively removes Cr (IV) from the aqueous medium (Singh et al., 2014). Porous silica functionalized by cobalt and mesoporous silica (SBA-15) both cause total oxidation. Between these two, the cobalt-containing porous silica is cost-effective and more stable, but both are promising options to remove volatile organic compounds (Popova et al., 2014).

7.5 Nanomaterials for Environmental Remediation

Unintentional and intentional accumulation of waste, unused, discarded products from domestic, industrial and other possible sources in the components of the environment results in pollution. Pollutants in the environment do not reintegrate into the environment effectively because of their physical and chemical conformation. The chemical composition and amount of these pollutants interfere or disrupt the structure and functionality of the components of the environment causing unwarranted impacts on its abiotic and biotic components. Urbanization and anthropogenic activities are among the major aspects of the increased amount and quality of the pollutants. Some of the current traditional modes of remediation include the use of activated carbon, reverse osmosis membrane, nanofiltration, activated sledge and so on. Conventional physicochemical treatment processes, such as oxidation, appear are ineffective to remediate widespread toxic compounds, micropathogens and complex additives, among others (Amin et al., 2014). The existing techniques either produce harmful by-products or are unsuccessful at degrading contaminants. These contaminants/pollutants have easy access to the water table, affecting the vulnerable underlying aquifers. Such pollutants and contaminants render water, air and land as problematic sites for difficult remediation and influence the quality of water, clean air and land (Salipira et al., 2007; Sipley et al., 2010).

Remediation is primarily dedicated to either removal or reduction of pollutants in the environment and provide an amicable macro- and ambient environment to the biota living therein. Existing conventional modes of disposal of waste and the treatment rendered are expensive, relatively ineffective and take a long duration to achieve the target (Salipira et al., 2007; Mehndiratta et al., 2013). Nanomaterials can be easily procured and suitably tailored with respect to applications such as sensing, detection and remediation, as needed for the environment and industries (Feng et al., 1997). Nanotechnology provides potential tools for monitoring the environment. Nanoforms of iron, titanium dioxide, silicon dioxide, zinc oxide, carbon nanotubes, dendrimers, nanopolymers and others are being utilized to decontaminate air, water and soil (Zhang, 2005; Das et al., 2015). It also provides a very effective 'breakthrough' in the fields of safe remediation of pollutants in the environment. These tools and the solutions from nanotechnology are eco-friendly and facilitate environmental sustainability (Corsi et al., 2018).

Fabricated nanomaterials are used in various modes to eliminate pollutants. The formulated nanomaterials and other metal oxides nanoparticles are utilized to remediate organic pollutants for the aquatic and other media. TiO_2-based nanomaterials exhibit specific features, such as abilities to convert energy and their semiconducting nature, to sense electric charge, gases and photocatalytic behavior. TiO_2 possesses semiconductor photocatalytic abilities utilized in different reduction processes at the interface between the semiconductor and the solution (Mansoori et al., 2008). The semiconductor materials get modulated when they come into contact with donor atoms that provide an electron for the conduction band, and at these bands, current can be carried. The semiconductor materials can readily accept atom which forms valance band and result in some positive charge (referred to as 'holes'). The specific energy levels of these donors and acceptors fall within the energy gap. Another feature of semiconductors' nanoparticles is their ability to bring a shift in their optical absorption spectra towards shorter wavelengths in comparison to the respective bulk forms, (Mansoori et al., 2008). The interaction of MgO nanostructure follows the pseudo-second-order kinetic model and adsorption isotherm model. Parameters, such as intra-particle diffusion, bulk diffusion, film diffusion mechanism and others, regulate the adsorption process involved in these interactions. The mode of a formulation is also one of the parameters that regulate the functionality of MgO nanostructure. The spontaneous interaction between MgO nanostructure and textile dyes is exothermic and physisorption (Nassar et al., 2017). Nanorods, hierarchical structures and nanoflakes are the fabricated nanoforms of MgO and show remarkable ability to remove organic dyes like malachite green and Congo red from an aqueous medium. MgO nanorods surface area 48.45 m^2/g, nanoflakes with the surface area around 147.5m^2/g while nano-hierarchical structures with 110.9 m^2/g were used to remove these dyes. Nanorods, nanoflakes and nano hierarchical structures removed malachite green from an aqueous medium 95%, 97.42% and 99.98%, respectively. The Congo red dye was separated from the water as 86.20%, 92.68% and 99.94%, respectively, by these nanomaterials. Thus, nano-hierarchical structures exhibited maximum capacity of removal of malachite green and Congo red by 1205.23 mg/g and 1050.82 mg/g, respectively (Dhal et al., 2015).

Nitroaromatic compounds like nitrobenzene, 2–4-dinitrotoluene, and 2–4–6-trinitrotoluene, among others, are common occurrences in aquatic waste and cause toxicity. The nanocrystalline form of titanium dioxide with its surface modified or functionalized with arginine removes such nitroaromatic pollutants. These also degrade nitrobenzene by the reduction process. The addition of methanol to the reaction mixture increases the degree of degradation of nitroaromatic compounds. The degree of coupling between nitrobenzene and functionalized TiO_2 is accomplished by the shift of photogenerated electrons from TiO_2. The photodecomposition of nitrobenzene (nitroaromatic compounds) is achieved with electron-donating chelating agents using surface-modified nanocrystalline TiO_2 (Makarova et al., 2000). Phenol is another aromatic toxic pollutant ubiquitously present in aquatic waste. Such toxic pollutants are eliminated by the wet oxidation process involving hydrothermal treatment of microemulsification and photocatalytic activities of anatase and rutile forms of TiO_2. During such interactions, the size of particles is affected by the ratio between the surfactants and water. The interaction between phenol and TiO_2 involves the formation of $OH^·$, H^+ and intermediate components. There are two microemulsions phases, namely, anatase and rutile forms of TiO_2. Of these two microemulsions, anatase-microemulsion exhibited a higher rate of photodegradation throughout the process while rutile microemulsion showed a higher rate of photodegradation during initial stages, but during the later period, it declined (Anderson et al., 2002). A reactor device involving a UV lamp, nano TiO_2 as a photocatalyst and source of 'hole electron pair') degrades organic pollutants in wastewater. This device is used as a combined photoelectrochemical system along with 'Rhodamine 6G' as an indicator of the degradation of organic waste in the water. These test samples were procured from the dyeing and printing units. The test water samples were treated for about half an hour. The observations showed a 93% reduction in 'chemical oxygen demand', a 87.6% reduction in 'biochemical oxygen demand', a 99.9% reduction in bacterial quantity and a 67.5% decline in ammonia nitrogen in the resultant water after the treatment (Chen et al., 2003). Fabricated 'carbon-black-modified-nanoTiO_2' (C-B-TiO_2) supported on the alumina sheets degrade organic matter. The comparative study of this fabrication and pristine TiO_2 showed that C-B-TiO_2 was 1.5 times more effective in degrading organic matter in wastewater than the pristine TiO_2. In this setup, ozone concentration played a very important role in the removal of 'total organic carbon' (TOC) from the wastewater under test. Thus, in this set up combined impact of three aspects were used namely photocatalysis, UV irradiation and ozonation (TiO_2/UV/O_3). Their impacts resulted in an enhanced percentage of removal TOC. During this interaction of total mineralization of catechols was accomplished by pseudo-zero-order kinetics and ozone (O_3) and the concentration of catechols regulated this interaction, but catechols' concentration did not influence the kinetics when UV/O_3 and TiO_2/UV/O_3 (O_2) processes were in progress. TiO_2/UV/O_3 (O_2) process gave a maximum yield of degradation of organic wastewater from an effluent understudy (Li et al., 2003).

Photoelectrochemical phenomena are good remedial modes for aquatic waste. Nanocrystalline colloid anatase TiO_2 doped with chromium (Cr) is derived using nanotubes involving hydrothermal ion-intercalation under acidic ambient medium. During photoelectrochemical interaction and under visible light, water splits into H_2 and O_2, involving Cr-doped TiO_2 in the form of thin film acting as a photoelectrode. The photoelectrochemical activity was relatively much higher when Cr-doped TiO_2 was used in comparison to undoped TiO_2; that is, chromium doping of TiO_2 elevated the impact of the photoelectrochemical activity. The Cr-doped TiO_2 influenced the redshift in the bandgap revealing the ability of Cr-doped TiO_2 to use a wide range of the visible spectrum (Tsai and Teng, 2008). The ability of photocatalytic degradation of titanium dioxide is enhanced after gets modified with asparagine, serine, phenylalanine and tyrosine involving the carboxyl group in the presence of high-pressure mercury irradiation. TiO_2 nanoparticles thus treated increase the rate of photocatalytic degradation of nitrobenzene into active aniline as compared to the use of pristine TiO_2. Asparagine, serine, phenylalanine, tyrosine and the ability of modified TiO_2 to donate electrons facilitate significantly the process of photoreduction in the case of nitrobenzene (Huang et al., 2010).

Sonocatalytic degradation of nitrobenzene/nitroaromatic compounds can be accomplished by functionalizing the surface of metal oxides like $Fe_3 O_4$, CuO, NiO and Co_3O_4 with amino acids like glycine, arginine, and glutamic acid. This sonocatalytic degradation of a nitroaromatic compound at pH above 7.0, arginine showed the highest degree of sonocatalytic degradation. This degradation is because of the promoted coupling through electrostatic attraction between negatively charged nitrobenzene molecules and positively charged arginine layer on the surface of these metal oxides. Hydrogen bonding, n-π and π-π bonding are involved during this attraction interaction (Yehia et al., 2015).

Nanocomposites are supportive materials and are most suitable for varied remedial options in the environment. Their compositional components within nano dimensions provide physicochemical properties that enhance their functionality and enable them to meet the requirements of material science in solving the existing problems (Ajayan et al., 2004). Nanocomposites are the structural and functional representation of the increased level of complexity involving nanocrystals, such as nanoparticles and nanorods; polymers, such as nanocopolymers and block copolymers; and small molecules, such as amphiphilic and mesogen (having mesophase). These undergo surface modification or association results in the formation of super-molecules. During this rearrangement, a super-lattice or directed nanocrystal assembly or hierarchical self-assembly takes place. In the recent past, the development in polymer nanocomposites includes the involvement of metals, CNTs and clay nanoplatelets as reinforce components. Most polymer matrixes utilize polystyrene, poly (ethylene oxide), poly (ε-caprolactone), polypropylene, various nylons and poly (ethylene terephthalate; Stefanescu et al., 2009).

Ultimately, waste is either organic or inorganic chemicals and is present beyond the permissible levels in all three aspects of the environment. These pollutants behave as per their physical and chemical nature. In order to deal with these pollutants, common techniques, such as coagulation, flocculation, adsorption, use of membrane filters, photo-Fenton oxidation and electrochemical reduction, are involved. These techniques are potential remedial processes to remove heavy metals, metal oxides and organic pollutants. Major sources of pollutants are the industrial effluent waste containing mostly persistent organic and inorganic pollutants. Such pollutants are commonly referred to as 'recalcitrant pollutants' because these are not easily eliminated. Advanced oxidation is very effective in such cases. The basic functional principle involves the production of highly reactive oxygen species (ROS). These ROS play a significant role in the catalytic breakdown of the contaminants.

Toxic gases such as NO, NO_2 and SO_2 are produced during fossil fuel combustion. Carbon nanotubes proved to be better adsorbents for these gases at room temperature in comparison to clay, zeolites and Fe-OOH-coated activated carbon. CNTs may not be suitable for adsorption of CO_2 gas. This problem can be solved when CNTs are treated with ethylamine, polyethyleneimine and 3-aminopropyltriethoxysilane are used to remove CO_2. These modified CNTs are effective mostly at a lower temperature (Long and Yang, 2001; Ian Sofian, et al., 2012).

Polyaromatic compounds and volatile organic compounds can be readily absorbed by using mesoporous magnesium oxide and chromium oxide (Sinha and Suzuki, 2007; Ian Sofian et al., 2012).

In the recent past, advanced oxygen has been empowered with processes such as ozonation, sonochemical degradation, electrochemical remediation, heterogeneous catalysis, Fenton oxidation and photo-Fenton oxidation (Gautam and Chattopadhaya, 2016). Ozonation is a process during which ozone is allowed to pass through the test sample of water. Ozone has a tendency to release one atom of oxygen that acts as a powerful oxidizing agent. This atom kills most of the pathogens and degrades the organic contaminants present. It produces very less dangerous by-products in comparison to chlorine and does not leave any residual disinfectant, unpleasant odor and taste. This process in combination with the denitrification biofilter process and biological aerated process gives remarkable results in the removal of the contaminant in aquatic discharge. After the treatment, values of chemical oxygen demand and total nitrogen are brought to permissible limits in the resultant discharge. Only ozonation can cause 94% removal of antibiotics, such as tetracycline, in the experimental sample water (Li et al., 2015).

Sonochemical degradation of contaminants involves 'sonochemistry'. The 'acoustic cavitation' reflects on the formation, growth and implosion of the bubble within the test liquid sample. These are induced when sound waves are irradiated on the study sample. The implosion or collapse of the bubble is instantaneous. These small bubbles formed under the irradiation of sound waves have very high temperatures and pressures. During the cavitation process, extreme physical and chemical localized conditions are caused. These bring about chemical degradation. Sound waves are successfully exploited in the varied range of chemical degradation process (Suslick, 1990). Sonochemical degradation of a toxic organic compound such as chlorinated and phenolic compounds, along organic dyes, is a successful mode of remediation of pollutants (Chowdhury and Viraraghvan, 2009).

Electrochemical remediation/degradation is accomplished by carrying out the oxidative reaction at the anode. At the anode, pollutants are converted into a nontoxic form. These toxic compounds either decompose or change their oxidative form. The contaminants or organic pollutants are either directly or indirectly subjected to electrochemical oxidation. During direct electrochemical oxidation, physically adsorbed oxygen is produced at anode, consuming only electrical energy, not the chemical. This aspect is referred to as 'electrochemical mineralization'. During indirect electrochemical oxidation, strong oxidants are produced at anode followed by chemical oxidation. Mostly highly reactive hydroxyl radicals are produced through a 'Fenton reagent'. The use of UV-A radiation is capable to accelerate the process of mineralization. This process is referred to as the 'Photoelectron-Fenton phenomenon' (Dimitrijevic et al., 2013). Polyaromatic hydrocarbons, such as naphthalene, fluoranthene and pyrene, undergo electrochemical oxidation, and these are degraded by oxidation at the anode (Muff and Sogaard, 2010).

Heterogeneous catalysis is a phenomenon related to catalysis that involves different phases of catalysts and reactants. It not only refers to solid, liquid versus gaseous phases, but it includes immiscible liquids such as oil and water (Swathi and Sebastian, 2008). Functionalized carbon nanotubes are able to catalyze n-butane to butenes and butadiene during oxidative dehydrogenation (Zhang et al., 2008).

Nanoscaled metals and metal oxides such as gold, silver, iron, and TiO2 iron oxide are of common use as sensors and remedial agents for environmental pollution. Silver nanoparticles act as antimicrobial, antifungal and antiviral agents. These are potential agents for removing these contaminants from the air, water and soil. TiO_2 exhibits photoactive antimicrobial action. Toxic chlorinated organic compounds, pesticides and inorganic mercury are readily removed from wastewater and air. Atmospheric contaminants like oxides of sulfur are eliminated by the combined form of Au-TiO_2 nanoparticles. TiO_2 nanoparticles are used to remediate and degrade metal toxins, dyes, halogenated compounds. Carbon nanomaterials like carbon nanotubes clusters are capable of sensing and eliminating bacteria, heavy metals such as cadmium and organic contaminants such as benzene and 1,2-dichlorobenzene from wastewater. Nanoparticles, when associated with host polymer materials like porous resins, cellulose and silica, cause relatively less harm to humans and biota of the environment. These fabricated nanomaterials are very effective in catching the pollutants from wastewaters. ZnO nanoparticles, when associated with polymer, become capable to degrade organochlorine pesticides, halogenated herbicides and azo-dyes (Khim et al., 2012).

7.6 Nanomaterials for Sensing and Detecting Pollutants

Advancements in nanotechnology and nanomaterials have been focused on the development of nanosensors and biosensors. Nanosensors are the need of the current scenario to ensure fast, reliable, accurate, sensitive and reproducible results of an analysis. Such an analysis of contaminants ensures appropriate monitoring of the changes in the environment. Such nanogadgets are aimed to be simple so that these are comfortably used by a semi- and/or nontechnical individuals in the laboratory and/or at the site or field (Hernandez-Ramirez et al., 2007; Lahir et al., 2016). These analyses detect the quality and quantity of the specific pollutant, contaminant in a given aspect of the environment, (Mehndiratta, et al., 2013). Miniaturization is of great application in the field of nanosensors. The nature, efficacy and compatibility of the nanosensors are significant parameters in the functional efficiency and reliability of nanosensors (Lahir et al., 2016). Nanomaterials are capable of being functionalized or modified as per the need or target. These can be either coated, dispersed or bound to the biomolecules or organic molecules or to polymeric complex. The nanosensors can be functionally specific. There are prepared, fabricated or manipulated with respect to the needs that are suitable to laboratory or field conditions (Lahir et al., 2016). The biocompatibility of the nanosensor/biosensor plays a prime role in the success of nanosensors. If the nanomaterials to be used in the biosensor or nanosensor are not biocompatible, these are likely to cause disturbance or derogative impacts in the biosystem or at the site of analysis. The biocompatibility of nanomaterials depends on their nature, ability to be modified, surface properties, morphology, degree of functionality of the sensors and so on. In the case of composite support/mesoporous silica support/materials the biocompatibility depends on the controlled surface properties and the morphology of the particles. On average, the mesoporous silica support possesses a surface area >700 m^2/g^{-1} and volume of the pore >lcm3/g^{-1}. The functionalization of the internal pore and external particle surface of the support are important parameters and influence the biocompatibility as well as the functionality of the mesoporous support (Trewyn et al., 2007).

Nanomaterials are the most suitable materials to be used in combination or otherwise as sensors to monitor environmental contamination in air, water and land. This concept is based on their specific features like electrical, optical, thermal, mechanical strength and catalytic behavior. Nanomaterials are used to sense and detect toxic metal ions, toxic gases, pesticides and hazardous industrial chemicals (Su et al., 2012). Nanomaterials are potential candidates for developing sensing devices and increasing their sensitivity or other similar devices. The size, shape and specific physicochemical features enhance the probability of nanomaterials to be structural and functional components in the field of sensing/detection devices. Nanomaterials like nanowires, semiconductor particles, allotropic forms of carbon and 'imprinted polymeric spheres' are used in the development of nanosensors (BelBruno, 2013). The components of nanosensors and transducers are fabricated so the complete unit functions in accordance with targeted sensitivity, accuracy and specificity of 'input' and 'output'. Depending on the type of transducers, nanosensors can be classified as amperometric, potentiometric, conductometric, ion-sensitive, acoustic or microcantilever (mass-sensitive) (San and Rogach, 2010; Monosik, $2012_{1\text{-}a}$; Reza et al., 2013; Lahir et al., 2016). Nanomaterials, such as metals and a metal oxide, nanowires, nanoporous materials, magnetic nanoparticles, optic fibers, carbonaceous nanomaterials, quantum dots and dendrimers are mostly used as nanosensors (Lahir et al., 2016).

Nanomaterials can disinfect water. TiO_2, ZnO and fullerol induce the formation of ROS during photocatalytic action. These nanomaterials damage the cell membrane of microbes and viruses. Peptides, chitosan, carboxy fullerenes, carbon nanotubes, ZnO and silver nanoparticles compromise the cell membrane of bacteria. Silver and aqueous fullerene nanoparticles interrupt energy transduction in biotic infectants. Chitosans, as fabricated nanomaterials, interfere with the enzyme activities and synthesis (Ian Sofian et al., 2012)

The process of quenching of photoluminescence is the functional basis in the detection of pesticides. During quenching, there is a 'transfer of energy to the substrate of an organic molecule and to the electron transfer for conduction band of silica nanocrystal to the vacant orbit of the quencher' (Stefano et al., 2005; Mehndiratta et al., 2013). In this case, it is an organic pesticide. This principle of quenching photoluminescence is very useful for the detection of organic pesticides. Porous silica crystal containing nanomaterial exhibits photoluminescence. The energy transfer takes place between silicon nanocrystal and the organic molecules and pesticides. With the porous silica technology, a pesticide can be detected up to 1-pmm concentration, (Stefano et al., 2005).

Monodisperse single-crystalline Cu_2O nanospheres are potential materials for sensors to sense and detect inflammable gases. Cu_2O nanospheres get associated in the form of a thin layer. The spaces are able to interact with gases that are present in ppm concentration. Cu_2O nanospheres are sensitive sensors for gases (Zhang et al., 2006). The working principle of heavy-metal-ion nanosensors involves chitosan and 11-mercaptoundecanoic acid. These two compounds bring about metal chelation. During this process, aggregation of nanoparticles takes place when heavy metal ions bind to the chelating agents and the absorption wavelength shifts, causing the change in color, that is, from red to blue. This basic functional principle may not be applicable to all heavy metal ions, but most of them, such as Pb, Hg and Cd, among others, are detected by such nanosensors (Sugunumm et al., 2005).

Semiconducting oxide, carbon nanotubes and polyaniline are suitable nanomaterials for chemical-resistant nanosensors. Quasi-one-dimensional materials such as nanorods and nanotubes are probably more appropriate for the fabrication of gas sensors (Backenkov and Sergeer, 2007). SnO_2 metal nanowires exhibit specific contact characteristics and varied resistant abilities when investigated for two- and four-probe measurements. The interaction between SnO_2 metal nanowires and gas exhibits the influence on its size and diameter and show specific electrical transductional features, specifically in the case of CO and varying degrees of humidity. These studies suggest that single SnO_2 nanowires are a good option as individual monocrystal to detect electrical transduction with respect to gas CO and can be used as sensing materials (Hernández-Ramírez et al., 2007). Nanomaterials as sensors are preferred options for sensing contaminants in the air, water and land. 'Nano strips' made of 'nylon-6-nanofibers net' is one such example. These nanomaterials can detect formaldehyde—a toxic air pollutant (domestic and industrial contaminant); a yellow sensor strip becomes red in the presence of formaldehyde in traces (Khim et al., 2012). Semiconductors oxidize organic molecules in the presence of light. A charge is transferred from the valence band to the conduction band in presence of appropriate intensity of light in this process. Nanosized semiconductor photocatalysts are fabricated with specific selectivity and reactivity are based on this principle. Such nanosized semiconductors are able to remove organic compounds like 1, 1-dichloroethane, cis-1, 2-dichloroethane, 1, 1, 1 trichloroethane, xylene and toluene. Benzene, toluene, ethylbenzene and toluene are removed from groundwater using these semiconductors. TiO_2, ZnO, Fe_2O_3, and WO_3 (tungsten oxide) are among some of the oxides that act as photocatalysts. These catalytic oxide forms perform better as detectors and remedial agents than their normal oxide forms (Liang and Noworty, 2010; Ian Sofian et al., 2012).

Ferritin is a specific protein that is of animal and plant origin. It has an affinity for iron and helps to store it. This protein is made of structurally similar 24 polypeptides. These get associated with a cage-like conformation. This ferritin cage permits the entry of iron molecules and changes them to ferrihydrite nanoparticles. This device can eliminate toxic metals and chloro-carbon from waste in the presence of solar radiation, and visible light is preferred in comparison to the normal iron catalyst (Ian Sofian et al., 2012).

Contaminants like 'As' (arsenic) are influenced by the complexity of the environment and biological origin and these features limit their detection. Thus, to meet such challenges, it is essential to enhance the degree of sensitivity and selectivity of the sensor. To meet such challenges, different electrode systems with surface modifications as per the targets are devised (Lahir et al., 2016). These devices include carbonaceous nanomaterials, metallic nanoparticles, metal nanotubes and even enzymes. These are cost-effective and easy to use, portable and have a facility of 'flow-injection-mode' (Sureshkumar et al., 2017). Fabricated selective electrochemical biosensors can detect 'As'. For this purpose, L-leucine and GO were associated, and this complex acted like a sensitive sensor. This sensor showed high and reproducible responses. This sensor could sense 'As' up to 30 μA ppm^{-1} cm^{-2} and could detect 'As' up to the limit of 0.5 ppm. It gave similar responses when detection was carried out in the field and in laboratory conditions (Sandeep et al., 2016). Copper (Cu) can be recovered from wastewater by employing low-cost stainless steel and potassium ferrocyanide and obtaining its refined grain form. An additive like 2, 2-dipyridine is useful and facilitates the grown forms. This technique of using additive-assisted electrodeposition offers a promising option to recover metal from wastewater (Lou et al., 2018).

Mesoporous ZnO nanomaterial is fabricated involving dielectrophoresis using mesoporous carbon and zinc nitrite as a precursor. The morphology of the resultant nanostructure is polycrystalline in nature. This device exhibited sensitivity up to 60% at 250°C with responses and recovery time is 80 min to 90 min for CO gas (Liu et al., 2009). One-dimensional ZnO nanoassemblies are made involving Al_2O_3 as substrate and 'plasma-enhanced chemical vapor deposition' technique. The resultant nanomaterial is referred to as the 'PE-CVD-ZnO nanosystem'. This system exhibited a very good response to sense H_2, CO and CH_4 at 100°C and 200°C (Barreca et al., 2010). ZnO nanoparticles and nanofibers fabricated using a modified sol-gel technique under supercritical conditions and electrospinning technique. The ability of ZnO nanoparticles (25 nm) is attributed to annealing temperature and shape of ZnO nanoparticles. The sensitivity and selectivity of CO sensors are dependent on annealing at high temperatures. The sensitivity NO_2 gas sensor is dependent on the method of synthesis of ZnO nanofibers (Hijiri et al., 2013). Gaudium (Ga)–doped ZnO nanopowder was prepared by the sol-gel technique. The product Ga-doped-ZnO nanomaterial formed is polycrystalline in nature and has a specific 'wurtzite structure'. When associated with an interdigitated alumina substrate as a thick layer, it acts as a very sensitive sensor for detecting CO gas that has more response and recovery (Hirjiri et al., 2015).

Improved electrochemical sensors are the need to ensure effective 'real-time' and 'robust' authentic analysis of pollutants/contaminations. These states of the art metal and metal oxide nanoparticles, carbon nanomaterials, polymers and others are utilized in devising efficient sensors. Electrochemical sensors provide a sufficient platform for sensing, detecting contaminants such as heavy metals, inorganic anion, phenolic compounds, pesticides and warfare reagents (Govindhan and Jin, 2017). There have been many attempts to fabricate electrochemical sensors. When the surface of exfoliated GO is treated electrochemically with lead ions it results in fabricating reduced GO–lead dioxide composite support. This composite nanomaterial can detect As when present in traces. The cause of this ability is the functional groups linked at the exfoliate GO that remain active in a wide range of pH (2–11pH). A very fine layer of lead dioxide enhances its efficacy that enables the detection of As and Pb in traces (Ramesh and Sampath, 2011). Hexavalent chromium is extracted from wastewater, and this removal is accomplished by involving a 'clean reagent', electrocoagulation, electrochemical reduction, electrodialysis and electro-deionization. Parameters such as pH, the material of electrodes, type and configuration of the cell and integration technique are involved in these techniques (Wei et al., 2016). Chromium (Cr) is detected and extracted from wastewater alkaline electrochemical advanced oxidation process-oxidation of chromium. Graphitized MWCNTs were modified by the formation of hydroxyl reactive species electrochemically. During *in situ* conditions, 2-electron oxygen species are created electrochemically and chromium-induced alkaline electro-Fenton-like reaction. Chromium could be oxidized 95% efficiently within 90 min of presence in wastewater. This approach is convenient and eco-friendly (Xue et al., 2017).

Graphene has specific properties, and its two forms, namely, GO and reduced graphene, find a variety of applications (Lahir, 2018). GO exhibits restricted electrical and thermal transport, it also exhibits oxygen functionalization. GO is used in those devices that are concerned with the storage of energy, monitoring nucleic acids. It gives specific electrochemical responses in cases where oxygenated groups are associated with it and mostly used in voltametric studies. Reduced graphene is an equally target-oriented component that dominates the electrochemical process, (Rowley-Neale et al., 2018; Lahir, 2018).

Two-dimensional (2-D) nanomaterials, such as boron nitride, nanosheets, graphite-carbon-nitride (g-C_3N_4) and various transition metal dichalcogenides (TMDs) are very suitable materials to monitor pollution in the environment. These 2-D nanomaterials are made of 2 or more elements and have a large surface area, greater chemical stability, easy functionalization, higher electrical conductivity and mechanical strength. This is because these 2-D nanomaterials are incorporating graphene or graphene-like structure. For example, boron-nitride nanosheets have higher thermal transport functionality. This functionality of such nanomaterials is enhanced by improving functional components like noble metal nanoparticles, metal oxides and others. Such fabrications are referred to as nanohybrids and perform better because of their synergistic impacts (Su et al., 2017).

The anodic-stripping voltametric technique is another sensing technique that detects heavy metals in wastewater. This mode is very sensitive, and detection becomes feasible due to the high reduction potential. Heavy metals like Hg++, Pb++, Cd++, Cu++, Zn++ and Fe++ can be sensed and detected if 2-D nanomaterials are used as an electrode while performing the 'anodic-stripping voltammetric technique'. Bismuth nanoparticles are readily deposited on to the surface of reduced graphene oxide sheets without involving any surfactant—'Hummer's method'. This results in the formation of 'reduced graphene oxide/bismuth' nanomaterials, (RGO/Bi). This nanomaterial exhibits relatively better interactive features with respect to the bismuth film electrode when used as an electrode in 'stripping voltametric analyses'. This modified electrode detects heavy metals such as Cd++, Pb++, Cu++ and Zn++. The sigma detection limits at different potentials for Cd++ was 2.8μg/L^{-1}; for Pb++, 0.55μg/L^{-1}; for Zn++, 17μg/L^{-1}; and for Cu++, 26μg/L^{-1} (Sahoo et al., 2013). Graphene nanosheets were exfoliated ultrasonically onto the glass-carbon electrode. These modified electrodes are used to study the signal responses for Cd++ and

Pb++. The presence of bismuth elevates the degree of signaling response. The synergistic impact of graphene sheets and bismuth showed better detection of these metals. Cd++ could be detected up to 0.35μg/L^{-1} while Pb++ could be detected up 0.045μg/L^{-1}. The synergistic combination is better novel electrochemical sensors for Cd++ and Pb++ (Wang et al., 2016).

7.7 Conclusion

Nanotechnology has undergone tremendous growth and development. It has revolutionized almost all industries. It provides potential remedial solutions to environmental problems such as global warming, water treatment, waste disposal, management and materials sciences, among others, that are effective not only at the macro level but also at nano levels. Nanosensors are the need of the current scenario and essential tools to monitor the pollutants/contaminants to facilitate the health and sustainability of the environment. The types and the quanta of pollutants/contaminants have escalated at the maximum rate and are the cause of 'worry'. In spite of strict regulations, the remedial efforts seem to be futile. The uncontrolled industrial, domestic and anthropogenic developmental activities taking place globally and release their waste indiscriminately in air, water and land. This is posing danger to life. Obviously, the sustainability of human life and other biosystems depends on the quality and quantity of water, air and soil available. The deteriorated state of these three components of the environment is a challenge to the current scenario. The major remedial efforts are being focused on the use of nanomaterials. Common modes of remediation are thermal, physicochemical and biological in nature. The nanomaterials have higher affinity, capacity and specific selectivity toward pollutants/contaminants. Nanomaterials have structural and functional superiority over their respective bulk forms. Moreover, mater at the nanoscale can be modified in accordance with the set applications. This feature makes them preferred options for application as remedial agents to rectify the dilapidated state of the environment. This review reflects on the potentialities of nanomaterials and nanosensors in tackling the varied issues of environmental pollution. The nanosensors are cost-effective and easy to use both in the laboratory and field. The specific features of nanomaterials add enormously to the durability in terms of mechanical strength, declined weathering ability, weight reduction and energy-saving ability, self-cleaning nature of nanosurfaces, functionalized textiles against bacteria, fungi and others. Due to their miniaturization ability, they exhibit better management of raw materials and facilitate functionalization and optimization. Nanomaterials utilized in the field of nonconventional energy, solar cells, batteries, treatment of groundwater, effluent from various anthropogenic sources, elimination of hazardous materials and anticorrosion materials ensure sustainability. The current reports are encouraging and suggest that in spite of derogative impacts the nanomaterials will rule remedial aspects of industries, domestic, biomedical, pharmaceutical, food technology, electronics, cosmetics and the bio-agro industry under the guidelines from the respective regulatory and legal authorities in the times to come.

ACKNOWLEDGMENT

The author gratefully acknowledges the encouragement from Dr. P M Dongre, head of the Department of Biophysics; Dr. A V Chitre, Department of Biophysics; and Dr. Archana Rath, head of the Department of Biotechnology, University of Mumbai, Mumbai, India.

REFERENCES

Ajayan, P. M., Schadler, L. S., and Braun, P. V., (2004) Nanocomposite science technology, Wiley-VCH Verlag Gmb H and Co KgaA, ISBN 9783527303595, http://dx.doi.org/10.1002/3527602127.

Alvarez-Ayuso, E., Garcia-Sanchez, A., and Querol, X., (2003) Purification of metal electroplating wastewater using zeolites, *Water Res*, 37: 4855–4862.

Amin, M. J., Alazba, A. A., and Manzoor, U., (2014) A review of removal of pollutants from Water/wastewater using different types of nanomaterials, *Adv Mater Sci Eng*, 2014(2014); article Id 825910, 24pg, http://dx.doi.org/10.1155/2014/825910.

Anderson, M., Osterlund, L., Ljungstrom, S., and Palmquist, A., (2002) Preparation of nanosize anatase and rutile TiO_2 by hydrothermal treatment of microemulsions and their activity for photocatalytic wet oxidation of phenol, *J Phys Chem B*, 106(41): 10674–10674; https://doi.org/10.1021/jp025715y.

Aorkas, M., Tsiourvas, D., and Paleo, C. M., (2003) Functional dendrimeric "nanosponge" for removal of polycyclic aromatic hydrocarbon from water, *Chem Mater*, 15(14): 2844–2847.

Ashasi, R., Morikawa, T., Ohwaki, T., Aoki, K., and Taga, Y., (2001) Visible-light photocatalysis in N-doped TiO_2, *Science*, 293(5528): 269–271.

Augustus, E. N., Allen, E. T., Nimibofa, A., and Donbebe, W., (2017) A review of synthesis, characterization and application of functionalized dendrimers, *Am J Polymer Sci*, 7(1): 8–18; https://doi.org/10.5923/j.ajps.20170701.02.

Babapour, A., Yang, B., Bahang, S., and Cao, W., (2012) Biocide Ag nanoparticles in two different silica-based coatings, American Institute of Physics – Conference Proceedings (1482,681, 2012); https://doi.org/10.1063/1.4757558.

Backenkov, V. E., and Sergeer, G. B., (2007) Nanomaterials for sensors, *Russ Chem Rev*, 76(11): 1084–1093; https://doi.org/10.1070/RC20076n11ABEH003735.

Bae, E., and Choi, W., (2003) Highly enhanced photoreductive degradation of perchlorinated Compounds on dye-sensitized metal/TiO_2 under visible light, *Environ Sci Technol*, 37(2): 147–152.

Barreca, D., Beckerman, D., Domini, E., Devi, A., Fischer, R. A., Gasparotta, A., Maccato, C., Sada, C., Sberveglieri, G., and Tondello, E., (2010) 1-D ZnO nanoassemblies by plasm-CVD-on chemical sensor for inflammable and toxic gases, *Sens Actuators B*, 149(1): 1–7; https://doi.org/10.1016/j.snb.2016.048.

BelBruno, J. J., (2013) Nanomaterials in sensors, *Nanomaterials*, 3(4): 572–573; https://doi.org/10.3390/nano3040572.

Bleyl, S., Kopinka, F. D., Georgi, A., and Mackenzie, K., (2013) (in German) Carbo-iron ein Maßgeschneidertes Reagenz zurin situ ground wasser sonierung, *Chemie Ingenier Technik*, 85(8): 1302–1311; (Special issue) Membrane und membrane processes; http://doi.org/10.1002/cite.201300009.

Brady-Estevez, A. S., Schnoor, M. H., Vecitis, C. D., Baleh, N. B., and Elimelech, M., (2010) Multiwalled carbon nanotubes filter: Improving viral removal at low pressure, *Langmuir*, 26(18): 14975–14982; https://doi.org/10.1021/la102783v.

Buzea, C., Pacheco, I. I., and Robbie, K. (2007) Nanomaterials and nanoparticles: Sources and toxicity, *Biointerphases*, 2(4); http://doi.org/10.1116/1.2815690.

Chen, C. Z. S., and Cooper, S., (2002) Interaction between dendrimers biocides and bacterial membrane, *Biomaterials*, 23(16): 3359–3368.

Chen, J., Chen, J., Lin, M., Zhang, L., Zhang, J., and Jin, L., (2003) Application of nano TiO_2 towards polluted water treatment combined with the electrochemical method, *Water Res*, 37(16): 3815–3820; https://doi.org/10.1016/S0043–1554(03)00332-y.

Cheng, X., Kan, A. T., and Tomson, M. B., (2004) Naphthalene adsorption and desorption from aqueous C60 fullerene, *J Chem Eng Data*, 49: 675–683.

Chiban, M., Saudani, A., Sinan, F., and Resin, M., (2012) Water treatment by batch adsorption Method on to microparticles of direct Withania frutescens plant a new adsorbent, *J Environ Manag*, 95(supplement): S61–S65.

Chowdhury, P., and Viraraghvan, T., (2009) Sonochemical degradation of chlorinated organic compounds, phenolic compounds, and organic dyes- A review, *Sci Total Environ*, 407(8): 2474–2492; https://doi.org/10.1016/j.scitotenv.2008.12.013.

Corsi, M., Winther-Nielsen, M., Sethi, R., Punta, C., Torre, C. D., Librato, G., Lofrano, G., Sabatini, L., Aiello, M., Fiordi, L., Cinuzzi, F., Caneschi, A., Pellegrini, D., and Buttino, I., (2018) Ecofriendly nanotechnologies and nanomaterials for environmental applications: Key issue and consensus recommendations for sustainable and eco-safe nanoremediations, *Ecotoxicol Environ Safety*, 154: 237–244; https://doi.org/10.1016/j.ecoenv.2018.02.037.

Das, S., Sen, B., and Deb-Nath, N., (2015) Recent trends in nanomaterials applications in environmental monitoring and remediation, *Environ Sci Pollut Res Int*, 22(23): 1833–1844; https://doi.org/10.1007/s11356-015-5491-6.

Defever, S. R., Geithner, K. M., Bhattchayra, P., Ding, F., Ke, C. P., and Sarupia, S., (2015) PAMAM Dendrimer and graphene materials for removing aromatic contaminants from the aqueous medium, *Environ Sci Technol*, 49(7): 4490–4497; https://doi.org/10.1021/es505518r.

Defriend, K. A., Wiesner, M. R., and Barron, A. R., (2003) Alumina and aluminate ultrafiltration membranes derived from alumina nanoparticles, *J Membrane Sci*, 224(1–2): 11–28.

Deliyanmi, E. A., Bakayannkis, D. N., Zouboutis, A. J., and Matis, K. A. (2003) Sorption of As (V) ion by Akaganeit-type – nanocrystals, *Chemosphere*, 50(1): 155–163.

Dhal, J. P., Sethi, M., Mishra, B. G., and Hota, G., (2015) MgO nanomaterials with different morphologies and their sorption capacity for removal of toxic dyes, *Mat Lett*, 141: 267–271; https://doi.org/10.1016/j.matlet.2014.10.055.

Dimitrijevic, S. B., Dimitrijevic, S. P., and Vukovic, M. D., (2013) Modern water treatment by electrochemical oxidation-A review, 17th international research /expert conference, trends in the development of machinery and associated technology, TMT-2013, Istanbul, Turkey, 10–11 September 2013; www.tmt.unze.ba/zbornik/TMT2013/017-TMT13-053.

Dror, T. D., Baram, D., and Berkowitz, B. (2005) Use of nanosized catalysts for the transformation of chloro-organic pollutants, *Environ Sci Technol*, 39(5): 1283–1290.

Farouq, R., and Yousef, N. S., (2015) Equilibrium and kinetics studies of adsorption of Cu (II) ions on natural biosorbent, *Int J Chem Eng Appl*, 6(5): 319–324.

Feng, X., Fryxell, G. E., Wang, L-Q., Kim, A. R., Liu, J., and Kemner, K. M., (1997) Functionalized monolayered mesoporous support, *Science*, 276(5314): 923–926; https://doi.org/10.1126/science.276.5314.923.

Fugetsu, B., Satoh, S., Shiba, T., Mizutani, T., Lin, Y. B., Terui, N., et al., (2004) Caged multiwalled carbon nanotubes as adsorbents for affinity-based elimination of ionic dyes, *Environ Sci Technol*, 38: 6890–6996.

Gautam, R. K., and Chattopadhaya, M. C., (2016) Advanced oxidation process based nanomaterials for remediation of recalcitrant pollutants, Chapter-3, In the book, Nanomaterials for wastewater remediation, Butter worth-Heinemann; https://doi.org/10.1016/B978-012-804609-8-00015-s.

Govindhan, M., and Jin, W., (2017) Nanomaterials based electrochemical sensors and biosensors a platform for environmental applications, *Environ Anal Chem*, 13: 10–23; https://doi.org/10.1016/j.teac.2017.02.001.

Hallman, A. M., and Bhattacharya, D. (2004) Pore assembled multilayered of a charged polypeptide in microporous membranes for ion separation, *Langmuir*, 20(13): 5418–5424.

Hamal, D. B., Haggstrom, J. A., Marchin, G. L., Ikenberry, M. A., Hohu, K., and Klabunde, K. J., (2010) A multifunctional biocide/sporocide and photocatalyst based on TiO_2 codoped with silver, carbon and sulfur, *Langmuir*, 26(4): 2805–2810; https://doi.org/10.1021/la902844r.

Hernández-Ramírez, F., Tarancon, A., Casals, O., Arbiol, J., Romano-Rodiguez, A., and Moranto, J. R., (2007) High response and stability in CO and humidity measure using a single SnO2 nanowire Sense, *Actuatotor B: Chem special issue: 25th Anniversary Sense Actuators B Chemical*, 121: 2–17; https://doi.org/10.1016/j.snb. 2006.09.015.

Hijiri, M., Mir, L., Leonardi, S. G., Donaton, N., and Neri, G., (2013) CO and NO_2 selective monitoring by ZnO based sensors, *Materials*, 3(3): 357–369; https://doi.org/10.3390/nano.3030357.

Hirjiri, M., Dhahri, R., ElMir, L., Leonordi, S. G., and Neri, G., (2015) Excellent CO gas sensor based on Ga-doped-ZnO nanoparticles, *J Material Sci, Materials in Electronics*, 26(8): 6020–6024; https://doi.org/10.1007/s10854-015-3178-x.

Huang, H., Zhou, J., Liu, H., Zhou, Y., and Feng, Y., (2010) Selective photoreduction of nitrobenzene to Aniline on TiO_2 nanoparticles modified with amino acids, *J Hazardous Mat*, 178(1–3): 994–998; https://doi.org/10.1016/jhazmat.2010.02.037.

Ian Sofian, Y., Harwin, Kuruniawan, A., Dandy, A., and Antonio, I., (2012) Nanotechnologies in water and air pollution treatment, *Environ Technol Rev*, 1(1): 136–148; https://doi.org/10.1080/21622515.2012.733966.

Ion, A. C., Culetu, A., Gherase, D., Florinela, S., and Ion, A., (2014) Environmental applications of carbon-based nanomaterials: Acetylcholinesterase biosensors for organophosphate pesticide analysis, In *Book: New Applications of Nanomaterials*, Eds: Ion, A. C., Dan, D., Gabriela, C., and Magdalena, K. C., Edition-Academic Book Publisher, Romaina, http://www.nanolink.net/mne/volume22/vol/2022/MNE.pdf.

Jayavarthanan, R., Nanda, A., and Bhatt, M. A., (2017) The impact of nanotechnology on environment, Source title: Integrating biologically inspired nanotechnology into Medical practice; https://doi.org/10.4018/978-1-5225-0160-2en007.

Kamtsikakis, A., Kevetsou, E., Chronaki, K., Kiosidou, E., Pavatou, E., Karana, A., Papaspyrides, C., Detsi, A., Karantonis, A., and Vouyiouka, S., (2017) Encapsulation of antifouling organic biocides in poly(lactic acid) nanoparticles, *Bioengineering*, 4(4): 81; https://doi.org/10.339o/bioengineering4040081.

Kelly, S. D., Kemmer, K. M., Fryxell, G. E., Liu, J., Mattgod, S. V., and Ferris K. F., (2001) X-ray absorption fine structure spectroscopy study of the self-assembled monolayer mesoporous support, *J Physchem B*, 105(27): 6337–6346.

Khim, M. M., Nair, A. S., Babu, J., Murugan, R., and Ramkeishna, S., (2012) A review on nanomaterials for environmental remediation, *Energy Environ Sci*, 5: 8075–8109; https://doi.org/10.1039/c2ee21818f.

Kwoka, M., Lyson-Sypien, B., Kulis, A., Muslyk, M., Borysoemicz, M. A., Kaminska, E., and Szuber, J., (2018) Surface properties of nanostructured, porous ZnO thin film prepared by direct current reactive magnetron sputtering, *Materials*, 11(1):131; https://doi.org/10.3390/ma11010131.

Lahir, Y. K., (2015) Role and adverse effects of nanomaterials in food technology, *J Toxicol Health*, 2(2); https://doi.org/10.7243/2056-3779-2-2.

Lahir, Y. K., (2017) Impacts of fullerene on a biological system, *Clin Immunol Metab Drugs*, 4(1): 48–57; https://doi.org/10.2174/221270700466617.

Lahir, Y. K., (2018) Graphene and graphene-based nano-drug delivery: A critical review, Chapter In *Applications of Targeted Nano-Drugs and Delivery Systems*, Eds: Mohapatra, S., Ranjan, S., Dasgupta, N., Mishra, R. K., and Thomas, S., Elsevier, Cambridge, MA.

Lahir, Y. K., Samant, M., and Dongre, P. M., (2016) The role of nanomaterials in the development of biosensors, *Global J Biosci Biotechnol*, 5(2): 146–163.

Lazaridis, N. K., Bokayannkis, D. N., and Deliyanni, E. A., (2005) Chromium (VI) sorptive removal from aqueous solution by nanocrystals Akaganeit, *Chemosphere*, 58(1): 68–75.

Li, L., Zhu, W., Zhang, P., Chen, Z., and Han, W., (2003) Photocatalytic oxidation and ozonation of catechol over carbon-black-modified-nano TiO_2 thin film supported on aluminum sheet, *Water Res*, 37(15): 3646–3651; https://doi.org/10.1016/S0043-1354 (03)00269-9.

Li, X., Shi, H., Li, K., and Zhang, L., (2015) Combined process of biofiltration and ozone oxidation, as the advanced treatment process for waste rescue, Frontiers of Environmental Sci and Engineering, 9(6): 1076–1083; https://doi.org/10.1007/s11783-015-0770-5.

Liang, H. C., and Noworty, J., (2010) Photocatalytic properties of TiO_2 nanotubes, *Solid State Phenomena*, 162: 295–328; doi:10.4028/www.scientific.net/SSP.162.295.

Lin, Y., Kiskum, S. K., Yantasee, W., Wu, H., Mattigod, S. V., Vorpargel, E., Feryxell, G. E., Raymond, K. N., and Xu, J. (2005) Incorporation of hydropiridinone ligands into a self assembly monolayered on mesoporous support for selective actinide sequestration, *Environ Sci Technol*, 39(5): 1332–1337.

Liu, C-Y., Chen, C-F., and Leu, J-P., (2009) Fabrication and CO sensing properties of meso-structured ZnO gas sensor, *J Electrochem Soc*, 156: j16–j19; https://doi.org/10.1149/1.3021044.

Long, R. Q., and Yang, R. T., (2001) Carbon nanotubes as a superior sorbent for nitrogen oxides, *Int Eng Chem Res*, 40: 4288–4291; https://doi.org/10.1021/ie/000976k.

Lou, W., Cai, W., Li, P., Su, J., Zheng, S, Zhang, Y., and Jin, W., (2018) Additive-assisted electrodepositing of fine spherical copper powder from the sulfuric acid solution, *Powder Technol*, 326: 84–88; https://doi.org/10.1016/J.powtec.2017.12.060.

Mackenzie, K., Bleyl, S., Georgi, A., and Kopinke, F-D., (2012) Carbo-iron – An Fe/AC composite as an alternative to nano-iron for groundwater treatment, *Water Res*, 46(12): 3817–3826; https://doi.org/10.1016/j.waters.2012.04.013.

Mackenzie, K., Bleyl, S., Kopinke, F-D., Doose, H., and Burns, J., (2016) Carbo-iron as improvement of the non-iron technology; from laboratory to the field test, *Sci Total Environ*, 563–564: 641–648; https://doi.org/10.1016/j.scitotenv.2015.07.107.

Magnum, C. L., Yue, Z. R., Economy, J., Maloney, S., Kemme, P., and Cropek, D. (2001) Adsorption of organic contaminants from water using tailored activated carbon fibers, *Carbon Chem Mater*, 13: 2356–2360.

Makarova, O. V., Rajh, T., Thurnauner, M. C., Martin, A., Kemme, P. A., and Cropek, D. (2000) Surface modification of TiO_2 nanoparticles for photochemical reproduction of nitrobenzene, *Environ Sci Technol*, 34(22): 4797–4803; http://dx.doi.org/10.1021/esDOI/1109.http://dx.doi.org/.

Mansoori, G. A., Rohani-Bastani, T., Ahmadpour, A., and Eshagi, Z. (2008) Environmental applications of nanotechnology, *Annual Rev of Nano Res*, 2: 1–73, chapter 2, World Scientific Publishing Company, ISBN 10:9812790225; ISBN 13:97819812790224.

Mehndiratta, P., Jain, A., Srivastava, S., and Gupta, N. (2013) Environmental pollution and nanotechnology, *Environ Pollut*, 2(2): 2013; http://dx.doi.org/10.5539/ep.v2n2p49.

Meyer, D. E., Wood, K., Bachas, L. G., and Bhattacharya, D., (2004) Degradation of chlorinated organic compound by means of membrane-immobilized nanosized metals, *Environ Prog*, 23(3): 232–242.

Mijangoes, S., Harnandez, R., and Martin, J., (2016) A review on the progress of polymer nanostructures with modulated morphologies and properties using a nonporous anodic aluminum oxide film template, *J Prog Polym Sci*, 54–55: 148–182.

Mireles, M., and Gabroski, T. R., (2017) Fabrication techniques enabling ultrathin nanostructured Membranes for separation, *Electrophoresis*, 1–14; https://doi.org/10.1002/elps.201700114.

Monosik, M., Stredansky, M., and Sturdik, E., ($2012_{1\text{-}a}$) Biosensors classification, characterization and new trends, *Acta Chimica Slovaca*, 5(1): 109–120.

Moreno, N., Qurol, X., and Ayora, C., (2001) Utilization of zeolite synthesized from coal fly ash for purification of acid mine water, *Environ Sci Technol*, 35: 3526–3534.

Muff, J., and Sogaard, E. G., (2010) Electrochemical degradation of PAH compounds in a process water: A kinetic study on model solutions and proof of concept study on runoff water from the harbor sediment purification, *Water Sci Technol*, 61(8): 2043–2051; https://doi.org/10.2166/wst.2010.129.

Nassar, M. Y., Mohamed, T. Y., Ahmed, I. S., and Samir, I., (2017) MgO nanostructure via sol-gel combustion synthesis method using different fuels: An efficient nano-adsorbent for removal anionic textiles dyes, *J Molecular Liquids*, 225: 730–740; https://doi.org/10.1016/j.molliq.2016.10.135.

Peng, X., Li, Y., Luan, Z., Di, C., Wang, H., and Tian, B. (2003) Adsorption of 1.2- dichlorobenzene from water to carbon nanotubes, *Chem Phys Lett*, 376(1): 154–158.

Peng, X., Luan, Z., Ding, J., Di, Z., Li, Y., and Tian, B. (2005) Ceria nanoparticles supported carbon nanotubes for the removal of arsenic in water, *Mater Lett*, 59: 399–403.

Popova, M., Ristic, A., Mavrodinova, V., Maucec, D., Mindizova, L., and Tusar, N. N., (2014) Design of cobalt functionalized silica with interparticle mesoporosity as a promising a catalyst for volatile organic compounds decomposition, *Catal Lett*, 144(6): 1096–1100; https://doi.org/10.1007/s10562-014-1246-0.

Rachid, T., (2011) Dendrons, dendrimers new material for environmental and science applications, *J Mater Environ Sci*, 2(30): 201–214.

Ramesh, G. K., and Sampath, S., (2011) In situ formation of graphene-lead-oxide composite and its use in trace 'As' detection, *Sen Actuators B Chem*, 160(1): 306–311; https://doi.org/10.1016/j.snb. 2011.07.053.

Reza, K. D., Azadeh, A., Maryam, N., Golnaz, R., and Morteza, A. A., (2013) Biosensors: Functions and applications, *J Biol Today's World*, 2(1): 53–61; www.biology.cnbjournals.com.

Rowley-Neale, S. J., Randviir, E. P., Abu Dena, A. S., and Bank, C.E., (2018) An overview of recent applications of reduced graphene oxide as a basis of electrochemical sensing platforms, *Appl Mater Today*, 10: 218–226; http://doi.org/10.1016/j.pml.2017.11.010.

Sahoo, P., Panighrahy, B., Sahoo, S., Satat, A., Li, D., and Bahadur, D., (2013) In situ synthesis and properties of reduced graphene oxide/Bi nanocomposites: As an electroactive material for analysis of heavy metals, *Biosens Bioelectron*, 43: 293–296: https://doi.org/10.1016/j.bios.2012.12.031.

Saini, R., Saini, S., and Sharma, S. (2010) Nanotechnology: The future medicine, *J Cutan Aesthet Surg*, 3(1): 32–33; https://dx.doi.org/10.4103/0974-2077.63301.

Salipira, K., Mamda, B. B., Kranse, R. W., Malfetse, T. J., and Durback, S. H., (2007) Carbon nanotubes and cyclodextrin polymers for removing organic pollutants from water, *Environmental Chem Lett*, 5: 13–17; https://dx.doi.org/101007/s10311-006-0057-y.

San, T., and Rogach, A. L., (2010) Nano-spherical noble metals nanoparticles: Colloid-chemical Synthesis and morphology control, *Adv Mater*, 22: 1781–1804.

Sandeep, K., Bhajana, G., Dilbaghi, N., Kumar, A., and Umar, A., (2016) Fabrication and characterization of highly sensitive and selective arsenic sensors based on ultra-thin graphene oxide nanosheets, *Sen Actuators B Chem*, 227: 29–34; https://doi.org/10.1016/j.snb. 2015.11.101.

Savage, N., and Diallo, M. S., (2005) Nanomaterials and water purification: Opportunities and challenges, *J Nanoparticles Res*, 7: 331–342; https://dx.doi.org/10.1007/s11051-005-7523-5.

Serenko, O., Strashnov, P., Kapustin, G., Kalini, M., Kuchinika, N., Serkova, E., Shirfrina, Z., and Mazafarov, A., (2017) Adsorption properties of pyridyphenylene dendrimers, *Royal Soc Chem Adv*, 7: 7870–7875.

Shahbasi, A., Younesi, H., and Badiei, A., (2013) Batch and fixed-bed column adsorption of Cu (II), Pb (II) and Cd (II) from aqueous solutions on to functionalized SBA-15 mesoporous silica, *Can J Chem Eng*, 91(4): 739–750.

Shirvastava, A., Shrivastava, O. N., Talpatra, S., Vajtai, R., and Ajayan, P. M., (2004) Carbon nanotubes filters, *Nature Mater*, 3(9): 610–614; https://dx.doi.org/10.1038/nmat1192.

Singh, N. B., Rai, S., and Agarwal, S., (2014) Polymer nanocomposites and Cr (IV) removal from water, *NanoSci Technol*, 1(1): 10; http://dx.doi.org/10.15226/2374-8141/1/100104.

Sinha, A. K., and Suzuki, K., (2007) Novel mesoporous chromium oxide for VOCs elimination, *Appl Catal B, Environ*, 70: 417–422; https://dx.doi.org/10.1016/j.apctb.2005.10.035.

Sipley, H. J., Engates, K. E., and Guttner, A. M., (2010) Study of iron oxide nanoparticles in the soil for remediation of arsenic, *J Nanoparticles Res*, 13: 2387–2397; http://dx.doi.org/10.1007/s11051-010-9999-x.

Sondi, I., and Salopek-Sondi, B., (2004) Silver nanoparticles as antimicrobial agent: A case study on E-coli as a model for Gram-negative bacteria, *J Coll Interf Sci*, 275(1): 177–182.

Srijanto, B. R., Reyyerer, S. T., Fowlkers, J. D., and Doktycz, M. J. (2010) Nanostructured silicon membranes for control of molecular transport, *J Vac Sci Technol B, Nanotechnol Microelectronics: Mat Proc, Measurement Phenomena*, 28: C6P48 (2010); https://dx.doi.org/10.1116/1.3518911.

Stanton, B. W., Harris, J. J., Miller, M. D., and Bruening, M. L., (2003) Ultra-thin multilayered polyelectrolyte film as nanofiltration membranes, *Langmuir*, 19(17): 7038–7042.

Stefano, L. D., Moretti, L., Rendina, I., and Rotirpti, L., (2005) Pesticide detection in water and humic solution porous silicon technology, *Sens Actuators-B, Chem*, 111–112(11): 522–525; https://doi.org/10.1016.j.snb.2005.03.007.

Stefanescu, E. A., Daranga, C., and Stefanescu, C., (2009) Insight into the broad field of polymer nanocomposites from carbon nanotubes to clay nanoplatelets via metal nanoparticles, *Materials*, 2(4): 2095–2153; https://10.3390/ma2042095.

Stoimenov, P. K., Klinger, R. L., Marchin, G. L., and Klabunde, K. J., (2002) Metal oxide nanoparticles as bactericidal agents, *Langmuir*, 18(7): 6679–6686.

Su, S., Chen, S., and Fan, C., (2017) Recent advances in 2-D nanomaterials-based electrochemical sensors for environmental analysis, *Green Energy Environ*, xx2017: 1–10; https://dx.doi.org/10.1016/j.gee.2017.08.005 (Science Direct).

Su, S., Wu, W., Gao, J., Lu, J., and Fan, C., (2012) Nanomaterials based sensors for applications in environmental monitoring, *J Mat Chem*, 22(35): 1801–18110; https://dx.doi.org/10.1039/czJM33284A.

Sugunumm, A., Thaachayanont, C., Dutta, C., and Hilborn, J. C., (2005) Heavy metal ion sensors using chitosan capped gold nanoparticles, *Sci Technol Adv Mat*, 6(3): 335–340; https://dx.doi.org/101.1016/j.stam.2005.03.007.

Sureshkumar, K., Deep, A., Kim, K-H., Sureshkumar, K., and Yoon, H-O., (2017) Nanomaterials based electrochemical sensors for as, a review, *Biosens Bioelectron*, 95: 106–116; https://doi.org/10.1016/j.bios.2017.04.013.

Survade, S. P., Smirnov, S. N., Vlasslouk, I. V., Unocic, R. R., Veith, G. N., Dai, S., and Mahurin, S. M., (2015) Mechanism of ion adsorption to aqueous interfaces: Graphene/water v/s air/water, *Nat Nanotechnol*, 10: 459–464.

Suslick, K. S., (1990) Sonochemistry, *Science*, 247: 1439–1445.

Swathi, R. S., and Sebastian, K. L., (2008) Molecular mechanism of heterogeneous catalysis, *Resonance*, 13(6): 548–560.

Thingavimoghaddam, J., Knowles, G. P., and Chaffec, A. L., (2013) Impact of the preparation method of SBA-15 supported low cobalt contents composites; St. and Catalytic activity, *J Mol Catal Chem*, 377: 115–122; https:// doi.org/10.1016/j.molcata.2013; 04031.

Thugittiplakorn, W., Cohen, C., and Lion, L. W., (2005) Engineered polymeric nanoparticles for Bioremediations of hydrophobic contaminants, *Environ Sci Technol*, 39(5): 1354–1358.

Thugittiplakorn, W., Lion, L. W., Cohen, C., and Kim, J. Y., (2004) Engineered polymeric nanoparticles for soil remediation, *Environ Sci Technol*, 38(5): 160601610.

Trewyn, B. G., Giri, S., Slowing, J. J., and Lin, V. V-Y., (2007) Mesoporous silica nanoparticles based controlled release, drug delivery, and biosensors system, *Chem Commun*, 3236–3245; https://dx.doi.org/10.1039/B701744H.

Tsai, C-C., and Teng, H., (2008) Chromium doped TiO_2 thin film photoanodes in visible-light-induced Water cleavage, *Appl Surf Sci*, 254(15): 4912–4918; https://dx.doi.org/10.1016/j.apsusc.2008.01.140.

UN-ESCAP, (2010) *Economic and Social Commission for Asia and Pacific Commission*, New Delhi.

Wang, J., Chen, X., Wu, K., Zhang, M., and Huang, W., (2016) Highly sensitive electrochemical sensor for Cd++ and Pb++ based on a synergistic enhancement of exfoliated graphene nanosheets bismuth, *Electroanalysis*, 28(1): 63–68; https://doi.org/10.1002/elan.201500447.

Wei, J., Du, H., Zhang, S., and Zhang, Y., (2016) Electrochemical processes for environmental remediation of toxic Cr (VI): A review, *Electro Chimica Acta*, 191: 1046–1053; https://doi.org/10.1016/j.elecacta.2016.01.130.

Xue, Y., Zhang, S., Sun, Z., Zhang, Y., and Jin, W., (2017) Alkaline electrochemical advanced oxidation process for chromium oxidation at graphitized MWCNTs, *Chemosphere*, 183: 156–163; https://dx.doi.org/10.1016/j. chemosphere.2017.05.115.

Yantasee, W., Lin, Y., Feryxell, G. E., Busche, B. J., and Brinbaum, J. C., (2003) Removal of heavy metals from aqueous solution using novel nanoengineered sorbents: Self-assembled carbamoyl-phosphonic acid on mesoporous silica, *Sci Technol*, 38(15): 3809–3825.

Yehia, F. Z., Eshaq, G., and El-Metwaley, A. E., (2015) Role of surface modification of some metal oxides with amino acids are upgrading the sonsocatalytic degradation of nitrobenzene, *Desalin Water Treat*, 56(8): 2160–2167; https://doi.org/10.1080/19443994.2014.958541.

Zaytseva, D., and Neumann, G., (2016) Carbon nanomaterials: Production, impact on plant development, agricultural and environmental applications, *Chem Biol Technol Agric*, 3: 17; https://dx.doi.org/10.1186/s40538-016-0070-8.

Zhang, J., Lin, X., Blume, R., Zhang, A., Schlogl, R., and Su, D. S., (2008) Surface modified carbon nanotubes catalyze oxidative dehydrogenation of n-butane, *Science*, 322(5898): 73–76; https://dx.doi.org/10.1126/science.1161916.

Zhang, J., Liu, J., Peng, Q., Wang, X., and Li, Y., (2006) Nearly monodisperse Cu_2O and CuO nano-spheres: preparation and applications for sensitive gas sensors, *Chem Mater*, 18(4): 867–871; https://dx.doi.org/10.11021/cm052256f.

Zhang, W. X., (2005) Nanoscale iron particles for environmental remediation, *J Nanopart Res*, 5: 323–332.

Zhao, H. T., and Nagy, K. L. (2004) Dodecyl sulfate hydrotalcite nanocomposites for tramping chlorinated organic pollutants in water, *J Coll Interf Sci*, 274(2): 613–624.

8

Climate Change Impact on Livestock Reproduction

S. Mondal, Ramanuj Banerjee, Mamoni Banerjee, S. Nandi, IJ Reddy, PSP Gupta, A. Mor and PK Agarwal

CONTENTS

8.1 Introduction

Climate change has recently become one of the most serious long-term formidable challenges to be faced by livestock owners and farmers of our country. Higher environmental temperatures are the major intriguing factor responsible for reduced fertility in farm animals (Ealy et al., 1995). Heat stress has been found to affect endometrial function (Malayer et al., 1988) besides compromised follicular growth (Badinga et al., 1985), hormonal secretion (Wolfenson et al., 1995), uterine blood flow (Roman-Ponce et al., 1978), embryonic function (Ealy et al., 1993) and oocyte developmental potential (Al-Katanani et al., 2002). Several reports indicated that heat stress caused a decrease in sperm motility along with increased production of dead sperm cells or sperm cells with morphological abnormalities (Abdul Niyas et al., 2015; Takahashi, 2012; Hansen, 2009; Marai et al., 2002; Setchell, 1998). Oocyte growth, protein synthesis or formation of transcripts required for subsequent embryonic development was affected by heat stress. The nuclear or cytoplasmic maturation process is disrupted by exposure to heat stress (Roth and Hansen, 2005). It has been reported that heat shock of 42°C for 30 minutes is able to alter the zona pellucida structure and the microvillus pattern of the oolemma of the bovine oocyte (Suzuki et al., 1998). Direct application of heat stress at 41°C for the first 12 hours of meiotic maturation reduced de novo protein synthesis by approximately 30%, coincident with 65% heat-induced reductions in blastocyst development (Edwards and Hansen, 1996). Similar exposure of murine oocytes to 43°C immediately after removal from ovarian follicles (i.e., at germinal vesicle [GV]–stage) reduced de novo protein synthesis by 30% to 80% in a duration-dependent manner (Curci et al., 1987). Moreover, heat stress of matured (i.e., metaphase II [MII]) murine oocytes at 43°C for 1 hour resulted in an approximately 85% reduction in de novo protein synthesis (Hahnel et al., 1986). Perturbations of this magnitude are concerning as protein synthesis is critical for meiotic maturation (i.e., progression from GV-stage to MII) and subsequent development after fertilization (Sirard et al., 1989). Environmental pollutants or environmental chemicals or endocrine disruptors or endocrine active compounds comprise a wide variety of synthetic or natural chemicals found in the environment arising from anthropogenic, industrial, agricultural and domestic sources that may interfere with the natural regulation of the endocrine system by either mimicking or blocking the function of endogenous hormones (sex steroids, e.g., androgens and estrogens), and they may act directly on gene expression by means of epigenetic modifications (Schug et al., 2011; Uzumcu et al., 2012). These chemicals have been shown to affect cell survival, follicle formation and growth, as well as steroidogenic functions in gonads. This chapter highlights the impact of climate change on sperm and oocyte function, embryo development and exposure to environmental chemicals on biodiversity and ecosystems, which are used to develop the paradigm for climate resilience in livestock.

DOI: 10.1201/9781003186304-8

8.2 Impact of Climate Change on Male Reproduction

High environmental temperatures disrupt reproductive processes in male animals, with the most pronounced consequences being reduced quantity and quality of sperm production. In most mammals, the testes are located outside the body cavity in the scrotum allowing a testis temperature between 2 and 8 degrees below the core body temperature (Alvarez and Storey, 1985; Takahashi, 2012). Different mechanisms exist for dissipating or increasing heat from the testis, such as the contraction or relaxation of the external cremaster muscle and tunica dartos, and the extensive vascularization next to the spermatic cord (Chenoweth and Lorton, 2014; Hansen, 2009). An increase of the testicular temperature can induce testicular degeneration and a reduction of sexual desire resulting in sub- or infertility (Abdul Niyas et al., 2015; Marai et al., 2002). Marai et al. (2002) and Setchell (1998) observed a decrease in testicular weight along with a decrease of the volume and sperm concentrations of the ejaculates due to acute or chronic heat stress. A decrease in sperm motility coupled with an increase in the production of dead sperm cells or sperm cells with morphological abnormalities in the ejaculate was also observed following heat stress (Abdul Niyas et al., 2015; Marai et al., 2002; Setchell, 1998). Hansen (2009) and Takahashi (2012) had observed an increase in abnormalities in chromatin structure in the ejaculates. However, Taylor et al. (1985) showed a minimal effect on sperm production following exposure to extreme temperatures (–24°C to −19°C and 27–32°C). The main reason for disruption in sperm production of male buffalo raised in the Amazon basin is heat stress, which results in testicular degeneration, decreased sperm motility and increased number of abnormal sperm in the ejaculated semen (Ohashi, Sousa and Vale et al., 1988). A reduction in scrotal circumference, testicular size and weight in the subtropics were found in buffalo during the summer as compared to those of the same breeds of buffalo in temperate climatic conditions (Yarny et al., 1990). The significant seasonal difference in semen characteristics on sperm quality has been reported in several studies (Cardozo et al., 2006; Balic et al., 2012; Mishra et al., 2013; Rahman et al., 2013; Bhakat et al., 2014). The seasonal changes in testicular volume, hormonal profiles, sexual behavior and semen quality have been reported by Cardozo et al. (2006). Semen quality parameters decreased in Bos taurus bulls following heat stress (Balic et al., 2012). As air temperature increased from 10°C to 18°C to more than 35°C, the membrane integrity of fresh spermatozoa in four different breeds of bulls (Crossbred, Red Sindhi, Haryana and Jersey) were affected significantly ($P<0.01$) (Mishra et al., 2013). The fertilization rate significantly decreased in heat stress (HS) spermatozoa as compared to non-HS or normal control spermatozoa (53.7% vs. 70.2% or 81.5%, respectively; Rahman et al., 2013). Bhakat et al. (2014) reported that optimal semen qualities during winter, poor during summer and intermediate during the rainy season and concluded that hot-dry or summer season adversely affect the various biophysical characteristics of semen in Karan Fries bulls.

8.3 Impact of Climate Change on Female Reproduction

a. Oocyte Maturation

Heat stress affects the fertility of various livestock species during extreme summer months. The impairment of reproduction is mainly because of failure of the animal to cope with heat stress, which can affect the functioning of the germ cells and the viability of an early developing embryo (Hansen, 2009). Heat-induced infertility has been observed in females of many species experiencing hyperthermia near the time of breeding when the oocyte is undergoing maturation (Stott and Williams, 1962). However, effects on fertility depend on the duration and severity of the heat stress imposed (Cavestany et al., 1985). Putney et al. (1989) reported that exposure of 42°C and 75% relative humidity for 10 hours to super-ovulated heifers at the onset of estrus increased respiration rates by more than 200% and rectal temperatures from 38.9°C to 41.3°C (Putney et al., 1989). This heat stress enhanced the proportion of retarded embryos and reduced embryo quality 7 days after artificial insemination but did not alter the fertilization rate (Putney et al., 1989). Similarly Dutt (1963) also reported the proportion of abnormal oocytes following exposure of sheep to high temperature and humidity (90°F to 95°F and 60 to 65% relative humidity) at the time of breeding. There was an increased arrest of oocytes at metaphase I (MI) and increased oocyte degeneration and fragmentation when mice were exposed to 35°C and 65% relative humidity for 15.5 hours (all of oocyte maturation) at the time of hCG administration (to induce resumption of meiosis). Exposure of the oocyte to heat stress at the germinal vesicle stage or during early stages of maturation results in deleterious effects on the internal organelles so that the capacity of the oocyte for fertilization and further embryonic development is reduced. The membrane integrity was not altered by exposure of oocytes to physiological heat stress (41°C) for 12 hours during maturation. Tseng et al. (2004) suggested an alteration in cytoskeleton structure in mature bovine oocytes and two-cell bovine embryos exposed to temperatures of 41°Cto 41.5°C. The Golgi system and the endoplasmic reticulum (ER) become fragmented under stress conditions, and the number of mitochondria and lysosomes decreases (Welch and Suhan, 1985), which leads to a dramatic drop in adenosine triphosphate (ATP) levels during heat stress (Lambowitz et al., 1983). Alteration in the ratio of protein to lipids and an increased fluidity of the membranes were observed (Kruuv et al. 1983). There was a drop in cytosolic pH and changes in ion homeostasis due to these membrane changes (Coote et al., 1991). It has been observed that cellular exposure to heat stress may induce oxidative stress. The mechanisms by which heat stress generates oxidative stress may include a reduction of antioxidant activity in the plasma of lactating cows (Harmon et al., 1997). A decrease in plasma lipid-soluble antioxidants (vitamin E and β-carotene) was observed in moderately heat-stressed cows (Calamari et al., 1999). Cellular exposure to heat stress has been found to increase the production of reactive oxygen species (ROS) and/or the promotion of cellular oxidation events (Skibba et al., 1991). ROS, such as superoxide anions ($O2^-$), hydroxyl radicals (OH^-), and hydrogen peroxide (H_2O_2) can cause lipid peroxidation and enzyme inactivation, resulting in cell damage by promoting hydroxyl radical formation (Ribarov and Benov, 1981).

Reduced embryo development was observed following exposure of oocytes to elevated temperature during in vitro maturation which may be due to a direct impact on the maturing oocytes. In addition, the timing of exposure to elevated temperature during oocyte maturation is important: The effects of culture at 41°C for 12 hours on blastocyst development after fertilization were more pronounced during the first half of maturation than the last 12 hours (Edwards and Hansen, 1996). Edwards et al. (2009) reported reduced blastocyst development (30%–65%) when the bovine oocytes were cultured at 41°C for the first 12 hours due to a reduced ability of embryos derived from heat-stressed oocytes to undergo compaction. The reductions in development are not due to an inability of embryos to undergo early cleavage divisions (Edwards and Hansen, 1996). The ability of embryos derived from heat-stressed oocytes to cleave decreased after exposure of maturing bovine oocytes to 41°C for the first 12 hours of maturation with simultaneous alterations in pH due to increased (7%) CO2 concentrations (Roth and Hansen, 2004a, 2004b).

b. Embryo Development

Environmental stress has a lethal effect on the embryos both in vivo and in vitro. The demise of preimplantation embryos increased following exposure of heat stress to the pregnant mammals. The development of mammalian pre-implantation embryos was impaired by elevated ambient temperatures (Zavy, 1994). There was a reduction in the embryonic developmental rate and hatched blastocyst rate reduced when embryos were exposed to high temperatures in vitro. The proportion of cultured two-cell embryos that developed to the blastocyst stage was reduced by heat shock than heat shock of four- to eight-cell embryos. Early embryos (<8- to 16-cell stage) would be more susceptible to heat shock because these embryos are transcriptionally quiescent and unable to produce protective molecules such as heat-shock protein 70 (HSP70) in response to heat shock. No deleterious effects were observed on subsequent blastocyst formation after in vitro fertilization when bovine oocytes were exposed to heat shocking for 1 hour at 40°C to 42°C had (Ju et al., 1999). However, the blastocyst formation rate decreased when heat shock is sustained for 12 hours at 41°C. The developmental competence of the treated oocytes was severely reduced when the temperature was increased to 43°C for 45 minutes (Ju et al., 1999). No effect on further development was observed for two-cell bovine embryos following a 3-hour heat shock at 40°C, but developmental competence decreased when heated at 41°C to 42°C for a longer period (Ealy et al., 1995). A negative relationship between embryo survival to term and pregnancy rate with increased rectal temperatures (37.5–40°C) was observed (Ulberg and Burfening, 1967). A positive correlation between rectal temperatures and percentage of retarded or abnormal embryos was reported when bovine embryos were recovered on day 7 after insemination, (Putney et al., 1988). A negative regression coefficient with elevated rectal temperatures (39–41°C) was found between the survival and implantation rates of early embryos of maternally stressed mice (Elliott and Ulberg, 1971).

Heat-induced reductions in embryo development do not appear to be due to an inability of the oocyte to mature at the nuclear level. The ability of oocytes to progress to MII was not altered by exposure of bovine oocytes to 41°C for 12 or 24 hours of maturation (Dorado et al., 2001). However, polar body extrusion was reduced by exposure of bovine oocytes to 41°C during maturation (Lenz et al., 1983). Heat stress had been found to reduce the ability of oocytes to progress to MII after heat stress due to increased arrest at MI (Roth and Hansen, 2005). Exposure of murine oocytes to 41°C (4°C above normal) for 17 hours had been found to inhibit completely meiotic maturation while temperatures of 40°C or higher reduced normal chromosome spread and increased abnormal ploidy (Fiorenza and Mangia, 1992). A decrease in type II and increase in type III cortical granules was observed in bovine ova exposed to 41°C during the first 12 hours of maturation as compared to controls at 24 hours (Edwards et al., 2005). Abnormal chromosome morphology and number were observed following exposure of heat stress to murine oocytes at 40°C during maturation (Fiorenza and Mangia, 1992). Meiotic spindle abnormalities (Roth and Hansen, 2005) were also observed which affect the progression of meiosis I for chromosome segregation and maturation-promoting factor (MPF) inactivation (Brunet et al., 2003). Polar bodies were absent in matured bovine oocytes after heat stress (Lenz et al., 1983) due to an inhibition of meiosis. However, the staining of nuclear material with Hoechst showed that heat-stressed bovine ova had not only the same rate of meiotic progression as controls but reached MI and MII earlier than controls (Edwards et al., 2005). When heat-stressed bovine ova (41.0°C, during the first 12 hours of maturation) were fertilized earlier, there was an increase in the number of embryos that developed into blastocysts, with the best development occurring when IVF was performed 18 hours after the onset of maturation (Schrock et al., 2007). Heat stress has been found to induce elevations in cytoplasmic ROS to reduce embryo development. Blastocyst yield was reduced by exposing maturing oocytes to elevated temperature by 20% as compared to controls and supplementation of retinol-inhibited heat-induced reductions in blastocyst development (Lawrence et al., 2004). But the levels of an intracellular antioxidant, glutathione, were unchanged in heat-stressed in vivo matured murine oocytes despite heat-induced reductions in blastocyst development (Matsuzuka et al., 2004). Edwards and Hansen (1996) reported reduced de novo protein synthesis in bovine oocytes by 40% following exposure of elevated temperature for the first 12 hour of maturation. A reduction in protein synthesis was observed following exposure of GV-stage and matured (MII) murine oocytes to elevated temperatures (Hahnel et al., 1986). Edwards and Hansen (1996) observed a reduction of HSP70, a thermoprotective molecular chaperone. An examination of global protein populations in matured bovine oocytes after 12 hours' exposure to 41°C at the onset of maturation revealed a differential abundance of four proteins, some of which should have declined during maturation but failed to do so after heat stress (West-Rispoli et al., 2006).

8.4 Climate Change and Endocrine Disruption

Climate change and exposure to environmental chemicals are two of the most serious anthropogenic threats to biodiversity and ecosystems, and to justify this, case studies of mammals and birds have been described. It was reported that climate change was likely to pose additional stress to individuals, and because different endocrine systems are important for enabling animals to respond adequately to environmental stress, environmental chemicals may interfere with adaptation to increased stress situations. Climate change alters the location and frequency of rainfall resulting in floods in some areas, and drought in others. Floods are associated with the release of sediment-bound chemicals, including environmental chemicals, into the water column (Wölz J et al., 2008), whereas droughts have led to the development and implementation of stormwater and wastewater recycling technologies that have been associated with the introduction of environmental chemicals into publicly utilized water supplies (Westerhoff et al., 2005). So both drought and flood increase human and domestic animal exposure to environmental chemicals (Brevini T et al., 2005). Currently, little is known about the effects of environmental chemicals on animal or human health, and even less is known about the consequences of exposure to waterborne environmental chemicals (Kolpin W et al., 2002).

The Environmental Protection Agency has defined environmental chemicals as "an exogenous agent that interferes with synthesis, secretion, transport, metabolism, binding action, or elimination of natural blood-borne hormones that are present in the body and are responsible for homeostasis, reproduction, and developmental process" (Diamanti-Kandarakis et al., 2009), and the International Programme for Chemical Safety (IPCS) defines endocrine-disrupting chemicals (EDCs) as "exogenous substances or mixtures that alter the function(s) of the endocrine system and consequently cause adverse health effects in an intact organism, or its progeny, or (sub) populations".

Existing human data neither confirm nor refute the hypothesis that human reproductive health is adversely affected by exposure to environmental chemicals (Falconer I R et al., 2006). It is, however, recognized that potential sources of exposure are through contaminated food and groundwater, and environmental chemicals have been found in humans (Koch et al., 2004; Younglai et al., 2004) and domestic animals. The evidence for human exposure to EDCs resulting in testicular dysgenesis syndrome is weak, primarily because studies have not been conducted rather than because there are any data to the contrary, and studies regarding environmental chemicals effects on female reproductive health are particularly sparse. Although adverse human health effects have clearly occurred after high exposures to environmental chemicals associated with poisoning incidents or other accidents, data on low, environmentally relevant levels of exposure, either to single or combined environmental chemicals, are lacking. It is these lower levels of exposure, however, that present significant risks because even small changes in fertility could potentially have a large impact at the population level (Brevini et al., 2005).

Environmental chemicals can act at multiple sites via multiple mechanisms of action (Brevini et al., 2005; Gandolfi et al., 2002), and although receptor-mediated mechanisms have been studied extensively, little is known about other equally important mechanisms such as hormone synthesis. In contrast to many other toxicants, numerous environmental chemicals do not show linear dose-response relationships (Caserta et al., 2008). Low-dose exposure sometimes causes effects opposite to the ones seen after high-dose exposure (Andrade et al., 2006), and the in vitro methodological approach we are proposing will allow elucidation of these effects. For studies of environmental chemicals, effects or mechanisms of action, farm animals are crucial and often superior to laboratory rodents (Magnusson and Persson, 2015). We showed previously that endocrine-disrupting heavy metals and pesticides caused fertilization failure or low cleavage rates in domestic buffalos (Nandi et al., 2008; Selvaraju et al., 2009) and that human reproductive cells were also susceptible to the endocrine-disrupting activity of agricultural chemicals (Young et al., 2005) and other environmental contaminants (Young et al., 2008).Ref_670_FILE150644130008

The fertilizing ability of a sperm is a complex process involving many molecular and cellular events, such as the acrosome reaction, which has to be completed before spermatozoa can fuse with the plasma membrane of the oocyte (egg; Mantovani et al., 2005; Fraser et al., 2006). Sperm morphology and motility are also good indicators of fertilization ability (Mantovani et al., 2005). The ability of environmental chemicals–exposed sperm to fertilize an egg and form an embryo in vitro (IVF) can be tested using sheep gametes but not human. We therefore propose to assess the morphology, motility and ability to undergo the acrosome reaction in both human and sheep sperm (Fraser et al., 2006) and relate these data to the sheep IVF outcomes. The DNA integrity of spermatozoa and the level of ROS are important factors in normal embryo development (Aitken, 2010), and we will therefore apply the COMET test is often used to characterize sperm DNA integrity and fragmentation.

Bisphenol A (BPA) is used in the manufacture of polycarbonate plastics and epoxy resins from which food and beverage containers and dental materials are made (Saal et al., 2007). The previous studies showed that phthalates and BPA interfered with the estrogen receptor (ER), androgen receptor (AR) and arylhydrogen receptor (AhR) in mammalian cell lines (Ghisari et al., 2009). Octylphenol (OP) and 4-n-nonylphenol (NP) are common surfactants with reproductive toxicity (Ferguson et al., 2000; Nielsen et al., 2000), and Cd and Hg also cause toxicity in reproductive systems (Yang et al., 2006; Siu et al., 2009). Steroid hormones have known effects on reproduction (Gandolfi et al., 2002; Caserta et al., 2008). Plasticizers and detergent by-products have been found in sewage effluents at concentrations that cause endocrine disruption, affecting fertility (Inglis et al., 2008; Noyes et al., 2009).Ref_594_FILE150644130008

Throughout the previous century, the production, use and, as a result, presence of chemicals in the environment increased enormously. Consequently, humans and animals are exposed to a wide variety of environmental chemicals of which some possess the ability to disrupt the endocrine system in the body, thereby they are denominated as "EDCs" or "endocrine disruptors". Because the reproductive system is a target organ for endocrine disruption, environmental chemicals are postulated as one of the possible causes of human subfertility (Petro et al., 2012). For studies of environmental chemicals, effects or mechanisms of action, farm animals are crucial and often superior to laboratory rodents (Magnusson and Persson, 2015). Several solid experimental evidences both in vivo and in vitro, demonstrate clinical effects and explain mechanisms of action of man-made endocrine disruptors on farm animal reproduction (Magnusson and Persson, 2015). In the context of the many other physical, chemical, and biological stressors that alter with climate change, it is important to identify those species and populations that are especially vulnerable to climate-pollutant interactions. It is important to develop early warning signals or biomarkers that convincingly reflect adverse biological responses toward anthropogenic environmental toxins even in minute concentrations, so it may trigger or accelerate synergistic interactions between climate change and contaminant exposures and (Kaviraj et al., 2014). A *biomarker* is defined as "[a]biological tool that responses to molecular or cellular reactions which changes physiological response to behavioral response after exposure to toxic environmental chemicals" (Peakall and Walker, 1994). Recent developments in molecular biology and biotechnology have developed novel and more sensitive biomarkers that have a high susceptibility to the adverse effects of pollutants (Galloway, 2006). Biomarkers have been a promising technique in the assessment of pollutants (Kaviraj et al., 2014). This evidence is adequate to prove that precautionary actions against excess exposure to environmental chemicals specifically active on hormonal homeostasis (Magnusson and Persson, 2015; Caserta et al., 2008). The gaps in knowledge include identifying compounds able to act as reproductive toxicants and then developing reliable and sensitive assays to test for them; ascertaining the health effects; understanding the impact of age, dose, length of exposure, timing of exposure and genetics on health effects; and learning more about the interactions between multiple synthetic chemicals and how they react in the body and the environment (Deepthi and Nandi S, 2016).

8.5 Environment Pollutants

Environmental pollutants are a threat to farm animal health, welfare and productivity (Rhind, 2005). There is a possibility that certain chemicals in the environment can adversely affect the various systems of animals. Such environmental chemicals (endocrine disrupters or xenobiotics) may be synthetic (man-made) or natural (originating in plants or microorganisms). It is unclear whether observed adverse effects linked to certain environmental pollutants are a matter of isolated incidents or part of a widespread environmental problem. As the endocrine system controls hormonal functions that influence growth, development, reproduction, and the body's chemical equilibrium, this question merits further investigation. Although these chemicals have effects on immune and thyroid dysfunctions, studies have mainly focused on development and reproduction. The effects observed in various studies on wildlife, aquatic and laboratory animals might serve as the background for raising concerns about domestic animal and human health effects. Hence, an urgent need is to address the effects of some of the most common and potent environmental pollutants on the reproductive system.

8.6 Environment and Pollution

Environment is the whole context of life—the physical, chemical and biological settings of where and how plants and animals live. Pollution is an undesirable change in the physical, chemical, or biological characteristics of an ecosystem that poses risks to animal health (and other living organisms). Pollution is the presence of a substance (i.e., pollutant) in the environment that because of its chemical composition or quantity prevents the functioning of natural processes and produces undesirable environmental and health effects. Pollutants are chemicals (inorganic, organic or radioactive) containing toxic elements or compounds above animal tolerance levels (damage of body tissues, cancer, chemical imbalance, etc.). Exposure to environmental pollutants can occur from a number of different sources: Humans and animals can be exposed involuntarily to pollutants as a result of drinking contaminated water, breathing contaminated air, ingesting food, or contacting contaminated soil. Pollutants can impact natural resources, public health and the economy. Most enter the environment from industrial and commercial facilities; oil and chemical spills; nonpoint sources such as roads, parking lots and storm drains; wastewater treatment plants; and sewage systems. Many hazardous waste sites and industrial facilities have been contaminated for decades and continue to impact the environment. Pollutants that resist breakdown and accumulate in the food chain are of greatest concern because they are consumed or absorbed by fish and animals and wildlife, which, in turn, are consumed by humans.

8.7 Endocrine System—Normal Functions and Disruption

Some major endocrine glands are the pituitary, thyroid, pancreas, adrenal, hypothalamus, testes and ovaries (Boerjan et al., 2002), and these glands have major physical functions, including reproduction, growth, and development; maintenance of normal glucose levels; blood pressure; and metabolism, that also produces hormones that travel through the bloodstream to organs and tissues and modify their functions. One of the major functions of hormones is bind to protein molecules called receptors respond to hormone signals (Janosek et al., 2006). This specific binding further responses in tissue according to its function. The binding of hormones to their receptors is also affected by certain chemicals, which shows competitive binding to their ligands with receptors and therefore alter or interfere with the normal binding of a hormone to its corresponding receptors, which ultimately leads to normal, abnormal or no function.

It has been well documented in experimental and wildlife animals that environmental pollutants have adverse effects on the normal development of male and female reproductive organs. The available data also reveal that chemical exposure may interfere with reproductive function and cause infertility, pregnancy loss, growth retardation, intrauterine fetal demise, birth defect and ovarian failure. But the actual mechanism for how toxicants can affect reproductive functions is not well understood. These effects may be due to disruption at site of action, including the hypothalamus–pituitary system and the gonads, which results in disruption of the normal pattern of gonadotrophin secretion and the destruction of gamete viability. The time of exposure of reproductive toxicants in the reproductive lifespan and in the developmental stages is also an important factor to be considered. Environmental chemicals cause destruction in germ cells during fetal development and in adult cyclic females causes temporary infertility whereas prenatal exposure to toxicants can cause anomalies in sexual differentiation that become evident only later in development. From these studies, it can be concluded that the developing embryo and fetus are more sensitive than adults to endocrine disruption and alterations in reproductive development observed at concentrations far below the one able to induce toxicity in the adult animals.

8.8 Environmental Endocrine Disruptors

Environmental endocrine disruptors (EEDs) are substances that can cause adverse effects by interfering in some way with the body's hormones or chemical messengers. Like hormones, small amounts of these chemicals (parts per billion or even trillion) affect the endocrine systems of animals. EEDs fall into three broad classes (Pocar et al., 2003; Brevini et al., 2005): (; a) synthetic chemicals used in industry, agriculture and consumer products; (b) synthetic chemicals used as pharmaceuticals drugs; and (c) phytoestrogen. Chemicals capable of acting as reproductive toxicants are ubiquitous in our environment:

- Pesticides (DDT metabolites, endosulfan, mancozeb, chloropyriphos, atrazine, nitrofen, tributyltin, etc.)
- Household products (breakdown products of detergents and associated surfactants, including nonylphenol and octylphenol)
- Plastics (bisphenol A, phthalates)
- Pharmaceuticals (drug estrogens—birth control pills, diethyl stilbesterol (DES), cimetidine)
- Industrial chemicals (polychlorinated biphenyls [PCBs], dioxin, benzo[a]pyrene)
- By-products of incineration, paper production, and fuel combustion
- Heavy metals (cadmium, lead, mercury)
- Food products (soybeans, legumes, flax, yams, clover)
- Plants (phytoestrogens are chemicals naturally found in plants that can act as endocrine disruptors and are present in fruits, veggies, beans and grasses)

Endocrine disruptors can exert their effects in many ways (Danzo, 1998; Goettlich, 2005): (a) They can either bind to the hormone's receptor and mimic the hormone; (b) they block the action of the hormone; (c) they can stimulate or inhibit the enzymes responsible for the synthesis or clearance of a hormone and thereby give rise to an increased or decreased action of the hormone. A majority of endocrine disruptors are lipophilic, resistant to metabolism and able to bioconcentrate up the food chain. These substances are stored in body fats and body fluids even in follicular fluid and seminal plasma (Kamarianos et al., 2003a, 2003b) and can be transferred to the developing offspring. These persistent EEDs are carried in air and water over several hundred miles, and so even animals living far away from where these substances are used are under significant threat. Although some of EEDs are banned and monitored for limited usage, exposure to those EEDs can still arise from a number of sources, including the mobilization of existing body burdens, from continued illegal usage, from sites contaminated in the past and from continued usage elsewhere.

Steroids are necessary for the development and function of the male and female reproductive systems, but when pollutants disrupt the normal mechanisms of steroid hormone action, this leads to reproductive problems in animals. Various pollutants act as agonists (acting as an estrogen or androgen) or antagonists (interfering with steroid hormone action), and these pollutants act through steroid-dependent mechanisms and interact with the estrogen receptor, androgen receptor or with certain steroid-binding proteins, androgen-binding protein or sex hormone–binding globulin. Steroid hormone acts through its binding to a receptor or binds to a protein followed by gene activation or suppression. Both males and females have androgens and all four binding proteins. Therefore, a given pollutant could potentially interfere with reproductive tract development in either sex. Evidence of demasculinizing effects of pollutants in wildlife species led investigators to assume that pollutants were estrogenic or worked through the estrogen receptor.

8.9 Environmental Pollutants and Female Reproduction

The reproductive cycle is a complex process characterized by gametogenesis and embryogenesis which is under the control of steroid hormones and is therefore a potential target for endocrine disruptors. The feedback mechanisms related of hypothalamus–pituitary–gonadal axis affects ovarian function and follicles, which targets female reproductive function because pollutants directly target these glands sites, such as the hypothalamus, the pituitary, the ovary and the reproductive tract. Therefore disruption of any of these sites disrupts endocrine balance, causing a reduction in estrogen and progesterone and an elevation in Follicle Stimulating Hormone (FSH) and Luteinizing Hormone (LH) and extensive oocyte destruction and ultimately infertility. The developmental stage at which exposure occurs determines the impact that the exposure will have on reproduction. Oocytes contained in primordial and primary follicles are likely to have a delayed effect on reproduction until such a time when recruitment of growing and antral follicles can no longer be supported whereas selective damage of large growing or antral follicles generally disrupts reproductive function only temporarily because these follicles can be replaced by recruitment from the greater pool of primordial follicles. The level and dose of chemicals required to cause ovarian damage are other key factors. In studies, it has been found that ovotoxic chemicals rarely affect animals, and these effects can usually be detected and evaluated, but it is difficult to evaluate chronic exposure to low levels of toxicants that silently damage the reproductive function (Sweeney, 2002). Prenatal exposure to persistent organic pollutants, such as dioxins, causes premature reproductive aging because it has been seen that the number of germ cells in the ovaries decreased 40% to 50% compared to control figures and appeared in all stages.

Oocyte destruction by environmental chemicals requires that these compounds reach the ovary. Studies also indicated a correlation between the presence of environmental pollutants and fertilization failure or low cleavage rates. One of the earlier studies investigated the effect of three widely used pesticides DDT, hexachlorocyclohexane and methoxyclor on bovine oocytes and embryos. A clear dose-dependent response on ovum growth was observed for all three compounds with an incidence of DNA degeneration. The most interesting observation was that the lowest dose affected the embryonic development. The ovary could also be a target organ for dioxin toxicity. Exposure of bovine oocytes during the maturation process to the mixture of polychlorinated biphenyls (PCBs), Aroclor-1254 (A-1254) was studied (Pocar et al., 2005). Concentrations as low as 0.01 μg/ml of A-1254 significantly decreased the percentage of oocytes that reached MII, reduced the fertilization ability of oocytes and caused a significant decrement in the proportion of cleaved embryos reaching the blastocyst stage. Exposing pig oocytes to an organochlorine mixture during in vitro maturation negatively affected the maturation rate without increasing the degeneration rate. Developmental competence of the exposed oocytes was also affected, reducing blastocyst rate and quality in a dose-dependent manner. Furthermore, there is a decrease in the quality and viability of cumulus cells, which may account for reduced maturation and developmental competence. Alkylphenolic compounds, such as ocytlphenol and others, were found to have detrimental effects on oocyte maturation and developmental competence. These substances are a widespread family of xenoestrogens present at high concentrations in sewage sludge that is spread on arable land and pasture as fertilizer. We demonstrated that chlorpyrifos and endosulfan had a negative impact on oocytes at 0.02 and 0.1 lg/mL levels, respectively. These pesticides reduced the oocyte nuclear maturation by a direct effect on oocytes, cumulus cell-mediated action, and by blocking the action of hormones. Chlorpyrifos was found to be more ovotoxic and embryotoxic than endosulfan (Nandi et al., 2010). Similarly, the effect of heavy metals on oocyte development was studied in our laboratory (Nandi et al., 2010). We demonstrated that the developmental block, degeneration and asynchronous divisions were higher in embryos exposed to cadmium than in those exposed to lead. Cadmium and lead were found to lower the viability and development of buffalo oocytes but at a concentration higher than that estimated in the body fluids and environment. Cadmium was found to be more ovotoxic than lead (Nandi et al., 2010).

Synthetic chemical substances are not the only threat to farm animals' reproductive efficiency, but natural toxins found in grazing plants or produced by molds contaminating livestock feedstuffs have also been recognized as a serious danger. More recently, in vitro studies have been used to define the toxic potential of plant alkaloids (steroidal glycoalkaloids from potatoes) and micotoxin (deoxynivalenol and zearalenol) on oocytes and embryos in cows and pigs, respectively.

Most of the studies described evaluated the toxic effects of only one class of pollutants at a time, an experimental approach essential for understanding the mechanism of action of each substance. However, an environmentally relevant mixture of more than 15 organochlorines (including PCBs, DDT, its metabolite DDE, and lindane) was used to investigate the effects of these chemicals on in vitro maturation of pig oocytes (Campagna et al., 2002). This study has a particular ecological significance as mixtures, and not individual compounds, are used industrially and accumulated in the environment. An environmentally relevant mixture provides a more realistic representation of possible exposure because it contains components that can interact with each other in an additive or non-additive fashion (synergistic or antagonistic) and can interact with different receptors and molecular pathways.

Steroids are known as important factors on the route of oocytes development and cumulus-oocyte complexes as well as follicular granulosa cells (GCs) are suggested to be involved in steroidogenesis. Fungicides were found to disturb steroidogenesis in GCs and during in vitro maturation (IVM) in cumulus-oocyte complexes. Dioxin is capable of disrupting ovarian steroid production by acting directly on the follicular granulosa cells, reducing estradiol secretion and increasing apoptosis. Furthermore, dioxin may alter cell membrane permeability, disrupting steroid secretion. The effect of dioxin on steroidogenesis may be attributable to downregulation of the gonadotrophin receptor. Data on farm animals show that dioxin decreases both progesterone and estradiol secretion by granulosa cells in pigs, in a dose-dependent manner, by altering the activity of enzymes involved in the steroid biosynthesis cascade. Two studies investigating the effects of PCBs in follicular cells in pigs have revealed that PCBs 126 and 153 have dioxin-like and non-dioxin-like activity, respectively. Both PCBs 153 and 126 alter steroid secretion in follicular cells (Wojtowicz et al., 2005). PCB 153 decreases estradiol secretion and increases progesterone secretion, indicating that it causes the disruption of the aromatization process. In contrast, PCB 126 is able to induce only the increase of progesterone with no effect on estradiol secretion by GCs. This finding differs from that previously observed for dioxin, indicating the presence of a different mechanism of action. Mixtures of pesticides, herbicides and fungicides increased the percentage of apoptosis in exposed embryos. Industrial chemicals caused ovotoxicity, and induced follicle loss occurs via a cell death process involving elevated expression of Bax, a proapoptotic Bcl-2 family member, and increased caspase-3-like activity.

In many case-control studies, it has been found that there is distinct association between adipose tissue concentrations of pollutants and udder/breast, an ovarian cancer risk. Exposure of diethylstilbestrol to rats produced several changes in the mammary gland, such as the proliferative activity being drastically altered, the cell cycle of mammary epithelial cells being perturbed and telomeric length being attenuated (Kawaguchi et al., 2009). However, studies suggest that AhR can be the target for the development of drugs and endometrial cancer and has an overall antitumorigenic effect because dioxins may activate an inhibitory effect between the AhR and the estrogen receptor in udder/breast, uterus and endometrial cancer cells. Moreover, it was also demonstrated that different pollutants have the ability to inhibit gap-junctional intercellular communication in normal human breast epithelial cells, pointing to a possible tumor-promoting potential of EDs in human breast tissue.

8.10 Pollutants and Reproductive Problems—Mechanism of Actions

Various interactions of environmental pollutants with signaling pathways of receptors were described and include either direct binding of pollutants to receptors or indirect effects mediated via the modulation of associated signaling pathways (Pocar et al., 2003; Brevini et al., 2005; Lee et al., 2013). Many receptors are physiologically activated by low-molecular-weight ligands (steroid hormones, vitamin A derivatives, thyroid hormones). Many low-molecular-weight ligands such as steroid hormones, vitamin A derivatives, thyroid hormones often display substantial structural similarities to many environmental contaminants such as polychlorinated dibenzodioxins and dibenzofurans (PCDD/Fs), PCBs, polycyclic aromatic hydrocarbons (PAHs) or phthalates. The receptors of these ligands become highly susceptible to nonphysiological modulations by contaminants, and the resulting disruption of signaling pathways was related to numerous in vivo effects, such as immunosuppressions, endocrine disruption, carcinogenesis, developmental toxicity and others.

Estrogen receptors: Environmental pollutants like phthalates were shown to bind nonphysiologically to estrogen receptors and activate their function. On the other hand, other environmental contaminants, for example, some hydroxylated PCBs, can act as "antihormones" by competitive binding to active site without the activation of estrogen receptors, thus inhibiting the function of natural estrogens. The environmental relevance of the estrogen-like or anti-estrogen biochemical mechanisms was experimentally proved by causal linking to several in vivo effects, such as male feminization or reproduction disorder, in various organisms. In laboratory experiments and field studies, it has been found that exposure to specific chemicals such astoxaphene, alkylphenol ethoxylates, phthalates, some pharmaceuticals, hydroxylated chlorinated biphenyls, methoxychlor, o,p′-DDT, some PCBs, PAHs or natural flavonoids and phytoestrogens or PCDD/Fs, PCBs elicits estrogenity or anti-estrogenity and causes adverse health effects, for example, reproduction toxicity, male testis and uterus tumors, delayed male puberty, decreased semen quality and volume and increases in developmental anomalies of the male reproductive system (Veeramachaneni, 2000; Pflieger Bruss et al., 2004),Ref_655_FILE150644130008including reduced secondary sex organ size, hypospadias, cryptochordism and enhanced susceptibility to seminomas.

AhRs: For better comparison of AhR-mediated effects and risk assessment purposes, toxicity equivalency factor (TEF) concept was accepted which represents the toxic potency of a particular compound to induce AhR-mediated effect related to the reference substance-TCDD. 2,3,7,8-tetrachloro-dibenzo-p-dio xin (TCDD) is the most potent Ah-receptor ligand which is a coplanar aromatic compound. The TEFs of TCCD and some other highly toxic congeners of polychlorinated dibenzo-p-dioxins and dibenzofurans were set to 1.0. The toxic potencies of coplanar PCBs correspond to TEF values ranging from 10^{-5} to 10^{-1}.

ARs: Both androgen-like and anti-androgenic actions of pollutants may lead to significant adverse effects. Pollutants like metabolites of fungicides like vinclozoline and las anzat, DDT and its metabolites (o,p′-DDT or p,p′-DDE), bisphenol-A, 3-hydroxy-phenylphenol or 4-hydroxy-phenylphenol, endosulfan, mirex, some PAHs, PCBs, tris-(4-chlorophenyl)-methanol act through androgen receptors.

Retinoid receptors: Pollutants such as PCBs and dioxins, act through an adverse effect of retinoid homeostasis.

Thyroid receptors: Many pollutants have been shown to cause adverse effects at multiple levels of thyroid signaling, both in vitro and in vivo. Only a few compounds, such as tetrabromo and tetrachlorobisphenol A, have been shown to bind directly to thyroid receptors. A large group of pesticides, such as DDT and its metabolites, dieldrin, pentachlorophenol, alachlor or toxaphene, also act through thyroid receptors.

8.11 Conclusion

The world is facing major crucial challenges from feeding the huge livestock population to tackling severe environmental crises, including climate change. High environmental temperatures have been found to compromise follicular dynamics, hormonal secretion, oocyte and embryonic function. Heat-induced disruption of microtubules and microfilaments altered the distribution of cellular structures, such as cortical granules and mitochondria, besides affecting chromosome segregation during fertilization and cell division of oocytes. The cellular and molecular mechanisms involved in heat-stressed oocyte death or survival following are not deciphered. The identification of novel genes involved in morula to blastocyst transition and to hatching and the analysis of their expression patterns under heat stress will explore the compromised embryonic development under environmental stress. Ensuring good animal welfare will be paramount in addressing these challenges; breeds suited to the local environment are often more robust and resilient than industrially farmed breeds.

ACKNOWLEDGMENTS

We are thankful to National Agricultural Science Fund, Ministry of Agriculture, ICAR, New Delhi for providing financial support. We thank Director, NIANP, for providing the necessary facilities for the preparation of the chapter. Special thanks are due to Dr. Asit Chakraborty, Bangalore, for taking much pain in editing the chapter. The help rendered by Uzma, Shanmugapriya, Deepa and other lab Mates is duly acknowledged.

REFERENCES

Abdul Niyas, P., Chaidanya, K., Shaji, S., Sejian, V., Bhatta, R., Bagath, M., Rao, G., Kurien, E., and V. Girish. 2015. Adaptation of livestock to environmental challenges. *Journal of Veterinary Science and Medical Diagnosis* 4:1–7.

Aitken, J. 2010. Whither must spermatozoa wander? The future of laboratory seminology. *Asian Journal of Andrology* 12:99–103.

Al-Katanani, Y.M., and P.J. Hansen. 2002. Effect of season and exposure to heat stress on oocyte competence in Holstein cows. *Journal of Dairy Science* 85:390–396.

Alvarez, J.G., and B.T. Storey. 1985. Spontaneous lipid peroxidation in rabbit and mouse epididymal spermatozoa: dependence of rate on temperature and oxygen concentration. *Biology of Reproduction* 32:342–351.

Andrade, A.J., Grande, S.W., Talsness, C. E., Grote, K., and I. Chahoud. 2006. A dose-response study following in utero and lactational exposure to di-(2-ethylhexyl)-phthalate (DEHP): Non-monotonic dose-response and low dose effects. *Toxicology* 227:185–192.

Badinga, L., Collier, R.J., Thatcher, W.W., and J. Wilcox. 1985. Effects of climatic and management factors on conception rate in subtropical environment. *Journal of Dairy Science* 68:78–85.

Balic, I.M., Milinkovic-Tur, S., Samardzija, M., and S. Vince. 2012. Effect of age and environmental factors on semen quality, glutathione peroxidase activity and oxidative parameters in Simmental bulls. *Theriogenology* 782:423–431.

Bhakat, M., Mohanty, T.K., Gupta, A.K., and M. Abdullah. 2014. Effect of season on semen quality of crossbred (Karan Fries) bulls. *Advanced in Animals Veterinary Sciences* 11:632–637.

Boerjan, M., Freijnagel, S., Rhind, S., and G. Meijer. 2002. The potential reproductive effects of exposure of domestic ruminants to endocrine disrupting compounds. *Animal Science* 74:3–12.

Brevini, T., Cillo, F., Antonini, S., and F. Gandolfi. 2005. Effects of endocrine disrupters on the oocytes and embryos of farm animals. *Reproduction in Domestic Animals* 40:291–299.

Brunet, S., Pahlavan, G., Taylor, S., and B. Maro. 2003. Functionality of the spindle check point during the first meiotic division of mammalian oocytes. *Reproduction* 126:443–450.

Campagna, C., Guillemette, C., Paradis, R., Sirard, M.A., Ayotte, P., and J.L. Bailey. 2002. An environmentally relevant organochlorine mixture impairs sperm function and embryo development in the porcine model. *Biological Reproduction* 1:80–87.

Cardozo, J., Fernández-Juan, M., Forcada, F., Abecia, A., Muiño-Blanco, T., and J.A. Cebrián-Pérez. 2006. Monthly variations in ovine seminal plasma proteins analyzed by two-dimensional polyacrylamide gel electrophoresis. *Theriogenology* 4:841–850.

Caserta, D., Maranghi, L., Mantovani, A., Marci, R., Maranghi, F., and M. Moscarini. 2008. Impact of endocrine disruptor chemicals in gynaecology. *Human Reproduction Update* 59–72.

Cavestany, D., el-Wishy, A.B., and R.H. Foote. 1985. Effect of season and high environmental temperature on fertility of Holstein cattle. *Journal of Dairy Science* 68:1471–1478.

Chenoweth, P.J., and S.P. Lorton, 2014. Animal andrology. *Theories and Applications, CABI* 53:1–568.

Coote, P.J., Cole, M.B., and M.V. Jones. 1991. Induction of increased thermotolerance in saccharomyces-cerevisiae may be triggered by a mechanism involving intracellular PH. *Journal of General Microbiology* 137:1701–1708.

Curci, A., Bevilacqua, A., and F. Mangia. 1987. Lack of heat-shock response in preovulatory mouse oocytes. *Developmental Biology* 1:154–160.

Danzo, B. 1998. The effects of environmental hormones on reproduction. *Cellular and Molecular Life Sciences* 54:1249.

Deepthi, K.B., and S. Nandi. 2016. Environmental chemicals as reproductive toxicants. *International Journal Applied Natural Science* 3:9–12.

Diamanti-Kandarakis, E., Bourguignon, J.P., Giudice, L.C., Hauser, R., Prins, G.S., Soto, A.M., Zoeller, R.T., and A.C. Gore. 2009. Endocrine-disrupting chemicals: An Endocrine Society scientific statement. *Endocrine Reviews* 30:293–342.

Dorado, C.M., McCann, L.M., Miller, J.A., and J.L. Edwards. 2001. Nuclear status of bovine oocytes exposed to elevated temperature after resumption of meiosis. *Theriogenology* 55:469.

Dutt, R.H. 1963. Critical period for early embryo mortality in ewes exposed to high ambient temperature. *Journal of Animal Science* 22:713–719.

Ealy, A.D., Drost, M., and P.J. Hansen. 1993. Developmental changes in embryonic resistance to adverse effects of maternal heat stress in cows. *Journal of Dairy Science* 76:2899–2905.

Ealy, A.D., Howell, J.L., Monterroso, V.H., Arechiga, C.F., and P.J. Hansen. 1995. Developmental changes in sensitivity of bovine embryos to heat shock and use of antioxidants as thermoprotectants. *Journal of Animal Sciences* 73:1401–1407.

Edwards, J.L., and P.J. Hansen. 1996. Elevated temperature increases heat shock protein 70 synthesis in bovine two-cell embryos and compromises function of maturing oocytes. *Biology of Reproduction* 2:340–346.

Edwards, J.L., Bogart, A.N., Rispoli, L., Saxton, A.M., and F.N. Schrick. 2009. Developmental competence of bovine embryos from heat-stressed ova. *Journal of Dairy Science* 92:563–570.

Edwards, J.L., Saxton, A.M., Lawrence, J.L., Payton, R.R., and J.R. Dunlap. 2005 Exposure to a physiologically relevant elevated temperature hastens in vitro maturation in bovine oocytes. *Journal of Dairy Science* 881:4326–4333.

Elliott, D.S., and L.C. Ulberg. 1971 Early embryo development in the mammal. I. Effects of experimental alterations during first cell division in the mouse zygote. *Journal of Animal Sciences* 33:86–95.

Falconer, I.R., Chapman, H.F., Moore, M.R., and G. Ranmuthugala. 2006. Endocrine-disrupting compounds: A review of their challenge to sustainable and safe water supply and water reuse. *Environmental Toxicology* 21:181–191.

Ferguson, S.A., Flynn, K.M., Delclos, K.B., and R.R. Newbold. 2000. Maternal and offspring toxicity but few sexually dimorphic behavioral alterations result from nonylphenol exposure. *Neurotoxicol Teratol* 4:583–591.

Fiorenza, M.T., and F. Mangia. 1992. Hyperthermia specifically inhibits bivalent chromosome following hyperthermic stress during meiotic maturation. *Experimental Cell Research* 132:359–366.

Fraser, L.R., Beyret, E., Stuart, R., Susan, M., and A. Adeoya-Osiguwa. 2006. Effects of estrogenic xenobiotics on human and mouse spermatozoa. *Human Reproduction* 21:1184–1193.

Galloway, T.S. 2006. Biomarkers in environmental and human health risk assessment. *Pollution Bulletin* 12:606–613.

Gandolfi, F., Pocar, P., Brevini, T.A.L., and Fischer, B. (2002) Impact of endocrine disrupters on ovarian function and embryonic development. *Domestic Animal Endocrinology* 23:189–201.

Ghisari, M., Long, M., Tabbo, A., Eva, C., and B. Jorgensen. 2009. Effects of currently used pesticides and their mixtures on the function of thyroid hormone and aryl hydrocarbon receptor in cell culture. *Toxicology Letter* 25:67–77.

Goettlich, P. 2005. What are endocrine disruptors? In M. Friedman (ed.), *Fundamentals of Naturopathic Endocrinology: A Complementary and Alternative Medicine Guide*, CCNM Press, Toronto.

Hahnel, A.C., Gifford, D.J., Heikkila, J.J., and G.A. Schultz. 1986. Expression of the major heat-shock protein (Hsp70) family during early mouse embryo development. *Teratogen Carcinogen Mutagenesis* 6:493–510.

Hansen, P.J. 2009. Effects of heat stress on mammalian reproduction. *Philosophical Transactions of the Royal Society B: Biological Sciences* 364:3341–3350.

Harmon, R.J., Lu, M., Trammel, D.S., and B.A. Smith. 1997. Influence of heat stress and calving on antioxidant activity in bovine blood. *Journal of Dairy Sciences* 80:264.

Inglis, D., et al. 2008. Improving the E-Screen: Effect of FCS on ZR75–1 cell proliferation & estrogen receptor in 5th Society for Environmental & Chemical Toxicology World Congress, Sydney.

Janosek, J., Hilscherová, K., Bláha, L., and I. Holoubek. 2006. Environmental xenobiotics and nuclear receptors – interactions, effects and in vitro assessment. *Domestic Animal Endocrinology* 23:203–209.

Ju, J.C., Parks, J.E., and X. Yang. 1999. Thermotolerance of IVM-derived bovine oocytes and embryos after short term heat shock. *Molecular Reproduction and Development* 53:336–340.

Kamarianos, A., Karamanlis, X., Goulas, P., Theodosiadou, E., and A. Smokovitis. 2003a. The presence of environmental pollutants in the follicular fluid of farm animals (cattle, sheep, goats, and pigs). *Reproductive Toxicology* 17:185–190.

Kamarianos, A., Karamanlis, X., Theodosiadou, E., Goulas, P., and A. Smokovitis. 2003b. The presence of environmental pollutants in the semen of farm animals (bull, ram, goat, and boar). *Reproductive Toxicology* 17:439–445.

Kaviraj, A., Unlu, E., Gupta, A., and El Nemr, A. 2014. Biomarkers of environmental pollutants. *BioMed Research International Volume*, Article ID 806598.

Kawaguchi, H., Umekita, Y., Souda, M., Gejima, K., Kawashima, H., Yoshikawa, T., and H. Yoshida. 2009. Effects of neonatally administered high-dose diethylstilbestrol on the induction of mammary tumors induced by 7,12-Dimethylbenz[a] anthracene in female rats H. *Veterinary Pathology* 46:142–150.

Koch, H.M., et al., 2004. Internal exposure of nursery-school children and their parents and teachers to di (2-ethylhexyl) phthalate (DEHP). *International Journal of Hygiene and Environmental Health* 207:5–22.

Kolpin, D.W., Furlong, E.T., Meyer, M.T., Thurman, E.M., Zaugg, S.D., Barber, L.B., and H.T. Buxton. 2002. Pharmaceuticals, hormones, & other organic wastewater contaminants. *Environmental Science and Technology* 36:1202–1211.

Kruuv, J., Glofcheski, D., Cheng, K.H., Campbell, S.D., Alqysi, H.M.A., Nolan, W.T., and J.R. Lepock. 1983. Factors influencing survival and growth of mammalian-cells exposed to hypothermia. Effects of temperature and membrane lipid perturbers. *Journal of Cell Physiology* 115:179–185.

Lambowitz, A.M., Kobayashi, G.S., Painter, A., and G. Medoff. 1983. Possible relationship of morphogenesis in pathogenic fungus, histoplasma-capsulatum, to heat-shock response. *Nature*, 303:806–808.

Lawrence, J.L., Payton, R.R., Godkin, J.D., Saxton, A.M., Schrick, F.N., and J.L. Edwards. 2004. Retinol improves development of bovine oocytes compromised by heat stress during maturation. *Journal of Dairy Science* 87:2449–2454.

Lee, H.R., Jeung, E.B., Cho, M.H., Kim, T.H., Leung, P. C. K., and K.C. Choi. 2013. Molecular mechanism(s) of endocrine-disrupting chemicals and their potent oestrogenicity in diverse cells and tissues that express oestrogen receptors. *Journal of Cellular and Molecular Medicine* s1:1–11.

Lenz, R.W., Ball, G.D., Leibfried, M.L., Ax, R.L., and N.L. First. 1983. In vitro maturation and fertilization of bovine oocytes are temperature-dependent processes. *Biology of Reproduction* 29:173–179.

Magnusson, U., and S. Persson. 2015. Endocrine disruptors in domestic animal reproduction: A clinical issue? *Reproduction in Domestic Animals* 50:15–19.

Malayer, J.R., Hansen, P.J., and W.C. Buhi. 1988. Effect of day of the oestrous cycle, side of the reproductive tract and heat shock on in-vitro protein secretion by bovine endometrium. *Journal of Reproduction and Infertility* 84:567–578.

Mantovani, A., and F. Maranghi. 2005. Risk assessment of chemicals potentially affecting male fertility. *Contraception* 72:308–313.

Marai, I.F.M., Habeeb, A.A.M., and A.E. Gad. 2002. Rabbits' productive, reproductive and physiological performance traits as affected by heat stress: A review. *Livestock Production Science* 78:71–90.

Matsuzuka, T., Ozawa, M., Hirabayashi, M., Ushitani, A., and Y. Kanai. 2004. Developmental competence and glutathione content of maternally heat-stressed mouse oocytes and zygotes. *Animal Science Journal* 75:117–124.

Mishra, S.R., Kundu, A.K., and A.P.K. Mahapatra. 2013. Effect of ambient temperature on membrane integrity of spermatozoa in different breeds of bulls. *The Bioscan* 1:181–183.

Nandi, S., Girish, K., Manjunatha, B.M., Ramesh, H.S., and P.S.P. Gupta. 2008. Follicular fluid concentrations of glucose, lactate and pyruvate in buffalo and sheep, and their effects on cultured oocytes, granulosa and cumulus cells. *Theriogenology* 69:186–196.

Nandi, S., Gupta, P.S.P., Selvaraju, S., Roy, S.C., and J.P. Ravindra. 2010. Effects of exposure to heavy metals on viability, maturation, fertilization, and embryonic development of buffalo (Bubalus bubalis) oocytes in vitro. *Archives of Environmental Contamination and Toxicology* 58:194–204.

Nielsen, M., Høyer, P.E., Lemmen, J.G., van der Burg, B., and A.G. Byskov. 2000. Octylphenol does not mimic diethylstilbestrol-induced oestrogen receptor-alpha expression in the newborn mouse uterine epithelium. *Journal Endocrinology* 167:29–37.

Noyes, P.D., McElwee, M.K., Miller, H.D., Clark, B.W., Van Tiem, L.A., Walcott, K.C., Erwin, K.N., and E.D. Levin. 2009. The toxicology of climate change: Environmental contaminants in a warming world. *Environmental International* 6:971–86.

Ohashi, O.M., Sousa, J.S., and W.G. Vale. 1988. Aspecto reprodutivo no macho bubalino. In *Bubalinos: fisiologia e Patologia da Reprodução*. Fundação Cargill, São Paulo, pp. 69–86.

Peakall., Walker. 1994. The role of biomarkers in environmental assessment (3). *Ecotoxicology* 3:173–179.

Petro, E.M., Leroy, J.L., Van Cruchten, S.J., Covaci, A., Jorssen, E.P., and P.E. Bols. 2012. Endocrine disruptors and female fertility: Focus on (bovine) ovarian follicular physiology. *Theriogenology* 9:1887–1900.

Pflieger Bruss, S., Schuppe, H.C., and W.B. Schill. 2004. The male reproductive system and its susceptibility to endocrine disrupting chemicals. *Andrologia* 36:337–345.

Pocar, P., Brevini, T.A.L., Fischer, B., and F. Gandolfi. 2003. The impact of endocrine disruptors on oocyte competence. *Reproduction* 125:313–325.

Pocar, P., Nestler, D., Risch, M., and B. Fischer. 2005. Apoptosis in bovine cumulus – oocyte complexes after exposure to polychlorinated biphenyl mixtures during in vitro maturation. *Reproduction* 130:857–868.

Putney, D.J., Drost, M., and W.W. Thatcher. 1988. Embryonic development in superovulated dairy cattle exposed to elevated ambient temperatures between days 1 to 7 post insemination. *Theriogenology* 30:195–209.

Putney, D.J., Mullins, S., Thatcher, W.W., Drost, M., and T.S. Gross. 1989. Embryonic development in superovulated dairy cattle exposed to elevated ambient temperatures between the onset of estrus and insemination. *Animal Reproduction Science* 19:37–51.

Rahman, M.B., Kamal, M.M., Rijsselaere, T., Vandaele, L., Shamsuddin, M., and A.V. Soom. 2013. Altered chromatin condensation of heat stressed spermatozoa perturbs the dynamics of DNA methylation reprogramming in the paternal genome after in vitro fertilisation in cattle. *Reproduction in Fertilization in Development* 8:1107–1116.

Rhind, S.M. 2005. Are endocrine disrupting compounds a threat to farm animal health, welfare and productivity? *Reproduction of Domestic Animals* 4:282–290.

Ribarov, S.R., and L.C. Benov. 1981. Relationship between the hemolytic action of heavy-metals and lipid-peroxidation. *Biochimica et Biophysica Acta* 640(3):721–726.

Roman-Ponce, H., Thatcher, W.W., Canton, D., Barron, D.H., and C.J. Wilcox. 1978. Thermal stress effects of uterine blood flow in dairy cows. *Journal of Animal Science* 46:175–180.

Roth, Z., and P.J. Hansen. 2004a. Sphingosine 1-phosphate protects bovine oocytes from heat shock during maturation. *Biology of Reproduction* 71:2072–2078.

Roth, Z., and P.J. Hansen. 2004b. Involvement of apoptosis in disruption of developmental competence of bovine oocytes by heat shock during maturation. *Biology of Reproduction* 71:1898–1906.

Roth, Z., and P.J. Hansen. 2005. Disruption of nuclear maturation and rearrangement of cytoskeletal elements in bovine oocytes exposed to heat shock during maturation. *Reproduction* 129:235–244.

Saal, F.S.V., and C. Hill. 2007. Bis phenol A expert panel consensus statement: Integration of mechanisms, effects in animals & potential to impact human health. *Reproduction in Toxicology* 24:131–138.

Schrock, G.E., Saxton, A.M., Schrick, F.N., and J.L. Edwards. 2007. Early in vitro fertilization improves development of bovine ova heat stressed during in vitro maturation. *Journal of Dairy Science* 90:4297–4303.

Schug, T.T., Janesick, A., Blumberg, B., and J.J. Heindel. 2011. Endocrine disrupting chemicals and disease susceptibility. *Journal of Steroid Biochemistry and Molecular Biology* 5:204–215.

Selvaraju, S., Reddy, I.J., Nandi, S., Rao, S.B., and J.P. Ravindra. 2009. Influence of IGF-I on buffalo (Bubalusbubalis) spermatozoa motility, membrane integrity, lipid peroxidation and fructose uptake in vitro. *Animal Reproduction Science* 113:60–70.

Setchell, B.P. 1998. The Parkes lecture: Heat and the testis. *Journal of Reproduction and Fertility* 114:179–194.

Sirard, M.A., Florman, H.M., Leibfried-Rutledge, M.L., Barnes, F.L., Sims, M.L., and N.L. First. 1989. Timing of nuclear progression and protein synthesis necessary for meiotic maturation of bovine oocytes. *Biology of Reproduction* 40:1257–1263.

Siu, E.R., Mruk, D.D., Porto, C.S., and C.Y. Cheng. 2009. Cadmium-induced testicular injury. *Toxicology in Applied Pharmacology* 3:240–9.

Skibba, J.L., Powers, R.H., Stadnicka, A., Cullinane, D.W., Almagro, U.A., and J.H. Kalbfleisch. 1991. Oxidative stress as a precursor to the irreversible hepatocellular injury caused by hyperthermia. *International Journal of Hyperthermia* 5:749–761.

Stott, G.H., and R.J. Williams. 1962. Causes of low breeding efficiency in dairy cattle associated with seasonal high temperatures. *Journal of Dairy Science* 45:1369–1375.

Suzuki, H., Ju J.C., Parks, J.E., and X. Yang. 1998. Surface ultrastructural characteristics of bovine oocytes following heat shock. *Journal of Reproductive Development* 44:345–351.

Sweeney, T. 2002. Is exposure to endocrine disrupting compounds during fetal/post-natal development affecting the reproductive potential of farm animals? *Toxicol in Vitro* 20:18–37.

Takahashi, M. 2012. Heat stress on reproductive function and fertility in mammals. *Reproductive Medicine and Biology* 11:37–47.

Taylor, J.F., Bean, B., Marshall, C.E., and J.J. Sullivan. 1985. Genetic and environmental components of semen production traits of artificial insemination Holstein bulls. *Journal Dairy Sciences* 68:2703–2722.

Tseng, J.K., Chen, C.H., Chou, P.C., Yeh, S.P., and J.C. Ju. 2004. Influences of follicular size on parthenogenetic activation and in vitro heat shock on the cytoskeleton in cattle oocytes. *Reproduction in Domestic Animal* 39:146–153.

Ulberg, L.C., and P.J. Burfening. 1967. Embryo death resulting from adverse environment on spermatozoa or ova. *Journal of Animal Sciences* 26:571–577.

Uzumcu, M., Zama, A.M., and E. Oruc. 2012. Epigenetic mechanisms in the actions of endocrine-disrupting chemicals: Gonadal effects and role in female reproduction. *Reproduction in Domestic Animals* 47:338–347.

Veeramachaneni, D.N.R. 2000. Deteriorating trends in male reproduction: Idiopathic or environmental? *Animal Reproduction Science* 61:121–130.

Welch, W.J., and J.P. Suhan. 1985. Morphological-study of the mammalian stress response – characterization of changes in cytoplasmic organelles, cytoskeleton, and nucleoli, and appearance of intranuclear actin-filaments in rat fibroblasts after heat-shock treatment. *Journal of Cell Biology* 4:1198–1211.

Westerhoff, P., Yoon, Y., Snyder, S., and E. Wert. 2005. Fate of endocrine-disruptor, pharmaceutical, & personal care product chemicals during simulated drinking water treatment processes. *Environmental Science of Technology* 17:6649–6663.

West-Rispoli, L.A., Rozanas, C.R., Stoming, T.A., and J.L. Edwards. 2006. Proteomic analysis of bovine oocytes during maturation using 2-D fluorescence difference gel electrophoresis (DIGE). *Biology of Reproduction* 152:359.

Wojtowicz, A., Tomanek, M., Augustowska, K., and E.L. Gregoraszczuk. 2005. Aromatic hydrocarbon receptor (AhR) in the porcine theca and granulosa cells: effect of TCDD, PCB 126 and PCB 153 on the expression of AhR. *Endocrine Regulations* 4:109–118.

Wolfenson, D., Thatcher, W.W., Badinga, L., Savio, J.D., Meidan, R., Lew, B.J., Braw-Tal, R., and A. Berman. 1995. Effect of heat stress on follicular development during the estrous cycle in lactating dairy cattle. *Biology of Reproduction* 52:1106–1113.

Wölz, J., Engwall, M., Maletz, S., OlsmanTakner, H., van Bavel, B., Kammann, U., Klempt, M., Weber, R., Braunbeck, T., and H. Hollert. 2008. Changes in toxicity and ah receptor agonist activity of suspended particulate matter during flood events at the rivers Neckar and Rhine. *Environmental Sciences Pollutants Research* 15:536–553.

Yang, K., Julan, L., Rubio, F., Sharma, A., and H. Guan. 2006. Cadmium reduces 11 beta-hydroxysteroid dehydrogenase II activity & expression in human placental trophoblast cells. *American Journal of Physiology -Endocrinology Metabolism* 290:E135–E142.

Yarny, T.A., Sanford, L.M., and W.M. Palmer. 1990. Pubertal development of ram lambs bodyweight and testicular size measurements as indices of post pubertal reproductive function. *Canadian Journal of Animal Science* 17:139–149.

Young, F., Micklem, J., and A.R. Humpage. 2008. Effects of blue green algal toxin cylindrospermopsin (CYN) on human granulosa cells in vitro. *Reproduction of Toxicology* 25:374–380.

Young, F.M., Menadue, M.F., and T.C. Lavranos. 2005. Effects of the insecticide amitraz, an α2-adrenergic receptor agonist, on human luteinized granulosa cells. *Human Reproduction* 20:3018–3025.

Younglai, E.V., Holloway, A.C., Lim, G.E., and W.G. Foster. 2004. Synergistic effects between FSH & P,P' DDE on human granulosa cell aromatase activity. *Human Reproduction* 19:1089–1095.

Zavy, M.T. 1994. Embryonic mortality in cattle. In M.T. Zavy and R.D. Geisert (eds.), *Embryonic Mortality in Domestic Species*. CRC, Boca Raton, pp. 99–140.

9

Biosurfactants and Their Applications in Environmental Cleanup

M Devasena, A. Thenmozhi and Shatirah Akib

CONTENTS

9.1 Introduction

Surfactants are surface-active compounds that create self-assembled molecular clusters called micelles at the interface and reduce the surface tension of single-phase systems such as water or oil or interfacial tension of two-phase or multi-phase systems such as gas–solid, gas–liquid, and liquid–solid. A good and effective surfactant can reduce the surface tension of water from 72 to 30 mN/m (Volkering et al., 1998). Synthetic surfactants such as Tween 80, sodium dodecyl sulfate (SDS), sodium lauryl sulfate (SLS) and polysorbate 80 are produced from petroleum compounds whereas living cells such as bacteria, fungi and yeast produce biosurfactants. When bacteria, fungi or yeast have an affinity at the interface between immiscible fluids and significantly reduce the interfacial tension, they are simply called biologically produced surfactants or biosurfactants. The complex structure of surfactants with hydrophilic (showing affinity to water) and hydrophobic (not showing affinity to water) moieties determines the unique amphiphilic characteristic of such molecules.

The hydrophobic moiety is either saturated or unsaturated, linear or branched fatty acid and tends to concentrate at interfaces and reduces the free surface energy of the system. The hydrophilic moiety is either an amino acid or a peptide or mono-, di- or polysaccharide that will make the surfactant soluble in water. Both together constitute a surfactant molecule generally with an alkyl chain of 8 to 22 carbons (Nakama, 2017). Figure 9.1 shows the head and tail portions of a surfactant. Depending on the nature of the polar head group carrying a positive or negative charge, surfactants can be classified as cationic or anionic surfactants, respectively. If the head contains both, then surfactant is termed as zwitterionic and nonionic if the head does not contain any charge.

Biosurfactants are grouped according to their chemical composition, properties and microbial isolates. They are also classified as low and high-molecular-weight compounds, with glycolipids, lipopeptides and lipoproteins, phospholipids and neutral lipids belonging to the former while polymeric and particulate biosurfactants belong to the latter. While low-molecular-weight biosurfactants purely reduce the surface tension, high-molecular-weight compounds act as emulsifiers and stimulate bioremediation and removal of hydrophobic substances from environment. Different species of Acinetobacter produces high-molecular-weight bioemulsifiers, such as *alasan*, and helps in solubilizing and mineralizing polyaromatic hydrocarbons (PAHs; Barkay et al., 1999). Bacterial species such as *Pseudomonas aeruginosa* and *Bacillus subtilis* are low-molecular-weight biosurfactants and are used to remediate hydrocarbon and metal contaminated soil samples.

Other bacterial species including *Corynebacterium*, *Nocardia* and *Rhodococcu sp.*, *Bacillus subtilis*, *Bacillus licheniformis* and *Arthrobacter paraffineus* also produce surfactants. Fungi

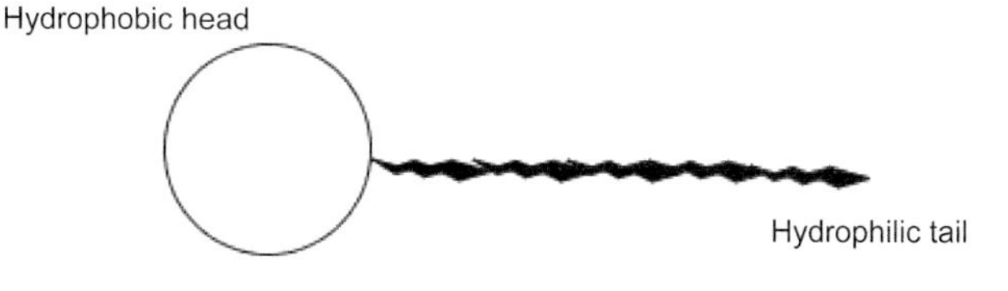

FIGURE 9.1 Typical surfactant molecule.

DOI: 10.1201/9781003186304-9

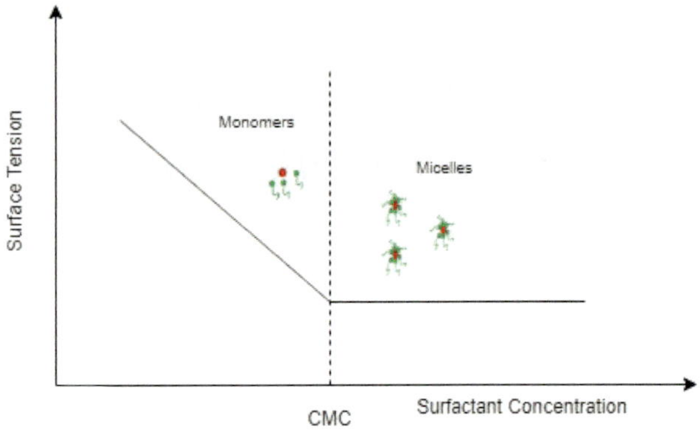

FIGURE 9.2 CMC value and micelle formation.

involved in surfactant production include the yeasts *Torulopsis sp.* and *Candida sp.* Another kind of classification is based on the hydrophile–lipophile balance (HLB; Pacwa-Płociniczak et al., 2016), which indicates whether surfactants form water in oil or oil in water emulsion. A surfactant is lipophilic (hydrophobic) and is used as an emulsifier when the HLB value is low, whereas surfactants with high HLB value are lipophobic (hydrophilic) that are used as cleansers (Franzetti et al., 2008).

The distinctive feature of surfactants is their tendency to adsorb at interfaces and alter the surface chemistry of the system. Depending on the concentration and orientation at interfaces, the surface energy varies in a desirable manner. Until a critical point called critical micelle concentration (CMC) is reached, the surface energy decreases with an increase in surfactant concentration. Above CMC, surface or interfacial tension no longer reduces. Typically, surfactants exist as monomers at the CMC, and when the saturation limit has been reached, large colloidal aggregates called micelles are formed with the hydrophobic part of the surfactant in the interior of the aggregate and the hydrophilic portion at the exterior (Figure 9.2).

Surfactant structure, surfactant concentration, pH, ionic strength and temperature of the solution influence the CMC of any surfactant. CMC values are specific to each and every surfactant; in principle, an effective surfactant will have a low CMC and will lower the surface tension between two dissimilar phases, increase the contact area and improve the bioavailability. CMC of alkylbenzene sulfonate is 590 mg/L; SLS, 2000 to 2900 mg/L; and CMC of biosurfactants, such as rhamnolipid and surfactin, are 20 mg/L and 11 mg/L, respectively. The CMC values denote that biosurfactants are better than synthetic surfactants in terms of efficiency (Sobrinho et al., 2008).

9.2 Production of Biosurfactants

Biosurfactants are synthesized by a variety of microorganisms such as *Pseudomonas, Bacillus, Arthrobacter, Mycobacterium, Acinetobacter, Thiobacillus* and different other isolates obtained from contaminated soil, seawater, sediments, sludges, agro-wastes and oil fields, including petrochemical industries. They are produced as secondary metabolites during the stationary growth or exponential growth phase. They are also produced under growth-limiting conditions, such as nitrogen or trace element limitations. Immobilized cells obtained from a surfactant-producing culture can contribute to an increase in surfactant production by almost five- to sixfold. The production, quantity and composition of biosurfactants depend not only on the bacterial strain, but the carbon and nitrogen source of the substrate, buffers, trace elements of phosphorus, magnesium, iron and manganese ions are also equally important (Ron and Rosenberg, 2001). Vegetable oil, crude oil, mahua oil and waste frying oils produce biosurfactants.

Few hydrophilic sources such as glycerol, sucrose and glucose are used for biosurfactant production, but the yield of biosurfactants would be less (Nitschke et al., 2011). The nitrogen sources include yeast extract, peptone, ammonium sulfate, ammonium nitrate, potassium nitrate, sodium nitrate and malt extracts. It has been reported that potassium nitrate can decrease the surface tension considerably and give high emulsification index when compared to other nitrogen sources (Chi et al., 2008). An appropriate control of temperature and pH is also important for their production. However, the cost of biosurfactant production is at least five to ten times more than producing a synthetic surfactant due to high raw material cost and operational costs; therefore, the production of biosurfactant is still challenging in the industrial level.

The most commonly used biosurfactants, called rhamnolipids, are mainly produced as secondary metabolites by *Pseudomonas sp.* and, in particular, by *P. aeruginosa* and *P. fluorescens* during their stationary growth phase. *P. aeruginosa* produces several biomolecules with alkyl chains of 8 to 12 carbon atoms and are grouped as monorhamnolipids and dirhamnolipids. *P. chlororaphis* produces only monorhamnolipid (Gunther et al., 2005). *P. nautica, P. maltophilia, P. marginalis* are a few more *Pseudomonas* isolates that produce biosurfactants. Mata-Sandoval et al. (2001) investigated the production of *P. aeruginosa* using both hydrophobic and hydrophilic substrates and found that rhamnolipid production was at least onefold higher with corn oil when compared to glucose and succinic acid as carbon sources. Similarly, *Bacillus subtilis* produce the biosurfactants surfactin, a lipopeptide that has found its application in various environmental sectors and has proved to be a better hydrocarbon degrader than the other isolates.

Hartoto and Manguwidjaja (2002) studied the ability of *Bacillus sp.* BMN14 using different carbon sources, such as glucose, fructose and sucrose, each at different concentrations and found that the highest biosurfactant production was obtained with glucose as the carbon source. Surfactin yields are generally higher under nitrogen limitation conditions. Davis et al. (2001) observed that *Bacillus sp.* showed increased biosurfactant production only under nitrate-limiting conditions. *Bacillus firmus* and *Bacillus alvei* are few other Bacillus isolates capable of producing biosurfactants. Biosurfactants and microbial isolates are shown in Table 9.1. Rhamnolipids produced by the gram-positive *P. aeruginosa* are used in various applications as they can be produced from numerous substrates. Sophorolipids, which are produced by yeast, differ in the length of alkyl chain, degree of unsaturation and the number of acyl groups.

TABLE 9.1
Biosurfactants: Sources and Applications

Group	Biosurfactant	Organisms	Applications
Glycolipids	Rhamnolipids	*Pseudomonas aeruginosa* *Pseudomonas chlororaphis Serratia rubidea*	Bioremediation, antimicrobial and biocontrol properties, degradation of hydrocarbons; emulsification of hydrocarbons and vegetable oils; removal of metals from soil
	Trehalose lipids	*Rhodotorula glutinis Rhodotorula graminis Mycobacterium sp.*	Enhancement of the bioavailability of hydrocarbons, dissolution of hydrocarbons
	Sophorolipids	*Candida bombicola* *Candida antartica* *Candida Stellata* *Candida bogoriensis*	Recovery of hydrocarbons, removal of heavy metals from sediments, enhanced oil recovery, antimicrobial and antiviral properties
	Emulsan Glycolipopeptide	*Acinetobacter calcoaceticus*	Microbially enhanced oil recovery
Lipopeptides	Surfactin/iturin	*Bacillus subtilis* *Bacillus amyloliquefaciens*	Antimicrobial properties, biodegradation of hydrocarbons and chlorinated pesticides, removal of heavy metals
	Lichenysin	*Bacillus licheniformis*	Microbially enhanced oil recovery
	Viscosin	*P. Fluorescens* *Leuconostoc mesenteriods*	Surface-active lipopeptides
	Subtilisin	*Bacillus subtilis*	Antimicrobial properties
	Peptide lipids	*B. licheniformis*	Antimicrobial properties
	Serrawettin	*Serratia marcescens*	Chemorepellent
	Gramicidin	*Brevibacterium brevis*	Antibiotic, disease control
	Polymixin	*Bacillus polymyxa*	Bactericidal and fungicidal activity
	Emulsan	*A.calcoaceticus*	Microbially enhanced oil recovery
Polymeric biosurfactants	Liposan	*C. Tropicalis*	Bioemulsion
	Biodispersan	*Acinetobacter calcoaceticus A2*	Dispersion of limestone in water
	Alasan	*A.radioresistens*	Biodegradation of polyaromatic compounds
	Ornithine lipids	*Pseudomonas sp.* *Thiobacillus thiooxidans, Agrobacterium sp.*	Bioemulsifiers
Fatty acids, phospholipids and neutral lipids	Corynomycolic acid	*Corynebacterium lepus*	Enhancement of bitumen recovery
	Spiculisporic acid	*Penicillium spiculisporum*	Heavy metal sequestrants
	Phosphati-dylethanolamine	*Acinetobacter* sp., *Rhodococcus erythropolis*	Heavy metal removal
	Aminoacids lipids	*Bacillus sp.*	Antimicrobial properties
	Phospholipids	*Acinetobacter sp.*	Bioremediation
Particulate biosurfactants	Vesicles	*Acinetobacter calcoaceticus, Pseudomonas marginalis*	Degradation and removal of hydrocarbons
	Whole microbial cells	*Cyanobacteria*	

9.3 Environmental Applications

A number of subsurface contaminants such as chlorinated methane and ethane, nitroaromatics, energetic compounds, uranium, heavy metals such as cadmium, chromium and lead, hydrocarbons such as non–aqueous phase liquids (NAPLs), PAHs and Benzene, Toluene, Ethylbenzene, Xylene (BTEX) compounds prevail and hence contaminated subsurface sites require effective remediation. Remediation techniques, although site-specific, can either be thermal treatment technology, physio-chemical removal or bioremediation. The time required for the removal of such compounds is usually high due to their sorption potential and recalcitrant nature. Ex-situ treatment methods, such as pump and treat or soil excavation, are economically high and difficult in real practice whereas *in situ* treatment methods, such as soil washing, mobilization, stabilization or soil flushing, have earned an increase in interest in recent times.

Biosurfactants play a significant role in enhancing subsurface remediation by solubilization and mobilization of contaminants. Solubilization increases the dissolved concentration of the contaminant through micellar partitioning and aids in remediation. Surfactants reduce the interfacial tensions and the capillary trapping of the contaminants, especially for NAPLs, such as oil/tetrachloroethylene (PCE)/tricholoroethylene (TCE). The contaminant thereby is readily extracted with the groundwater. Currently, biosurfactants help in degrading hydrocarbons, degrading explosive residues and in bioremediation of soil and water (Volkering et al., 1998) and are expected to extend their use as smart biomaterials in diverse fields.

9.4 Enhanced Oil Recovery

Petroleum engineering has a long history of research in understanding the mechanisms of capillary trapping of oil and its recovery from the subsurface by mobilization either with water or surfactants. Accidental spills and improper disposals of petroleum components, such as crude oil, gasoline and other hydrocarbons, are the most frequent causes of subsurface contamination. They are sparingly soluble in water, migrate under the influence of gravity forces and get trapped within the pore spaces due to capillary forces. Under such scenarios, water flooding is the most common process used in which oil is recovered as a result of displacement by water. Even after primary and secondary recovery, about 50% to 60% of the oil still remains in the subsurface as residuals. Shrinking oil reserves and burgeoning costs for exploring new discoveries makes oil recovery itself as one of the economical oil production processes. Only the diminution of capillary forces can mobilize/displace the trapped oil and aid in recovering it.

The interfacial tension can be reduced by flooding with either steam or hot water, alcohol or surfactant. The effectiveness of using hot water or steam in enhanced oil recovery at the field level is still unknown. Flushing with surfactants and cosolvents generally reduces the interfacial tension and mobilizes the trapped oil thereby promoting enhanced oil recovery (Taylor et al., 2001; Pennell et al., 1996; Gupta and Mohanty, 2001). This complex process involving water, brine and oil requires the addition of surfactants to form a micelle solution. The technique creates low interfacial tension (IFT), in which an especially very low IFT (0.001mN/m) between the surfactant and the oil is a requirement in order to mobilize the residual oil. Such microemulsions decrease the IFT and increase the mobility of residual oil. However, care must be taken to avoid the formation of microemulsions, as microemulsions may plug the soil pores.

Certain microbes such as *Bacillus subtilis* and *Pseudomonas aeruginosa* are widely reported to use crude oil and hydrocarbons present in oil reservoirs as carbon sources and possibly help in reducing the surface tension at the interface and thereby increase oil production in reservoirs. Biosurfactants, even with low purity, have proved their success in the enhancement of oil recovery especially in the presence of rhamnolipids. However, screening of the surfactants is important as certain surfactants tend to cause macro emulsions and plug the void spaces of the subsurface.

9.5 Petroleum Hydrocarbons

Subsurface contamination with organic hydrocarbons such as dense non–aqueous phase liquids (DNAPLs), light non–aqueous phase liquids and polyaromatic hydrocarbons remains a persistent problem due to capillary trapping of these contaminants in the subsurface zone. Pennell et al. (1996) studied the influence of viscous and buoyancy forces on DNAPL mobilization during surfactant flushing. Because surfactants are amphiphilic in nature, the hydrophobic part tends to accumulate at the interface and reduce the interfacial tension between the wetting (water) and nonwetting fluid (DNAPL). In their study, the authors were able to bring down the interfacial tension of DNPAL–water from 47.8 dyn/cm to 5 dyn/cm, then from 5 dyn/cm to 0.58 dyn/cm and finally to 0.09 dyn/cm by injecting different surfactant solutions. Jeong and Corapcioglu (2005) used etched glass micromodels to understand the physical forces acting on TCE removal during surfactant and surfactant foam flushing.

During surfactant flushing, capillary forces dominated and residual TCE either retained in the pore space or solubilized, whereas during surfactant foam flushing, viscous forces were dominated and the trapped TCE was flushed out. Although both surfactant and surfactant foam flushing were applied under the same conditions, the results obtained were dissimilar. Surfactant foam is a homogeneous mixture of surfactant and air. Viscosity of foam was nearly 100 times greater than the surfactant and therefore resulted in high TCE reduction. Inoculants of biosurfactant strains can be applied in order to improve the efficiency of *in situ* phytoremediation technology for contaminated soils. However, toxic effects, if any on plants must be studied.

9.6 Biodegradation of Nitroaromatics

Contamination of soil and groundwater with nitroaromatics is extensive due to military operations and any other industrial waste disposal. Nitroaromatics include compounds such as 2,4, 6-trinitrotoluene (TNT) and 2,4-dinitrotoluene (DNT), among many others, and are classified as hazardous compounds. *In situ* remediation technologies, such as incineration, phytoremediation and chemical oxidation, have their limitations. Nitroaromatics can be treated by bioremediation wherein the addition of synthetic surfactants or biosurfactants can be used in order to enhance the bioavailability of these compounds (Thenmozhi and Devasena, 2019).

Pseudomonas sp., *Desulfovibrio sp.*, *Bacillus sp.* And *Staphylococcus sp.* are a few isolates of bacteria that can transform nitroaromatic compounds. With the bioavailability of nitro compounds being a challenging factor, surfactants/biosurfactants facilitate desorption of nitroaromatics from the soil matrix and increase the bioavailability. *P. aeruginosa* is capable of producing rhamnolipids and transforming DNT/TNT into nontoxic metabolites by means of reduction of the nitro groups. Boopathy (2002) studied TNT degradation and desorption with synthetic surfactant Tween 80. It was observed that both TNT and its metabolite 4-amino-2,6-dinitrotoluene were degraded in a span of 35 days. Sadani et al. (2016) have reported biotransformation of the explosive pentaerythritol tetranitrate (PETN) using rhamnolipids. Rhamnolipids increased the solubility of PETN, enhanced the dispersion in an aqueous solution and prompted easy removal from the soil. Amin et al. (2017) investigated simultaneous degradation of TNT and PETN using rhamnolipids.

9.7 Heavy Metals

The application of biosurfactants in the recovery of heavy metals involves mobilization of the metals through desorption and solubilization. Biosurfactants are quick in replacing ethylenediamine tetraacetic acid (EDTA), acids and chelating agents that cause secondary consequences to the soil affecting its fertility. In a study by Elouzi et al. (2012), rhamnolipids obtained from *P. aeruginosa* removed multiple metals cadmium, lead, nickel and zinc from contaminated soil. Maity et al. (2013) used rhamnolipids and saponin to remove chromium and arsenic from soils. Ozturk et al. (2012) produced rhamnolipids from *P. aeruginosa* and evaluated the removal of chromium from soil. Sarubbo et al. (2018) investigated the yeast *Candida guilliermondii* for producing biosurfactant where the production medium was supplemented with used 4% soya bean oil, 2.5% sugar cane molasses and 2.5% corn steep liquor.

The biosurfactant produced reduced the surface tension from 50.0 mN/m to 28.0 mN/m with a CMC of 0.42. The soil sample which contained 3018 g/L of lead, 1877 g/L of iron and 1470 g/L of zinc underwent removal efficiency of 99.9%, 96.8% and 93.5%, respectively. Adopting biosurfactants' usage in permeable reactive barriers reduces the soil permeability and aids in removing many groundwater pollutants such as petroleum by-products and heavy metals. Simultaneous removal of mixed heavy metals, such as lead and cadmium from soil, was studied by Wan et al. (2015), who obtained good desorption rates with the aid of rhamnolipids mixed with citric acid. Rhamnolipids consist of one or two rhamnose molecules in the hydrophilic side and three molecules of fatty acids on the hydrophobic side. The carboxyl group present in rhamnolipids helped in desorption of metals and mobilization and solubilization of hydrocarbons trapped in the subsurface.

The sophorolipids produced by *Starmella bombicolla* exhibited greater removal efficiency of heavy metals than the chemical surfactants (Tween 80 and SDS) according to the observations by Qi et al. (2018). With concentrations of crude oil ranging from 0.1% to 8%, the removal efficiency of cadmium and lead were 5.28% to 79.75% and 0.47% to 30.28%, respectively. With a crude oil concentration of 8%, the removal efficiency of cadmium raised to 83% and 45%, a threefold increase for lead. Singh et al. (2012) found good chromium removal potential in dirhamnolipid, which showed a 92% chromium removal in a period of 30 hours. The authors found this most convincing result with 0.1% dirhamnolipid, which showed no lethal effects to the soil microbes, thus maintaining the stability and fertility of the soil. Biosurfactants form complexes with heavy metals that are strongly bonded to the soil. Sangeetha et al. (2020) removed lead with simultaneous biosurfactant production by the *P. aeruginosa* isolated from lead-contaminated soil. Lead removal by biosurfactants was investigated with the addition of palm oil and gingelly oil, respectively. Batch adsorption experiments showed 96% to 99.6% of lead adsorption following the Langmuir isotherm model. In the presence of biosurfactants, enhanced desorption of 62.3% was observed.

Removal of heavy metals from the soil is initiated with the biosurfactant getting sorbed to the soil surface and forming complex with the metal, followed by desorption of the metal–biosurfactant complex from the soil and finally getting trapped within the micelles through electrostatic interactions and recovered by means of precipitation of biosurfactant from the complex. Anionic surfactants create very strong nonionic complexes with metals, which are later desorbed by the previously described mechanism due to reduction in interfacial tension, whereas cationic surfactants undergo ion exchange processes and replace similarly charged metal ions alone. Falciglia et al. (2016) studied the removal of mercury from marine sediments by using nonionic surfactants, typically Tween 80 in combination with a biodegradable complexing agent (EDTA). This combination resulted in a synergetic effect in remediating mercury-contaminated areas. In another work by Tang et al. (2017), Tetrasodium of N, N-bis (carboxymethyl) glutamic acid was combined with rhamnolipid, which led to good removal efficiencies of metals.

9.8 Agriculture

Biosurfactants are applied in many other industries such as agrochemicals, fertilizers, foods, mining, beverages and cosmetics. Biosurfactants remove plant pathogens and increase the bioavailability of nutrients for microbes associated with plants. They improve soil quality in agriculture through soil rehabilitation. Rhizosphere-associated microbes and plants are capable of producing biosurfactants and controlling the interaction between plants and microbes. *Pseudomonas* and *Bacillus* isolated from rhizomes control the soft rot caused by *Pectobacterium* and *Dickeya sp.* Rhamnolipids inhibit many plant pathogens and act as an insecticidal compound. Zoospores of pathogens are inhibited by P. aeruginosa and *P. putida*.

Biosurfactants also target the fungal pathogens, such as *Pythium ultimum* that causes plant root rot, *Fusarium oxysporum* that causes wilting of crops and *Phytophthora cryptogea* that causes rotting of fruits and flowers (Sachdev and Cameotra, 2013). Their potential role in the promotion of plant growth, plant-pathogen inhibition and other related agricultural applications needs more attention and investigation. Bioavailability of nutrients can be increased when biosurfactants are used as soil-remediating agents. Biosurfactants improve agricultural soil quality when used for soil remediation. Conventional methods of microbial population screening for the production of biosurfactants are currently utilized. New biosurfactants from the soil biosphere need to be researched using advanced techniques.

9.9 Conclusion

Biosurfactants act as substitutes for synthetic surfactants because of their low toxicity, better biodegradability and ecological acceptability. They are microbially synthesized compounds that can be produced *in situ* and are effective in biodegradation and bioremediation when compared to artificial and chemical surfactants. *In situ* treatment of heavy metals, nitroaromatic compounds and chlorinated hydrocarbons can be effective and fast with biosurfactants. Field-level studies are important in order to capture the dynamic interaction of all interfering factors, such as pollutant concentration, type of microbes, carbon and nitrogen sources and optimum environmental conditions. Although biosurfactants are thought to be eco-friendly, enhanced concentrations might be toxic to microorganisms and inhibit the biodegradation process. Nevertheless, the production of tailor-made biosurfactant compositions will aid in the remediation of polluted sites.

9.10 Funding

The authors wish to thank the financial support provided by SERB—Department of Science and Technology, New Delhi—SB/FTP/ETA-0199/2013.

REFERENCES

Amin, M., H. Khanahmad, F. Teimouri, M. Sadani and R. Karami. 2017. Improvement of biodegradability of explosives using anaerobic-intrinsic bioaugmentation approach. *Bulg. Chem. Commun* 49 (3): 735–741.

Barkay, T., S. Navon-Venezia, E. Z. Ron and E. Rosenberg. 1999. Enhancement of solubilisationzation and biodegradation of polyaromatic hydrocarbons by the bioemulsifier alasan. *Appl. Environ. Microbiol* 65: 2697–2702.

Boopathy, R. 2002. Effect of food-grade surfactant on bioremediation of explosives-contaminated soil. *J. Hazard. Mater* 92 (1): 103–114.

Chi, M., L. Naeun, K. Minju, K. Mia and L. Sangjoon. 2008. Heterologous production of pseudomonas aeruginosa EMS1 biosurfactant in Pseudomonas putida. *Bioresour. Technol* 99: 2192–2199.

Davis, D. A., H. C. Lynch and J. Varley. 2001. The application of foaming for the recovery of Surfactin from B. subtilis ATCC 21332 cultures. *Enzyme Microb. Technol* 25: 346–354.

Elouzi, A. A., A. A. Akasha, A. M. Elgerbi, M. El-Baseir and B. A. El Gammudi. 2012. Removal of heavy metals contamination by biosurfactants (Rhamnolipids). *J. Chem. Pharm. Res* 4(9): 4337–441.

Falciglia, P. P., D. Malarbì and G. A. V. Federico. 2016. Removal of mercury from marine sediments by the combined application of a biodegradable non-ionic surfactant and complexing agent in enhanced-electrokinetic treatment. *Electrochim. Acta,* 222: 1569–1577.

Franzetti, A., G. Bestetti, P. Caredda, P. La Colla and E. Tamburini. 2008. Surface-active compounds and their role in the access to hydrocarbons in Gordonia strains. *FEMS Microbiol. Ecol* 63: 238–248.

Gunther, N. W., A. Nuñez, W. Fettand and D. K. Solaiman. 2005. Production of rhamnolipids by pseudomonas chlororaphis, a nonpathogenic bacterium. *Appl. Environ. Microbiol* 71 (5): 2288–2293.

Gupta, D. K. and K. M. Mohanty. 2001. A laboratory study of surfactant flushing of DNAPL in the presence of macroemulsion. *Environ. Sci. Technol* 35: 2836–2843.

Hartoto, L. and D. Mangunwidjaja. 2002. Production of lipopeptide biosurfactant by indigenous isolated of Bacillus sp. BMN 14 in a Batch bioreactor. *2nd World Engineering Congress Sarawak, Malaysia*, pp. 22–25.

Jeong, S. W. and M. Y. Corapcioglu. 2005. Force analysis and visualization of NAPL removal during surfactant-related floods in a porous medium. *J. Hazard. Mater* A126: 8–13.

Maity, J. P., Y. M. Huang, C. W. Fan, C. C. Chen, C. Y. Li, C. M. Hsu, Y. F. Chang, C. I. Wu, C. Y. Chen and J. S. Jean. 2013. Evaluation of remediation process with soapberry derived saponin for removal of heavy metals from contaminated soils in Hai-Pu, Taiwan. *J. Environ. Sci* 25: 1180–1185.

Mata-Sandoval, J. C., J. Karns and A. Torrents. 2001. Effect of nutritional and environmental conditions on the production and composition of rhamnolipids by P. aeruginosa UG2 *Microbiol. Res* 155 (4): 249–256.

Nakama, Y. 2017. Chapter 15 – surfactants, in cosmetic science and technology theoretical principles and applications. *Amsterdam, Elsevier Inc*: 231–44.

Nitschke, M., S. G. V. A. O. Costa and J. Contiero. 2011. Rhamnolipids and PHAs: Recent reports on pseudomonas – derived molecules of increasing industrial interest. *Process Biochem* 46: 621–630.

Ozturk, S., T. Kaya, B. Aslim and S. Tan 2012. Removal and reduction of chromium by pseudomonas spp. and their correlation to rhamnolipid production. *J. Hazard Mater* 231: 64–69.

Pacwa-Płociniczak, M., T. Płociniczak, J. Iwan, M. Żarska, M. Chorążewski, M. Dzida and S. Z. Piotrowska. 2016. Isolation of hydrocarbon-degrading and biosurfactant-producing bacteria and assessment their plant growth-promoting traits. *J. Environ. Manage* 168: 175–184.

Pennell, K. D., G. A. Pope and L. M. Abriola. 1996. Influence of viscous and buoyancy forces on the mobilization of residual tetrachloroethylene during surfactant flushing. *Environ. Sci. Technol* 30: 1328–1335.

Qi, X., X. Xu., C. Zhong, T. Jiang, W. Wei and X. Song. 2018. Removal of cadmium and lead from contaminated soils using sophorolipidsfrom fermentation culture of starmerella bombicola CGMCC 1576 fermentation. *Int J Environ Res Public Health* 15 (11): 2334.

Ron, E. Z. and E. Rosenberg. 2001. Natural roles of biosurfactants, *Environ. Microbiol* 3: 229–236.

Sachdev, D. P. and S. S. Cameotra. 2013. Biosurfactants in agriculture. *Appl Microbiol Biotechnol* 97: 1005–1016.

Sadani, M., M. A. Karami, N. Mirzaei, F. Teimouri and M. M. Amin. 2016. Enhanced biodegradation of pentaerythritol tetranitrate (PETN) anaerobic/aerobic biological treatment by biosurfactants, *Bulg. Chem. Commun* 48 (D): 50–54.

Sangeetha, V., A. Thenmozhi and M. Devasena. 2020. Enhanced removal of lead from soil using biosurfactant derived from edible oils. *Soil Sediment Contam* 30 (2): 135–147.

Sarubbo, L. A., P. F. Pedro, B. Gabriel, N. M. Silveira, J. M. Lunaa, R. D. Rufino, A. Valdemir and dos Santosa. 2018. Application of a low cost biosurfactant in the removal of heavy metals in soil. *Chem. Eng. Trans* 64: 433–438.

Singh, S. K., A. Juwarkar, A. Moodhu and K. V. Dubey. 2012. Remediation of chromium contaminated sites using pseudomonas aeruginosa BS2. *TAET* 6 (2): 100–105.

Sobrinho, H. B. S., R. D. Rufino, J. M. Luna, A. A. Salgueiro and L. A. Sarubbo. 2008. Utilization of two agro industrial by – products for the production of a surfactant by Candica sphaerica UCP 0995. *Process Biochem* 43: 912–917.

Tang, J., J. He, T. Liu, X. Xin and H. Hu. 2017. Removal of heavy metal from sludge by the combined application of a biodegradable biosurfactant and complexing agent in enhanced electro kinetic treatment. *Chemosphere* 189: 599–608.

Taylor, T. P., K. D. Pennell, L. M. Abriola and J. H. Dane. 2001. Surfactant enhanced recovery of tetrachloroethylene from a porous medium containing low permeabilitylenses,1-Experimental studies. *J. Contam. Hydrol* 48: 325–350.

Thenmozhi, A. and M. Devasena. 2019. Remediation of 2,4,6-trinitrotoluene persistent in the environment – a review. *Soil Sediment Contam.* Doi: 10.1080/15320383.2019.1664394.

Volkering, F., A. M. Breure and W. H. Rulkens, 1998. Microbiological aspects of surfactant use for biological soil remediation. *Biodegradation* 8: 401–417.

Wan, J., D. Meng, T. Long, R. Ying, M. Ye, M. S. Zhang, Q. Li, Y. Zhou, Y. Lin and J. Mao. 2015. Simultaneous removal of lindane, lead and cadmium from soils by rhamnolipids combined with citric acid. *PLoS One* 10: e0129978. Doi: 10.1371/journal.pone.0129978.

10

Biobased Fuel and Chemicals from Lignocellulosic Biomass—Prospects and Challenges

R. Pontes, A. Romaní, M. Michelin, L. Domingues, J. Nunes and J. Teixeira

CONTENTS

10.1 Introduction

During the last decades, the current demand for energy, fossil fuel depletion, increase in oil prices, and environmental constraints, such as the greenhouse effect, global warming and others, have driven scientists and technologists to consider renewable, sustainable and cheap energy resources such as biofuels.

The paradigm in biofuel research forced countries to investigate an efficient biomass source, starting from first-generation biofuels to fourth generation, with associated advantages and disadvantages. The utilization of lignocellulosic materials, as renewable biomass, provides the basis for a sustainable development of a biorefinery, based on the selective separation of the major fractions to enable their efficient utilization. However, the lignocellulosic biomass structure is very recalcitrant, forming a three-dimensionless matrix that requires pretreatments for an efficient bioconversion.

In this context, the integrated production of biofuels with the production of power, heat and value-added chemicals are a major step in the development of a biorefinery. These processes are favorable from an environmental and economic point of view, reduce nonrenewable fuel consumption while simultaneously stimulating the expansion of the lignocellulosic biomass industry.

10.2 Current Status of Biofuels

Nowadays, the world's energy requirements are largely met by fossil resources. These nonrenewable sources are used for the manufacturing of fuels, fine chemicals, pharmaceuticals, detergents, synthetic fiber, plastics, pesticides, fertilizers, lubricants, solvents, waxes, coke and asphalt, among others to meet the needs of the growing population (Yu et al., 2011). Furthermore, the use of fossil resources contributes to global warming, increasing the level of carbon dioxide in the atmosphere and leading to the emission of greenhouse gases (Naik et al., 2010). During the last three decades, greenhouse gas emissions have already made the global temperature rise 0.6°C, according to the Intergovernmental Panel on Climate Change (Bessou et al., 2011).

The economic development around the globe and the increasing shortage of petroleum reserves coupled with these environmental problems, greenhouse gas emission and global warming, have boosted the demand of sustainable resources

DOI: 10.1201/9781003186304-10

for the production renewable energy (John et al., 2009). Currently, only 10% of the total energy is supplied by bioenergy, including solar, wind and other renewable energy sources (Bessou et al., 2011).

Fossil fuels represent 80% of primary energy consumed, of which 58% is used in the transport sector (Nigam and Singh, 2011). Therefore, the production of biofuels (such as biohydrogen, biomethane, bioethanol, biomethanol, biobutanol) is one of the key priorities for achieving sustainable growth (Kumari and Singh, 2018). The use of biofuels in the transport sector will reduce the fossil fuel dependency and greenhouse gas emissions since biofuels are not accompanied by an increase in the same emissions during the production and transformation of the biomass (Bessou et al., 2011). Biofuels are called "CO_2 neutral" because the plants consume CO_2 through photosynthesis and the CO_2 released in burning in the biorefinery is equal to the CO_2 caught by the plant. For this reason, the CO_2 in the atmosphere does not increase as shown in Figure 10.1.

First-, second- and third-generation biofuels are classified according to the origin and production technology; the fourth-generation biofuels, still at the basic research level, are an emerging trend and make use of novel synthetic biology tools.

10.2.1 First-Generation Biofuels

First-generation biofuels are produced from sugars, grains or seeds. The major feedstocks (corn, wheat, barley, sugarcane) used for first-generation biofuels are sources of food that have a negative impact on food security (Dutta et al., 2014). Among the first generation of fuels biodiesel, bioethanol and biogas are the most important.

Biodiesel is produced from a large range of oilseed crops, mainly rapeseed or canola, soybean and sunflower and palm oil, among others. These oils and fats contain a mixture of triglycerides, free fatty acids, phospholipids, sterols, water, odorants and other impurities. So, there are different types of biodiesels depending on the composition of the oils (Bessou et al., 2011). Biodiesel is produced by the transesterification of vegetable oils or animal fats with an alcohol to produce esters. The by-products, as glycerin from glycerol recovery, are used in cosmetics or as a substrate for bioprocessing. However, the rapid expansion of biodiesel has already saturated the market of glycerin in Europe, undercutting its ability to reduce the biodiesel price (Yazdani and Gonzalez, 2007). Empirical studies have reported that the energy and greenhouse gas balances of rapeseed-based biodiesel are obviously positive, but there is no information about the overall environmental balance (Zabaniotou et al., 2008). Therefore, it is important to find new, efficient alternatives based on both renewable and conventional technologies.

Bioethanol, unlike biodiesel, is a single-compound biofuel whose final composition does not vary with the type of feedstock and is obtained by fermentation. The raw materials are sugar and starch crops; the main sugar crops are sugarcane and

FIGURE 10.1 The fully integrated forest–biofuel–biomaterial–energy cycle for sustainable technologies at a biorefinery.

sugar beet, and the main starch crops used are maize and wheat. Sugar crops show better ethanol yields than starch crops, as they present higher sugar amounts per hectare compared with starch crops, and sugar can be directly fermented while starch crops have to be hydrolyzed beforehand (Bessou et al., 2011). About 60% of global ethanol production comes from sugarcane (Dufey, 2006). According to the renewable fuels association statistics (World Fuel Ethanol Production, 2005), the production of bioethanol has increased rapidly during recent years. In the United States, by 2009, the production reached 10.600 million of gallons, which represents 54% of the worldwide production. In 2011, operating productions generated a total of 13.900 million of gallons, 24% higher than in 2009. Considering the transport sector, although ethanol produced from corn and sugar represent promising substitutes to gasoline, these are not enough to replace the fossil fuel consumed each year worldwide (1 trillion gallons; Limayem and Ricke, 2012).

On the other hand, biogas that is produced through the anaerobic digestion by bacteria of biodegradable matter, methanization, is less considered as transportation biofuels, due to gas storage constraints (Bessou et al., 2011). The feedstock used for its production can be municipal solid or agricultural waste, liquid slurry, solid manure or maize silage. The greater biogas yield relies on the drier matter and fatty acid contents in the substrate.

First-generation biofuels, such as biodiesel, bioethanol and biogas, became useful for improving domestic energy security, and they have the ability to blend with petroleum-based fuels (Naik et al., 2010). However, the use of these biofuels poses some concerns about the supply of feedstocks, including the impact that may have on biodiversity, carbon balances and competition with food crops.

10.2.2 Second-Generation Biofuels

The European Directive (EU) 2009/28 has established a substitution of 10 % of the final consumption of energy by biofuels in the transport sector by 2020. Moreover, the European Union in this directive regulates the indirect land use and limits to 7% the contribution of fuels produced from food crops or first-generation feedstocks. Therefore, the directive (EU) 2015/1513 incentives the second-generation biofuels based on lignocellulosic biomass using environmentally friendly treatments.

Bioethanol based on second-generation feedstocks results in lower greenhouse emissions than first-generation bioethanol, reducing the environmental impact. By 2012, there were already more than 10 demonstration-scale facilities for second-generation bioethanol production, with the potential to grow to a commercial scale. Moreover, some commercial-scale facilities have come up, and an increase in the number of these ethanol plants is expected. However, its large-scale commercialization is still an important challenge due to the difficulty of biomass fractionation. In this sense, more details related to the lignocellulose-to-ethanol process are addressed in the following sections.

10.2.3 Third- and Fourth-Generation Biofuels

Third-generation biofuel, also called algae fuel or oilgae, is a biofuel based on algal biomass. Algae do not compete with food and can be produced throughout the year. However, they have some limitations in terms of ecological footprint, economic viability, sunlight dependence and geographical location and thus are not sufficient to replace fossil fuels. The search for the use of microalgae to produce fuel has received renewed attention as an ideal alternative third-generation biofuel feedstock because of their higher photosynthesis and fast growth rate as compared to terrestrial plants (Dutta et al., 2014).

Biodiesel produced from microalgae as an alternative to conventional crops is being investigated. Commonly, microalgae produce more oil, take less space, can be grown on land unsuitable for agriculture and are estimated to be able to produce 10 to 20 times more biodiesel than rapeseed. However, they need 55 to 111 times more nitrogen fertilizer—8 to 16 tons/ha/year (Demirbas, 2011).

Microalgae can be cultured in an open pond system (raceways) or using photobioreactors. The first is more acceptable in terms of a commercial point of view. The latter, however, is easy to control, less prone to contamination and can achieve high productivity.

The major advantage of the use of microalgae is the high growth rate, making it possible to satisfy the significant demand for biofuels and the use of limited land resources and less water consumption than land crops.

Several researchers have focused on the search of a perfect combination of algal species with a high lipid content and optimal growth conditions have been reported. Generally, fast-growing algae have a low lipid content. The highest average lipid content and biomass were obtained in *Chlrorella,* but it has a low triglyceride content. On the other hand, *Botryococcus braunii* and *Schizochytrium sp.* Produce high percentage of triglycerides with low biomass yields (Dutta et al., 2014).

Therefore, metabolic engineering of algae for production of biofuel, called fourth generation—photobiological solar fuels and electrofuels—has been considered. It is a direct conversion of solar energy into fuel using raw materials that are limitless, low-priced and widely available. This is an emerging field and is expected to bring fundamental advances on biofuels (Aro, 2016).

In this context, genetic modification/metabolic engineering could be a promising alternative to maximize lipid content and biomass yield to make it a more cost-effective and sustainable source of biofuel.

Fourth-generation biofuels include (1) photosynthetic microorganisms to produce photobiological solar fuels; (2) electrobiofuels, combining photovoltaics and microbial fuel production; or (3) synthetic cell factories or synthetic organelles designed to produce high-value chemicals and biofuels (Aro, 2016).

In the future, biofuels will definitely have an important role in meeting the world´s energy needs, sourced from plants and photosynthetic microorganisms for a carbon-neutral bioeconomy. Each generation of biofuel has come with its own set of advantages, disadvantages and challenges. For profitable commercial production of biofuel, the availability of feedstock, cost-effective processes and high-yield conversion systems are needed.

Currently, great efforts are devoted to the feasibility of second generation of biofuels. The main drawbacks of large-scale production are related to the technical difficulty of biomass processing. Therefore, this chapter focuses on the main aspects involved on the development of processes for second-generation biofuels and chemicals production following the biorefinery scheme.

10.3 Lignocellulosic Biomass for Biobased Fuels and Chemicals Production within Biorefinery Scheme

Lignocellulosic biomass (LCB) is composed of cellulose (a crystalline, linear polymer of D-glucose molecules bound together by β-1,4-glycosidic linkages with a high polymerization degree), hemicellulose (a set of branched, amorphous heteropolymers made from different sugars, which can be substituted by uronic acid moieties, acetyl groups or esterified phenolic acids, with a polymerization degree several times lower than cellulose) and lignin (a three-dimensional, amorphous polymer of complex structure, made of different phenyl-propane units). Additionally, LCB contains nonstructural components (including ashes, extractives, pectins and proteins) in lower proportions (Ares-Peón et al., 2013; Romaní et al., 2010). So, this feedstock is made up 50% to 60% carbohydrates (cellulose and hemicellulose) and 10% to 30% of lignin, together with nonstructural components (González-Muñoz et al., 2012).

LCB represents an abundant and renewable resource potentially suitable for the sustainable production of ethanol, chemicals, paper, pharmaceutical and biomaterials following the biorefinery concept (Gullón et al., 2012; Ruiz et al., 2011). The utilization of LCB can be carried out using two different approaches: (1) utilization as a whole (e.g., combustion, gasification or pyrolysis) or (2) the utilization of methods based on the selective separation and further valorization of its main components (cellulose, hemicellulose, lignin), known as biorefinery (Gullón et al., 2012). However, the efficient separation of LCB's main components is one of the major drawbacks for its utilization (Ruiz et al., 2011) because their complex structure hampers enzymatic hydrolysis (Yánez et al., 2014). This separation process can follow several steps, namely, starting with the separation of easily recovered fractions (extractives and hemicellulose) from the more resistant ones (cellulose, lignin), which could be further fractioned with more aggressive treatments. Therefore, LCB is a difficult raw material to convert into chemical products due to its complex polymeric structure, requiring multidisciplinary approaches to achieve its integrated benefit (González-Muñoz et al., 2012).

The integrated biorefinery is capable of providing a robust, self-dependent and sustainable alternative for the production of energy and high value-added products that could meet the demand for energy and chemicals (Alvarado-Morales et al., 2009). Generally, a biorefinery includes several processes from biological, enzymatic to chemical technologies, where biomass feedstock is fractionated into main constituting components for the conversion of a broad range of valuable products with applications in many industrial sectors, likewise to petroleum refineries (Carvalheiro et al., 2008; Yánez et al., 2014). Nevertheless, in contrast to petroleum refinery, a biorefinery uses renewable resources to produce fuels and chemicals that have less impact on the environment (Fernando et al., 2006). Biorefineries work as a "catch and release" carbon method, and the development of new manufacturing concepts enables the management of environmental challenges, namely, the decrease in greenhouse gas emissions and the mitigation of global warming (González-Muñoz et al., 2012). However, in order to become a viable alternative to petroleum refinery, a biorefinery must be competitive and cost-effective (Ruiz et al., 2011) and must follow sustainable development principles, such as (1) utilization of green processing technologies, (2) efficiency in the utilization of feedstocks (use of raw material as a whole avoiding waste generation) and (3) limitation of energy consumption and environmental impact (Gullón et al., 2012).

There is a wide range of feedstock that can be used and the main factors that play an important role in its selection are geographical location, availability, transport costs and feedstock composition (Alvarado-Morales et al., 2009; Carvalheiro et al., 2008).

In a biorefinery, biomass can be converted into useful products using a combination of technologies and processes, that produce high-value low-volume (HVLV) and low-value high-volume (LVHV) products through a series of unit operations (Fernando et al., 2006). Moreover, it can maximize the economic value of the biomass used while, simultaneously, reduce the waste streams produced (Kaparaju et al., 2009). In this field, the profitability prospect from both (1) high value-added compounds in limited amounts and (2) lower added-value compounds at a large scale fits in the scope of biorefineries (Gullón et al., 2012).

Consequently, a wide range of products can be made with different values and markets. Most emphasis to date has been on fuel production, but it is clear that there are many opportunities for chemical production, which will become increasingly significant in the near future (Alvarado-Morales et al., 2009).

10.3.1 Fractionation of Lignocellulosic Biomass: Pretreatment as First Step of Biorefinery

For the development of lignocellulosic biorefineries, suitable fractionation making use of environmentally friendly pretreatments is necessary. One of the main objectives is the removal of lignin/hemicellulose to increase the cellulose accessibility for hydrolytic microorganisms or enzymes (Karimi and Taherzadeh, 2016).

LCBs have external and internal surfaces, where the total accessible surface area is the sum of both. The external surface could be increased by size reduction, according to a typical physical pretreatment, because it depends on the size and shape. The internal surface depends on the pore size and distributions, and it decreases especially after drying. Thus, a suitable pretreatment should significantly increase the internal surfaces (Karimi and Taherzadeh, 2016).

Therefore, the pretreatment is considered the first step in a biorefinery since it allows the disruption of the LCB complex structure and facilitates the enzymatic access to polysaccharides. Enzymatic saccharification of polysaccharides produces fermentable sugars able to be used for bioconversion of fuels and chemicals. Hence, two steps should be optimized, namely, pretreatment and enzymatic hydrolysis (González-Muñoz et al., 2012).

Nevertheless, there is no consensus on the best type of pretreatment, although some representative examples are reported in literature according to hemicelluloses solubilization or delignification (Romaní et al., 2010).

A suitable pretreatment should fulfill the following requirements: (1) allow high carbohydrate recovery; (2) produce substrates with high digestibility towards the subsequent enzymatic hydrolysis; (3) avoid the formation of by-products (phenolic acids, furfural or 5-hydroxymethylfurfural) that cause inhibition in the subsequent hydrolysis and/or fermentation, reducing or suppressing detoxification steps; (4) suitable for operation at high solid concentration, enabling the production of hydrolysates with a high concentration of sugars; (5) to be cost-effective; (6) limited energy consumption; (7) employ green solvents; and (8) limited generation of wastes (Gregg and Saddler, 1996; Hama et al., 2015). The multiple methods proposed for LCB pretreatment can be grouped in different categories: physical, chemicals, biological or a combination of them. Table 10.1 resumes the main pretreatments, and their advantages and disadvantages. During the pretreatment some sugars are converted into toxic inhibitors of microbial growth or enzymes, such as furan derivatives (mostly furfural and 5-hydroxymethylfurfural)

TABLE 10.1

Pretreatments: Physical, Chemical, Biological or a Combination of These

Pretreatment		Method	Advantages	Disadvantages
Physical (Chen et al., 2017; Kumar and Sharma, 2017; Kumari and Singh, 2018)	Mechanical	Dry crushing, wet crushing, vibrating ball mill grinding and comprehension	Reduce the particle size to expand the contact surface	Cannot remove lignin and hemicellulose High cost
	Microwave	Acceleration of ions and collision with other molecules with an alternating electric field	Improve cellulose accessibility and reactivity Enhance ability of lignocellulose feedstock to enzymes Simple operation; Energy-efficient Heat in short time	High cost of equipment
	Ultrasound	Cell disruption technique	Improve accessibility and chemical reactivity of cellulose Decompose hemicellulose	Limited effect on fine structure of cellulose Negative effect on enzymatic hydrolysis
	High-energy electron radiation	Emission of rays at the raw material	Reduction of the degree of polymerization of cellulose Loose cellulose structure Improve hydrolysis and conversion rates Increase moisture Less pollution Environmentally friendly	High degree of cellulose structure damage High cost with high-energy electron radiation Difficult large-scale industrial production
	High-temperature pyrolysis	Include pyrolysis and liquid hot water decomposition	Decompose cellulose rapidly Promote hydrolysis of raw material to oligosaccharides High hydrolysis efficiency	Limited microbial fermentation Low solid content Large energy consumption
Chemical (Chen et al., 2017; Kumar and Sharma, 2017; Kumari and Singh, 2018)	Concentrated acid pretreatment	Acid concentration >30%; 100°C; 2–10 h; 1 atm	High sugar conversion rates	Highly toxic and corrosive acids High operational and maintenance costs
	Diluted acid pretreatment	Acid concentration <10 %; 100–240°C; several minutes; > 10 atm	Fast reaction Suitable for continuous production (no need to recycle acid).	High temperature and pressure Negative influence of degradation products on fermentation
	Alkali pretreatment	NaOH, KOH $Ca(OH)_2$ and ammonium hydroxide, seconds to days, room temperature, atmospheric pressure	Room temperature and natural atmospheric pressure Increase internal area Decrease polymerization degree and crystallinity Break the chemical bonds between lignin and carbohydrates Destroy lignin structure	Less sugar degradation compared with acid pretreatment

(*Continued*)

TABLE 10.1 (Continued)

Pretreatment		Method	Advantages	Disadvantages
	Oxidation pretreatment	Degradation of lignocellulose by oxidant (ozonolysis, wet oxidation and photocatalysis)	Ozonolysis: Does not produce toxic substances and inhibitory compounds Remove lignin at room temperature and normal pressure Wet oxidation: Effective treatment compared with other methods Photocatalysis: Cannot affect the final products	Ozonolysis: High cost with large amount of ozone needed Wet oxidation: High temperature and high pressure Photocatalysis: Reduce reaction time.
	Ionic Liquid pretreatment	Cellulose "green solvent", no explosive or toxic gases are formed.	Large temperature range Break noncovalent interactions of lignocellulose Minimize products degradation Environmentally friendly	High cost
	Organosolv pretreatment	Organic solvents to degrade lignin and, to break and decompose the internal chemical bonds of cellulose and hemicellulose	Obtain pure lignin, cellulose and hemicellulose	Expensive organic solvents Effect on environment and fermentation
Physicochemical (Chen et al., 2017; Kumar and Sharma, 2017; Kumari and Singh, 2018)	Steam explosion pretreatment	Hydrothermal pretreatment with high-temperature steam; combines chemical effects and mechanical forces.	Lower environmental impact Lower requirement for reaction conditions Lower cost investment Fewer hazards of chemical reagent; Complete sugar recovery.	High temperature and pressure Degradation compounds could affect enzymatic hydrolysis
	AFEX pretreatment	Combination of steam explosion with alkali treatment in liquid anhydrous ammonia with high temperature	Increase surface area of cellulose Improve enzyme accessibility Remove hemicellulose and lignin partly Absence of inhibitory substances	High cost Not efficient for raw material with high lignin content
	CO_2 explosion treatment	Addition of CO_2 to steam explosion	Improve enzymatic hydrolysis Remove lignin Does not produce inhibitory compounds Non-toxic Nonflammable Easy recovery after extraction Cost-effective	High cost (higher than steam explosion and lower than AFEX)
	SO_2 explosion treatment	Similar to CO_2 explosion	Low optimal temperature Partial cellulose hydrolysis	High requirements of the equipment Large amount of degradation compounds
	Electrical catalysis pretreatment	Use electrode material, namely, titanium-based SnO_2 and CeO_2	Increase surface area Remove lignin Does not produce inhibition compounds Cleanliness; Cost-effective	Lower efficiency
Biological		Use fungi to degrade lignin (wood-rotting fungi)	High delignification efficiency Mild conditions Low energy consumption Absence of pollution	Need long cycle Fungi use cellulose and hemicellulose to growth Low activity of ligninolytic enzymes

and several phenolic compounds (namely vanillin) and from heavily acetylated polymers, such as acetic acid, which is the frequent inhibitor present in plant-biomass hydrolysates (Duarte et al., 2013).

Considering LCB biorefineries, the key instrument is the efficient utilization of the major feedstock components and the development of viable technologies that allow efficient and clean fractionation to produce multiple commercial products and develop the so-called bioeconomy (Menon and Rao, 2012; Nunes et al., 2012).

10.3.2 Second-Generation or Lignocellulosic Biofuels

The second generation of biofuels are produced from lignocelluloses materials, namely, forest biomass and agro-residues (e.g., corn stover, wheat, barley straws), agro-processing by-products (e.g., corn fiber, sugarcane bagasse, seed cake, etc.), energy crops (e.g., switchgrass, poplar, banagrass, miscanthus, etc.) and wood industry residues (bark wood, sludge paper; Cherubini, 2010). These biofuels can be obtained by different methods including biochemical or thermochemical processes. Thermochemical fuels include methanol, refined Ficher-Tropsch liquids and dimethyl ether. Lignocellulosic biomass is firstly submitted to gasification, liquefaction, combustion or pyrolysis for the production of energetically liquid and gaseous products. These processes require extreme temperatures and pressures.

On the other hand, the production of liquid fuels as bioethanol or butanol, includes chemical or biological processes. The first step is a pretreatment for the fractionation of the recalcitrant LCB structure to improve the access toward cellulose.

Bioethanol is the most important biofuel produced worldwide (Balat, 2011). The main steps required for bioethanol production are (1) mechanical and physic-chemical pretreatment to decompose the complex structure of material, (2) enzymatic hydrolysis of cellulose to monomer sugars, (3) fermentation of sugars to ethanol and (4) product recovery and concentration usually by distillation (Galbe and Zacchi, 2002).

In short, the overall ethanol yield and ethanol production rate rely on the sugar yield and the fermentability of the solution, which is influenced by the concentration of the soluble substances released during pretreatment, present in the original raw material or formed in the pretreatment step.

Great efforts have been devoted to improving the cost-efficient production of ethanol. In this sense, one main objective is to reduce costs related to the process. Accordingly, operating at high solid loadings in the pretreatment and saccharification and fermentation is mandatory (Koppram et al., 2014). The increase of substrate loading up to 15% implies a high glucose concentration that allows reaching ethanol concentration of 4% (w/w). Moreover, the saccharification and fermentation processes may be carried out simultaneously or separately. Simultaneous saccharification and fermentation (SSF) show several advantages as glucose produced is immediately consumed by the yeast, which reduces the product inhibition of enzymes. Moreover, the use of only one reactor results in overall cost reduction of equipment. However, the SSF process is carried out in nonoptimal conditions for yeast and enzymes (Domínguez et al., 2017). In this sense, thermotolerance of yeast is a desired feature for the development of integrated and intensified processes.

Another strategy for cost reduction aims to use the whole slurry (liquid and solid phases) obtained from pretreatment. This approach shows several advantages as the increase of fermentable sugars derived from hemicellulose, which allows to achieve higher ethanol concentration. The use of hemicellulosic fraction as liquid phase avoids washing steps and saves water consumption, which have positive effects on production costs (Romaní et al., 2014a). Nevertheless, these operating conditions are very demanding for microorganisms and enzymes involved in the process, causing their inhibition and lower product yields. In fact, the presence of degradation compounds derived from physical or chemical treatment is considered one of the main drawbacks related to lignocellulosic fermentations. These degradation compounds can be classified into weak acids (acetic acid, levulinic acid, formic acid), furan compounds (furfural, hydroxymethylfurfural) and phenolic compounds. In this sense, the selection of the microorganisms used to obtain a high ethanol yield is of great importance. Recent scientific advances are focused on the search for more tolerant yeast and enzyme (Pereira et al., 2014; Romaní et al., 2014b). Isolated strains from industrial environments have shown great potential for lignocellulosic ethanol process due to a higher capacity to detoxify furan-inhibitory compounds (Pereira et al., 2014). Another important challenge of whole-slurry fermentation is pentose consumption, required also for a cost-efficient ethanol production. Xylose is the main sugar present in lignocellulosic hydrolysates obtained from the processing of agricultural wastes and hardwoods. Nevertheless, the most used yeast for ethanol production, *Saccharomyces cerevisiae*, does not naturally consume xylose. Therefore, metabolic engineering tools have been used for the xylose consumption and for the increase of tolerance to inhibitory compounds (such as acetic acid and furan compounds; Costa et al., 2017; Cunha et al., 2018, 2015).

Moreover, the reduction of the global enzyme loading by enzymes recycling (Gomes et al., 2016) and the addition of additives (such as bovine serine albumin or recombinant CBM-carbohydrate-binding module) to improve saccharification yields in the presence of lignocellulosic hydrolysates (Oliveira et al., 2018) have been proposed as improvements on enzymatic saccharification of lignocellulosic biomass.

The coproduction of value-added compounds and biofuels from lignocellulosic materials so far has been based on single feedstocks. Despite the low cost of lignocellulosic materials, some problems related to seasonality and lower sugars concentration than first-generation feedstocks hinder large-scale exploitation of second-generation bioethanol. Therefore, another trend on lignocellulosic biofuels is to use a multi-waste valorization approach including the mixture of renewable resources as cheese whey (containing high amounts of lactose) with pretreated lignocellulosic biomass (Cunha et al., 2018). The mixture of simple sugars as lactose or sucrose with second-generation feedstocks increases the ethanol concentration (>90 g/L), which resolves one of the limitations of lignocellulosic ethanol fermentation in which the final concentration is limited to cellulose content (Cunha et al., 2018). On the other hand, the mixture of forest and marginal land resources from s determined geographical area aims to resolve problems related to availability, seasonality and fire risk of the uncultivated area (Pontes et al., 2018).

Therefore, several strategies have been developed to reduce operation costs related to lignocellulosic-to-ethanol processes. Nevertheless, finding an effective solution is still the main challenge for a real cost-effective exploitation of lignocellulosic biofuels. In this sense, the coproduction of value-added compounds from hemicellulosic fractions (such as xylitol or furfural) or valorization of lignin could improve the global process contributing to creation of new jobs and industries.

10.3.3 Chemical Platform from Lignocellulosic Biomass: Lactic Acid, Furan Derivatives, Phenolic Compounds and Xylitol

The chemical platform aims to use the three main products (cellulose, hemicellulose, lignin), with full valorization of all fractions, which significantly contributes to a better economy of biorefinery operations. The products include specialty sugars, such as food additives, as well as chemicals and products with multiple processing routes; see Figure 10.2.

10.3.3.1 Lactic Acid/Polylactic Acid

LCBs are potential feedstocks for the production of organic acids, such as lactic acid (LA). Both cellulose and hemicellulose can be hydrolyzed into fermentable monosaccharides, such as glucose and xylose, and converted to LA (Maas et al., 2006). Lactic acid production from renewable biomass is an economic and environmentally friendly alternative when compared to the LA production from fossil resources, considered unsustainable due to resources depletion and the accumulation of environmentally hazardous chemicals (John et al., 2009).

LA presents two optically active configurations L(+) and D(–), the first one is produced in humans, other mammals and bacterial systems, the second one is only produced in bacterial systems (Garlotta, 2002).

LA is industrially produced either by chemical synthesis or by microbial fermentation. However, 90% of commercially produced LA is made by bacterial fermentation, whereas only 10% is produced synthetically through the hydrolysis of lactonitrile (Hofvendahl and Hahn-Hägerdal, 2000; John et al., 2007; Schmidt and Padukone, 1997).

LA bacteria (LAB) ferment sugars through different pathways as homo-, hetero-, or mixed acid fermentation. Homofermentation, used by homolactic acid bacteria, produces only LA as the end product of glucose metabolism (Maas et al., 2006). Heterofermentation produces equimolar amounts of LA, carbon dioxide and ethanol or acetate from glucose through the phosphoketolase pathway, as is shown in Figure 10.3. The ratio of ethanol and acetate formed, by facultative heterofermenters, rely of the redox potential of the system (Hofvendahl and Hahn-Hägerdal, 2000).

The conventional method to produce lactic acid is an anaerobic fermentation in batch reactors, and to keep the fermentation, the acid produced is neutralized with alkali or removed from the fermentation system (Ghaffar et al., 2014).

LA is an interesting end product because of (1) high market price, (2) growing demand, (3) high fermentation yield (theoretical product yield in homolactic fermentations) and (4) wide range of applications (Moldes et al., 2000).

The most important application of LA may yet be as a raw material for producing the polymer polylactic acid (PLA). However, the use of LA has a wide range of applications, such as acidulant, flavoring and preservative in the food, pharmaceutical, leather and textile industries, and for the production of other base chemicals (Hofvendahl and Hahn-Hägerdal, 2000; John et al., 2007; Schmidt and Padukone, 1997).

PLA is one of the most important commercial biodegradable thermoplastic polymer. It has a favorable melting point (about 160–180°C, below the temperatures at which natural fibers start to degrade), versatility (many applications in several areas), suitability for melt processing (injection molding), biodegradability, thermoplasticity and high strength (González et al., 2010; John et al., 2009). Naturworks LLC, the major producer of PLA, has

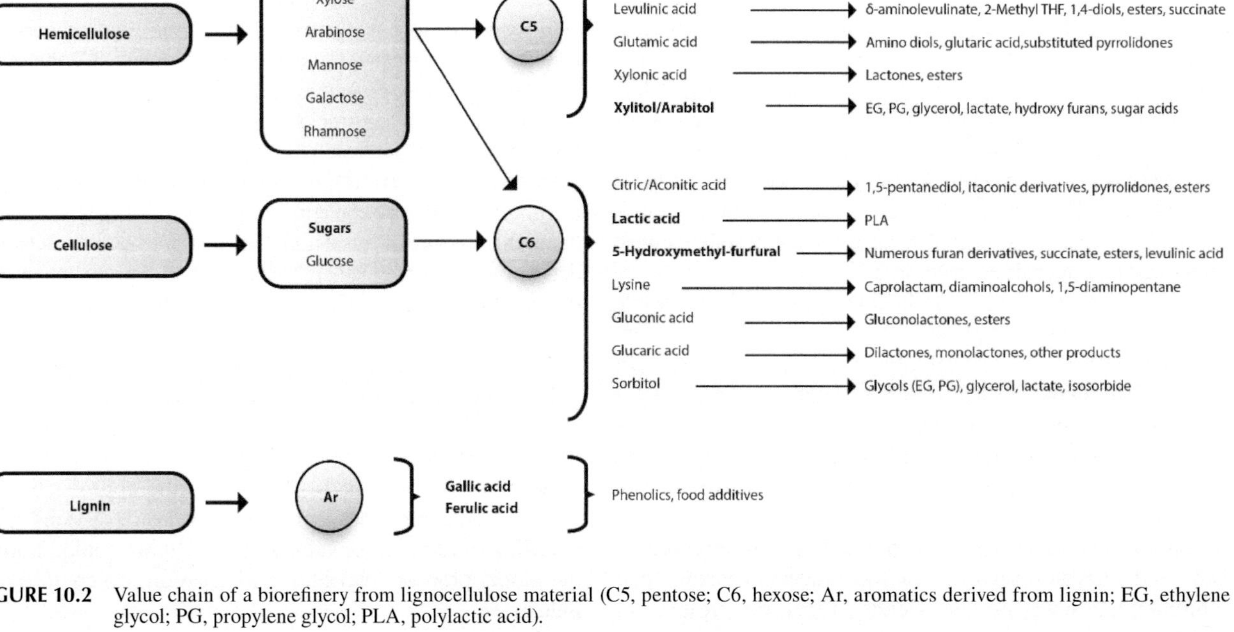

FIGURE 10.2 Value chain of a biorefinery from lignocellulose material (C5, pentose; C6, hexose; Ar, aromatics derived from lignin; EG, ethylene glycol; PG, propylene glycol; PLA, polylactic acid).

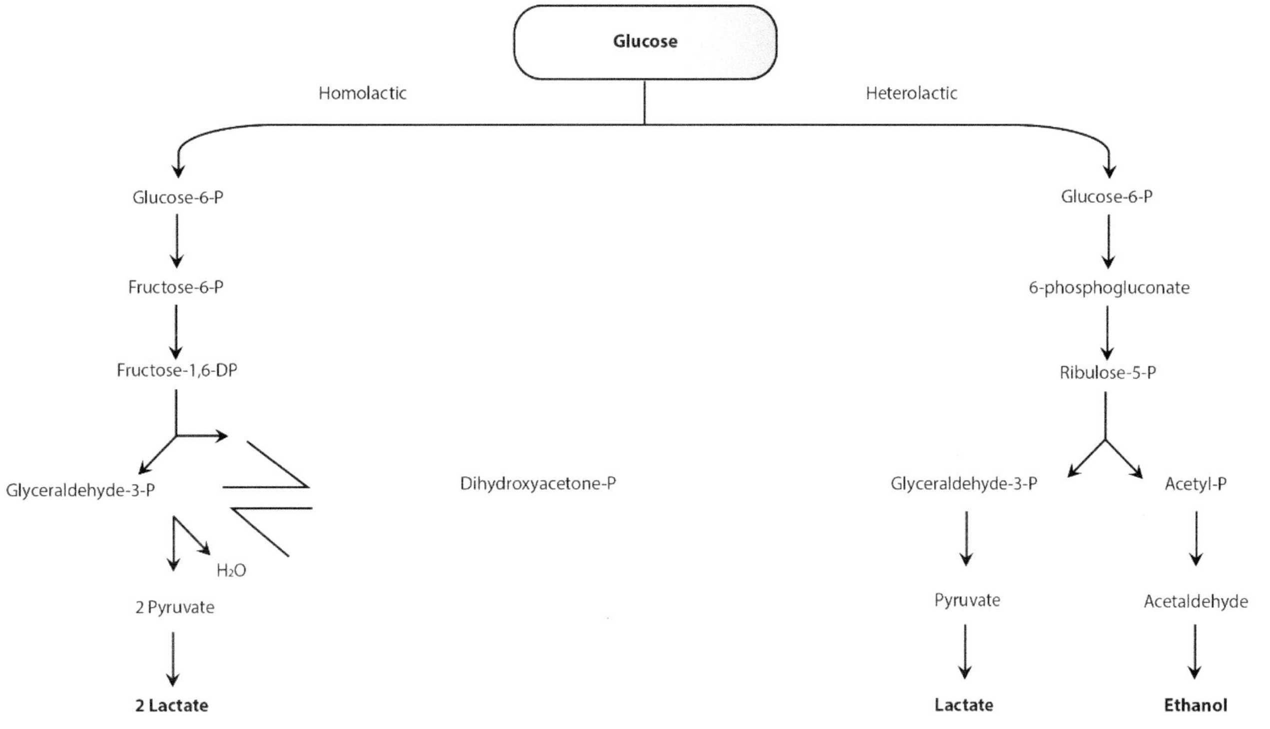

FIGURE 10.3 Generalized scheme for the fermentation of glucose to lactic acid bacteria (adapted from Cubas- Cano et al. (2018).

considerably increased the production capacity of its PLA plant in Blair (Nebraska, USA) reaching 140,000 metric tons/year. Environmental concerns and legislative incentives, particularly in the European Union and the United States, tend to promote the use of biodegradable plastics (González et al., 2010).

Highly purified L(+)–LA anhydrous monomer is required for the production of the biodegradable polymer PLA (Maas et al., 2006). The synthesis of lactic acid into PLA can be done in two different polymerization methods as shown in Figure 10.4, in which lactic acid can be polymerized to yield a low-molecular-weight polymer, that through coupling agents will increase the molecular weight of the PLA. The second route is to collect, purify and ring-open polymerize lactide to yield high molecular weight PLA (Garlotta, 2002).

10.3.3.2 Furan Derivatives

According to pretreatment, toxic compounds are generated, and their amounts depend on the raw material and the harshness of pretreatment. These coproducts can be divided into carboxylic acids, furan derivatives and phenolic compounds.

Among degradation products, the main furan derivatives are furfural and 5-hydroxymethylfurfural (HMF) derived from pentoses and hexoses degradation, respectively. These compounds have been considerably undesirable due to their inhibition effect on enzymes and microorganisms, limiting polysaccharides hydrolysis and sugar fermentation to value-added products. Currently, to improve the potential of lignocellulosic biorefineries, several studies are focusing on the development of a furan derivatives platform (furfural and HMF) since they are considered important chemical building blocks able to be transformed in several commodities (Jean-Paul et al., 2011; Rosatella et al., 2011).

In this sense, hemicellulose could be depolymerized to oligomer and monomers by an acid and hydrothermal treatment, and these sugars can be isomerized and dehydrated to furfural and HMF. Hence, hemicellulose fraction is a potential substrate for furan production, which is formed with different sugars units, namely, glucose, galactose, mannose, xylose and arabinose. The rate of hydration depends on the sugar type, showing the decrease order: xylose, mannose and glucose.

Furfural is typically produced by the dehydration of pentose in the presence of an acidic catalyst but can also be formed through thermal decomposition of xylose wherein the organic acids formed act as catalysts.

Quaker Oats in 1921 was the first company that produced furfural in a batch process. Currently, most furfural is produced by continuous process (Westpro-modified Huaxia Technology), 280,000 tonnes per year, where China is the main exporter (200,000 tones), followed by small producers in the Dominican Republic, South Africa, Thailand and Spain (Karinen et al., 2011). Some furfural production processes are shown schematically in Figure 10.5.

Furfural has suitable properties to be used in different industrial fields as food (flavor-enhanced), chemical (solvent, gasoline additive, resin, lubricants, fuel), plastic (polymers) and pharmaceutical industries.

HMF is very useful because it is reactive and can be an intermediate for the production of the biofuel dimethylfuran (DMF) can be also converted to levulinic acid, 2,5-furandicarboxylic acid (FDA), 2,5-diformylfuran (DFF), dihydroxymethylfuran and 5-hydroxy-4-keto-2-pentenoic acid.

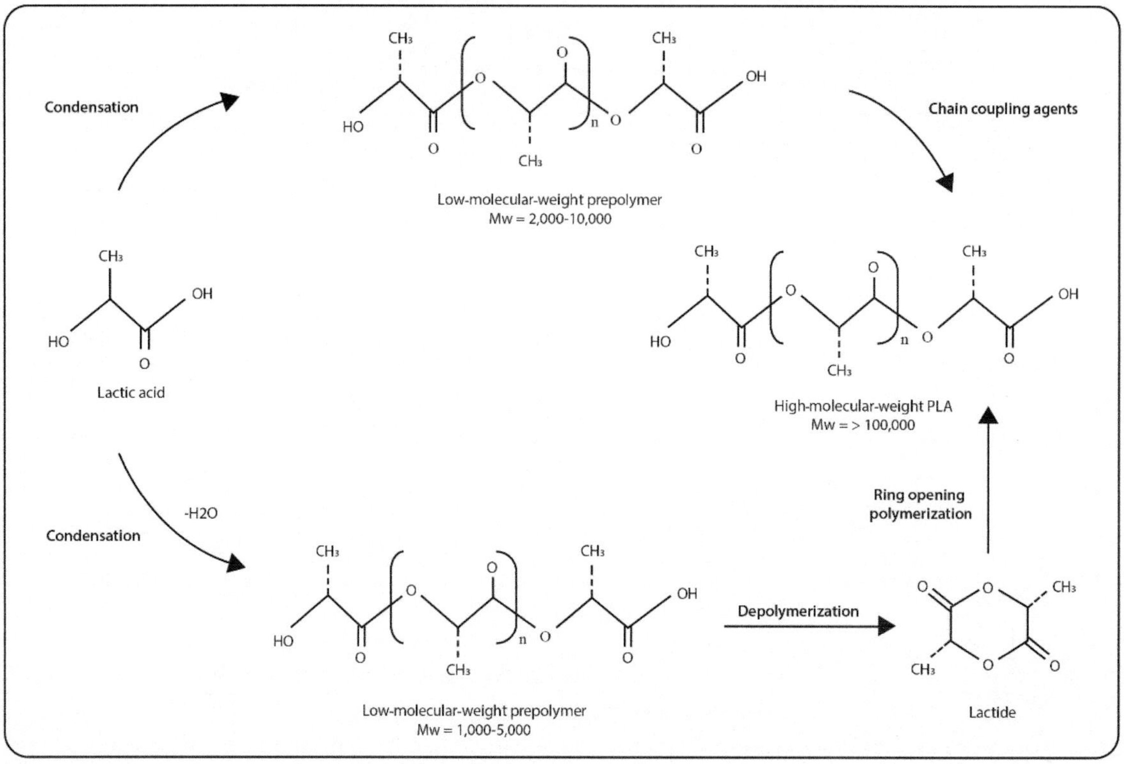

FIGURE 10.4 Synthesis routes of PLA (adapted from Garlotta et al., 2002).

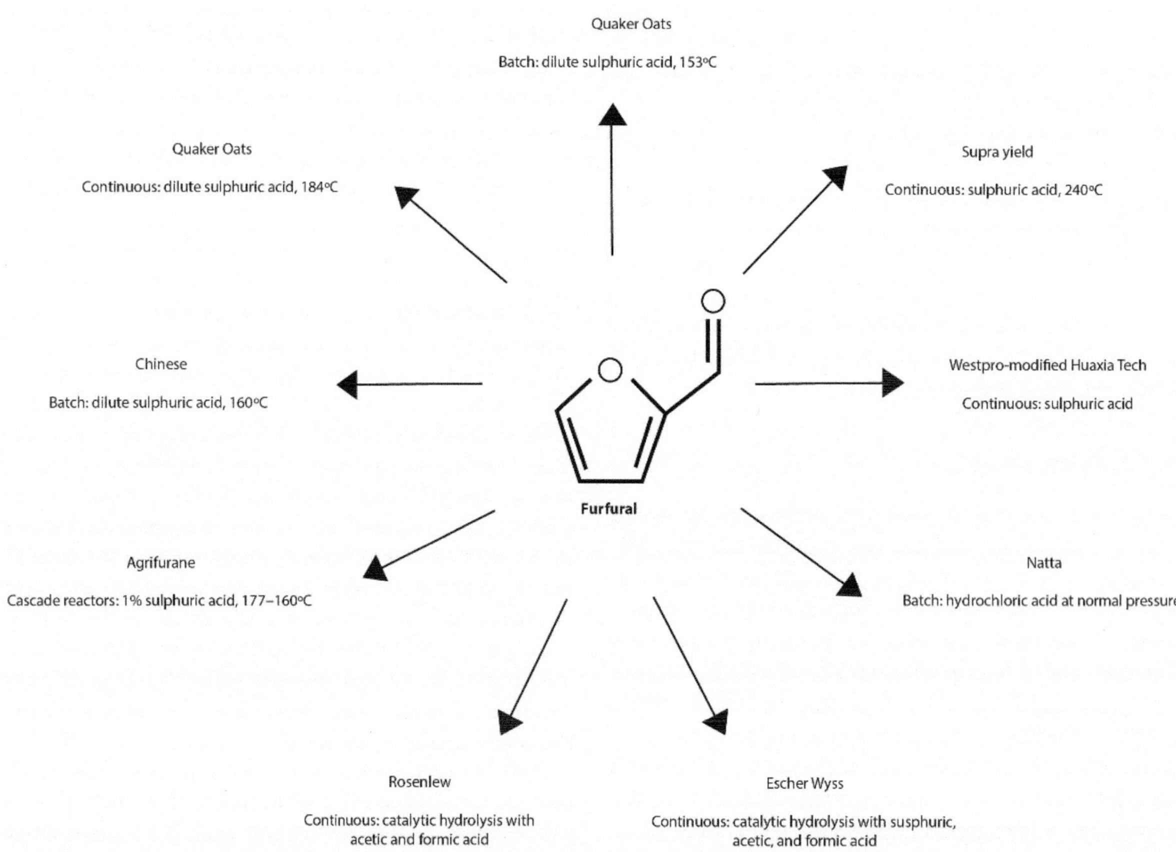

FIGURE 10.5 Furfural production processes (based on Karinen et al., 2011).

10.3.3.3 Xylitol

Other important chemical obtained from lignocellulosic biomass is xylitol, considering the current search for sugar-free and low-calorie food. Xylitol is a product of great commercial value as a food additive, with a lot of health benefits (insulin-independent metabolism, anticariogenic, pharmacological properties). It has sweetening power similar to sucrose and has high market value, making it an important coproduct in biorefinery.

Xylitol, naturally found in small amounts in many fruits, vegetables, yeasts, lichens, seaweeds and mushrooms, can be extracted through solid/liquid extraction; nonetheless, they have low xylitol content and a high cost in raw materials. Thus, several studies have been done regarding the xylitol production by chemical or biotechnological processes (D. et al., 2015; Rafiqul and Sakinah, 2013)

Currently, xylitol is industrially produced by a chemical approach with a catalytic reduction of pure D-xylose, valorizing the fraction of pentoses (C5) of hydrolyzed biomass.

The chemical process comprises four steps: (1) pretreatment of lignocellulosic (acid or enzymatic hydrolysis), (2) purification and separation of hydrolysate to obtain pure xylose from a variety of sugars (xylose, arabinose, glucose, galactose and mannose), (3) catalytic hydrogenation of xylose to xylitol at high pressure (40–70 bar) and temperature (80–140°C) with a nickel catalyst and (4) crystallization and separation of xylitol. The recovery of xylose solution with a high degree of purity and high xylitol yield for commercial scale is fundamental, although it makes the process very expensive.

As an alternative to the conventional method, there are two biotechnological processes: the microbial process and the enzymatic approach. The microbial process relies on bacteria, fungi and yeast for xylitol production from pure xylose or hemicellulose hydrolyzate. There are several studies regarding xylitol production, and the best xylitol producers among microorganisms are considered to be yeast, namely, from the genus *Candida* (Nigam and Singh, 1995; Parajó et al., 1998; Winkelhausen and Kuzmanova, 1998).

The enzymatic approach uses isolated enzymes, namely, xylose reductase and xylitol dehydrogenase, isolated from yeast, under oxygen-limited or anaerobic conditions. However, there are few articles regarding the commercial enzymatic conversion of xylose to xylitol and the high cost for large-scale production can be responsible for that. Thus, more research is required to develop a cost-effective way, as well as an optimized method, for enzyme-based xylitol production, which may provide an alternative for chemical process. Cheaper than the chemical process, the microbial process does not require extensive xylose purification. Moreover, microbial method has low xylitol productivity, and it requires large amounts of sterilized distilled water and long residence times.

10.3.3.4 Lignin-Derived Compounds

The chemical structure has made lignin a potential feedstock for the production of biorenewable fuels and chemicals. Higher in energy content per unit mass when compared to carbohydrates, lignin is the largest source of renewable aromatics on Earth (Azadi et al., 2013).

Lignin presents physicochemical properties that enable its use for a wide range of products, namely, (1) compatibility with an extensive range of industrial chemicals; (2) presence of aromatic rings, allowing a wide range of chemical transformations; (3) the presence of other reactive functional groups, which enable easy preparation of graft copolymers; (4) good rheological and viscoelastic properties; (5) good film-forming ability; (6) small particle size and (7) hydrophilic or hydrophobic character according to its origin, allowing a wide range of blends to be produced (Kleinert and Barth, 2008).

Nevertheless, lignin is not usually used because of its structure, low degree of order and high degree of heterogeneity. Thus, the major disadvantages to the development of lignin-based biorefining technologies are, among others, the existence of a variety of different interunit linkages, a high affinity for the formation of a better condensed structure when thermochemically processed and subpar product selectivity. (Azadi et al., 2013).

The method of extraction, as wells as its source, has a strong impact on lignin properties. The reactivity and physicochemical properties of lignins are dependent to certain extent, on their molecular weight distribution (Kleinert et al., 2008); therefore, functional groups present in lignin extracts depend significantly on method of isolation (Pandey and Kim, 2011)

The environmental waste regulations, increasingly stringent, at the national and European level, mean that all "wastes" must be dealt, as well as lignin, and integrated into new and established chemical industry subsectors, such as polymers, resins, adhesives and others (Stewart, 2008).

Lignin might be used as a phenol substitute in phenol formaldehyde resins. However, its chemical heterogeneity is a limiting factor that can be overcome by a biochemical modification of lignin, the addition of filler agents and the use of lignins derived from novel process (lignins derived from acetosolv, acid hydrolysis and organosolv processes). From an environmental point of view, the use of lignin in replacement of phenol-formaldehyde resin decreases formaldehyde emission during the process (Doherty et al., 2011).

Epoxy-resin adhesives, with the substitution of phenol by lignin, produce an inexpensive and renewable adhesive, although, this transformation is strongly affected by the type of lignin used. The epoxide resin market is economically interesting, with specific reference to phenol-epoxy resins, which could be a good opportunity for lignin (Stewart, 2008).

Low-molecular-weight aromatic compounds are commonly used in a variety of different industries as solvents or chemical reagents. Concerning the use of lignin-derived aromatics, there are two different options: (1) the production of the most commonly used aromatics such as BTX (benzene, toluene, xylene), which are easy to incorporate into the existing industrial processes and products, and (2) the production of alkylated phenols with a high degree of similarity with the monolignols and attempting to modify the existing industrial processes and products. In both cases, product purification adds value to lignin-derived product but imposes additional energy and material loss. The degree of use of mixed products, such as lignin depolymerized mixtures, can be tolerated in the following decreasing order, according to the cascade value: fuels, solvent and chemical reagent.

In biomass conversion processes, lignin-derived phenolic monomers, such as propyl guaiacol, can be used to extract biomass compounds from aqueous acid solutions. It can serve as organic phase in biphasic catalytic reactors in which carbohydrates are transformed into useful chemical platforms and as a medium for the selective hydrogenation of these platform compounds to fuels and fuel blend, such as gamma-valerolactone from aqueous solutions containing mineral acids. Furthermore, these alkylated phenols obtained from lignin are stable in the operating conditions usually utilized in biphasic reactors and the downstream hydrogenation steps. In summary, the high-boiling alkylated phenols obtained from depolymerization of lignin seem to offer major technical and economic advantages if adapted to be utilized within the lignocellulosic biorefineries (Azadi et al., 2013).

However, there are lignin utilizations not readily categorized, and all the applications previously reported are dependent on the source and purity of the lignin. Usually, the less lignin is heterogeneous, the fewer problems are found in the application or end product. This can be explained, due to the removal of impurities which decreases its heterogeneity, chemically and structurally, making lignin behave in a more uniform manner during any subsequent process (Stewart, 2008).

Although there are prospects of macromolecular lignin being utilized as raw material in many industrial processes (e.g., copolymer and polymer, metal recovery, cosmetic, resins, adhesives, fuels and aromatics chemicals), the most prominent lignin utilization so far appears to be lower-molecular-weight compounds derived from its depolymerization. Lignin can be gasified to produce H_2 and CO (syngas) which in turn can be synthesized into methanol and various other chemicals through known technologies. Pyrolysis oil can be obtained from lignin in a way similar to biomass pyrolysis. However, pyrolysis oil produced from lignin pyrolysis is highly oxygenated and unstable during storage and hence requires further upgrading (Pandey and Kim, 2011).

The efficient use of lignin can potentially reduce greenhouse gas emissions because it can directly produce biorenewable fuels and chemicals with a small carbon footprint with a minimal impact on food. The ever-tightening environmental regulation requires a smart and efficient use of all fractions of biomass, including lignin, in order to reduce the overall emissions and achieve more sustainable economics. Additionally, the conversion of lignin into value-added products will improve the viability of the lignocellulosic biorefineries and potentially enable the conversion of cellulose to biofuels (Azadi et al., 2013).

10.4 Conclusion Including Future Directions

Current trends in energy consumption show the urgency of the search for alternative clean fuel. The shortage of petroleum-based fuels that is foreseen for the near future, associated with the negative environmental impact, has driven the use of lignocellulosic biomass to produce alternative sources of biofuels. The high demand of the transport sector for fuels have led, for example, Brazil to the successful introduction of Flexi-fuel vehicles that use bioethanol (first-generation bioethanol from sugarcane) as fuel. The production of bioethanol from lignocellulosic biomass (second-generation biofuel) could help avoid a forthcoming energy crisis. However, there are some difficulties in the viability of the production of second-generation bioethanol, such as the high cost of enzymes, problems related to biomass-processing technologies, the formation of inhibitory by-products of hydrolysis and/or fermentation, the corrosion of equipment and so on. Thus, a chemical platform for the valorization of the three main components of lignocellulosic biomass (i.e., cellulose, hemicellulose, lignin), aiming the production of biobased chemicals besides bioethanol, with high value, can enable to reduce nonrenewable fuel consumption while, simultaneously, could contribute significantly to bioethanol production and for the viability of the biorefining industry.

REFERENCES

Alvarado-Morales, M., Terra, J., Gernaey, K.V., Woodley, J.M., Gani, R., 2009. Biorefining: Computer aided tools for sustainable design and analysis of bioethanol production. *Chem. Eng. Res. Des*. 87, 1171–1183. https://doi.org/10.1016/j.cherd.2009.07.006.

Ares-Peón, I.A., Romaní, A., Garrote, G., Parajó, J.C., 2013. Invasive biomass valorization: Environmentally friendly processes for obtaining second generation bioethanol and saccharides from Ulex europaeus. *J. Chem. Technol. Biotechnol*. 88, 999–1006. https://doi.org/10.1002/jctb.3963.

Aro, E.-M., 2016. From first generation biofuels to advanced solar biofuels. *Ambio*. 45, 24–31. https://doi.org/10.1007/s13280-015-0730-0.

Azadi, P., Inderwildi, O.R., Farnood, R., King, D.A., 2013. Liquid fuels, hydrogen and chemicals from lignin: A critical review. *Renew. Sustain. Energy Rev*. 21, 506–523. https://doi.org/10.1016/J.RSER.2012.12.022.

Balat, M., 2011. Production of bioethanol from lignocellulosic materials via the biochemical pathway: A review. *Energy Convers. Manag*. 52, 858–875. https://doi.org/10.1016/j.enconman.2010.08.013.

Bessou, C., Ferchaud, F., Gabrielle, B., Mary, B., 2011. Biofuels, greenhouse gases and climate change. A review. *Agron. Sustain. Dev*. 31, 1–79. https://doi.org/10.1051/agro/2009039.

Carvalheiro, F., Duarte, L.C., Gírio, F.M., 2008. Hemicellulose biorefineries: A review on biomass pretreatments. *J. Sci. Ind. Res*. (India). 67, 849–864.

Chen, H., Liu, J., Chang, X., Chen, D., Xue, Y., Liu, P., Lin, H., Han, S., 2017. A review on the pretreatment of lignocellulose for high-value chemicals. *Fuel Process. Technol*. https://doi.org/10.1016/j.fuproc.2016.12.007.

Cherubini, F., 2010. The biorefinery concept: Using biomass instead of oil for producing energy and chemicals. *Energy Convers. Manag*. 51, 1412–1421. https://doi.org/10.1016/j.enconman.2010.01.015.

Costa, C.E., Romaní, A., Cunha, J.T., Johansson, B., Domingues, L., 2017. Integrated approach for selecting efficient Saccharomyces cerevisiae for industrial lignocellulosic fermentations: Importance of yeast chassis linked to process conditions. *Bioresour. Technol*. 227, 24–34. https://doi.org/10.1016/j.biortech.2016.12.016.

Cubas-Cano, E., González-Fernández, C., Ballesteros, M., Tomás-Pejó, E., 2018. Biotechnological advances in lactic acid production by lactic acid bacteria: Lignocellulose as novel substrate. *Biofuels, Bioprod. Biorefining*. 12, 290–303. https://doi.org/10.1002/bbb.1852.

Cunha, J.T., Aguiar, T.Q., Romaní, A., Oliveira, C., Domingues, L., 2015. Contribution of PRS3, RPB4 and ZWF1 to the resistance of industrial Saccharomyces cerevisiae CCUG53310 and PE-2 strains to lignocellulosic hydrolysate-derived inhibitors. *Bioresour. Technol*. 191, 7–16. https://doi.org/10.1016/j.biortech.2015.05.006.

Cunha, J.T., Costa, C.E., Ferraz, L., Romaní, A., Johansson, B., Sá-Correia, I., Domingues, L., 2018. HAA1 and PRS3 overexpression boosts yeast tolerance towards acetic acid improving xylose or glucose consumption: unravelling the underlying mechanisms. *Appl. Microbiol. Biotechnol*. 102, 4589–4600. https://doi.org/10.1007/s00253-018-8955-z.

D., M.A., R., K.K., C., K.A., 2015. The role of process integration in reviewing and comparing biorefinery processing routes, in: *Process Design Strategies for Biomass Conversion Systems*. Wiley-Blackwell, pp. 309–327. https://doi.org/10.1002/9781118699140.ch12.

Demirbas, M.F., 2011. Biofuels from algae for sustainable development. *Appl. Energy*. 88, 3473–3480. https://doi.org/10.1016/J.APENERGY.2011.01.059.

Doherty, W.O.S., Mousavioun, P., Fellows, C.M., 2011. Value-adding to cellulosic ethanol: Lignin polymers. *Ind. Crops Prod*. 33, 259–276. https://doi.org/10.1016/J.INDCROP.2010.10.022.

Domínguez, E., Romaní, A., Domingues, L., Garrote, G., 2017. Evaluation of strategies for second generation bioethanol production from fast growing biomass Paulownia within a biorefinery scheme. *Appl. Energy*. 187, 777–789. https://doi.org/10.1016/j.apenergy.2016.11.114.

Duarte, G.V., Moura, A.I., Moreira, R., Nunes, J., Figueiredo, M.M., Carvalho, M.G., 2013. Evaluation of several forest residues as potential raw material for bioethanol production in Portugal. *J. Bioprocess Eng. Biorefinery*. 2, 73–78. https://doi.org/10.1166/jbeb.2013.1035.

Dufey, A., 2006. *Biofuels Production, Trade and Sustainable Development: Emerging Issues*. International Institute for Environment and Development, London.

Dutta, K., Daverey, A., Lin, J.-G., 2014. Evolution retrospective for alternative fuels: First to fourth generation. *Renew. Energy*. 69, 114–122. https://doi.org/10.1016/J.RENENE.2014.02.044.

Fernando, S., Adhikari, S., Chandrapal, C., Murali, N., 2006. Biorefineries: Current status, challenges, and future direction. *Energy Fuels*. 20, 1727–1737. https://doi.org/10.1021/ef060097w.

Galbe, M., Zacchi, G., 2002. A review of the production of ethanol from softwood. *Appl. Microbiol. Biotechnol*. 59, 618–628. https://doi.org/10.1007/s00253-002-1058-9.

Garlotta, D., 2002. A literature review of poly (Lactic Acid). *J. Polym. Environ*. 9, 63–84. https://doi.org/10.1023/A:1020200822435.

Ghaffar, T., Irshad, M., Anwar, Z., Aqil, T., Zulifqar, Z., Tariq, A., Kamran, M., Ehsan, N., Mehmood, S., 2014. Recent trends in lactic acid biotechnology: A brief review on production to purification. *J. Radiat. Res. Appl. Sci*. 7, 222–229. https://doi.org/10.1016/j.jrras.2014.03.002.

Gomes, D., Domingues, L., Gama, M., 2016. Valorizing recycled paper sludge by a bioethanol production process with cellulase recycling. https://doi.org/10.1016/j.biortech.2016.06.004.

González, D., Campos, A.R., Cunha, A.M., Santos, V., Parajó, J.C., 2010. Utilization of fibers obtained by peroxyformic acid processing of broom as reinforcing agents for biocomposites. *BioResources*. 5, 2591–2610. https://doi.org/10.15376/biores.5.4.2591-2610.

González-Muñoz, M.J., Alvarez, R., Santos, V., Parajó, J.C., 2012. Production of hemicellulosic sugars from Pinus pinaster wood by sequential steps of aqueous extraction and acid hydrolysis. *Wood Sci. Technol*. 46, 271–285. https://doi.org/10.1007/s00226-011-0408-0.

Gregg, D.J., Saddler, J.N., 1996. Factors affecting cellulose hydrolysis and the potential of enzyme recycle to enhance the efficiency of an integrated wood to ethanol process. *Biotechnol. Bioeng*. 51, 375–383. https://doi.org/10.1002/(SICI)1097-0290(19960820)51:4<375::AID-BIT1>3.0.CO;2-F.

Gullón, P., Romaní, A., Vila, C., Garrote, G., Parajó, J.C., 2012. Potential of hydrothermal treatments in lignocellulose biorefineries. *Biofuels, Bioprod. Biorefining*. 6, 219–232. https://doi.org/10.1002/bbb.339.

Hama, S., Mizuno, S., Kihara, M., Tanaka, T., Ogino, C., Noda, H., Kondo, A., 2015. Production of d-lactic acid from hardwood pulp by mechanical milling followed by simultaneous saccharification and fermentation using metabolically engineered Lactobacillus plantarum. *Bioresour. Technol*. 187, 167–172. https://doi.org/10.1016/j.biortech.2015.03.106.

Hofvendahl, K., Hahn-Hägerdal, B., 2000. Factors affecting the fermentative lactic acid production from renewable resources(1). *Enzyme Microb. Technol*. 26, 87–107. https://doi.org/10.1016/S0141-0229(99)00155-6.

Jean-Paul, L., Evert, van der H., Jeroen, van B., Richard, P., 2011. Furfural – a promising platform for lignocellulosic biofuels. *ChemSusChem*. 5, 150–166. https://doi.org/10.1002/cssc.201100648.

John, R.P., Anisha, G.S., Nampoothiri, K.M., Pandey, A., 2009. Direct lactic acid fermentation: Focus on simultaneous saccharification and lactic acid production. *Biotechnol. Adv*. 27, 145–152. https://doi.org/10.1016/j.biotechadv.2008.10.004.

John, R.P., Madhavan Nampoothiri, K., Pandey, A., 2007. Production of L(+) lactic acid from cassava starch hydrolyzate by immobilized Lactobacillus delbrueckii. *J. Basic Microbiol*. 47, 25–30. https://doi.org/10.1002/jobm.200610208.

Kaparaju, P., Serrano, M., Thomsen, A.B., Kongjan, P., Angelidaki, I., 2009. Bioethanol, biohydrogen and biogas production from wheat straw in a biorefinery concept. *Bioresour. Technol*. 100, 2562–2568. https://doi.org/10.1016/j.biortech.2008.11.011.

Karimi, K., Taherzadeh, M.J., 2016. A critical review on analysis in pretreatment of lignocelluloses: Degree of polymerization, adsorption/desorption, and accessibility. *Bioresour. Technol*. 203, 348–356. https://doi.org/10.1016/J.BIORTECH.2015.12.035.

Karinen, R., Vilonen, K., Niemelä, M., 2011. Biorefining: Heterogeneously catalyzed reactions of carbohydrates for the production of furfural and hydroxymethylfurfural. *ChemSusChem*. 4, 1002–1016. https://doi.org/10.1002/cssc.201000375.

Kleinert, M., Barth, T., 2008. Towards a lignincellulosic biorefinery: Direct one-step conversion of lignin to hydrogen-enriched biofuel. *Energy & Fuels*. 22, 1371–1379. https://doi.org/10.1021/ef700631w.

Koppram, R., Tomás-Pejó, E., Xiros, C., Olsson, L., 2014. Lignocellulosic ethanol production at high-gravity: Challenges and perspectives. *Trends Biotechnol.* 32, 46–53. https://doi.org/10.1016/j.tibtech.2013.10.003.

Kumar, A.K., Sharma, S., 2017. Recent updates on different methods of pretreatment of lignocellulosic feedstocks: A review. *Bioresour. Bioprocess*. https://doi.org/10.1186/s40643-017-0137-9.

Kumari, D., Singh, R., 2018. Pretreatment of lignocellulosic wastes for biofuel production: A critical review. *Renew. Sustain. Energy Rev.* 90, 877–891. https://doi.org/10.1016/J.RSER.2018.03.111.

Limayem, A., Ricke, S.C., 2012. Lignocellulosic biomass for bioethanol production: Current perspectives, potential issues and future prospects. *Prog. Energy Combust. Sci.* 38, 449–467. https://doi.org/10.1016/j.pecs.2012.03.002.

Maas, R.H.W., Bakker, R.R., Eggink, G., Weusthuis, R.A., 2006. Lactic acid production from xylose by the fungus Rhizopus oryzae. *Appl. Microbiol. Biotechnol.* 72, 861–868. https://doi.org/10.1007/s00253-006-0379-5.

Menon, V., Rao, M., 2012. Trends in bioconversion of lignocellulose: Biofuels, platform chemicals & biorefinery concept. *Prog. Energy Combust. Sci.* 38, 522–550. https://doi.org/10.1016/j.pecs.2012.02.002.

Moldes, A.B., Alonso, J.L., Parajó, J.C., 2000. Multi-step feeding systems for lactic acid production by simultaneous saccharification and fermentation of processed wood. *Bioprocess Eng.* 22, 175–180. https://doi.org/10.1007/s004490050028.

Naik, S.N., Goud, V. V., Rout, P.K., Dalai, A.K., 2010. Production of first and second generation biofuels: A comprehensive review. *Renew. Sustain. Energy Rev.* 14, 578–597. https://doi.org/10.1016/J.RSER.2009.10.003.

Nigam, P., Singh, D., 1995. Processes of fermentative production of Xylitol – a sugar substitute. *Process Biochem.* 30, 117–124. https://doi.org/10.1016/0032-9592(95)80001-8.

Nigam, P.S., Singh, A., 2011. Production of liquid biofuels from renewable resources. *Prog. Energy Combust. Sci.* 37, 52–68. https://doi.org/10.1016/J.PECS.2010.01.003.

Nunes, J., Fierro, J., Freitas, H., 2012. Sustainable management of forest ecosystems and lignocellulosic biomass use: A case study from Portugal. 20th Eur. Biomass Conf. Exhib.

Oliveira, C., Romaní, A., Gomes, D., Cunha, J.T., Gama, F.M., Domingues, L., 2018. Recombinant family 3 carbohydrate-binding module as a new additive for enhanced enzymatic saccharification of whole slurry from autohydrolyzed Eucalyptus globulus wood. *Cellulose*. 25, 2505–2514. https://doi.org/10.1007/s10570-018-1722-6.

Pandey, M.P., Kim, C.S., 2011. Lignin depolymerization and conversion: A review of thermochemical methods. *Chem. Eng. Technol.* 34, 29–41. https://doi.org/10.1002/ceat.201000270.

Parajó, J.C., Domínguez, H., Domínguez, J., 1998. Biotechnological production of xylitol. Part 1: Interest of xylitol and fundamentals of its biosynthesis. *Bioresour. Technol.* 65, 191–201. https://doi.org/10.1016/S0960-8524(98)00038-8.

Pereira, F.B., Romaní, A., Ruiz, H.A., Teixeira, J.A., Domingues, L., 2014. Industrial robust yeast isolates with great potential for fermentation of lignocellulosic biomass. *Bioresour. Technol.* 161, 192–199. https://doi.org/10.1016/j.biortech.2014.03.043.

Pontes, R., Romaní, A., Michelin, M., Domingues, L., Teixeira, J., Nunes, J., 2018. Comparative autohydrolysis study of two mixtures of forest and marginal land resources for co-production of biofuels and value-added compounds. *Renew. Energy*. 128, 20–29. https://doi.org/10.1016/j.renene.2018.05.055.

Rafiqul, I.S.M., Sakinah, A.M.M., 2013. Processes for the production of xylitol – a review. *Food Rev. Int.* 29, 127–156. https://doi.org/10.1080/87559129.2012.714434.

Romaní, A., Garrote, G., Alonso, J.L., Parajó, J.C., 2010. Bioethanol production from hydrothermally pretreated Eucalyptus globulus wood. *Bioresour. Technol.* 101, 8706–8712. https://doi.org/10.1016/j.biortech.2010.06.093.

Romaní, A., Ruiz, H.A., Pereira, F.B., Teixeira, J.A., Domingues, L., 2014a. Integrated approach for effective bioethanol production using whole slurry from autohydrolyzed Eucalyptus globulus wood at high-solid loadings. https://doi.org/10.1016/j.fuel.2014.06.061.

Romaní, A., Ruiz, H.A., Pereira, F.B., Teixeira, J.A., Domingues, L., 2014b. Effect of hemicellulose liquid phase on the enzymatic hydrolysis of autohydrolyzed Eucalyptus globulus wood Time needed to reach CGC max /2 (h). *Biomass Conv. Bioref.* 4, 77–86. https://doi.org/10.1007/s13399-013-0093-3.

Rosatella, A.A., Simeonov, S.P., Frade, R.F.M., Afonso, C.A.M., 2011. 5-Hydroxymethylfurfural (HMF) as a building block platform: Biological properties, synthesis and synthetic applications. *Green Chem.* 13, 754. https://doi.org/10.1039/c0gc00401d.

Ruiz, H.A., Ruzene, D.S., Silva, D.P., Da Silva, F.F.M., Vicente, A.A., Teixeira, J.A., 2011. Development and characterization of an environmentally friendly process sequence (autohydrolysis and organosolv) for wheat straw delignification. *Appl. Biochem. Biotechnol.* 164, 629–641. https://doi.org/10.1007/s12010-011-9163-9.

Schmidt, S., Padukone, N., 1997. Production of lactic acid from wastepaper as a cellulosic feedstock. *J. Ind. Microbiol. Biotechnol.* 18, 10–14. https://doi.org/10.1038/sj.jim.2900339.

Stewart, D., 2008. Lignin as a base material for materials applications: Chemistry, application and economics. *Ind. Crops Prod.* 27, 202–207. https://doi.org/10.1016/J.INDCROP.2007.07.008.

Winkelhausen, E., Kuzmanova, S., 1998. Microbial conversion of d-xylose to xylitol. *J. Ferment. Bioeng.* 86, 1–14. https://doi.org/10.1016/S0922-338X(98)80026-3.

World Fuel Ethanol Production., 2005. RFA: Renewable Fuels Association. http://ethanolrfa.org/pages/World-Fuel-Ethanol-Production

Yánez, R., Gómez, B., Martínez, M., Gullón, B., Alonso, J.L., 2014. Valorization of an invasive woody species, Acacia dealbata, by means of Ionic liquid pretreatment and enzymatic hydrolysis. *J. Chem. Technol. Biotechnol.* 89, 1337–1343. https://doi.org/10.1002/jctb.4207.

Yazdani, S.S., Gonzalez, R., 2007. Anaerobic fermentation of glycerol: A path to economic viability for the biofuels industry. *Curr. Opin. Biotechnol.* 18, 213–219. https://doi.org/10.1016/J.COPBIO.2007.05.002.

Yu, Q., Zhuang, X., Yuan, Z., Qi, W., Wang, Q., Tan, X., 2011. The effect of metal salts on the decomposition of sweet sorghum bagasse in flow-through liquid hot water. *Bioresour. Technol.* 102, 3445–3450. https://doi.org/10.1016/j.biortech.2010.10.084.

Zabaniotou, A., Ioannidou, O., Skoulou, V., 2008. Rapeseed residues utilization for energy and 2nd generation biofuels. *Fuel*. 87, 1492–1502. https://doi.org/10.1016/J.FUEL.2007.09.003.

11

Impact of CRISPR-Based Gene Editing in Environmental Biotechnology

Saurabh Srivastav, Mahino Fatima, Mir Hilal Ahmad and Amal Chandra Mondal*

* Laboratory of Cellular and Molecular Biology, School of Life Sciences, Jawaharlal Nehru University, New Delhi, India-110067
E-mail: acmondal@mail.jnu.ac.in Telephone: +91-11-2670-4505

CONTENTS

11.1 Introduction

Environmental biotechnology is a multidisciplinary branch of science that encompasses wide aspects of daily life, in which challenges associated with environmental problem is one of the major concern encountered by environmental biotechnologists (Gavrilescu M, 2010). The environment is capable of maintaining its equilibrium by recycling its components. However, with the advent of industrialization and an ever-increasing population demand, the natural healing capacity of the environment is getting compromised, thereby imposing a serious threat to its optimized functioning. The heavy industrialization and monstrous population increase in the last few decades have adversely affected land, air and water quality (Gavrilescu M, 2010). Industrialization and technological advancement have improved quality of life, but simultaneously, it has also imposed several adversaries to the environment, for example, emission of hydrocarbons (oil spills), disposal of electronic/electrical/plastic wastes, uncontrolled use of antibiotics, exposure of hazardous materials, contamination of rivers, lakes, estuaries and coastal waters, among others (Dai *et al.*, 2018; Yuan *et al.*, 2016; Zhao *et al.*, 2013; Liu *et al.*, 2012; Tacão *et al.*, 2012). Furthermore, the agriculture sector is facing several issues such as soil erosion, excess use of pesticides, limited availability of fertile land, limitations of conventional plant breeding and others, in order to overcome the demand of increasing population (Garnett *et al.*, 2013; Brussaard *et al.*, 2010). The majority of the adversaries are the outcome of anthropogenic activities, which is leading to the widespread disruption of environmental balance (Ansari and Matondkar, 2014). However, a lot of effort has been initiated and proposed to combat global issues by strategizing policies and developing more promising scientific methods (Reid *et al.*, 2010).

Recent advancement in the field of biotechnology has opened doors for sustainable development of the environment to provide a balanced and healthier life. In this milieu, genetic engineering has fascinated environmental biotechnologists since its introduction. The utilization of genetic engineering in bioremediation, phytoremediation, metabolic engineering and other methods offers immense potential in the sustainable development of the environment (Basu *et al.*, 2018; Daniell *et al.*, 2016; Fingerman, 2016; Azubuike *et al.*, 2016; Weber *et al.*, 2015).

DOI: 10.1201/9781003186304-11

However, genetic engineering had several limitations until the advent of CRISPR-Cas (clustered regularly interspaced short palindromic repeats–CRISPR associated) technique (Doudna and Charpentier, 2014). The CRISPR-Cas-based genome engineering technique has already revolutionized the life sciences research by providing a versatile and handy platform for researchers to manipulate genetic content with ease (Ratner *et al.*, 2016). The CRISPR-Cas technique is the latest addition to genome editing technique, which has immense potential to be utilized for the prevention or treatment of various human diseases (Charpentier E, 2015). However, the utilization of CRISPR is not limited to biomedicine, but this revolutionary technology has opened doors for scientists to develop methods that can help protect or repair the environment against the damage done by the various anthropogenic activities. The CRISPR-Cas is being utilized for a wide variety of applications, and the present chapter discusses its impact on various environmental challenges, including biofuels, bioplastic, bioremediation, metabolic engineering, agriculture, biosensing, greenhouse gas emissions, pesticide reduction, nitrogen fixation and others, that are being encountered by the environmental biotechnologists.

11.1.1 Overview of CRISPR-Cas

The origin of CRISPR-Cas is from bacteria (roughly 50%) and archaea (roughly 90%) (Makarova and Koonin, 2015), and where it exists as an ancient immune system, it provides defense against a wide variety of foreign elements, including bacteriophages and other mobile genetic elements (Charpentier E, 2015). The CRISPR-Cas system, unlike other cellular defenses, functions via a mechanism analogous to the adaptive immunity of vertebrates. It helps form a genetic memory of previous infections, which further elicits a quick and heightened response toward the next round of exposure to the same foreign entity (Wright *et al.*, 2016). The CRISPR-Cas in a bacterial genome is present as the CRISPR array consisting of a series of direct repeats (~20–50 base pairs) separated by 'spacers' that are of similar length and are unique (Kunin *et al.*, 2007; Jansen *et al.*, 2002). This array of direct repeat and spacer sequence is preceded by the AT-rich sequence known as a 'leader' sequence (Kunin *et al.*, 2007; Jansen *et al.*, 2002).

It is worth mentioning that the discovery of CRISPR dates back more than 25 years when it was first identified in an *Escherichia coli* genome as an array of short and repetitive DNA sequences that were interspaced with nonrepetitive sequences (Ishino *et al.*, 1987). However, the importance and function of these array of sequences were unknown until a decade after its discovery when it was found that these arrays of sequences were present in various bacteria and archaea (Mojica *et al.*, 2005; Bolotin *et al.*, 2005; Pourcel *et al.*, 2005). The unique spacer sequence was derived from different viral genomes and other mobile genetic elements, such as plasmids and transposons, as a representative of earlier infection (Mojica *et al.*, 2005; Bolotin *et al.*, 2005; Pourcel *et al.*, 2005). The sequences of foreign DNA, which is similar to the unique spacer sequence of the host, is referred to as 'protospacers' (Wright *et al.*, 2016). Furthermore, it was found that the spacer sequence similar to the phage genome tends to provide immunity to the host against this phage and that the exposure to the corresponding phage causes the CRISPR array to expand further by incorporating the new spacers specific for the corresponding phage genome (Gasiunas *et al.*, 2012; Westra *et al.*, 2012; Barrangou *et al.*, 2007).

Furthermore, research during the last 8 to 10 years in this area has increased our understanding of the underlying mechanism of action for the CRISPR-Cas system, which involves RNA-directed interference (Rath *et al.*, 2015; Barrangou and Marraffini, 2014; Makarova *et al.*, 2011). The mechanism of interference by CRISPR-Cas involves three different sequential stages: (a) spacer acquisition, (b) CRISPR RNA (crRNA) transcription and maturation and (c) interference via target identification and cleavage. The first step is spacer acquisition, which involves identifying foreign DNA followed by its processing and integration into the CRISPR array as a new spacer (Van der Oost *et al.*, 2014; Heler *et al.*, 2014). Once, the spacer acquisition is complete, the next stage is crRNA biogenesis, involving transcription of a CRISPR locus into a pre-crRNA followed by its subsequent processing to form mature crRNAs. Each crRNA contains a portion of repetitive sequence with a single spacer, which will help in the identification of the foreign nucleic acid having significant sequence complementarity to the spacer sequence of the crRNA. The third stage, or the interference stage involves the effector complex which mainly contains a crRNA complexed with Cas protein(s). This complex identifies the target and binds to it based on spacer sequence complementarity, thereby carrying out its target cleavage (Rath *et al.*, 2015; Barrangou and Marraffini, 2014; Makarova *et al.*, 2011). The overview of the steps involved in CRISPR-Cas system is illustrated in Figure 11.1.

The accomplishment of the previously mentioned three steps of CRISPR-Cas system primarily involves the role of Cas proteins encoded by the 'cas' gene located adjacent to the CRISPR array (Makarova *et al.*, 2011). The specific component of cas gene varies, and based on it, the CRISPR-Cas system has been categorized into six types: Type I, II, III, IV, V, VI (Makarova *et al.*, 2011). Among the six types of CRISPR-Cas systems, Types I through III are the best studied (Makarova and Koonin, 2015). However, the most popular subtype is Type II CRISPR-Cas systems, and it requires a single protein to carry out the maturation of crRNA, target recognition followed by interference steps of the CRISPR-Cas based immunity. The different types of CRISPR-Cas systems have been illustrated in Figure 11.2.

Furthermore, the CRISPR-Cas system is a unique defense system whose hallmark is specificity in terms of self- and non-self-recognition, which is accomplished via distinct mechanisms in different systems to minimize the off-target effects (Marraffini and Sontheimer, 2010). Minimizing the off-target is very important as binding of crRNA to fully or partially complementary sequences may result in the self-targeting of the host genome, thereby posing a lethal threat for the host (Ratner *et al.*, 2016; Wright *et al.*, 2016). In most of the CRISPR-Cas system, the selection of target protospacer (sequence of foreign DNA complementary to the spacer sequence of the host), relies on the identification of 2 to 5 bp nucleotide sequences adjacent to protospacer sequence known as the "protospacer adjacent motif" (PAM; Mojica *et al.*, 2009; Deveau *et al.*, 2008). The selection of PAM-specific protospacer ensures that the crRNA driven interference targets the foreign DNA with PAM while

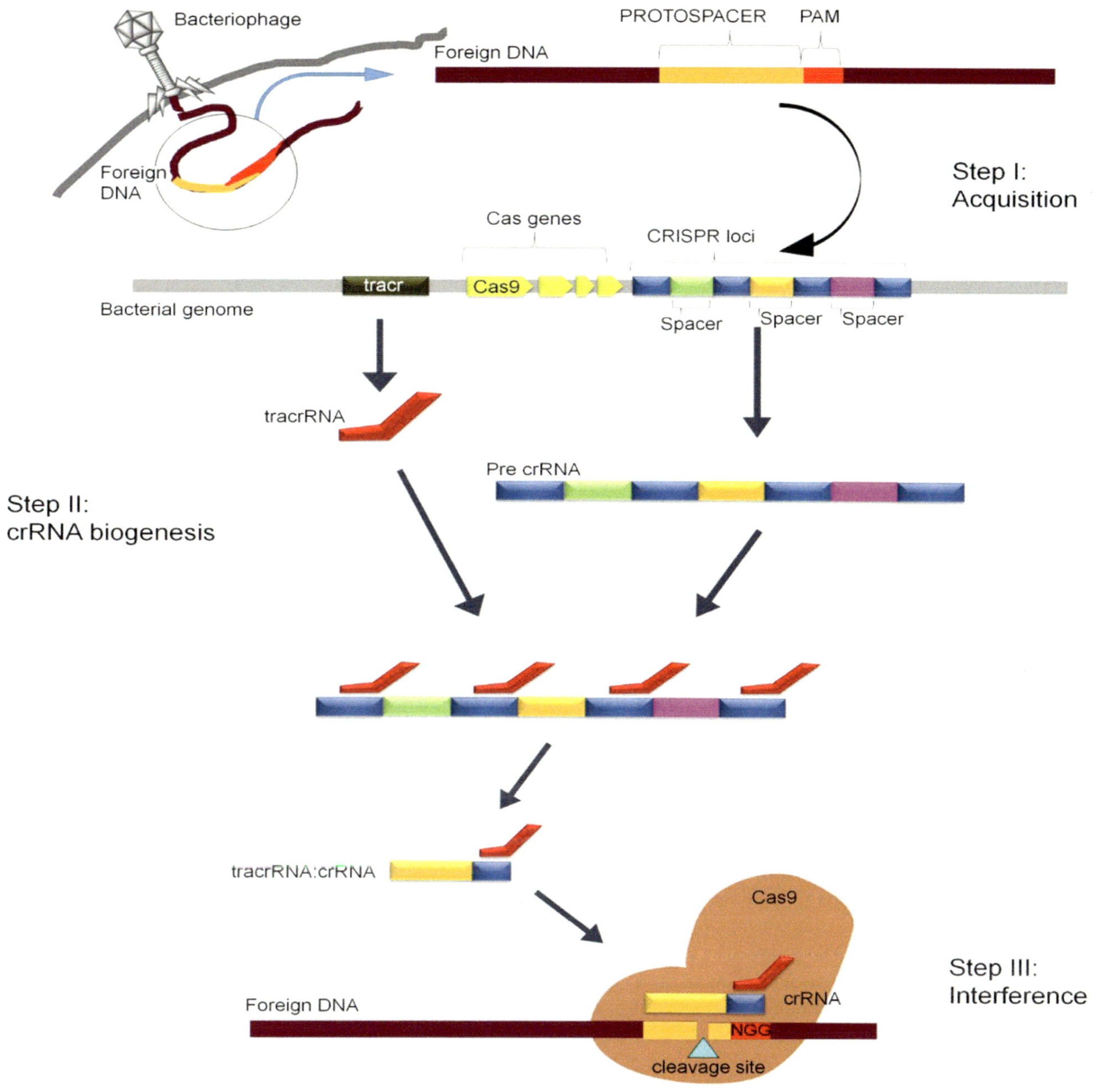

FIGURE 11.1 Schematic illustration of the CRISPR-Cas system in bacteria. The CRISPR-Cas system involves three steps mechanism: acquisition, crRNA biogenesis and interference in sequential order. The CRISPR-Cas systems are mainly composed of Cas operon (shown as cas genes in the bacterial genome in yellow) and the CRISPR array known as CRISPR loci. The CRISPR loci contain identical repeat sequences (blue rectangles within CRISPR loci) that are interspersed by spacers (different colored rectangles beside repeat sequences which are shown as a blue rectangle). In the acquisition step, a sequence of invading foreign DNA known as protospacer gets incorporated into CRISPR array by Cas1-Cas2 complex (not shown). Acquisition of protospacer relies on the identification of 2–5 bp nucleotide sequences adjacent to protospacer sequence known as PAM. The next step is the biogenesis of crRNA that involves a transcription of the CRISPR array into a single precursor crRNA (Pre crRNA), which is further processed by other Cas to form mature crRNA. The mechanism of biogenesis shown in the figure is representative of Type II CRISPR-Cas system, which involves tracrRNA (shown in red) to form mature crRNA. In the interference step, the effector complex mainly contains a crRNA complexed with Cas protein(s) (here shown as tracrRNA:crRNA duplex complexed with Cas9) identifies the target foreign DNA and binds to its protospacer sequence based on spacer sequence complementarity thereby carrying out the cleavage of target foreign DNA.

ignoring the PAM-free spacer found in the CRISPR array of the bacterial host (Wright *et al.*, 2016; Heler *et al.*, 2015; Marraffini and Sontheimer, 2010). Furthermore, the Type II system is one of the most studied types of CRISPR-Cas systems and involves Cas9 protein, which binds to the duplex of crRNA and tracrRNA (trans-activating crRNA, a small trans-encoded RNA within the CRISPR–Cas locus, which is unique to Type II systems) forming a Cas9: crRNA:tracrRNA complex (Jinek *et al.*, 2014; Nishimasu *et al.*, 2014). Once the Cas9 binds to the RNA, it undergoes conformational changes that will favor the subsequent degradation of its target (Jinek *et al.*, 2014; Nishimasu *et al.*, 2014). Cas9 then scans the target DNA associated with PAM, once the PAM-associated protospacer sequence is identified, the Cas9 interacts with the PAM by its

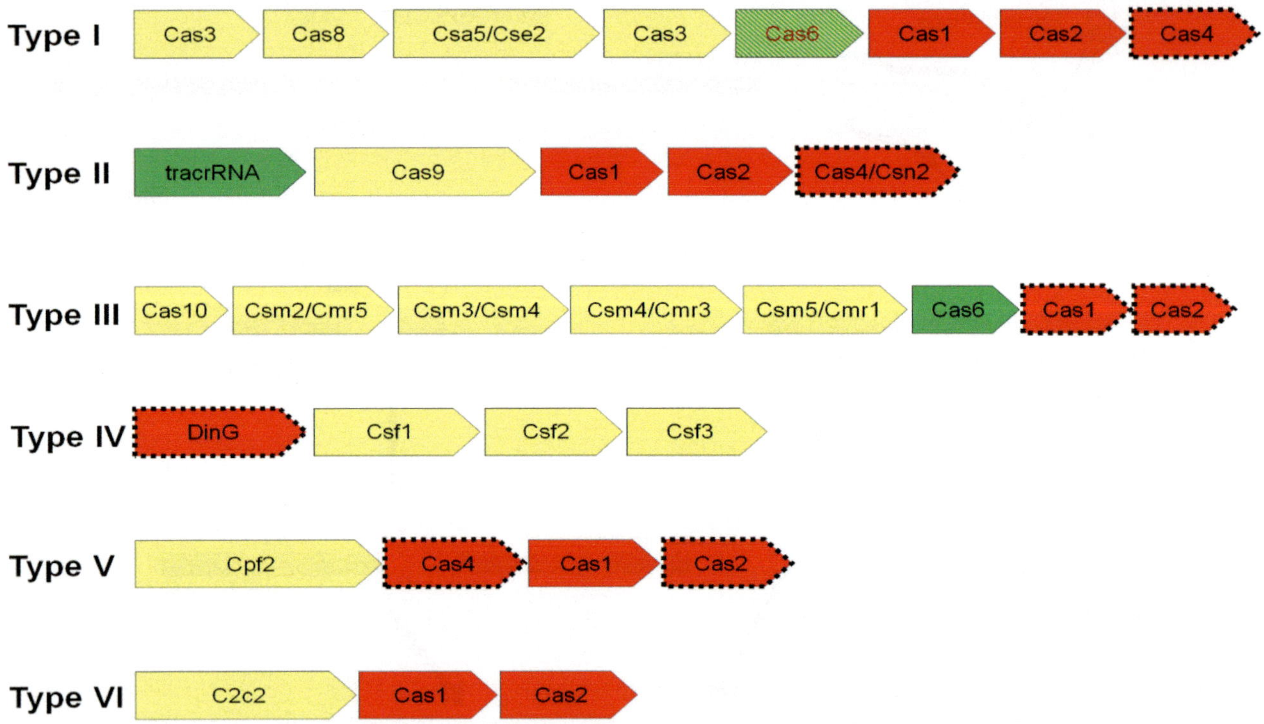

FIGURE 11.2 Illustration of the representative operons of different types of CRISPR-Cas system. Genes involved in the acquisition of CRISPR-Cas system are shown in red, while those that are involved in the biogenesis of crRNA are shown in green. The genes involved in the interference are shown in yellow. The genes shown with a dashed line are the one present only in some subtypes of the given type of CRISPR-Cas system. The image has been adapted and modified (Wright *et al.*, 2016).

PAM interacting domain. This interaction results in the destabilization of the adjacent double-stranded DNA (dsDNA), thereby facilitating the binding of the target sequence to the seed region of the crRNA (Jinek *et al.*, 2014; Nishimasu *et al.*, 2014). The target sequence with near-full complementary with the crRNA results in the denaturation of DNA followed by the formation of the RNA:DNA heteroduplex. This unwinding of the DNA and formation of the RNA:DNA hybrid initiates the cleavage of the RNA-bound DNA strand at the site three nucleotides upstream to the associated PAM sequence by the endonuclease domain (Anders *et al.*, 2014; Jinek *et al.*, 2014; Nishimasu *et al.*, 2014).

Therefore, the extensive study done on CRISPR-Cas system during the past decade has provided a platform for genetic engineers to utilize it as a valuable tool for gene editing. However, the use of spacer sequence of choice was a big question until the synthetic double-stranded RNA, also known as guide RNA (gRNA), similar to the crRNA: tracrRNA duplex, was engineered, which is capable of associating with the Cas9 to drive the CRISPR-Cas–mediated gene editing (Jinek *et al.*, 2012). The introduction of gRNA facilitated researchers to easily modify the spacer sequence as needed against a wide variety of targets (Deltcheva *et al.*, 2011; Jinek *et al.*, 2012), thereby helping pave a way for the efficient, cost-effective and revolutionary technique of gene editing. Henceforth, the CRISPR-Cas system has opened doors for a wide variety of applications, including human disease and therapeutic, agriculture, bioremediation and other applications. The technique is imparting an impact to every sector of science. Moreover, the art of CRISPR methodology is also making it feasible for scientist to design potential strategies against a variety of complex environmental challenges existing on account of anthropogenic activities. The current chapter discusses the impact of CRISPR on environmental biotechnology.

11.2 Impact of CRISPR on Various Environmental Challenges

11.2.1 Sustainable Agriculture

The growth of agriculture sector is quintessential for meeting the demand of the ever-increasing global population. Modern agricultural practices have made significant contributions toward meeting global demand, although still more efforts are needed for equilibrating the agricultural productivity with global population demand. During the last 50 years, crop productivity and quality have been significantly improved. However, there still exist several limitations, and it is unlikely for the conventional plant breeding method to cope up with the ever-increasing food demand along with the associated environmental challenges (Gao C, 2018). The advent of CRISPR has revolutionized the gene-editing technique, and its use in the agriculture sector is not only going to improve the breeding method of plants but will also produce results similar to conventional plant-breeding methods. Moreover, the use of CRISPR approach will not only provide a more predictable and quick method for breeding but also will be cost effective.

The best-studied and extensively used CRISPR-Cas system in plants are the CRISPR-Cas9 (Type II) and CRISPR-Cpf1 (Type V) systems (Zetsche *et al.*, 2015; Makarova *et al.*, 2015; Jinek *et al.*, 2012). The delivery of the CRISPR elements to the plant cells is done as either DNA, RNA or protein-RNA that gather as an active site-directed nuclease (SDN), whereby it carries out the generation of double-strand breaks (DSBs) by cleaving the target DNA sequences (Yin *et al.*, 2017; Arora and Narula, 2017). The plant cells utilize three approaches to repair these DSBs: SDN-1, SDN-2 and SDN-3 editing. The SDN-1 editing involves annealing of the DNA ends, thereby often resulting in the generation of gene knockout by causing small changes in the sequence (Gao C, 2018). The SDN-2 editing involves the integration of the separate piece of DNA at the site of DSB resulting in the replacement of short sequence (<20 nucleotides) (Gao C, 2018). The SDN-3 editing is similar to SDN-2 with the only exception that it involves the insertion or replacement of the longer sequence (Gao C, 2018).

11.2.1.1 Development of Improved Crops

The use of CRISPR-Cas technique in the improvement of different crops and medicinal plants has already shown some promising results (Liu *et al.*, 2017; Rodríguez-Leal *et al.*, 2017). The improvements in crop plants such as maize, tomato and wheat had already been initiated, thereby depicting immense potential of CRISPR-Cas technique (Rodríguez-Leal *et al.*, 2017; Yin *et al.*, 2017). The yield in tomato plant was boosted by applying the SDN-1 editing method for producing a mutation in the regulatory regions of the gene associated with yield. The approach resulted in the increase in genetic variation, thereby resulting in a greater yield of tomato. Moreover, the time taken to achieve was much less when compared to the conventional breeding approach (Rodríguez-Leal *et al.*, 2017). Wheat crop productivity is hampered by powdery mildew fungus and the utilization of the SDN-1 editing has imparted resistance to wheat crop against such a devastating fungal disease (Zhang *et al.*, 2017). Furthermore, the yield of maize during the drought stress was enhanced using the CRISPR-Cas technique by using the SDN-3 approach to edit the promotor of *Argos8* gene, thereby resulting in the constitutive expression of the endogenous gene (Yin *et al.*, 2017).

The CRISPR-Cas approach has been used in a wide variety of plants, including rice, soybean, wheat, maize, tobacco, potato, tomato, orange and many more (Liu *et al.*, 2017). The results obtained so far are very promising, but there exist a few hurdles that need to be addressed in order to achieve the best result from such revolutionary technique. The major hurdles are being the appropriate delivery of the CRISPR-Cas machinery to the specific plant cells with subsequent regeneration of the CRISPR modified plant. Although tissue culture–based methods have several limitations and, hence, direct delivery of CRISPR-Cas component to the pollen grains or apical meristem of the plant may help researchers obtain CRISPR-edited plants without the utilization of a tissue culture approach. The current approaches being utilized for the delivery of vector-carrying CRISPR-Cas component involves transformation mediated by Agrobacterium, particle bombardment and polyethylene glycol (PEG)–based protoplast transfections (Liu *et al.*, 2017). However, every current approach has its own advantages and associated limitations, and more refinement in the current methods is required to design a more potential use of CRISPR-Cas system.

The use of CRISPR tools in the development of crop improvement has certainly opened a new door for a more refined approach for sustainable agriculture. Although the technique offers a cost-effective and less tedious approach, but it exhibits far more potential, which can be applied to provide different, feasible innovative approaches. The technique can not only be utilized for the improvement of domesticated crops but it can also be utilized for domesticating those plants that are well adapted to varying environmental conditions but remain neglected (Gao C, 2018). This can be achieved by the addition of high-value traits such as improving oil and meal quality in such neglected plants (Gao C, 2018). Furthermore, the ability exclusive to certain plants can be utilized to enhance its benefit to agriculture productivity. One of the examples is nitrogen-fixation ability of legumes, whereby the use of CRISPR tools can help with designing approaches to extract maximum benefit for sustainable agriculture. Henceforth, more research in agriculture sector is required to get the most benefit from the CRISPR-Cas tool.

11.2.1.2 Reduce the Use of Pesticides

The use of pesticides in agriculture is considered to be one of the factors implicated in increasing crop productivity in order to quench the ever-increasing demand of growing human population. However, the extensive use of pesticides is adversely affecting both the environment and human health (Lechenet *et al.*, 2017). Therefore, there is an urgent need for the development of different strategies that will not only help in reduction of pesticide usage but simultaneously will not have any adverse effect on the quality and productivity of crops as well (Barzman *et al.*, 2015; Lamichhane *et al.*, 2016). Although, reducing pesticide use without affecting crop productivity is a debatable topic. However, with the advancement in plant research, it has been observed that the resistance of plants toward pest relies mainly on the content of plant's secondary metabolite profile, which includes alkaloids, terpenes, cyanogenic glycosides, phenylpropanoids, glucosinolates and flavonoids (Douglas A.E., 2018). The understanding of the chemistry of secondary metabolites and gene editing tool offers a potential approach toward reduction of pesticide usage. The CRISPR-based molecular tools of genome editing have made it feasible to design strategy for the development of pest-resistant variety of plants. Recently, with the use of CRISPR-Cas tools, researchers have developed a virus-resistant variety of cucumber nontransgenically without affecting the plant's development (Wolf *et al.*, 2016). A fungal disease–resistant (powdery mildew) variety of grapes and a bacterial disease–resistant (fire blight) variety of apples have been developed using CRISPR-Cas tools (Malnoy *et al.*, 2016). The CRISPR-Cas tool is being extensively utilized in the development of pest-resistant varieties of crops, thereby minimizing the usage of pesticides (Douglas *et al.*, 2018; Lechenet *et al.*, 2017; Malnoy *et al.*, 2016). The CRISPR-Cas method has shown immense potential for the development of resistant varieties of crops. However, there exist several issues which require attention in order to achieve maximum benefit with minimum risk by using the CRISPR tool. The biggest challenge for the utilization of CRISPR-Cas tool is less associated with scientific limitation; rather, it's more associated with convincing the public and government regulatory bodies. This will require a serious and strategic political acceptance in order to clarify the scope and use of this technique in a direction for sustainable growth of agriculture.

11.2.1.3 Improvement of Efficiency of Water Use

The agriculture sector requires a majority of freshwater content for its fruitfulness, and with the growing challenge of water scarcity and drought, the global goal of sustainable agriculture seems far away. However, the existence of drought-tolerant plants offer hope for developing approaches that will enable researchers to develop drought-tolerant crops without losing the yield content of the crop. This will also ensure efficient and minimal use of water for increasing crop productivity in an abiotic-stress condition involving water-deficit and high-temperature stress (Kawakami *et al.*, 2010; Hays *et al.*, 2007). In this milieu, the very first step requires the unraveling of the physiological process and underlying metabolism that impart drought tolerance to plants. However, the underlying mechanism is yet not delineated fully, but it has been shown that the gaseous plant hormone "ethylene" seems to be one of the key regulatory players involved in the response toward water-deficit and high-temperature stress (Kawakami *et al.*, 2010; Hays *et al.*, 2007). It has been shown that depreciating the ethylene biosynthesis (Habben *et al.*, 2014) or by reducing the sensitivity of plants towards ethylene (Shi *et al.*, 2015) enhances the grain yield in the maize plants exposed to drought stress. Furthermore, it was found that *Argos8* in maize is a negative regulator of ethylene, and when overexpressed, it imparts drought tolerance to the maize plant (Shi *et al.*, 2015). Therefore, the CRISPR-Cas tool was utilized to develop the novel variant of the *Argos8*, and it was found the variant generated by the CRISPR-Cas approach was found to enhance the grain yield by five bushels per acre of land under stress conditions as compared to its wild type (WT) form (Shi *et al.*, 2017). Thus CRISPR-Cas approach clearly depicts its potential in the development of novel allelic variation which will impart drought tolerance to the crops (Shi *et al.*, 2017). Henceforth, CRISPR technique can potentially be utilized for the efficient use of water in agriculture by developing drought-stress-tolerant crops for enhancing crop productivity.

11.2.2 Biological Nitrogen Fixation

Nitrogen deficiency in the soil hampers agricultural crop productivity and that nitrogen runoff is one of the agriculturally associated environmental concerns. Plants naturally depend on soil as their source of nitrogen, although not all plants can directly uptake major forms of nitrogen available in the soil. Certain plants, such as legumes, have developed an association with bacteria that helps the plant with nitrogen fixation. However, a majority of plants are unable to acquire ample amounts of nitrogen and thus rely on the additional nitrogen made available to them in the form of synthetic fertilizer or manure. These synthetic nitrogen sources in excess may run off fields, thereby contaminating the water sources that may result in the formation of an aquatic dead zone. Therefore, to overcome the problem of inefficient nitrogen fixation, the understanding of the genes in legume host that are involved in the symbiotic nitrogen fixation (SNF) is required. There exist hundreds of SNF-associated genes involved in a variety of functions, which include bacterial partner recognition, nodule development, nodule signaling pathway, regulation of nodule number, maintenance of efficient nitrogen fixation within the nodule, nodule number and nodule senescence (Wang *et al.*, 2017b). The utilization of the CRISPR-Cas technique in SNF is in progress, and studies done so far have provided new insights toward the understanding of functional genomics and SNF of legumes (Wang *et al.*, 2013). Future advances in this area may enable molecular breeding and help in the development of CRISPR-Cas–modified legume and nonlegume crops having increased efficiency of nitrogen usage, thereby reducing the use of synthetic nitrogen fertilizer and manure (Wang *et al.*, 2017b).

11.2.3 Bioremediation

Bioremediation is a process of degrading and/or detoxifying the toxicant and/or pollutants from the environment using microorganisms and plants. Heavy industrialization, urbanization and exponential population growth has resulted in the increased level of anthropogenic activity. Such activities have resulted in the formation of recalcitrant pollutants (hydrocarbons, oil spills, heavy metals, hazardous radioactive components, plastic, etc.), whereby the global environmental homeostasis is getting disturbed. Remediation involving physical and chemical approaches are not sufficient enough to cope up with the existing rate of pollutants introduction to the environment (Dangi *et al.*, 2019; de Lorenzo V, 2018; Dvořák *et al.*, 2017). Therefore, bioremediation approaches involving the utilization of microorganisms offer more diverse, eco-friendly and socially acceptable alternatives for removing pollutants (Dangi *et al.*, 2019; de Lorenzo V., 2018; Dvořák *et al.*, 2017). Several microbes have been isolated and characterized by their potential in bioremediation (Dangi *et al.*, 2019; de Lorenzo V., 2018; Dvořák *et al.*, 2017; Glass, 2005). However, in most cases, the complete degradation of pollutants/toxicants are often not accomplished, possibly due to the presence of mixed pollutants (Dangi *et al.*, 2019; de Lorenzo V., 2018; Dvořák *et al.*, 2017). Therefore, there is a need for new bioremediation approaches that will involve the utilization of environmental microbiomes engineered using modern gene editing technique involving CRISPR-Cas tool.

The selection of microbes specific for a given pollutant (e.g., plastic) will be capable of degrading the pollutant (Sen and Raut, 2015; Yang *et al.*, 2014). In this context, a CRISPR-Cas tool can be utilized to enhance the activity of the associated genetic pathways involved in degrading the pollutant. This will ease the removal of pollutants in a more efficient manner. Genetically engineered microbes could be utilized to help degrade plastics. Certain species of bacteria and fungi have evolved naturally to degrade compounds found in plastics. CRISPR-Cas could be utilized to enhance the activity of the associated genetic pathways. There exist different bioremediation approaches in which a CRISPR-Cas tool can be very useful in editing microbes and plants to efficiently uptake heavy metals, carry out the cleaning of oil spills, help in wastewater treatment and so on. However, the area of bioremediation can be heavily revolutionized by the use of CRISPR-Cas tool, and it requires a lot of future research in defining and selecting specific microbes based on pollutant type, genetic pathway involved and its variation by CRISPR-Cas tool. This area will have a greater impact on stabilizing environmental homeostasis for better and healthier life.

11.2.4 Generation of Biofuel

The growing appetite for energy has increased our dependency on fossil fuels, which is a nonrenewable source of energy having limited supplies. In addition, the burning of fossil fuels has been one of the main reasons for the rise in atmospheric carbon dioxide (CO_2) content (Meinshausen *et al.*, 2009). The increase in the atmospheric CO_2 content is a serious environmental concern as CO_2 is one of the greenhouse gases associated with the rise in global warming, thereby posing a threat to environmental homeostasis (Bose B.K., 2010). Furthermore, the heavy use and dependency on limited supply of fossil fuels have not only adversely affected the environment but have also resulted in the constraint on economic growth (Nel and Cooper, 2009). Therefore, there exists a global demand for alternative and renewable source of energy that will relieve pollution from the environmental and economic growth constraints. Wind energy and solar energy are very good sources of alternative, renewable and nonpolluting energy but are associated with geological constraints and high initial investment (Desai *et al.*, 2015).

The other alternative source of energy is 'biofuels' derived from biological sources, which include bacteria, algae and plants (Lü *et al.*, 2011; Lee *et al.*, 2008). The biofuels are obtained by the conversion of cellulosic biomass into simple sugars, and then these sugars are converted into different type of biofuels, which includes alcohols, terpenes, alkanes and fatty acid alkyl esters (Zhang *et al.*, 2011). The production of biofuel by nonfood oil plants, such as *Jatropa*; microorganisms; and algae offer several advantages. However, to make biofuel more cost-effective and increase its commercial availability, modern techniques of metabolic engineering will help achieve the goal (Desai *et al.*, 2015). The metabolically engineered saccharomyces, cyanobacteria and other microbes have provided successes in the production of biofuels (Oliver *et al.*, 2016; Runguphan and Keasling, 2014; Yang *et al.*, 2015). Recently, the role of metabolically engineered cyanobacteria in the production of biofuel has gained a lot of attention as the biofuel can be produced by utilization of CO_2 (Sharma *et al.*, 2008; Desai *et al.*, 2015). This approach is a carbon-neutral approach that will help in consuming CO_2 from the environment for biofuel production, thereby contributing in the utilization of greenhouse gas for beneficial purpose (Desai *et al.*, 2015). To enhance the biofuel productivity by cyanobacteria and marine algae, the CRISPR-Cas-mediated metabolic engineering can be applied. This will require exploring the genetic pathways involved in the production of biofuel and thus exploiting those pathways by CRISPR-Cas will provide a cost-effective approach for biofuel production.

Recently, it was shown that altering metabolic pathways via CRISPR-Cas tool in cyanobacteria results in the high titer of succinate as compared to its wild-type form (Li *et al.*, 2016). Furthermore, the construction of metabolic pathways in *E. coli* for biofuel production will enable commercialization of the production of biofuel (Wang *et al.*, 2017a). The use of the CRISPR-Cas approach in exploiting metabolic pathways to develop metabolic engineered strain of *E. coli* for biofuel production may provide a big commercial success in near future. Furthermore, the use of the multiplex CRISPR approach in Saccharomyces cerevisiae has shown that the fermentation rate of cellobiose is increased by tenfold (Ryan *et al.*, 2014). The increase in cellobiose fermentation will pave way for improved biofuel production. Henceforth, utilizing CRISPR-Cas tools I editing the metabolic pathways involved in biofuel production, especially in Cyanobacteria, marine algae and other commercially utilized bacterial species, will provide an environmentally friendly alternative source of energy, thereby helping maintain environmental homeostasis and relieving global stress.

11.2.5 Biosensing

One of the major challenges encountered by the environmental biotechnologist is to develop approaches for ensuring environmental security. Due to the increased rate of anthropogenic activities, the amount of toxicants introduced to the environment is very high. However, several approaches are being utilized to minimize the disposal of toxicants into the environment, but the success of these approaches relies much on the monitoring and detection of toxicants' status in the environment. Therefore, effective monitoring and detection of the pollutants/toxicants in the environment will enable us to carry out timely and effective remediation (Wanekaya *et al.*, 2008). There exist physical, chemical and biosensing approaches for carrying out effective monitoring and detection of toxicants' status in the environment (Wanekaya *et al.*, 2008; Yagi K, 2007). Based on the integrated property, portability, selectivity and sensitivity, with the rapid response time of the biosensors, it is considered to be the most effective approach for monitoring and detecting environmental toxicants (Gavrilescu *et al.*, 2015; Wanekaya *et al.*, 2008).

Biosensing is carried by the use of a 'biosensor', which is defined as an integrated device composed of a biological recognition species component (immuno-based, enzymatic, DNA or whole-cell) that is in direct contact of the second component referred as a transduction element (optical, electrochemical, thermal-based or mass-based) and a third component known as a signal-processing system (Figure 11.3; Nigam and Shukla, 2015; Wanekaya *et al.*, 2008). Biosensors have been developed for detection of heavy metals, such as zinc, cadmium, mercury and nickel; copper from different samples, such as soil, water and effluents (Pal *et al.*, 2009; Gruzina *et al.*, 2007; Claude *et al.*, 2007); pesticide detection (Mostafa GA, 2010; Rodríguez-Monaz *et al.*, 2006); and herbicide detection (Hayat and Marty, 2014). Furthermore, the different varieties of biosensors are being developed for detecting environmental pollutants, but the approach offers several limitations, one of which being the selection of biological recognition species components. The use of bacteria, algae and cyanobacteria has been shown to work as a potential biological

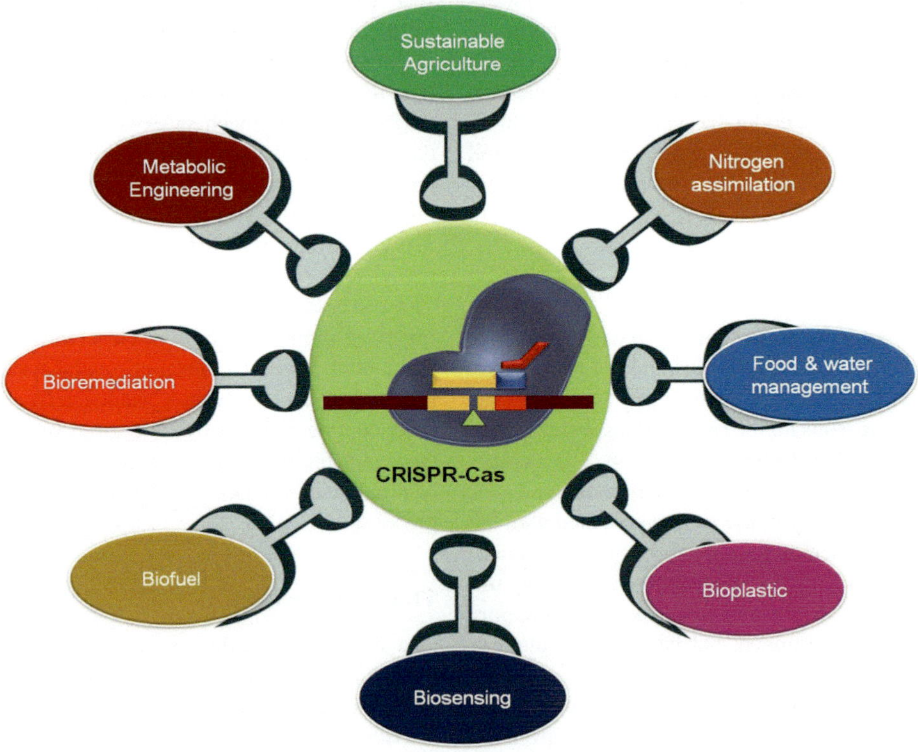

FIGURE 11.3 Illustration of the potential role of the CRISPR-Cas system in different sectors of environmental biotechnology.

recognition species component of the biosensor (Nigam and Shukla, 2015). The development of gene-editing technology offers novel methods of designing a desired biological recognition species component. In this milieu, bacteria can be engineered to be used as bacterial biosensors. This can be accomplished by fusion of a DNA-response element associated to stress-response proteins with the different types of reporter genes, thereby enabling the development of bacterial biosensors that will be used for a wide variety of analyses (Nigam and Shukla, 2015; Yagi, 2007). However, the development of such cell-based biosensors can be made feasible by the use of the CRISPR-Cas technique. Recently, a CRISPR-Cas-based biosensor known as SHERLOCK (Specific High-sensitivity Enzymatic Reporter unLOCKing) has been developed that is capable of detecting a single molecule and a single base–pair discrimination in the nucleic acid (Zuo *et al.*, 2017).

The sensitivity of SHERLOCK is very high and is capable of detecting femtomolar concentration (Zuo *et al.*, 2017). Furthermore, it has been shown that SHERLOCK-based method detected approximately 1000 viral particles of dengue virus per ml of solution, and it also detected about 2000 Zika virus copies in given clinical isolates (Zuo *et al.*, 2017). The degree of sensitivity in such biosensing is of immense importance as it will provide a tool for detecting infectious diseases. Another advantage is its cost-effectiveness, whereby the complete reagent can be incorporated into a single strip costing about US$0.61. Although current SHERLOCK-based biosensing is more of a diagnostic tool, it has opened immense possibilities for the CRISPR-Cas system in the development of highly sensitive and cost-effective biosensors. Therefore, the designing of CRISPR-Cas-based biosensors for the use of detecting and monitoring environmental pollutants will have a greater impact on environmental protection. Such biosensing is done by using microbes and plants that can be enhanced by unraveling the complex multigene network in the microbes and plants. This seems feasible by the use of the CRISPR-Cas technique. Furthermore, biosensing, when clubbed with bioremediation, will have a huge impact against existing environmental problems and will facilitate to combat unseen future environmental challenges.

11.2.6 Other Avenues

The application area of a CRISPR-Cas tool is continuously increasing, and with the advancing research, it will have a sustainable impact on environmental biotechnology. CRISPR-Cas tools are being further utilized in other environmentally beneficial projects, which include biosensing, bioplastic production, wastewater management, minimizing food wastage, a reduction of greenhouse gases and the like. The production of bioplastic from certain bacteria and algae will provide a biodegradable and non-pollutant form of plastic. The utility can be further enhanced by using the CRISPR approach in its production. Biosensing is another very important area in which the use of the CRISPR technique can have greater impact in environment protection. Biosensing, when clubbed with bioremediation, will have a huge impact on rising environmental problems.

Biosensing is done by using microbes and plants that can be enhanced by unraveling the complex multigene network in microbes and plants. This seems feasible through the use of the CRISPR-Cas technique. Furthermore, the management of wastewater from industries and households is a big challenge for environmental biotechnologists. However, the use of CRISPR-Cas tools in near future will enable researchers to develop microbial-based approaches to efficiently perform wastewater management. Hence, there exist a plethora of possibilities in which the revolutionizing CRISPR-Cas tool can have a huge impact, with an aim of achieving a healthy environment for a better life. However, there exist both scientific and social limitations that need to be sorted out for utilizing the full potential of the technique. The major concern is the public acceptability and consideration by governmental regulatory bodies from each country so that the practical applicability of the CRISPR-Cas tool can be utilized. To accomplish this, political willingness will be required so as to provide a clear thought over the efficient use of CRISPR technology. The use of CRISPR-Cas-mediated gene-editing technique in the field of environmental biotechnology is summarized in Figure 11.3.

11.3 Ethical Concerns about the Use of the CRISPR Technique

The CRISPR-Cas technique has received a lot of attention in research because of its easy-to-use, cost-effectiveness, time-saving and widely applicable properties. However, every potential approach has associated risks and the CRISPR-Cas system is no exception to the rule. There are several ethical concerns regarding the use of CRISPR, which are heavily being debated worldwide. One of the ethical concerns involves a balance between the benefit and the risk of the CRISPR technique, in which the benefit has to be on the upper hand. Besides, the risk assessment is of greater concern as the CRISPR technique involves 'off-target' risk, which means that the DNA sequence apart of target sequence may also be subjected for modification if it gets identified by the CRISPR-Cas system as target (Fu *et al.*, 2013; Yang *et al.*, 2014). This off-targeting may lead to deleterious mutation or transformation, thereby imposing severe risk (Zhang *et al.*, 2015; Fu *et al.*, 2013). Therefore, the risk assessment and minimizing the off-target effect may help in overcoming the ethical concern associated with the CRISPR-Cas tool. Furthermore, the use of CRISPR-Cas also imposes the risk of causing ecological disequilibrium (Rodriguez E, 2016).

The phenomenon of gene drive operates in living beings, and thus, there exists a risk of transfer of modified gene sequences among the population that may result in passing the negative trait to related organisms (Esvelt *et al.*, 2014). This may have severe consequences on ecological equilibrium, and hence, there exist a need for proper assessment for the use of the CRISPR-Cas tool (Rodriguez E, 2016). One of the major ethical concerns are the use of CRISPR-Cas in human germline. The utilization of the gene-editing technique in somatic cells is ethically acceptable provided informed consent is given prior to its use and the benefit/risk ratio has been taken into account (Rodriguez E, 2016). However, the scenario for germline cells are different as any changes to germline will be transmitted to future generations; therefore, the use of CRISPR-Cas in germline cells raises severe ethical concerns. Therefore, the immense potential of the CRISPR-Cas technique has to overcome these serious ethical issues, which impose a threat to humans and the environment.

11.4 Conclusion

The advent of the CRISPR-Cas tool has revolutionized genome editing research by making it highly efficient, cost-effective and widely applicable. Every sector of life science research is reaping a benefit from the CRISPR-Cas technique. The utilization of the CRISPR tool has begun to create a huge impact on the field of environmental biotechnology. The tool is being utilized in almost every sector of environmental biotechnology. It has been shown that CRISPR offers a platform that will enable researchers and scientists to tackle global environmental challenges, such as global warming, sustainable agriculture, water management, food production and others. Although CRISPR-Cas offers immense potential, certain scientific limitations and public concerns have to be minimized and sorted out for its efficient use.

ACKNOWLEDGMENTS

ACM highly acknowledges the partial financial supports from DBT (BT/PR32907/MED/122/227/2019), and institutional umbrella support grants under the following: DST PURSE-(Phase-II) (PAC-JNU-DST-PURSE-462), UGC DRS-II from School of Life Sciences, Jawaharlal Nehru University, New Delhi-110067, India. Dr. S. Srivastav gratefully acknowledges the financial support from SERB-NPDF Project (PDF/2016/000892).

REFERENCES

Anders, Carolin, et al. "Structural basis of PAM-dependent target DNA recognition by the Cas9 endonuclease." *Nature* 513.7519 (2014): 569–573.

Ansari, Z. A., and S. G. P. Matondkar. "Anthropogenic activities including pollution and contamination of coastal marine environment." *Journal of Ecophysiology and Occupational Health* 14.1/2 (2014): 71.

Arora, Leena, and Alka Narula. "Gene editing and crop improvement using CRISPR-Cas9 system." *Frontiers in Plant Science* 8 (2017): 1932.

Azubuike, Christopher Chibueze, Chioma Blaise Chikere, and Gideon Chijioke Okpokwasili. "Bioremediation techniques – classification based on site of application: Principles, advantages, limitations and prospects." *World Journal of Microbiology and Biotechnology* 32.11 (2016): 1–18.

Barrangou, Rodolphe, and Luciano A. Marraffini. "CRISPR-Cas systems: Prokaryotes upgrade to adaptive immunity." *Molecular Cell* 54.2 (2014): 234–244.

Barrangou, Rodolphe, et al. "CRISPR provides acquired resistance against viruses in prokaryotes." *Science* 315.5819 (2007): 1709–1712.

Barzman, Marco, et al. "Eight principles of integrated pest management." *Agronomy for Sustainable Development* 35.4 (2015): 1199–1215.

Basu, Supratim, et al. "Engineering PGPMOs through gene editing and systems biology: A solution for phytoremediation?" *Trends in Biotechnology* 36.5 (2018): 499–510.

Bolotin, Alexander, et al. "Clustered regularly interspaced short palindrome repeats (CRISPRs) have spacers of extrachromosomal origin." *Microbiology* 151.8 (2005): 2551–2561.

Bose, Bimal K. "Global warming: Energy, environmental pollution, and the impact of power electronics." *IEEE Industrial Electronics Magazine* 4.1 (2010): 6–17.

Brussaard, Lijbert, et al. "Reconciling biodiversity conservation and food security: Scientific challenges for a new agriculture." *Current opinion in Environmental Sustainability* 2.1–2 (2010): 34–42.

Charpentier, Emmanuelle. "CRISPR-Cas9: How research on a bacterial RNA-guided mechanism opened new perspectives in biotechnology and biomedicine." *EMBO Molecular Medicine* 7.4 (2015): 363–365.

Claude, Durrieu, et al. "Whole cell algal biosensors for urban waters monitoring." *Novatech* 7.3 (2007).

Dai, Zhenfei, et al. "Occurrence of microplastics in the water column and sediment in an inland sea affected by intensive anthropogenic activities." *Environmental Pollution* 242 (2018): 1557–1565.

Dangi, Arun Kumar, et al. "Bioremediation through microbes: Systems biology and metabolic engineering approach." *Critical Reviews in Biotechnology* 39.1 (2019): 79–98.

Daniell, Henry, et al. "Chloroplast genomes: Diversity, evolution, and applications in genetic engineering." *Genome Biology* 17.1 (2016): 1–29.

DeLorenzo, V. "Biodegradation and bioremediation: An introduction." Consequences of microbial Interactions with hydrocarbons, oils, and lipids: Biodegradation and bioremediation. Handbook of hydrocarbon and lipid microbiology. Springer, Cham, Switzerland (2018): 1–21.

Deltcheva, Elitza, et al. "CRISPR RNA maturation by trans-encoded small RNA and host factor RNase III." *Nature* 471.7340 (2011): 602–607.

Desai, Trunil, Vaishali Dutt, and Shireesh Srivastava. "Systems biology and metabolic engineering of marine algae and cyanobacteria for biofuel production." *Mar Bioenergy: Trends Dev* 163 (2015).

Deveau, Hélène, et al. "Phage response to CRISPR-encoded resistance in Streptococcus thermophilus." *Journal of Bacteriology* 190.4 (2008): 1390–1400.

Doudna, Jennifer A., and Emmanuelle Charpentier. "The new frontier of genome engineering with CRISPR-Cas9." *Science* 346.6213 (2014).

Douglas, Angela E. "Strategies for enhanced crop resistance to insect pests." *Annual Review of Plant Biology* 69 (2018): 637–660.

Dvořák, Pavel, et al. "Bioremediation 3.0: Engineering pollutant-removing bacteria in the times of systemic biology." *Biotechnology Advances* 35.7 (2017): 845–866.

Esvelt, Kevin M., et al. "Emerging technology: Concerning RNA-guided gene drives for the alteration of wild populations." *eLife* 3 (2014): e03401.

Fingerman, M. (Ed.). "*Bioremediation of aquatic and terrestrial ecosystems*." CRC Press (2016).

Fu, Yanfang, et al. "High-frequency off-target mutagenesis induced by CRISPR-Cas nucleases in human cells." *Nature Biotechnology* 31.9 (2013): 822–826.

Gao, Caixia. "The future of CRISPR technologies in agriculture." *Nature Reviews Molecular Cell Biology* 19.5 (2018): 275–276.

Garnett, Tara, et al. "Sustainable intensification in agriculture: Premises and policies." *Science* 341.6141 (2013): 33–34.

Gasiunas, Giedrius, et al. "Cas9 – crRNA ribonucleoprotein complex mediates specific DNA cleavage for adaptive immunity in bacteria." *Proceedings of the National Academy of Sciences* 109.39 (2012): E2579-E2586.

Gavrilescu, Maria, et al. "Emerging pollutants in the environment: Present and future challenges in biomonitoring, ecological risks and bioremediation." *New Biotechnology* 32.1 (2015): 147–156.

Gavrilescu, Maria. "Environmental biotechnology: Achievements, opportunities and challenges." *Dynamic Biochemistry, Process Biotechnology and Molecular Biology* 4.1 (2010): 1–36.

Glass, David J. "Commercial use of genetically modified organisms (GMOs) in bioremediation and phytoremediation." In M. Fingerman and R. Nagabhushanam, (eds.), *Bioremediation of Aquatic and Terrestrial Ecosystems* (2005), 41–96.

Gruzina, T. G., et al. "A bacterial multisensor for determination of the contents of heavy metals in water." *Journal of Water Chemistry and Technology* 29.1 (2007): 50–53.

Habben, Jeffrey E., et al. "Transgenic alteration of ethylene biosynthesis increases grain yield in maize under field drought-stress conditions." *Plant Biotechnology Journal* 12.6 (2014): 685–693.

Hayat, Akhtar, and Jean L. Marty. "Aptamer based electrochemical sensors for emerging environmental pollutants." *Frontiers in Chemistry* 2 (2014): 41.

Hays, Dirk B., et al. "Heat stress induced ethylene production in developing wheat grains induces kernel abortion and increased maturation in a susceptible cultivar." *Plant Science* 172.6 (2007): 1113–1123.

Heler, Robert, et al. "Cas9 specifies functional viral targets during CRISPR – Cas adaptation." *Nature* 519.7542 (2015): 199–202.

Heler, Robert, Luciano A. Marraffini, and David Bikard. "Adapting to new threats: The generation of memory by CRISPR-Cas immune systems." *Molecular Microbiology* 93.1 (2014): 1–9.

Ishino, Yoshizumi, et al. "Nucleotide sequence of the iap gene, responsible for alkaline phosphatase isozyme conversion in Escherichia coli, and identification of the gene product." *Journal of Bacteriology* 169.12 (1987): 5429–5433.

Jansen, Ruud, et al. "Identification of genes that are associated with DNA repeats in prokaryotes." *Molecular Microbiology* 43.6 (2002): 1565–1575.

Jinek, M., F. Jiang, D. W. Taylor, S. H. Sternberg, E. Kaya, et al. "Structures of Cas9 endonucleases reveal RNA-mediated conformational activation." *Science* 343(6176) (2014): 1247997.

Jinek, M., K. Chylinski, I. Fonfara, M. Hauer, J. A. Doudna, et al. "A programmable dual-RNA – guided DNA endonuclease in adaptive bacterial immunity." *Science* (2012): 1225829.

Kawakami, Eduardo M., Derrick M. Oosterhuis, and John L. Snider. "Physiological effects of 1-methylcyclopropene on well-watered and water-stressed cotton plants." *Journal of Plant Growth Regulation* 29.3 (2010): 280–288.

Kunin, Victor, Rotem Sorek, and Philip Hugenholtz. "Evolutionary conservation of sequence and secondary structures in CRISPR repeats." *Genome Biology* 8.4 (2007): 1–7.

Lamichhane, Jay Ram, et al. "Toward a reduced reliance on conventional pesticides in European agriculture." *Plant Disease* 100.1 (2016): 10–24.

Lechenet, Martin, et al. "Reducing pesticide use while preserving crop productivity and profitability on arable farms." *Nature Plants* 3.3 (2017): 1–6.

Lee, Sung Kuk, et al. "Metabolic engineering of microorganisms for biofuels production: From bugs to synthetic biology to fuels." *Current Opinion in Biotechnology* 19.6 (2008): 556–563.

Li, H., C. R. Shen, C. H. Huang, L. Y. Sung, M. Y. Wu, et al. "CRISPR-Cas9 for the genome engineering of cyanobacteria and succinate production." *Metabolic Engineering* 38 (2016): 293–302.

Liu, Liang-Ying, et al. "Polycyclic aromatic hydrocarbons (PAHs) in continental shelf sediment of China: Implications for anthropogenic influences on coastal marine environment." *Environmental Pollution* 167 (2012): 155–162.

Liu, Xuan, et al. "Application of CRISPR/Cas9 in plant biology." *Acta Pharmaceutica Sinica B* 7.3 (2017): 292–302.

Lü, Jing, Con Sheahan, and Pengcheng Fu. "Metabolic engineering of algae for fourth generation biofuels production." *Energy & Environmental Science* 4.7 (2011): 2451–2466.

Makarova, Kira S., and Eugene V. Koonin. "Annotation and classification of CRISPR-Cas systems." *CRISPR* (2015): 47–75.

Makarova, Kira S., et al. "An updated evolutionary classification of CRISPR – Cas systems." Nature Reviews Microbiology 13.11 (2015): 722–736.

Makarova, Kira S., et al. "Evolution and classification of the CRISPR – Cas systems." *Nature Reviews Microbiology* 9.6 (2011): 467–477.

Malnoy, Mickael, et al. "DNA free genetically edited grapevine and apple protoplast using CRISPR/Cas9 ribonucleoproteins." *Frontiers in Plant Science* 7 (2016): 1904.

Marraffini, Luciano A., and Erik J. Sontheimer. "Self versus non-self discrimination during CRISPR RNA-directed immunity." *Nature* 463.7280 (2010): 568–571.

Meinshausen, Malte, et al. "Greenhouse-gas emission targets for limiting global warming to 2 C." *Nature* 458.7242 (2009): 1158–1162.

Mojica, Francisco J. M., et al. "Short motif sequences determine the targets of the prokaryotic CRISPR defence system." *Microbiology* 155.3 (2009): 733–740.

Mojica, Francisco J. M., Jesús García-Martínez, and Elena Soria. "Intervening sequences of regularly spaced prokaryotic repeats derive from foreign genetic elements." *Journal of Molecular Evolution* 60.2 (2005): 174–182.

Mostafa, Gamal A. E. "Electrochemical biosensors for the detection of pesticides." *The Open Electrochemistry Journal* 2.1 (2010).

Nel, Willem P., and Christopher J. Cooper. "Implications of fossil fuel constraints on economic growth and global warming." *Energy Policy* 37.1 (2009): 166–180.

Nigam, Vinod Kumar, and Pratyoosh Shukla. "Enzyme based biosensors for detection of environmental pollutants-a review." *Journal of Microbiology and Biotechnology* 25.11 (2015): 1773–1781.

Nishimasu, Hiroshi, et al. "Crystal structure of Cas9 in complex with guide RNA and target DNA." *Cell* 156.5 (2014): 935–949.

Oliver, Neal J., et al. "Cyanobacterial metabolic engineering for biofuel and chemical production." *Current Opinion in Chemical Biology* 35 (2016): 43–50.

Pal, Priyabrata, et al. "The detection of mercury, cadium, and arsenic by the deactivation of urease on rhodinized carbon." *Environmental Engineering Science* 26.1 (2009): 25–32.

Pourcel, Christine, Gregory Salvignol, and Gilles Vergnaud. "CRISPR elements in Yersinia pestis acquire new repeats by preferential uptake of bacteriophage DNA, and provide additional tools for evolutionary studies." *Microbiology* 151.3 (2005): 653–663.

Rath, Devashish, et al. "The CRISPR-Cas immune system: Biology, mechanisms and applications." *Biochimie* 117 (2015): 119–128.

Ratner, Hannah K., Timothy R. Sampson, and David S. Weiss. "Overview of CRISPR – Cas9 biology." *Cold Spring Harbor Protocols* 2016.12 (2016): pdb-top088849.

Reid, Walter V., et al. "Earth system science for global sustainability: Grand challenges." *Science* 330.6006 (2010): 916–917.

Rodriguez, E. "Ethical issues in genome editing using Crispr/Cas9 system." *Journal of Clinical Research & Bioethics* 7.2 (2016): 1–4.

Rodríguez-Leal, Daniel, et al. "Engineering quantitative trait variation for crop improvement by genome editing." *Cell* 171.2 (2017): 470–480.

Rodriguez-Mozaz, Sara, Maria J. Lopez de Alda, and Damià Barceló. "Biosensors as useful tools for environmental analysis and monitoring." *Analytical and Bioanalytical Chemistry* 386.4 (2006): 1025–1041.

Runguphan, W., and J. D. Keasling. "Metabolic engineering of Saccharomyces cerevisiae for production of fatty acid-derived biofuels and chemicals." *Metabolic Engineering* 21 (2014): 103–113.

Ryan, Owen W., et al. "Selection of chromosomal DNA libraries using a multiplex CRISPR system." *eLife* 3 (2014). e03703.

Sen, Sudip Kumar, and Sangeeta Raut. "Microbial degradation of low density polyethylene (LDPE): A review." *Journal of Environmental Chemical Engineering* 3.1 (2015): 462–473.

Sharma, Ranjana, et al. "Diversity and distribution of commensal fecal Escherichia coli bacteria in beef cattle administered selected subtherapeutic antimicrobials in a feedlot setting." *Applied and Environmental Microbiology* 74.20 (2008): 6178–6186.

Shi, Jinrui, et al. "ARGOS 8 variants generated by CRISPR-Cas9 improve maize grain yield under field drought stress conditions." *Plant Biotechnology Journal* 15.2 (2017): 207–216.

Shi, Jinrui, et al. "Overexpression of ARGOS genes modifies plant sensitivity to ethylene, leading to improved drought tolerance in both Arabidopsis and maize." *Plant Physiology* 169.1 (2015): 266–282.

Tacão, M., A. Correia, and I. Henriques. "Resistance to broad-spectrum antibiotics in aquatic systems: Anthropogenic activities modulate the dissemination of bla CTX-M-like genes." *Applied and Environmental Microbiology* 78.12 (2012): 4134–4140.

Van Der Oost, John, et al. "Unravelling the structural and mechanistic basis of CRISPR – Cas systems." *Nature Reviews Microbiology* 12.7 (2014): 479–492.

Wanekaya, Adam K., Wilfred Chen, and Ashok Mulchandani. "Recent biosensing developments in environmental security." *Journal of Environmental Monitoring* 10.6 (2008): 703–712.

Wang, Chonglong, Brian F. Pfleger, and Seon-Won Kim. "Reassessing Escherichia coli as a cell factory for biofuel production." *Current Opinion in Biotechnology* 45 (2017a): 92–103.

Wang, Haoyi, et al. "One-step generation of mice carrying mutations in multiple genes by CRISPR/Cas-mediated genome engineering." *Cell* 153.4 (2013): 910–918.

Wang, Longlong, et al. "Use of CRISPR/Cas9 for symbiotic nitrogen fixation research in legumes." *Progress in Molecular Biology and Translational Science* 149 (2017b): 187–213.

Weber, Tilmann, et al. "Metabolic engineering of antibiotic factories: New tools for antibiotic production in actinomycetes." *Trends in Biotechnology* 33.1 (2015): 15–26.

Westra, Edze R., et al. "CRISPR immunity relies on the consecutive binding and degradation of negatively supercoiled invader DNA by Cascade and Cas3." *Molecular Cell* 46.5 (2012): 595–605.

Wolf, D., et al. "The plant breeders' dream come true: editing non-transgenic cucumber for broad virus resistance using CRISPR/Cas9 technology." Cucurbitaceae 2016, XIth Eucarpia Meeting on Cucurbit Genetics & Breeding, July 24–28, 2016, Warsaw, Poland. Cucurbitaceae 2016 Organizing Committee, 2016.

Wright, Addison V., James K. Nuñez, and Jennifer A. Doudna. "Biology and applications of CRISPR systems: Harnessing nature's toolbox for genome engineering." *Cell* 164.1–2 (2016): 29–44.

Xiaolei, Zuo, Fan Chunhai, and Chen Hong-Yuan. "Biosensing: CRISPR-powered diagnostics." *Nature Biomedical Engineering* 1.6 (2017).

Yagi, Kiyohito. "Applications of whole-cell bacterial sensors in biotechnology and environmental science." *Applied Microbiology and Biotechnology* 73.6 (2007): 1251–1258.

Yang, Jun, et al. "Evidence of polyethylene biodegradation by bacterial strains from the guts of plastic-eating waxworms." *Environmental Science & Technology* 48.23 (2014): 13776–13784.

Yang, Xiaorui, Mengmeng Xu, and Shang-Tian Yang. "Metabolic and process engineering of Clostridium cellulovorans for biofuel production from cellulose." *Metabolic Engineering* 32 (2015): 39–48.

Yin, Kangquan, Caixia Gao, and Jin-Long Qiu. "Progress and prospects in plant genome editing." *Nature Plants* 3.8 (2017): 1–6.

Yuan, Yuan, et al. "The impact of anthropogenic activities on marine environment in Jiaozhou Bay, Qingdao, China: A review and a case study." *Regional Studies in Marine Science* 8 (2016): 287–296.

Zetsche, Bernd, et al. "Cpf1 is a single RNA-guided endonuclease of a class 2 CRISPR-Cas system." *Cell* 163.3 (2015): 759–771.

Zhang, Fuzhong, Sarah Rodriguez, and Jay D. Keasling. "Metabolic engineering of microbial pathways for advanced biofuels production." *Current Opinion in Biotechnology* 22.6 (2011): 775–783.

Zhang, Xiao-Hui, et al. "Off-target effects in CRISPR/Cas9-mediated genome engineering." *Molecular Therapy-Nucleic Acids* 4 (2015): e264.

Zhang, Yunwei, et al. "Simultaneous modification of three homoeologs of Ta EDR 1 by genome editing enhances powdery mildew resistance in wheat." *The Plant Journal* 91.4 (2017): 714–724.

Zhao, Guangju, et al. "Soil erosion, conservation, and eco-environment changes in the Loess Plateau of China." *Land Degradation & Development* 24.5 (2013): 499–510.

12

Biotechnological Approaches for Increasing the Production and Utilization of Natural Fibers for a Cleaner and Greener Environment

Laxmi Sharma, Suman Roy, Soham Ray, Amit Bera, Debabrata Sarkar, Surendra Kumar Pandey and Pratik Satya

CONTENTS

DOI: 10.1201/9781003186304-12

12.1 Introduction

The impact of human civilization on earth, the only known inhabitable planet for biological organisms, can clearly be envisaged from the recent estimation of biomass distribution on earth (Bar-On *et al.*, 2018; Elhacham *et al.*, 2020). The overall estimate of human biomass (0.06 Gt carbon) is much higher than the combined biomass of all other wild mammals (0.007 Gt carbon). Not only that, the biomass of domesticated animals and poultry is much higher compared to their wild counterparts. Furthermore, the rate of energy consumption for sustaining this biomass is ever increasing; global energy demand has increased by 2% in 2017, along with a 1.4% increase in energy-related carbon dioxide (CO_2) emissions (International Energy Agency, 2018). The impact of the accumulation of such huge biomass and its typical high energy-consumption pattern is quite evident—mass extinction of animals, rising environmental pollution, global warming and rapid depletion of water and other natural resources. These issues are currently in the focus of attention among researchers, policymakers, environmentalists, farming community and many other groups. The rising concerns on environmental degradation have insisted nations neutralize carbon emissions, inclining them toward targeting reduced use of fossil fuel and synthetic polymers and promoting biobased technologies and products. In the Paris Agreement (UNFCC, 2016), 195 nations have joined hands to reduce the greenhouse gas (GHG) emissions for a better earth to live on. For this, goals have been set to reduce global temperature and to develop clean and sustainable socio-economy with energy security. However, most of the synthetic earth-polluting technologies are comparatively cheaper than their environmentally friendly alternatives, which makes the large-scale adoption of clean- and green-energy technologies a daunting task. Intensive research and technological developments are thus necessary to meet the two contrasting challenges of supplementing the required energy with a reduction in carbon emissions at a competitive cost.

To reduce the use of fossil fuel and synthetic polymers, alternate sources of renewable energy are required that are economically viable considering carbon budget. An alternate source is plant biomass, which has long been used to provide fuel and energy. However, the major component of plant-based fuel and energy comes from deforestation, which, again, is a major contributor to environmental pollution. For example, in the United States, the paper industry uses 68 million trees each year for pulping (www.greenamerica.org). A renewable energy source, such as agricultural crop biomass, is a far better alternative as a source of paper pulp than forest trees. However, the choice of renewable sources in this context is a crucial decision as it raises the issue like shifting of agricultural land to harvest energy at the cost of food production. Natural fiber–producing crops fit well in this frame, with the advantage of higher fixation of CO_2, environment cleanup and simultaneous production of green energy (Figure 12.1). Here, in this chapter, we concentrate more on natural fiber–producing plant species, which can play a significant role in reduction of both GHG emissions and the use of synthetic polypropylenes, with an occasional introduction to the potential of two animal-derived natural fibers, silk and wool, that also contribute to a significant reduction of polyethylene- and polypropylene-based products.

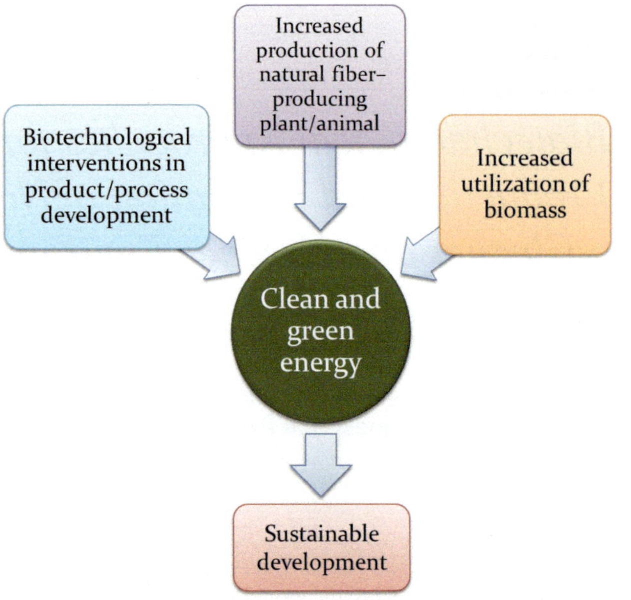

FIGURE 12.1 Sustainable development through biotechnological approaches from natural fibers.

12.2 Natural Fibers: Classification, Properties and Biotechnological Development

Natural fibers are produced from both animal and plant species (Figure 12.2) and vary greatly in composition, appearance and use. Plant-based natural fibers are either cellulosic (cotton, kapok, ramie) or lignocellulosic (jute, flax, hemp, kenaf, sisal). These fibers have several applications in textile and nontextile sectors, with a high potential for the development of various clean- and green-energy products. A number of plants producing natural fibers also produce high biomass, such as kenaf, hemp, jute and ramie (Satya and Maiti, 2013). The biomass itself is predominantly lignocellulosic, composed primarily of cellulose (47%–68%), hemicellulose (14%–24%), lignin (10%–30%) and pectin (5%–30%), which has a high potential for the production of paper pulp, biocomposites and biofuel.

12.2.1 Cotton

Cotton (*Gossypium hirsutum*, *G. barbadense*, *G. raimondii*, family: Malvaceae, 2n = 52) is the most important fiber-producing plant species, occupying the first position in terms of area (31.8 m ha; International Cotton Advisory Committee, 2017) and production (26 million tonnes; FAOSTAT, 2016). Cotton fiber is purely cellulosic (90% cellulose), originating from the seed coat. Due to its economic importance, several attempts have been made to improve the production and quality of cotton through biotechnological approaches, particularly transgenic technology. A number of transgenic hybrid cotton

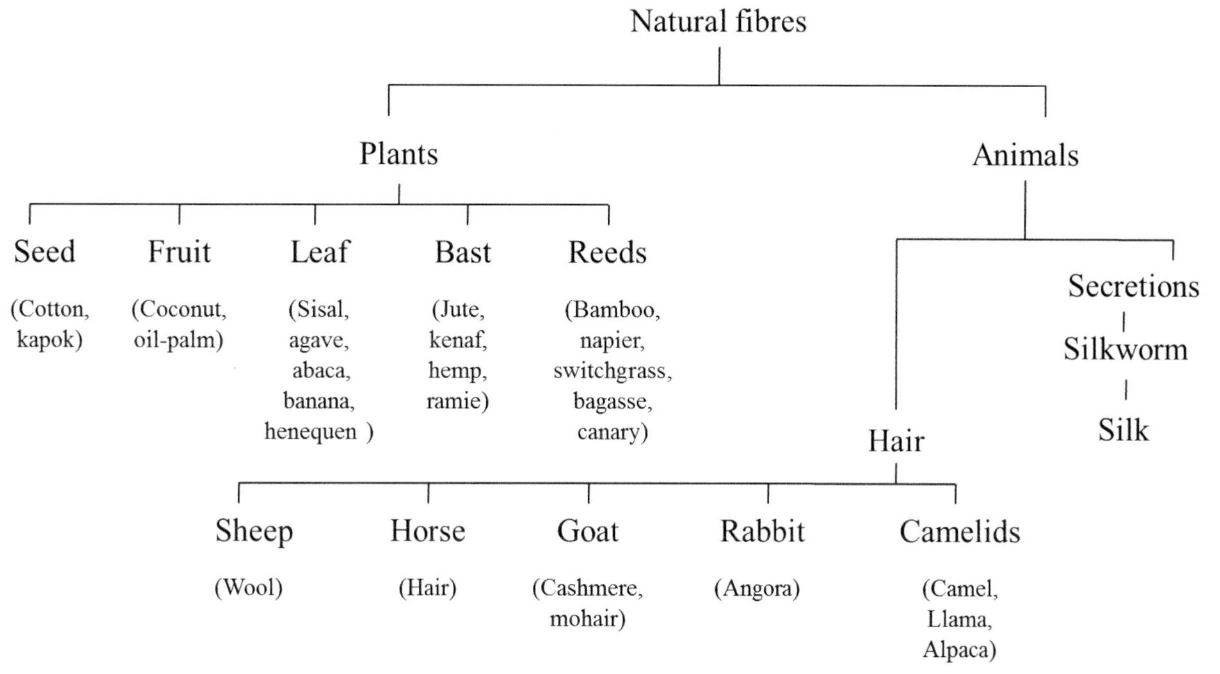

FIGURE 12.2 Classification of natural fibers.

varieties are commercially cultivated worldwide over an area of 24.9 million ha (Liu *et al.*, 2020), having an insect resistance trait along with a high yield. In addition, successful transgenic cultivars have also been developed for other traits, including fiber yield and quality (Walford *et al.*, 2011) and herbicide tolerance (Keller *et al.*, 1997). To further aid biotechnological interventions in this crop, several resources, such as genome and transcriptome sequences, genetic map, physical map, transformation and regeneration systems, have also been developed.

12.2.2 Flax

Flax (*Linum usitatissimum*, family: Linaceae, 2n = 30) is one of the oldest domesticated fiber crops, grown primarily in Canada, Kazakhstan, China, Russia, Belgium and many more countries, over an area of 0.22 million ha with a production of 44.3 thousand tonnes (FAOSTAT, 2016). Flax fiber, popularly known as linen, has a high textile value. It is also used for the production of tents, towels, sails, canvas and other materials. The biomass yield of flax is about 15 to 20 t/ha, while the seed yield of linseed, another ecotype of flax grown for oilseed production, is about 530 to 580 kg ha^{-1} (Jana *et al.*, 2018). The flax genome has been sequenced, and a high-density genetic map is available. Besides, a lot of genomic, transcriptomic, proteomic and metabolomic resources are available in this crop that have been very helpful in biotechnological manipulation of this crop (Huis *et al.*, 2012). Robust genetic transformation protocols have aided in the development of several transgenic lines in flax, although commercial cultivation of transgenic flax is yet to be realized.

12.2.3 Industrial Hemp

Hemp (*Cannabis sativa*, family: Cannabaceae, 2n = 20) has been a traditionally important fiber crop. So far, it has been cultivated as industrial hemp. There is another ecotype with a high-cannabinoid profile, which is used for pharmaceutical purposes. The potential of hemp metabolite (cannabinoids) is well recognized in the field of medical science (Ahmed *et al.*, 2015). The recent popularization of hemp for fiber, as well as other industrial uses, has spread its cultivation rapidly throughout the world. Currently, it is cultivated over an area of 0.44 m ha, with the production of 71.46 thousand tonnes (FAOSTAT, 2016). Industrial hemp has a high biomass of 8 to 15 tonnes ha^{-1} year^{-1} and is also capable of producing a fiber quality comparable to flax. The complete genome and transcriptome sequence of hemp have been deciphered (Van Bakel *et al.*, 2011), and a few reports on molecular breeding for improving fiber yield and quality are also available (Van der Broeck *et al.*, 2008).

12.2.4 Jute

Jute (*Corchorus olitorius*, *C. capsularis*, family: Malvaceae, 2n = 14) is the second-most important natural fiber crop after cotton, grown mainly in India and Bangladesh followed by China, and others over an area of 1.4 m ha, with a production of 331 thousand tonnes (FAOSTAT, 2016). The plant is cultivated mainly for coarser fiber suitable for packaging purposes. This crop has biomass production potential (10–25 tonnes ha^{-1} year^{-1}; Satya and Maiti, 2013) and has diversified applications in various sectors. A number of genetic maps and quantitative trait loci for fiber yield, wood yield, green biomass, fiber fineness, and tensile strength in *C. oiltorius* jute are available (Sarkar *et al.*, 2016; Satya and Chakraborti, 2015). However, it is recalcitrant to *in vitro* regeneration, and only a few transformation protocols are reported. A number of transcriptomic resources have been developed in jute. Whole-genome sequences of two jute species are also available (Islam *et al*., 2017; Sarkar *et al*., 2017), which open up avenues toward understanding the mechanisms underlying fiber development and quality and cell wall development through modulation of lignin and hemicellulose content.

12.2.5 Kenaf, Ramie and Sisal

Kenaf (*Hibiscus cannabinus*, *H. sabdariffa*, family: Malvaceae, 2n=36), ramie (*Boehmeria nivea*, family: Urticaceae, 2n = 28) and sisal (*Agave sisalana*, family: Asparagaceae, 2n = 150) are also important members of the natural fiber family. The former two are popular in Asian countries while the latter one is produced mainly in Brazil and other American countries. The fiber obtained from kenaf is allied with jute. However, ramie and sisal give a very high-quality fiber with a huge potential in the textile and paper industries, respectively. The biomass potential of these crops has been recognized, and their cultivation is being recognized for diversified purposes. Despite their immense economic importance, the basic and molecular information regarding genetic makeup, as well as growth and development, is still in its formative stage. A draft genome of ramie has been sequenced recently (Luan *et al*., 2018). Limited physical and genetic maps; transcriptomes, proteomes and metabolomes; and the transformation and regeneration protocols, among others, are designed for kenaf, ramie and sisal. Nevertheless, the availability of promising biotechnological tools provides a lot of scope for enriching the database for improving crop quality.

12.2.6 Silk

Silk is one of the most luxurious textiles across the world. The mulberry silkworm (*Bombyxmori*, family: Bombycideae, 2n = 28) is a domesticated insect and has assisted in silk production for the last 5000 years (Barber, 1991). Currently, global silk production is 177.53 thousand tonnes, and the major producers are China, India, Uzbekistan and Brazil (International Sericulture Commission, 2017). Studies on *B. mori* have significantly contributed to classical genetics and molecular biology (Willis *et al*., 1995). With its significance in sericulture, the sequencing and annotation of its genome have opened up novel targets for genetic manipulation. The current understanding of molecular biology has helped to utilize its larvae as live bioreactors. Furthermore, transgenic approaches have been used to improve silk yield (Zhang *et al*., 2017) and quality (Saviane *et al*., 2018). In addition, the silk glands of *B. mori* is a cost-efficient platform for the commercial production of recombinant proteins (Ma *et al*., 2014).

12.2.7 Wool

Wool, another important textile of animal origin, is produced from the hair follicles of the skin epidermis mainly from sheep. In addition, wool production from other animals, such as goats, rabbits, camelids and others, is also popular. Currently, 1.16 million kg of clean raw wool is produced by more than 1.163 billion sheep (International Wool Textile Organization, 2015). The fine quality of wool plays a critical role in textile sector and cloth industry, which mainly depend on traits like wool fiber length, fineness and density (Li *et al*., 2017). Several efforts have been put to improve the woolen fiber quality and length using various biotechnological tools. Recently, genome modification through transgenic (Liu *et al*., 2013) as well as genome editing technologies of sheep (Li *et al*., 2017; Hu *et al*., 2017) and goat (cashmere; Wang *et al*., 2016) have successfully produced longer wool length with better quality.

12.3 Importance of Natural Fibers in Environmental Protection

Natural fiber has supported mankind for cloth, shelter and energy since the dawn of human civilization. However, synthetic polymers, such as polyester, acrylic, nylon and others, in past five or six decades have replaced natural fibers due to their low cost and abundant availability. The production of these synthetic fibers adds carbon and pollutants to the environment, aggravating the global temperature. Fortunately, agreements from United Nations Framework Convention on Climate Change have created public awareness for mitigating GHG emissions and using eco-friendly products. More than 60 countries have completely banned plastics (CBS, 2018) and reduced dependence on nonrenewable fossil reserves. Fiber crops are versatile in nature and alternatively can be dedicated for bioenergy with sustainable clean development due to their lignocellulosic biomass as nonfood crops, renewable and biodegradable nature, high growth rate and biomass within a short period of 120 to 150 days, capacity to provide cloth and shelter, climate resilience in nature and requirement of low-cost input and can be grown in marginal land. They also provide benefits of carbon sequestration, phytoremediation and water use efficiency, lower cost of bioprocessing due to soft and herbaceous biomass and utilization as short-rotation crops.

Natural fibers have diversified industrial uses, such as paper, composites, fuel, woven and nonwoven textiles and other value-added products. Recently, a restriction on the cultivation of industrial hemp has been lifted in the United States; considering its potential for multiple end uses, the identification of its potential has led to the adoption of this crop within a short span. Similarly, the potential of kenaf has been identified in Malaysia for meeting energy crisis needs recently (Luthfi *et al*., 2017). Presently, sisal, hemp and kenaf are being used to generate high-quality papers at the commercial level to save forest reserves. Biocomposites made from hemp shives, commercially known as 'hempcrete', is used as a building material instead of cements in the United States. Henry Ford's hemp car, designed long ago in the 1940s, ran on fuel generated from hemp. Car interiors of popular brands, such as Mercedes Benz, are now made of hemp biocomposites. In addition, jute and kenaf biocomposites also have been quite popular in the construction of automobile interiors and exteriors. Biofuels are successfully produced at a commercial scale from many natural fiber crops. The potential of these bark and leaf fiber crops as biofuel sources has been well explained by Satya and

Maiti (2013) and has been explored for bioethanol production (Duque *et al.*, 2021).

Apart from sustainable energy production, natural fibers also protect the environment against global warming and reduce the depletion of fossil reserves, as well as remediate the environment. These are carbon-negative or carbon-neutral crops with higher atmospheric CO_2-fixing capacity. For example, a jute crop consumes 15 tonnes CO_2 ha^{-1} $year^{-1}$ in 120 days, producing 49.7 g m^{-2} biomass (Palit, 1993) whereas kenaf absorbs 21.89 tonnes CO_2 ha^{-1} $year^{-1}$ (Santoso *et al.*, 2015). The carbon sequestration potential of kenaf ranges from 2.9 to 12.1 tonnes CO_2 ha^{-1} $year^{-1}$ (Santoso *et al.*, 2015) whereas that of hemp ranges from 8.9 to 13.4 tonnes CO_2 ha^{-1} $year^{-1}$ (Vosper, 2011). Again, coupling phytoremediation and carbon sequestration to bioenergy production would be a better strategy for improving environmental health. Growing fiber crops for dual targets allows diversification of the contaminated area, enhances soil quality and CO_2 sequestration and increases the earth's greenness. The unutilized contaminated land can be used for growing fiber crops, solving the issue of land dedication for food versus energy crops as well as preventing the flow of leachate to adjacent areas. Fibre crops, such as hemp and flax, are good accumulators of Cu, Cd, Pd and Zn (Angelova *et al.*, 2004) and are used for biodiesel production in cadmium-contaminated soil (Shi and Cai, 2009). The phytoremediation potential of other fiber crops has also been reported (Table 12.1). Environmental cleanup in many countries is practiced by growing bioenergy crops at contaminated sites (Pandey *et al.*, 2016). Furthermore, integration of 'omics' approaches opened up additional avenues for combining the potential of soil microbial community with plant metabolism and consequently increasing the phytoremediation capacity. In flax, the association of mycorrhizal fungi with plants has improved the remediation of cadmium (Gao *et al.*, 2011) and nickel (Amna *et al.*, 2015) from contaminated soil. Heavy metal–tolerant bacteria supporting phytoremediation of cadmium, lead, zinc and copper are also reported in ramie rhizosphere (Jiang *et al.*, 2017). As a whole, growing these crops for 4 to 5 months sequesters at least 10 tonnes CO_2 ha^{-1} $year^{-1}$ with simultaneous removal of pollutants and production of bioenergy. However, a major constraint of these fiber crops (except cotton) is their industrially driven demand in an inconsistent manner. In addition, these are cultivated in small landholdings and used for local consumption (Satya and Maiti, 2013). Many authors have reviewed the potential of second-generation lignocellulosic biomasses dedicated to bioenergy that are otherwise sources of food/feed. However, discussion projecting the importance of natural fibers as second-generation lignocellulose biomass and their diversification uses is limited. Hence, the popularization and adoption of these fiber crops in a sustainable energy–demanding society

TABLE 12.1

Phytoremediation Potential of Natural Fiber Crops

Crop	Heavy Metal	Reference
Flax	Cd, Cu, Pb, Zn	Angelova *et al.*, 2004
	Ni	Amna *et al.*, 2015
	Cd	Gao *et al.*, 2011
Hemp	Cd, Cu, Pb, Zn	Angelova *et al.*, 2004
	Cd	Linger *et al.*, 2002; Shi *et al.*, 2012; Luyckx *et al.*, 2017
	Pb, Cu, Zn, Ni, Co, Cr	Malik *et al.*, 2010
	Ni	Linger *et al.*, 2002
	Cu	Bona *et al.*, 2007
Jute	Pb, Cr, Cu, Fe, Zn	Ahmed and Slima, 2018
	As	Nizamuddin *et al.*, 2016
Kenaf	Cd	Kurihara *et al.*, 2005; Deng *et al.*, 2017
	Cr	Ding *et al.*, 2016
	Cd, Zn	Arbaoui *et al.*, 2013
	Pb	Ho *et al.*, 2008
	Se	Banuelos *et al.*, 1997
	Cr, Pb, Cd	Bada, 2015
	Fe, Zn	Abioye *et al.*, 2012
	Fe, As	Meera and Agamuthu, 2012
	As	Nizamuddin *et al.*, 2016
	Mn, Fe, Zn, Cu, Cr	Taiwo *et al.*, 2016
Ramie	Cu, Zn, Pb, Cd	Jiang *et al.*, 2017
	Cd	Gong *et al.*, 2016; Li *et al.*, 2014
	Sb	Chai *et al.*, 2016

are the need of the hour for harnessing the previously mentioned environmental benefits.

12.4 Targets For Improving the Production of Natural Fiber–Producing Plants

12.4.1 Biomass Improvement

Genetic improvement and crop management practices of natural fiber crops have primarily targeted the traits associated with fiber yield and quality, resistance to diseases and pests and tolerance to abiotic stresses. These traits have also been manipulated by advanced biotechnological tools to attain higher productivity of fiber, the primary harvest obtained from these crops. However, from a different perspective, the potential of natural fibers are emerging in the industrial sector or as a bioenergy crop to support a clean and green environment. Improving biomass yield and quality (biomass recalcitrance) are major thrust areas that require the intervention of cutting-edge biotechnological tools. Biomass improvement in most crops has been achieved by targeting traits such as CO_2 fixation, photosynthetic efficiency, hormone metabolism, transcription factors expression, carbon metabolism and signal transduction at the shoot apical meristem (Table 12.2). The modulation of sucrose synthase from cotton (*GhSuSA1*), an important enzyme in carbon metabolism, showed significant association with plant biomass as well as fiber yield and quality. The downregulation of *GhSusA1* reduced the overall quality of fiber whereas its overexpression improved biomass at the vegetative stage (Jiang *et al.*, 2012). In another study, the overexpression of *GhCOBL9A*, responsible for cell elongation and cellulose crystallinity, increased the plant cell elongation and thickening and contributed to higher biomass (Niu *et al.*, 2018). The constitutive overexpression of *Arabidopsis thaliana GA20 oxidase* in kenaf under a constitutive promoter resulted in enhanced stem growth with a better cellulose content in fibers (Withanage *et al.*, 2015). Similarly, overexpression of the rate-limiting enzyme of cytokinin biosynthesis pathway (isopentenyl transferase, IPT) under senescence-associated promoter (SAG12) conferred a higher biomass in transgenic ramie through increased photosynthesis, chlorophyll content, height, stem diameter, stem and bast fresh weight, bast dry weight, resistance to exogenous ethylene and delayed senescence (An *et al.*, 2017).

Biomass recalcitrance has been successfully addressed in flax by manipulating the lignin biosynthesis pathway. The downregulation of cinnamyl alcohol dehydrogenase (*LuCAD*; Wróbel-Kwiatkowska *et al.*, 2007a) and caffeoyl coenzyme O-methyl transferase (*LuCCoAOMT*; Day *et al.*, 2009) have greatly reduced endogenous lignin content by 16% to 40% and 8% to 18%, respectively. Chantreau and coworkers in 2013 identified *lbf1*-mutated line with 35% high-lignin content, which increased the accumulation of lignin biosynthesis genes (*LuCCR, LuCOMT, LuCAD*) and the polymerization (peroxidases) gene. The role of cellulose synthase genes (*LuCes*A1 and *LuCes*A2) has been elucidated for the deposition of gelatinous layer (G-layer; Mokshina *et al.*, 2017) and bast fiber formation and thickening in flax (Chantreau *et al.*, 2013). A transgenic flax overexpressing bacterial PHB gene resulted in fibers with better elasticity (Wróbel-Kwiatkowska *et al.*, 2007b). The flax fiber thus obtained was used to prepare flax-polypropylene/polylactide composites, and these composites had better mechanical properties in terms of elasticity, flexibility and tensile strength (Gredes *et al.*, 2010; Kunert-Keil *et al.*, 2012). In hemp, recent studies have focused and characterized the metabolic processes of secondary wall growth and bast fiber development, the regulation of lignin biosynthesis and its association with phytohormones, particularly auxin and ABA (Behr *et al.*, 2017). Lignin-related genes are differentially expressed in tissue- and stage-specific manners (Behr *et al.*, 2018). A transcriptome signature from various parts of the stem at different developmental stages identified the snap point where the majority of cell wall–related enzymes are expressed during the transition phase from fiber elongation to fiber-thickening stage (Guerriero *et al.*, 2017a). Spatiotemporal expression of phenylpropanoid pathway genes has been investigated in kenaf (Zhao *et al.*, 2014). Comparison of transcriptome signatures of a lignin deficient mutant (deficient lignin phloem fiber; *dlpf*) and wild type in jute showed 16 and 10 differentially expressed genes for the shikimate-aromatic amino acid pathway and the monolignol biosynthesis pathway, respectively. Furthermore, the expression analysis of the lignin biosynthesis pathway from contrasting lignin content backgrounds at 30 days after germination showed the downregulation of 3-dehydroquinate dehydratase/shikimate dehydrogenase (*CcDHD-SHH*2), shikimate kinase1 (*CcSK*1), chorismite synthase1 (*CcCS*1), phenylalanineammonia-lyase1 (*CcPAL*1), 4-coumarate:CoA ligase2 (*Cc4CL*2), cinnamoyl-CoAreductase1 (*CcCCR*1), caffeoyl-CoA O-methyltransferase2 (*CcCCoAOMT*2) and cinnamyl alcohol dehydrogenase7 (*CcCAD*7) and upregulation of 3-deoxy-d-arabino-heptulosonate 7-phosphate synthase2 (*CcDAHPS*2), cinnamate 4-hydroxylase2 (*CcC4H*2), ferulate 5-hydroxylase1 (*CcF5H*1), caffeic acid O-methyltransferase1(*C cCOMT*1) (Chakraborty *et al.*, 2015). Moreover, gene expression analysis in kenaf showed upregulation of *HcPAL*, *HcC4H* and *Hc4CL* in the stem. On the other hand, chalcone synthase (*HcCHS*), chalcone isomerase (*HcCHI*) and flavones-3-hydroxylase (*HcF3H*) were highly expressed in flowers. Flavone synthase (*HcFLs*) showed the highest expression in leaves (Zhao *et al.*, 2014). The upregulation of cellulose synthase7 (*CcCes*7) and fascilin-like arabinogalatan6 (*CcFLA*6) with the downregulation of *CcCAD*7 at the early growth stage in jute revealed their involvement in fiber S-layer deposition (Chakraborty *et al.*, 2015). In hemp, the *FLA* genes were finely tuned by the transcriptional network in tissue- andstage-specific manners (Guerriero *et al.*, 2017b). Overexpression of *GhMYB*7 (R2-R3 MYB transcription factor) from cotton resulted in the deposition of lignin and cellulose through activation of a cascade of secondary wall biosynthesis-related genes. The phenotype of transgenic *Arabidopsis* showed small, curled, dark green leaves with shorter inflorescence stems (Huang *et al.*, 2016b).

12.4.2 Biotic and Abiotic Stress Tolerance

Studies on biotic and abiotic stress responses in crops reveal that plants undergo changes in the cascade of metabolic

TABLE 12.2

Possible Target Genes for Enhancement of Biomass in Natural Fibers

Gene	Function	Source	Expressed in	Phenotype	Reference
Photosynthesis and C-metabolism- and N-metabolism-related genes					
Phosphoenol pyruvate carboxylase (PEPcase) and Pyruvate orthophosphate dikinase (PPDK)	CO_2 fixation	Maize	Rice	Increased tiller number and biomass	Matsuoka *et al.*, 2001
PEPcase		Potato	Potato	High photosynthesis	Häusler *et al.*, 2001
Rubisco activase (RA)	Photosynthesis	*Arabidopsis*	*Arabidopsis*	Higher photosynthesis and increased vegetative growth	Kurek *et al.*, 2007
ZmPEPCase, Aspartate Amino transferase (GmAspAT), Glutamine synthetase (NtGS)	CO_2 fixation, photorespiration and ammonia assimilation	Maize, soybean, tobacco	*Arabidopsis*	Increased shoot biomass	Kaachra *et al.*, 2018
Sucrose synthase (SUS3)	Carbon partitioning and sucrose synthesis	Rice	Rice	Enhanced biomass saccharification and lodging resistance	Fan *et al.*, 2017
Hormonal biosynthesis and signaling genes					
Protein phosphatase 2C (PP2C)	Negative regulator of ABA signaling	*Arabidopsis*	*Arabidopsis*	Increased biomass yield with accelerated growth of inflorescence stem due to cell expansion and proliferation	Sugimoto *et al.*, 2014
GA20 oxidase (GA20OX)	GA biosynthesis	*Arabidopsis*	*Arabidopsis*	Increased biomass and fiber length	Eriksson *et al.*, 2000
GA20oxidase1 (GA20OX1)		Maize	Maize	Increased plant height and biomass	Voorend *et al.*, 2016
Isopentenyl transferase (IPT)	Cytokinin receptor	*Agrobacterium*	*Arabidopsis*	Increased biomass	Rupp *et al.*, 1999
More axillary growth2/Dwarf3/Ramosus4 (MAX2/D3/RMS4)	Strigolactone signaling	Switch grass	*Arabidopsis*	Higher tillering	Cheng *et al.*, 2018
Genes regulating shoot apical meristem (SAM) development and cell cycle					
FLOWERING LOCUS C (FLC), SUPPRESOR OF CONSTANS (SOC), FT (FLOWERING LOCUS T), FUL (FRUITFUL)	Regulation of flowering time and indetermancy of meristem	*Arabidopsis*	*Arabidopsis*	Delayed flowering and increased biomass	Salehi *et al.*, 2005; Melzer *et al.*, 2008
CYCD2	Cell division	*Arabidopsis*	*Arabidopsis*	35% increase in plant height and accelerated growth rate	Cockcroft *et al.*, 2000
Transcription factors					
Biomass yield 1 (BMY1) and biomass yield 3 (BMY3)	Regulation of photosynthesis and metabolism	Switchgrass	Switchgrass	Increased shoot and root biomass with increased sugar and starch	Ambavaram *et al.*, 2018
PLASTOCHRON1	Cytochrome P450 (CYP78A1)	Maize	Maize	Increased leaf elongation duration, biomass and yield	Sun *et al.*, 2017
CYP714DT	GA deactivating enzyme	Rice	Poplar	Increased growth, biomass and fibre length	Wang *et al.*, 2013
NAC1	Multiple pathways	*Arabidopsis*	*Arabidopsis*	Larger leaves with thicker stem and abundant roots	Xie *et al.*, 2000

processes. Literature shows different strategies to control the pest and diseases in fiber crops. However, imparting resistance through increasing the immunity of a crop is another way of addressing the problem. This is acquired by the accumulation of phytohormones, secondary metabolites and reactive oxygen species (ROS) (Feng and Shan, 2014; Wu *et al.*, 2014). This complex interaction between secondary metabolite from phenyl propanoid pathway (flavonoids and lignin), phytohormones (jasmonic acid and salicylic acid) and ROS is found in transgenic cotton harboring laccase and lignin polymerization gene for ENHANCED resistance against *Verticillium* wilt, bollworms and cotton aphids (Hu *et al.*, 2018) and *Hcm*1 (harpin protein from *Xanthomonas oryzae* pv. *Oryzicola* and cecropin A-melittin) for *Verticilium* and *Fusarium* wilt (Zhang *et al.*, 2016). The transcriptome analysis for *Fusarium* wilt in flax showed involvement of pathogenesis, ROS and cell wall–related genes to acquire resistance (Dmitriev *et al.*, 2017).

The molecular mechanism of abiotic stress response and tolerance in fiber crops are also well investigated. Differential metabolic process in various tissue, impaired cell elongation in hypocotyl and enhanced secondary wall/lignin biosynthesis in leaves under salinity has been studied at physiological as well as molecular level in hemp (Guerriero *et al.*, 2017c), kenaf (Kim *et al.*, 2013), ramie (An *et al.*, 2015) and flax (Yu *et al.*, 2016). This was also reflected in the differential metabolism response of contrasting ecotypes in hemp. Seed-type crops utilized fatty acid metabolism, amino acid metabolism and endoplasmic reticulum protein-processing pathways, while fiber-type crops used spliceosome and amino acid pathways as a stress response strategy (Liu *et al.*, 2016a). Drought-induced hemp seedlings showed the modulation of NAC, B3, peroxidases, expansin and inositol oxygenase as a stress response mechanism (Gao *et al.*, 2018). Moreover, a transgenic approach has been used to validate the response of bZIP (Huang *et al.*, 2016a) and auxin receptor (auxin-binding protein, ABP1) for leaf development (Zhang *et al.*, 2014) and abiotic stress tolerance in ramie and that of cotton ABA receptor, PYR/PYL/RCAR, for drought tolerance in transgenic *Arabidopsis* (Chen *et al.*, 2017). There is also a tight modulation of the mechanism observed under high temperature through micro-RNA and long noncoding RNA, calcium and heat-shock protein as well as the epigenetic regulation (Zahid *et al.*, 2016).

12.4.3 Improvement in Photosynthetic and Nutrient Use Efficiency

Regulation of CO_2-fixing genes and enhancement of nutrient use efficiency are important targets to improve biomass production of fiber crops. Somatic hybrid of *G. hirsutum* x *G. trilobum* showed a higher photosynthetic rate, stomatal conductance and transpiration rates (Yu *et al.*, 2012). The expression of photosynthesis-related genes photosystem I P700 chlorophyll a apoprotein A1 and A2, photosystem II protein D1, cytochrome c biogenesis protein, photosystem I assembly protein, NADH dehydrogenase subunit 5 were higher in the hybrid, suggesting interspecific hybridization enhances photosynthesis. A proteome analysis of ramie under nitrogen (N), phosphorus (P) and potassium (K) deficiency showed the differential expression of 32, 51 and 27 proteins, respectively (Deng *et al.*, 2014). These proteins were categorized into various metabolic pathways, such as photosynthesis, primary and energy metabolism, cell, protein destination and storage, signal transduction, secondary metabolite metabolism and cell and developmental processes. Most important, proteins related to photosynthesis like RUBISCO large subunit (RBCL) and ferredoxin-nicotinamide adenine dinucleotide phosphate (NADP) reductase were downregulated under N-deficient and proteins like RBCL, chlorophyll a/b binding protein, chloroplast photosynthetic water oxidation complex, ferredoxin-NADP reductase and phosphoribulokinase were downregulated under P- and K-deficient conditions. Moreover, the expression of energy metabolism–related genes, such as *hexokinase*, *2,3-bisphosphoglycerate-independentphosphoglycerate mutase*, *glyceraldehyde-3-phosphate dehydrogenase* and *malate dehydrogenase*, are modulated under nutrient deficiency. This dynamic regulation of carbon metabolism under nutrient deficiency further provides the scope for improving biomass in fiber crops. In Table 12.2, we describe a number of target genes for improving biomass.

Another factor, inherent biomass recalcitrance of lignocellulosic biomass increases the cost of bioproduct development. Modifications in plants thus require lignin content adjustment through the manipulation of new genes (Wang *et al.*, 2015). In a study, diversion of the flux out of shikimate pathway to synthesize protocatechuate through the overexpression of 3-dehydro shikimate dehydrogenase from bacteria reduced the lignin content in *Arabidopsis* and increased the enzyme digestibility of lignocelluloses (Eudes *et al.*, 2016). This type of approach may be used to divert the flux of shikimate pathway from lignin biosynthesis pathway to reduce the lignin content in natural fibers. Another approach would be incorporation of novel and nonconventional monomers embedded with functional groups to block polymerization of the monolignol backbone. The introduction of monomers lacking β-carbon or a monomer in which the hydroxyl group has been blocked will disrupt the coupling between monomers and 4-O position of oligomers. Therefore, the elongation of oligomer chain of lignin will be arrested (Van Acker *et al.*, 2013). This has been successfully applied in *Arabidopsis* (Eudes *et al.*, 2012) to arrest lignin polymerization. Furthermore, transcriptional and post-translational regulation of lignification has been achieved through modulation of gene expression (Shen *et al.*, 2013), RNAi-mediated downregulation (Coleman *et al.*, 2008) or genome-editing techniques employing the CRISPR/Cas9 system (Zhou *et al.*, 2015) in poplar. Biofuel production from transgenic poplar has been successful due to the increased availability of fermentable sugar even at the field level (Van Acker *et al.*, 2013). An miRNA-mediated reduction in lignin content also has been fruitful in many lignocellulosic biomass crops. Overexpression of miRNA (miR156, miR164, miR166, etc.) has reduced recalcitrance in bioenergy crop like poplar and switch grass (Trumbo *et al.*, 2015). A newer approach to enhance the delignification rate may be introduction of genes capable of deconstructing plant cell wall (either lignin or pectin component) by sensing specific signals during harvest or preharvest.

12.5 Utilization of Natural Fibers for a Clean and Green Environment

Development of sustainability out of biomass is the growing research interest to let lignocellulosic biomass attain its market potential and support a clean ecosystem. Here, we highlight the potential of natural fiber biomass derived from biotechnological interventions mainly through microbes/enzymes in manufacture of green products such as paper, biofuel, biocomposites and the like. The main enzymes used for research and development, as well as at industrial scale, are mentioned in Table 12.3.

12.5.1 Substituting Nonrenewable Forest Biomass with Renewable Fiber Biomass for Paper Pulp

Changing agricultural practices, environmental concerns and depleting forest reserves have increased the importance of natural fibers for the global paper and pulp industry (Ashori *et al.*, 2006). The paper industry is the fourth-largest consumer of energy and a significant contributor of GHGs (www.greenamerica.org). Therefore, to reduce the environmental impact and overall cost of production, enzyme-assisted processes are, by far, the most accepted method for bio-pulping and bio-bleaching of paper pulp. Obtaining a higher grade dissolving pulp from lignocellulosic biomass is a herculean task. Enzyme mediated processing has improved the quality and mechanical properties of paper and reduced the generation of toxic product like absorbable organic halides (AOXs). AOXs are produced due to chlorine dioxide (ClO_2)–mediated bleaching of pulp and is dose-dependent. The dosage of chlorine-based bleaching agent depends on the structure of feedstock, reaction temperature, pH and hexuronic acid (HexA) content (Nie *et al.*, 2016). HexA is produced during the cooking of pulp, and its presence increases the kappa number (the degree of brightness increases as the kappa number decreases) of paper (Costa and Colodette, 2007). Therefore, the delignification rate can be readily monitored, and it acts as an indicator for reducing the quantity of bleaching agent. Delignification of feed stocks using ligninases (laccase, lignin peroxidase, manganese peroxidase, versatile peroxidases) removes lignin and chromophores from the substrate, thereby assisting in biobleaching and reducing the dosage of ClO_2. The by-product of the reaction is water, and no toxic pollutant is produced (Pei *et al.*, 2016). Enzymes, such as cellulase, xylanase and ligninase altogether improve the quality and mechanical properties of pulp, reduce the cost of bleaching, arrest the flow of pollutants to the environment, reduce overall cost of production and finally enhance the quality of paper (Sindhu *et al.*, 2016). Another biobleaching system, laccase mediator system (LMS)–assisted bleaching using synthetic or natural coupler aided in better delignification, with improved paper quality in sisal (Aracri *et al.*, 2011), flax (Fillat *et al.*, 2012), kenaf (Andreu *et al.*, 2013), cotton (Spina *et al.*, 2016) and jute (Sahin *et al.*, 2016). It improved the bleaching performance, viscosity, reduced the kappa number, tensile strength, tear strength, cellulose reactivity and other intrinsic paper properties in sisal (Aracri and Vidal, 2011), flax (Fillat

TABLE 12.3

Major Enzymes Used in Bioprocessing of Natural Fibers

Enzyme	Major Enzymes	Function	Environmental Impact	Reference
Cellulase	Endo-glucanases, Cellobio-hydrolases, B-glucanases	• Improves the performance of dissolving pulp, mechanical properties and viscosity of paper pulp • Enzymatic hydrolysis for bioethanol production • Biostoning and biopolishing of cotton fabrics	Reduces energy consumption	Duan *et al.*, 2017; Singh *et al.*, 2015; Sindhu *et al.*, 2016; Araujo *et al.*, 2008
Hemi-cellulases	Endo-xylanases, Endo-mannanases		Reduces the load of pollutants from wastewater by reducing the generation of adsorbable organic halides	Nie *et al.*, 2016; Sindhu *et al.*, 2016; Anish *et al.*, 2007
Ligninases	Laccases, Lignin peroxidases, Mn-peroxidases, Versatile peroxidases	• Removes lignin and chromophores from the substrate • Improves biobleaching for paper pulp and textile fabrics • Improves adhesion strength on fiber-matrix interface of biocomposites • Biostoning and bio-desizing of fabric	Reduces the load of pollutants and use of chlorine-based bleaching agents	Gonçalves *et al.*, 2015; Pei *et al.*, 2016; Ibrahim *et al.*, 2011; Pazarlioğlu *et al.*, 2005
Others	Pectinases, Acetyl xylan esterases, Amylases/ gluco-amylases		Reduces the load of pollutants to the effluents	Dehabadi *et al.*, 2011; Anish *et al.*, 2007; Das and Ramaswamy 2006
	Savinases, Resinases, Alcalases, Proteases	• Bioscouring, shrink poofing and carbonization of wool • Degumming of silk fibers	Reduces the load of pollutants to the effluents and restricts the use of chlorine and alkalis	Das and Ramaswamy 2006; Anghileri *et al.*, 2007

et al., 2012) and kenaf (Andreu *et al.*, 2013). A comparison of paper and pulp properties from natural fibers such as sisal, flax, hemp, abaca and jute pretreated with xylanases, cold alkali and endo-glucanase treatment revealed that sisal had better paper pulp qualities (Ibarra *et al.*, 2010). Bio-pulping of jute and kenaf fibers also reduced the kappa number and improved paper properties (Mohiuddin *et al.*, 2005). Moreover, degraded hemicellulose from degummed ramie liquor enhanced the beatability and mechanical strength of paper from jute stick pulp. Cellulase-free xylanase or its recombinant from *Bacillus* sp. also improved brightness of pulp and reduced the usage of bleaching agent (Ayyachamy and Vatsala, 2007). Biobleaching of xylan-type fiber crops, such as kenaf (with abundant HexA), with a combination of laccase- and xylanase-enhanced delignification, reduced HexA and increased cellulose reactivity in totally chlorine-free (TCF) pulp (Andreu and Vidal, 2014). Furthermore, a new promising enzyme, carbohydrase, in combination with xylanase, improved the cellulose reactivity in TCF sisal pulp and subsequently reduced the requirement of NaOH for further processing (Beltramino *et al.*, 2015). These promising results suggest that biobleaching of pulp from natural fiber crops can successfully replace a considerable part of wood pulp, thereby helping reduce deforestation.)

12.5.2 Biofuel

Lignocellulosic biofuel is a safer alternative of renewable energy for sustainable global development (Chandel *et al.*, 2015). Although these outcompete first-generation biofuel with respect to food verses fuel, the major problem with these fiber crops is biomass recalcitrance, which hinders pretreatments from accessing their substrates for efficient conversion to fermentable sugars. Mitigation strategies are better developed to overcome the problem at the postharvest biomass pretreatment level. Microorganism/enzyme-mediated pretreatment has been the safest and sustainable way among the available methods. Delignification, the first step of bioethanol production, using certain lignin modifying enzymes, especially laccases from basidiomycetes (white rot fungi), ensures lignin removal as well as scavenging toxic organic pollutants and xenobiotics. Few studies have employed LMS to delignify the cotton feedstock along with physico-chemical pretreatments (Fockink, 2015). Microbial delignification of kenaf using fungi (*Corilus versicolor, Trametes gallica*) and bacteria (*Mycobacterium* and *Streptomyces* sp.) have been successful (Brzonova *et al.*, 2014). Enzymatic saccharification (conversion of polysaccharides to fermentable sugars), the second step of bioethanol production, executed with a wide range of hydrolases, such as cellulase, xylanase, β-glucosidase, cellobiase, pectinase and others, have been the major area of research interest so far. Major industrial source of commercial cellulase till date is *Trichoderma reesei* RUTC 30 (Martinez *et al.*, 2008). Although the higher cost of saccharification enzyme is the major economic constraint, enzymatic hydrolysis has been favored over acid hydrolysis. Cellulases from *Trichoderma reesei* is the preferably used enzyme both at a commercial scale and a research and development level for hemp (Abraham *et al.*, 2014), jute (Singh *et al.*, 2018) and kenaf (Brzonova *et al.*, 2014). In addition, nonenzymatic pretreatment, in combination with downstream enzymatic saccharification and fermentation in natural fibers, improved biofuel production. For example, microwave-assisted $FeCl_3$ and γ-valuronate pretreatment in cotton (Singh *et al.*, 2017), steam, dilute acid, alkali in combination with dilute acid in hemp (Kuglarz *et al.*, 2014), dilute acid, alkali and surfactant pretreatment in jute (Singh *et al.*, 2018) in addition to enzyme-assisted saccharification and fermentation have been reported. Besides, a novel hydrolytic enzyme, glycoside hydrolases 61 (HG61), from *Gleophyllum trabeum* in combination with xylanase (Gt X110G) or cellulase (GtCel5B) from the same source improved the efficiency of bioconversion by 56% (Jung *et al.*, 2015). The addition of cobalt to this further accelerated bioconversion by 11%. A new concept of cellulosome, the complex of several cellulolytic or hemicellulolytic enzymes (together called holocellulolytic) that attaches to the surface of the substrate and degrades it into soluble products, is emerging (Artzi *et al.*, 2017). *Clostridium thermocellum* B8 from goat rumen, efficient in producing cellulosome, contains the enzymatic cocktails (cellulase/xylanases) and is a potent candidate for a consolidated process of saccharification and fermentation from cotton stalk (Leitão *et al.*, 2017). To improve the efficiency, stability and activity of cellulase, immobilization of these enzymes to a matrix is also helpful within a short time. Immobilized cellulase on nanoparticles resulted in 93% saccharification in 48 hours from hemp biomass (Abraham *et al.*, 2014).

The final step of biofuel production, that is, enzyme-mediated fermentation, is mainly carried out by *Saccharomyces cerevisiae*, and the major problem of this step is catabolic repression. The abundance of fermentable sugar represses the conversion efficiency catalyzed by *Saccharomyces cerevisiae* in bioreactors. A novel non-inhibitory *Saccharomyces cerevisiae* RRP-03N used for fermentation showed 76% glucose conversion efficiency (Christopher *et al.*, 2017). Protoplast fusion of C6 fermenting *Saccharomyces cerevisiae* andC5 fermenting yeast such as *Pachyosolen tannophilus*, *Candida shelate*, *Pichia stipitis* yielded higher ethanol from cotton hydrolysate (Kumari *et al.*, 2012). Moreover, cellulosomal proteins and cellobiose hydrolases are triggered by catabolic repressor genes early in the process. Therefore, the more complex the substrate is, the earlier the expression of cellulosomal complex genes for hydrolysis. The use of yeasts other than *Saccharomyces cerevisiae*, such as *Candida tropicalis* CCT1516, has also been reported (Damião Xavier *et al.*, 2018). Apart from bioethanol, another biofuel obtained from lipid feedstocks has also been obtained from hemp seed oil (Li *et al.*, 2010), revealing a high potential of using fiber crops for biodiesel production.

12.5.3 Biocomposites

The use of natural fiber composites is on the rise due to increasing environmental and geopolitical awareness. Natural fiber composites, due to their renewable and biodegradable nature, low-cost input, good mechanical properties, durability and comparable specific strength are being preferred over synthetic fiber–based composites. In India, the construction industry contributes 10% of total CO_2 emission, of which 4% (102MT) is emitted by cement manufacturing units (Jami *et al.*, 2016). A major advantage of green composites based on natural fibers is that these are carbon-neutral or carbon-negative in nature. Hempcrete (blends of hemp shives and lime) with 45% carbon is carbon negative (Jami, 2018) and 1 kg of hemp shives sequester 1.6 to 1.8 kg CO_2 (Jami *et al.*, 2016). A functional wall unit of dimension 1 m × 1 m × 0.3 m has the capacity to sequester 82.7 kg CO_2 (Ip and Miller, 2012). However, the green composites made out of natural fibers possess some manufacturing constraints due to their lignocellulosic nature. One of the major disadvantages of fiber-reinforced composite is a weak interfacial bonding between the fiber and the matrix that imparts mechanical strength to the composite. The hydrophilicity of fiber and the hydrophobicity of matrix weakens the adhesion strength on fiber–matrix interface. To overcome this problem, lignocellulosic fiber has to be subjected to surface modification so that physical adhesion of fiber–matrix complex leads to optimum surface energies of material, chemical bonding and mechanical interlocking on the surface (Guven *et al.*, 2016). Low-lignin fiber plants thus would be a better choice than woody plants that have higher lignin content.

Most of the surface modification strategies reviewed in the literature have engaged a physical and chemical approach. However, employing green technology in improving the interfacial adhesion using microbes and enzymes has additional environmental benefits. Enzymes use less water and energy in a milder environment and release fewer toxic effluents. The treatment of plant fibers with enzymes, such as hydrolases (glycosidase, proteases, lipases) and oxidoreductases (laccases, peroxidases, tyrosinases; Gübitz *et al.*, 2003), is well established. Enzymatic blends of hydrolases and oxidoreductases for improving the surface properties have been reported in jute (Karaduman *et al.*, 2013) and hemp (Väisänen *et al.*, 2016). It increases fiber aspect ratio, flexural rigidity, tenacity, break strength and elongation, in addition to surface adhesion of the fiber composite. Improvement in surface smoothness, softness and brightness has been reported in several fiber crops (Kalia and Vashistha, 2012). The use of pectinase (commercial name: Viscozyme L from Novozyme) in addition to commercial or noncommercial chelators also improved the mechanical strength and interfacial adhesion in flax (Foulk *et al.*, 2006) and kenaf (Lee *et al.*, 2020) and hemp (Väisänen *et al.*, 2016). However, the use of LMS adds environmental benefits with different mediators for hydrophobization of the fiber surface with the advantage of detoxifying wastes. LMS has been successful in flax (Acero *et al.*, 2014), jute (Dong *et al.*, 2018) and kenaf (Islam *et al.*, 2014). Surface adhesion and mechanical properties of jute–polypropylene composites have been significantly improved by grafting acrylamide (AAM)/tertiary-butylhydroxide (t-BHP; Dong *et al.*, 2014), dodecyl gallate (Ni *et al.*, 2017), alkylgallate (Ni *et al.*, 2015). The LMS also improves the thermal properties of fiber-reinforced composites. Besides, a new enzyme system employing horseradish peroxidase with H_2O_2 grafting of dodecyl gallate also tremendously improved the mechanical strength of jute–polypropylene composites (Liu *et al.*, 2016). The improved technologies with help of enzymes have thus led to the development of green composites with better quality.).

12.5.4 Biogas

The production of biogas through anaerobic digestion of biomass is an established strategy to meet environmental challenges (Jiang *et al.*, 2011). Besides, the residual product can be used as a soil nutrient supplement and for soil and water conservation (Surendra *et al.*, 2014). Natural fiber, due to its lignocellulosic nature, has been used for biogas production. However, its biomass recalcitrance, the major limitation, is currently addressed through physical, chemical and biological pretreatment or combination. Although biological pretreatments are environmentally friendly, the higher cost of application limits their usage at a commercial scale. Biogas fermentation requires different microorganisms at every step of its processing: (1) hydrolysis through hydrolytic microorganisms, (2) acidogenesis through fermentive bacteria, (3) acetogenesis through homoacetogenic or syntrophic bacteria and (4) methanogenesis through methanogenic *archae* (Wei *et al.*, 2016). Removing lignin from feedstock allows sufficient access to the substrate to support biogas fermentation steps. Pretreatments of the complex lignocellulosic feedstocks with enzymes or in combination with other pretreatments, such as alkali pretreatment, are frequently practiced for delignification. Transgenic flax shives with downregulated lignin biosynthesis gene, *LuCAD*, followed by pretreatment with sodium hydroxide showed a 7% increase in biogas production and a 30% reduction in lignin content (Wróbel-Kwiatkowska *et al.*, 2015). Again, the treatment of different biomass with laccase- and versatile peroxidase–assisted enzyme digestion led to an increased release of phenolics (Schroyen *et al.*, 2015). In hemp, treatment with these enzymes increased biogas production as well as substrate detoxification by scavenging phenols (Schroyen *et al.*, 2017). In another study, MC1 consortia containing *Clostridium straminisolvens* (CSK1), *Clostridium* sp. FG4b, *Pseudoxanthomonas* sp. strain M1–3, *Brevibacillus* sp. M1–5 and *Bordetella* sp. M 1–6 (Kato *et al.*, 2005) increased the biodegradability of cotton stalk with enhanced release of substrate for acidogenesis and subsequent steps (Yuan *et al.*, 2016).

Although anaerobic digestion (AD) of lignocellulosic biomass results in biogas production, its full chemical potential remains unutilized in the form of solid digestate. Therefore, a concept of linking the pyrolysis to AD has been suggested for efficient utilization and upgradation of every phase. Recent studies show that the presence of biochar under anaerobic conditions acts as a detoxifying agent and substrate for the adhesion of microorganisms (Xu *et al.*, 2015). It also accelerates the syntrophic conversion of alcohols and volatile fatty acid to acetate, which is usually a slow step under anaerobic conditions (Zhao *et al.*, 2015). Valorizing the residues at each step by linking them to pyrolysis would be a better approach for harvesting green energy.

12.5.5 Textile Biotechnology

The global textile industry contributes 20% of the registered levels of water pollution from countries such as Turkey, Indonesia, China, Romania and Bulgaria and 44% from Macedonia (Paraschiv *et al.*, 2015). Attempts to overcome existing environmental concerns associated with the textile industry using white biotechnology have been quite successful. The first microbial process exploiting amylase from *Bacillus subtilis* dates to the early 1950s (Diderichsen *et al.*, 1995). Since then, extensive scientific contributions with help of green chemistry from the textile industry are frequently reported. Out of 186 commercialized enzyme marketed in 2001, 31 are used by the textile industry in Europe (Schumacher *et al.*, 2001). Enzyme technology has revolutionized the textile sector in lowering the pollutant load. The bioprocessing of major textile fibers such as cotton, wool and silk are briefly discussed here. The cotton industry uses enzymes in every bioprocessing stage, such as biodesigning, bio-scouring, bio-polishing, denim bio-stoning and others. Other emerging sustainable processing of textile includes silk degumming, wool scouring and shrink-proofing, bio-dyeing and so on.

12.5.6 Cotton

The past few decades have recorded enzymatic desizing of cotton with α-amylases. Adding glucoamylases with α-amylases removes earth alkaline and cationic metal particles in addition to starch desizing (Dehabadi *et al.*, 2011). Bio-scouring (the removal of noncellulosic fraction from cotton) using cellulose-free xylanases and pectinases improves the wetability, brightness, polymerization degree, crystallinity, color depth, softness, absorbency and weight loss, as well as low-stress mechanical properties (Shanthi and Krishnabai, 2013). The addition of cellulose further improves the efficiency of pectinases and hence the scouring process. Furthermore, the bleaching step of cotton fabric is a major contributor of pollutants to the effluents. LMS/H_2O_2 has been widely used for biobleaching of cotton fibers (Tian *et al.*, 2012). Another enzyme, glucose oxidase, uses β-D-glucose as a substrate to generate H_2O_2, which bleaches the cotton fabric with comparable mechanical properties and better whiteness (Tzanov *et al.*, 2002). The bio-stoning of denims at a commercial scale has employed cellulase (Araujo *et al.*, 2008), but recently, laccase-mediated bleaching is getting popularized (Pazarlioğlu *et al.*, 2005). Bio-desizing and biobleaching have been successful with a combination of amylase, cellulase and laccases on denims (Maryan and Montazer, 2013). Furthermore, to remove the microfibrillar cotton fragments on fabric, an optimal bio-polishing treatment is set using optimized cellulase preparations (a mixture of endoglucanases, cellobiohydrase and cellobiase). The bio-polishing of denims is also better achieved with endoglucanase and xylanase treatment (Anish *et al.*, 2007). The fabrics thus obtained are softer, smoother and have better color brightness (Ibrahim *et al.*, 2011). Although these enzyme-mediated bioprocessing is effective, a combination of physico-chemical method further improves enzyme efficiency for desizing (Ul-Haq and Nasir, 2012), scouring (Nithya *et al.*, 2011), bleaching (Gonçalves *et al.*, 2014) and bio-polishing (Ulson de souza *et al.*, 2013).

12.5.7 Wool

Traditionally, wool scouring, carbonization and shrink-proofing have been a source of environmental contamination due to the usage of heavy chemicals. The use of xylanase, pectinase, savinase and resinase on wool and other fibers from llama, alpaca, mohair and camel (Das and Ramaswamy, 2006) has led to clean scouring. Wool carbonization is now practiced with cellulase and pectinase to remove residues from wool (Gouveia *et al.*, 2008). The incorporation of protease instead of chlorine for shrinkage resistance prevents the generation of AOX by-products. The pretreatment of fibers with plasma, corona discharge, lipases, cutinases, transglutaminases and chitosan before protease treatment has also improved the quality and mechanical properties of woolen fabrics (Shahid *et al.*, 2016).

12.5.8 Silk

Silk processing has become easier and cleaner with enzyme-mediated degumming, bio-polishing, bio-dyeing, printing, functional finishing and cold washing (Shahid *et al.*, 2016). Alkaline degumming of silk fibers to remove sericin has been the most efficient and adopted process. However, due to its high cost and release of alkalis and surfactants to the effluents (Kalantzi *et al.*, 2013), enzyme-assisted degumming is now the most preferred technology. Although alcalase, savinases (Kalantzi *et al.*, 2013) and several other alkaline proteases (Anghileri *et al.*, 2007) are known to remove sericin efficiently without impairing silk quality, smoothness, luster and mechanical strength, the combined effect of these enzymes, for example, alcalase/savinase (Arami *et al.*, 2007) or protease/lipase (Gulrajani *et al.*, 2000), shows better degumming efficiency.

12.5.9 Bio-Dyeing

Commercial chemical dyes are harmful to both fabric and the environment. Enzyme-based bio-dyes are extensively used in the textile industry (Shahid *et al.*, 2016). The potential of laccases and catalases for dyeing of wool (Shin *et al.*, 2001), cotton and flax (Kim *et al.*, 2008) has been already exploited. Recently, tyrosinase-catalyzed oxidation of ferulic acid also has been introduced to the bio-dye family (Sun *et al.*, 2015). An environmental assessment of woolen fabric bio-dyeing through comparative life-cycle analysis of conventional and bio-dyeing approaches showed that cost of production and energy consumption is reduced by 50% and 70%, respectively, with a 55% reduction in CO_2-equivalent emissions (Parisi *et al.*, 2015).

12.6 Natural Fiber Crops for Greener Pharmaceuticals

12.6.1 Production of Xylitol

Xylitol, a 5-carbon sugar alcohol, is an artificial sweetener used in the food and pharmaceutical industries. The traditional. production of xylitol is energy- and cost-intensive, and thus, xylan biomass serves as an alternate for producing xylose on hydrolysis. Lignocellulosic biomass is rich in xylan, for example, sisal, jute and kenaf. Generally, fermentation through *Saccharomyces cerevisiae* uses C6 sugars whereas the C5 sugars remain unutilized. Integration of C5 and C6 fermentation using single or different microorganisms in a process reduces the cost of production and produces a value-added product such as xylitol. Currently, Asia is the largest producer of xylitol, followed by Europe, the United States and Australia. For efficient production, pretreatment of lignocellulosic biomass to remove lignin is a critical step, ensuring a reduction of the substrate crystallinity. Fungi such as white rot, brown rot or soft rot, as well as bacteria, are generally used in the process. Moreover, enzymes such as laccase and peroxidases have been widely used to remove lignin and detoxify the substrate. Xylitol has been produced from sisal fibers simultaneously with ethanol using *Candida tropicalis* CCT1516 within 60 hours (Damião Xavier *et al.*, 2018).

12.6.2 Production of Bio-Succinate

Succinate is one of the leading value-added compound in the bioeconomy (Cok *et al.*, 2014). Traditionally, it was produced by paraffin oxidation, catalytic hydrogenation and electroreduction of maleic anhydride (Magnuson and Lasure, 2004). Recently, the demand for biosuccinate and its derived product in global market has been rising (Marketsandmarkets.com, 2015). Of these, 1,4-butanediol, tetrahydrofuran, gamma-butyrolactone and polybutylene succinate are highly demandable compounds with the potential to drive the future market (Cok *et al.*, 2014). Various U.S.-based companies have currently employed several bacterial and fungal strains for biosuccinate production (Cavani *et al.*, 2016. The biobased production of each liter of succinate uses 5 tonnes of CO_2 and thus promises to be a sustainable green approach for carbon capture (Cavani *et al.*, 2016). Bacterial strains such as *Mannheimia succiniciproducens, Actinobacillus succinogenes, Anaerobiospirillum succiniproducens* and recombinant *E. coli* and fungal strains such as *Enterobacter faecalis* (Moon *et al.*, 2004) and *S. cerevisiae* (Ahn *et al.*, 2016) efficiently produce biosuccinate from several crop residues. Natural fibers are rich in fermentable sugars and are the most potent candidates for the simultaneous production of succinate with other bioproducts. Kenaf and hemp have already been used to produce biosuccinate along with bioethanol employing *Actinobacillus succinogenes* (Luthfi *et al.*, 2016). High-biomass natural fibers may be thus used to capture further carbon to produce value-added succinate.

With all such commercially important and value-added bioproducts from natural fibers, it is also important to understand the postharvest processing problem and address it through lignin-specific action using microbial interventions and logistic upgradations. Enzymes are, by far, the most important tools, and thus, most of the improvement strategies are centered on them. Bio-prospecting the ecosystem to explore thermally stable, nonspecific, efficient, robust, lytic novel holocellulolytic and/or cellulolytic enzymes/microbes is the most important strategy for improving biomass processing. Tailoring enzymes for enhanced efficiency, thermal stability and tolerance to inhibitors through mutation or recombinant DNA technology will further improve the overall performance of enzyme. Most important, designing of bacterial cellulosome and secretomes with desired enzymes will also improve the biomass degradation efficiency with consolidated bioprocessing. In addition, the blending of enzymes or preparation of enzymatic cocktails in addition to physico-chemical process will allow attaining better delignification. To further improve, immobilization of enzymes/microbes to a matrix such as a nanoparticle is important for harnessing its efficiency, stability, extended half-life and recyclability. From the logistics point of view, a single-pot processing of various steps involved in biofuel and biogas production can be thus achieved with the previously mentioned improvisations. Moreover, moving toward a biorefinery concept will efficiently allow using and recycling each and every residual component of bioprocessing steps, especially biofuel and biogas production.

12.6.3 Lignocellulosic Biorefinery Concept: Future for Bioenergy

The environmental and economic pressure has led to rethinking the processing of biomass in a sustainable manner. Enzymatic conversion of lignocellulosic biomass for integrating remediation and resource recovery (biofuels and biomaterials) through a closed-loop bioprocess cascade defines the biorefinery. According to the International Energy Agency, biorefinery is the sustainable processing of biomass into a wide spectrum of marketable products and energy (Silva *et al.*, 2018). This holistic approach requires management of crop production, biomass harvesting and procurement, logistics, storage and handling, pretreatment and processing and, finally, supply to market (Fahd *et al.*, 2012). The biomass conversion to bioenergy and biomaterial in a single platform requires a high-throughput, low-cost and robust approach and has the potential to sustain the socioeconomic, environmental and energy securities. Most of the biorefineries are operational at demonstration, pilot or commercial scale in Europe, the United States and Brazil based on using agricultural residues, corn stover and sugarcane bagasse, wood chips and a few dedicated crops, such as switchgrass, *Miscanthus* and others (Silva *et al.*, 2018; International Energy Agency, 2017). In addition, natural fiber/lignocellulosic biorefineries (Figure 12.3) also possess the potential to enhance the economy of any nation. However, there are a few challenges associated with feedstock, as well as bioprocesses, that need to be addressed to achieve commercial robustness (Mehmood *et al.*, 2021). Biomass recalcitrance and non-availability of efficient low-cost enzyme/enzyme consortia/microbial consortia are the major hurdles limiting fiber biomass from achieving its full potential. To ensure this, synergistic deconstruction of lignocellulosic biomass, enzyme consortium of lytic polysaccharide monooxygenase, ligninase, cellobiose dehydrogenase, hemicellulase, cellulase and other protein molecules, such as swollenins and expansins, are required (Van Dyk and Pletschke, 2012). Input cost for enzymes represent a major portion of overall cost of production. Therefore, high-throughput, low-cost, robust and efficient pretreatment methods, as well as enzymes, are required to economize the lignocellulosic biorefinery concept. Moreover, integrated 'omics' approach is essential for understanding the fundamental microbial–plant interaction required for bioenergy production at its highest potential. Employing microbial consortia from an ecosystem to a new environment (lignocellulosic feedstock) is now easier through identifying taxonomic groups and enabling microbes

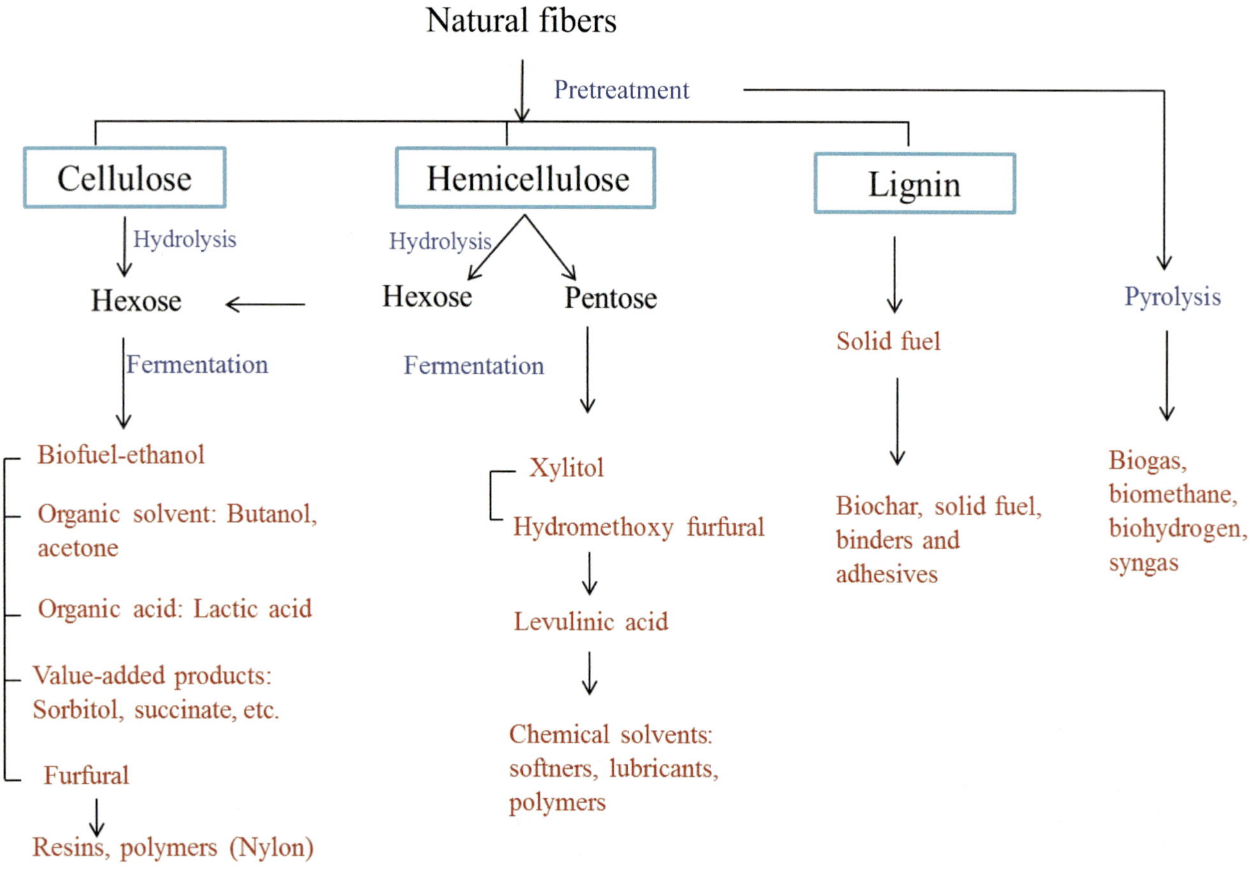

FIGURE 12.3 An outline of a natural fiber biorefinery.

to execute their substrate-specific action through a genomics and metagenomics approach (Silva *et al.*, 2018). This approach has been successfully employed on corn stover biomass and agricultural wastes using firmicutes and bacteroidetes. Novel glycoside hydrolases degrading rhamnogalacturon-II, one of the complex polysaccharides, have been identified from polysaccharide utilization loci of bacteroidetes (Ndeh *et al.*, 2017) and possess an immense potential for biorefinery application. Microbial secretomes and cellulosomes are also important aspects to be considered for degrading lignocellulosic biomass (Resch *et al.*, 2013).). Chip-based platforms, such as CAZymes (Carbohydrate Active enZymes) and ACPP (activity correlated quantitative proteomic platform) developed for an integrated approach, can be used to identify the compatibility of microbes with substrates (Ma *et al.*, 2017). These strategies will ensure the optimization of enzyme blends and the dosage required for enzymatic hydrolysis. Currently, biorefineries using natural fiber crops are limited despite their immense potential. Overall, the achievement of bioenergy production in a closed-loop biorefinery through green technology using natural fiber biomass will ensure a better substrate recovery and utilization with better environmental sustainability and energy security.

12.7 Conclusion

Building a biobased economy with a clean and green environment has been the global target, and developed nations are already set with this objective. To make it better, the use of lignocellulosic natural fiber biomass through recycling and recovering the products in a closed loop using green technology will be a better option. Green technology majorly employs the use of enzymes, and cost reductions of these enzymes are important. Moreover, efforts are also required to improve the inherent quality of biomass and its production. Integrating all these approaches will make it convenient to achieve a sustainable economy and environment.

ACKNOWLEDGMENT

We acknowledge financial support from Department of Science and Technology, Govt. of West Bengal under project no. 344(Sanc.)/ST/P/S&T/4G-01/2018.

REFERENCES

Abioye, O. P., P. Agamuthu, and A. R. Abdul Aziz. "Phytotreatment of soil contaminated with used lubricating oil using *Hibiscus cannabinus*." *Biodegradation* 23, no. 2 (2012): 277–286.

Abraham, Reinu E., Madan L. Verma, Colin J. Barrow, and Munish Puri. "Suitability of magnetic nanoparticle immobilised cellulases in enhancing enzymatic saccharification of pretreated hemp biomass." *Biotechnology for Biofuels* 7, no. 1 (2014): 1–12.

Acero, Enrique Herrero, Tukayi Kudanga, Andreas Ortner, Iwona Kaluzna, Stefaan De Wildeman, Gibson S. Nyanhongo, and

Georg M. Guebitz. "Laccase functionalization of flax and coconut fibers." *Polymers* 6, no. 6 (2014): 1676–1684.

Ahmed, Dalia A., and Dalia F. Slima. "Heavy metal accumulation by Corchorus olitorius L. irrigated with wastewater." *Environmental Science and Pollution Research* 25, no. 15 (2018): 14996–15005.

Ahmed, Safwat A., Samir A. Ross, Desmond Slade, Mohamed M. Radwan, Ikhlas A. Khan, and Mahmoud A. ElSohly. "Minor oxygenated cannabinoids from high potency *Cannabis sativa* L." *Phytochemistry* 117, (2015): 194–199.

Ahn, Jung Ho, Yu-Sin Jang, and Sang Yup Lee. "Production of succinic acid by metabolically engineered microorganisms." *Current Opinion in Biotechnology* 42, (2016): 54–66.

Ambavaram, Madana M. R., Aminat Ali, Kieran P. Ryan, Oliver Peoples, Kristi D. Snell, and Maria N. Somleva. "Novel transcription factors PvBMY1 and PvBMY3 increase biomass yield in greenhouse-grown switchgrass (*Panicum virgatum* L.)." *Plant Science* 273, (2018): 100–109.

Amna, Sajid Masood, Jabir Hussain Syed, Muhammad Farooq Hussain Munis, and Hassan Javed Chaudhary. "Phytoextraction of Nickel by *Linum usitatissimum* in Association with *Glomus intraradices*." *International Journal of Phytoremediation* 17, no. 10 (2015): 981–987.

An, Xia, Jie Chen, Jingyu Zhang, Yiwen Liao, Lunjin Dai, Bo Wang, Lijun Liu, and Dingxiang Peng. "Transcriptome profiling and identification of transcription factors in ramie (Boehmeria nivea L. Gaud) in response to PEG treatment, using illumina paired-end sequencing technology." *International Journal of Molecular Sciences* 16, no. 2 (2015): 3493–3511.

An, Xia, Jingyu Zhang, Yiwen Liao, Lijun Liu, Dingxiang Peng, and Bo Wang. "Senescence is delayed when ramie (*Boehmeria nivea* L.) is transformed with the isopentyl transferase (*ipt*) gene under control of the SAG 12 promoter." *FEBS Open Bio* 7, no. 5 (2017): 636–644.

Andreu, Glòria, Agustín G. Barneto, and Teresa Vidal. "A new biobleaching sequence for kenaf pulp: Influence of the chemical nature of the mediator and thermogravimetric analysis of the pulp." *Bioresource Technology* 130, (2013): 431–438.

Andreu, Glòria, and Teresa Vidal. "An improved TCF sequence for biobleaching kenaf pulp: Influence of the hexenuronic acid content and the use of xylanase." *Bioresource Technology* 152, (2014): 253–258.

Angelova, V., R. Ivanova, V. Delibaltova, and Kr Ivanov. "Bioaccumulation and distribution of heavy metals in fibre crops (flax, cotton and hemp)." *Industrial Crops and Products* 19, no. 3 (2004): 197–205.

Anghileri, Anna, Giuliano Freddi, Raffaella Mossotti, and Riccardo Innocenti. "Mechanical properties of silk yarn degummed with several proteases." *Journal of Natural Fibers* 4, no. 1 (2007): 13–23.

Anish, Ramakrishnan, Mohammad Safikur Rahman, and Mala Rao. "Application of cellulases from an alkalothermophilic *Thermomonospora* sp. in biopolishing of denims." *Biotechnology and Bioengineering* 96, no. 1 (2007): 48–56.

Aracri, Elisabetta, and Teresa Vidal. "Xylanase-and laccase-aided hexenuronic acids and lignin removal from specialty sisal fibres." *Carbohydrate Polymers* 83, no. 3 (2011): 1355–1362.

Aracri, Elisabetta, M. Blanca Roncero, and Teresa Vidal. "Studying the effects of laccase-catalysed grafting of ferulic acid on sisal pulp fibers." *Bioresource Technology* 102, no. 16 (2011): 7555–7560.

Arami, Mokhtar, Sharam Rahimi, Leila Mivehie, Firoozmehr Mazaheri, and Niyaz Mohammad Mahmoodi. "Degumming of Persian silk with mixed proteolytic enzymes." *Journal of Applied Polymer Science* 106, no. 1 (2007): 267–275.

Araujo, Rita, Margarida Casal, and Artur Cavaco-Paulo. "Application of enzymes for textile fibres processing." *Biocatalysis and Biotransformation* 26, no. 5 (2008): 332–349.

Arbaoui, Sarra, Aricia Evlard, Mohamed El Wafi Mhamdi, Bruno Campanella, Roger Paul, and Taoufik Bettaieb. "Potential of kenaf (*Hibiscus cannabinus* L.) and corn (*Zea mays* L.) for phytoremediation of dredging sludge contaminated by tracemetals." *Biodegradation* 24, no. 4 (2013): 563–567.

Artzi, Lior, Edward A. Bayer, and Sarah Moraïs. "Cellulosomes: bacterial nanomachines for dismantling plant polysaccharides." *Nature Reviews Microbiology* 15, no. 2 (2017): 83.

Ashori, Alireza, Jalaluddin Harun, Warwick D. Raverty, and Mohd Nor Mohd Yusoff. "Chemical and morphological characteristics of Malaysian cultivated kenaf (*Hibiscus cannabinus*) fiber." *Polymer-Plastics Technology and Engineering* 45, no. 1 (2006): 131–134.

Ayyachamy, M., and T. M. Vatsala. "Production and partial characterization of cellulase free xylanase by Bacillus subtilis C 01 using agriresidues and its application in biobleaching of nonwoody plant pulps." *Letters in Applied Microbiology* 45, no. 5 (2007): 467–472.

Bada, B. S. "Bioremediation of textile effluent polluted soil using kenaf (*Hibiscus cannabinus* Linn.) and composted market waste." *Journal of Applied Sciences and Environmental Management* 19, no. 4 (2015): 773–776.

Banuelos, G. S., H. A. Ajwa, Bruce Mackey, L. Wu, Charles Cook, S. Akohoue, and S. Zambruzuski. "Evaluation of different plant species used for phytoremediation of high soil selenium." *American Society of Agronomy, Crop Science Society of America, and Soil Science Society of America* 26, no. 3 (1997).

Barber, Elizabeth Jane Wayland. *Prehistoric textiles: The development of cloth in the Neolithic and Bronze Ages with special reference to the Aegean*. Princeton University Press, 1991. pp. 504.

Bar-On, Yinon M., Rob Phillips, and Ron Milo. "The biomass distribution on Earth." *Proceedings of the National Academy of Sciences* 115, no. 25 (2018): 6506–6511.

Behr, Marc, Kjell Sergeant, Céline C. Leclercq, Sébastien Planchon, Cédric Guignard, Audrey Lenouvel, Jenny Renaut, Jean-Francois Hausman, Stanley Lutts, and Gea Guerriero. "Insights into the molecular regulation of monolignol-derived product biosynthesis in the growing hemp hypocotyl." *BMC Plant Biology* 18, no. 1 (2018): 1–18.

Behr, Marc, Sylvain Legay, Jean-Francois Hausman, Stanley Lutts, and Gea Guerriero. "Molecular investigation of the stem snap point in textile hemp." *Genes* 8, no. 12 (2017): 363.

Beltramino, Facundo, Cristina Valls, Teresa Vidal, and M. Blanca Roncero. "Exploring the effects of treatments with carbohydrases to obtain a high-cellulose content pulp from a non-wood alkaline pulp." *Carbohydrate Polymers* 133, (2015): 302–312.

Bona, Elisa, Francesco Marsano, Maria Cavaletto, and Graziella Berta. "Proteomic characterization of copper stress response in *Cannabis sativa* roots." *Proteomics* 7, no. 7 (2007): 1121–1130.

Brzonova, Ivana, Evguenii Kozliak, Alena Kubátová, Michelle Chebeir, Wensheng Qin, Lew Christopher, and Yun Ji. "Kenaf biomass biodecomposition by basidiomycetes and actinobacteria in submerged fermentation for production of carbohydrates and phenolic compounds." *Bioresource Technology* 173, (2014): 352–360.

Cavani, Fabrizio, Stefania Albonetti, Francesco Basile, and Alessandro Gandini. *Chemicals and fuels from bio-based building blocks*. John Wiley & Sons, 2016.

CBS/AP. "Over 60 countries have introduced bans, fees-to cut single use plastic waste." *CBS News Blog*, 5 June 2018. https://www.cbsnews.com/news/over-60-countries-introduced-bans-fees-single-use-plastic-waste/ (accessed 02 May 2021).

Chai, Li-Yuan, Hussani Mubarak, Zhi-Hui Yang, Wang Yong, Chong-Jian Tang, and Nosheen Mirza. "Growth, photosynthesis, and defense mechanism of antimony (Sb)-contaminated *Boehmeria nivea* L." *Environmental Science and Pollution Research* 23, no. 8 (2016): 7470–7481.

Chakraborty, Avrajit, Debabrata Sarkar, Pratik Satya, Pran Gobinda Karmakar, and Nagendra Kumar Singh. "Pathways associated with lignin biosynthesis in lignomaniac jute fibres." *Molecular Genetics and Genomics* 290, no. 4 (2015): 1523–1542.

Chandel, Anuj K., Bruna C. M. Gonçalves, Janice L. Strap, and Silvio S. da Silva. "Biodelignification of lignocellulose substrates: An intrinsic and sustainable pretreatment strategy for clean energy production." *Critical Reviews in Biotechnology* 35, no. 3 (2015): 281–293.

Chantreau, Maxime, Sébastien Grec, Laurent Gutierrez, Marion Dalmais, Christophe Pineau, Hervé Demailly, Christine Paysant-Leroux et al. "PT-Flax (phenotyping and TILLinG of flax): development of a flax (*Linum usitatissimum* L.) mutant population and TILLinG platform for forward and reverse genetics." *BMC Plant Biology* 13, no. 1 (2013): 1–15.

Chen, Yun, Li Feng, Ning Wei, Zhi-Hao Liu, Shan Hu, and Xue-Bao Li. "Overexpression of cotton PYL genes in Arabidopsis enhances the transgenic plant tolerance to drought stress." *Plant Physiology and Biochemistry* 115, (2017): 229–238.

Cheng, Tingting, Donghua Wang, Yongfeng Wang, Shumeng Zhang, Chao Zhang, Shudong Liu, Yajun Xi, and Fengli Sun. "Identification and functional characterization of a MAX2 ortholog from switchgrass (*Panicum virgatum* L.)." *Plant Physiology and Biochemistry* 128, (2018): 106–114.

Christopher, Meera, Anil K. Mathew, M. Kiran Kumar, Ashok Pandey, and Rajeev K. Sukumaran. "A biorefinery-based approach for the production of ethanol from enzymatically hydrolysed cotton stalks." *Bioresource Technology* 242, (2017): 178–183.

Cockcroft, Claire E., Bart G. W. den Boer, J. M. Sandra Healy, and James A. H. Murray. "Cyclin D control of growth rate in plants." *Nature* 405, no. 6786 (2000): 575–579.

Cok, Benjamin, Ioannis Tsiropoulos, Alexander L. Roes, and Martin K. Patel. "Succinic acid production derived from carbohydrates: An energy and greenhouse gas assessment of a platform chemical toward a bio-based economy." *Biofuels, Bioproducts and Biorefining* 8, no. 1 (2014): 16–29.

Coleman, Heather D., Ji-Young Park, Ramesh Nair, Clint Chapple, and Shawn D. Mansfield. "RNAi-mediated suppression of p-coumaroyl-CoA 3′-hydroxylase in hybrid poplar impacts lignin deposition and soluble secondary metabolism." *Proceedings of the National Academy of Sciences* 105, no. 11 (2008): 4501–4506.

Costa, M. M., and J. L. Colodette. "The impact of kappa number composition on eucalyptus kraft pulp bleachability." *Brazilian Journal of Chemical Engineering* 24, no. 1 (2007): 61–71.

Damião Xavier, Franklin, Gustavo Santos Bezerra, Sharline Florentino Melo Santos, Líbia Sousa Conrado Oliveira, Flávio Luiz Honorato Silva, Aleir Joice Oliveira Silva, and Marta Maria Conceição. "Evaluation of the simultaneous production of xylitol and ethanol from sisal fiber." *Biomolecules* 8, no. 1 (2018): 2–13.

Das, Trina, and Gita N. Ramaswamy. "Enzyme treatment of wool and specialty hair fibers." *Textile Research Journal* 76, no. 2 (2006): 126–133.

Day, Arnaud, Godfrey Neutelings, Frédérique Nolin, Sébastien Grec, Anouk Habrant, David Crônier, Bouchra Maher, et al. "Caffeoyl coenzyme A O-methyltransferase down-regulation is associated with modifications in lignin and cell-wall architecture in flax secondary xylem." *Plant Physiology and Biochemistry* 47, no. 1 (2009): 9–19.

de Souza, A, A. Ulson, F. C. S. Ferreira, and S. M. A. Guelli U. Souza. "Influence of pretreatment of cotton yarns prior to biopolishing." *Carbohydrate Polymers* 93, no. 2 (2013): 412–415.

Dehabadi, Vahid Ameri, Klaus Opwis, and Jochen Gutmann. "Combination of acid-demineralization and enzymatic desizing of cotton fabrics by using industrial acid stable glucoamylases and α-amylases." *Starch-Stärke* 63, no. 12 (2011): 760–764.

Deng, Gang, Li Jun Liu, Xin Yue Zhong, Cheng Ying Lao, Hong Yang Wang, Bo Wang, Cong Zhu, Fahad Shah, and Ding Xiang Peng. "Comparative proteome analysis of the response of ramie under N, P and K deficiency." *Planta* 239, no. 6 (2014): 1175–1186.

Deng, Yong, Defang Li, Yumin Huang, and Siqi Huang. "Physiological response to cadmium stress in kenaf (*Hibiscus cannabinus* L.) seedlings." *Industrial Crops and Products* 107, (2017): 453–457.

Diderichsen, Børge. "A new desizing enzyme shows the changing nature of enzyme production." *Journal of Biotechnology* 38, no. 2 (1995): 196–197.

Ding, Han, Guodong Wang, Lili Lou, and Jinyin Lv. "Physiological responses and tolerance of kenaf (*Hibiscus cannabinus* L.) exposed to chromium." *Ecotoxicology and Environmental Safety* 133, (2016): 509–518.

Dmitriev, Alexey A., George S. Krasnov, Tatiana A. Rozhmina, Roman O. Novakovskiy, Anastasiya V. Snezhkina, Maria S. Fedorova, Olga Yu Yurkevich et al. "Differential gene expression in response to Fusarium oxysporum infection in resistant and susceptible genotypes of flax (*Linum usitatissimum* L.)." *BMC Plant Biology* 17, no. 2 (2017): 29–40.

Dong, Aixue, Jiugang Yuan, Qiang Wang, and Xuerong Fan. "Modification of jute fabric via laccase/t-BHP-mediated graft polymerization with acrylamide." *Journal of Applied Polymer Science* 131, no. 12 (2014).

Dong, Aixue, Xuerong Fan, Qiang Wang, Yuanyuan Yu, Ping Wang, Jiugang Yuan, and Artur Cavaco-Paulo. "Changes on content, structure and surface distribution of lignin in jute fibers after laccase treatment." *Journal of Natural Fibers* 15, no. 3 (2018): 384–395.

Duan, Chao, Xinqi Wang, YanLing Zhang, Yongjian Xu, and Yonghao Ni. "Fractionation and cellulase treatment for enhancing the properties of kraft-based dissolving pulp." *Bioresource Technology* 224, (2017): 439–444.

Duque, Aleta, Cristina Álvarez, Pablo Doménech, Paloma Manzanares, and Antonio D. Moreno. "Advanced bioethanol production: From novel raw materials to integrated biorefineries." *Processes* 9, no. 2 (2021): 206.

Elhacham, Emily, Liad Ben-Uri, Jonathan Grozovski, Yinon M. Bar-On, and Ron Milo. "Global human-made mass exceeds all living biomass." *Nature* 588, no. 7838 (2020): 442–444.

Eriksson, Maria E., Maria Israelsson, Olof Olsson, and Thomas Moritz. "Increased gibberellin biosynthesis in transgenic trees promotes growth, biomass production and xylem fiber length." *Nature Biotechnology* 18, no. 7 (2000): 784–788.

Eudes, Aymerick, Anthe George, Purba Mukerjee, Jin S. Kim, Brigitte Pollet, Peter I. Benke, Fan Yang et al. "Biosynthesis and incorporation of side-chain-truncated lignin monomers to reduce lignin polymerization and enhance saccharification." *Plant Biotechnology Journal* 10, no. 5 (2012): 609–620.

Eudes, Aymerick, Jose H. Pereira, Sasha Yogiswara, George Wang, Veronica Teixeira Benites, Edward E. K. Baidoo, Taek Soon Lee, Paul D. Adams, Jay D. Keasling, and Dominique Loqué. "Exploiting the substrate promiscuity of hydroxycinnamoyl-CoA: Shikimate hydroxycinnamoyl transferase to reduce lignin." *Plant and Cell Physiology* 57, no. 3 (2016): 568–579.

Fahd, S., G. Fiorentino, S. Mellino, and S. Ulgiati. "Cropping bioenergy and biomaterials in marginal land: The added value of the biorefinery concept." *Energy* 37, no. 1 (2012): 79–93.

Fan, Chunfen, Shengqiu Feng, Jiangfeng Huang, Yanting Wang, Leiming Wu, Xukai Li, Lingqiang Wang, et al. "AtCesA8-driven OsSUS3 expression leads to largely enhanced biomass saccharification and lodging resistance by distinctively altering lignocellulose features in rice." *Biotechnology for Biofuels* 10, no. 1 (2017): 1–12.

FAOSTAT, F. "FAOSTAT statistical database." 2016. http://www.fao.org/statistics/databases/en/ (accessed 28 July 2018).

Feng, Baomin, and Libo Shan. "ROS open roads to roundworm infection." *Science Signaling* 7, no. 320 (2014): pe10–pe10.

Fillat, Amanda, M. Blanca Roncero, and Teresa Vidal. "Elucidating the effects of laccase-modifying compounds treatments on bast and core fibers in flax pulp." *Biotechnology and Bioengineering* 109, no. 1 (2012): 225–233.

Fockink, Douglas Henrique, Marcelo Adriano Corrêa Maceno, and Luiz Pereira Ramos. "Production of cellulosic ethanol from cotton processing residues after pretreatment with dilute sodium hydroxide and enzymatic hydrolysis." *Bioresource Technology* 187, (2015): 91–96.

Foulk, Jonn A., Wayne Y. Chao, Danny E. Akin, Roy B. Dodd, and Patricia A. Layton. "Analysis of flax and cotton fiber fabric blends and recycled polyethylene composites." *Journal of Polymers and the Environment* 14, no. 1 (2006): 15–25.

Gao, Chunsheng, Chaohua Cheng, Lining Zhao, Yongting Yu, Qing Tang, Pengfei Xin, Touming Liu, Zhun Yan, Yuan Guo, and Gonggu Zang. "Genome-wide expression profiles of hemp (*Cannabis sativa* L.) in response to drought stress." *International Journal of Genomics* 2018, (2018).

Gao, Xiaopeng, Mario Tenuta, Donald N. Flaten, and Cynthia A. Grant. "Cadmium concentration in flax colonized by mycorrhizal fungi depends on soil phosphorus and cadmium concentrations." *Communications in Soil Science and Plant Analysis* 42, no. 15 (2011): 1882–1897.

Gonçalves, Idalina, Carla Silva, and Artur Cavaco-Paulo. "Ultrasound enhanced laccase applications." *Green Chemistry* 17, no. 3 (2015): 1362–1374.

Gonçalves, Idalina, Victor Herrero-Yniesta, Iratxe Perales Arce, Monica Escrigas Castañeda, Artur Cavaco-Paulo, and Carla Silva. "Ultrasonic pilot-scale reactor for enzymatic bleaching of cotton fabrics." *Ultrasonics Sonochemistry* 21, no. 4 (2014): 1535–1543.

Gong, Xiaomin, Yunguo Liu, Danlian Huang, Guangming Zeng, Shaobo Liu, Hui Tang, Lu Zhou, Xi Hu, Yaoyu Zhou, and Xiaofei Tan. "Effects of exogenous calcium and spermidine on cadmium stress moderation and metal accumulation in *Boehmeria nivea* (L.) Gaudich." *Environmental Science and Pollution Research* 23, no. 9 (2016): 8699–8708.

Gouveia, Isabel C., José M. Fiadeiro, and João A. Queiroz. "Enzymatic removal of plant residues from wool: Application of experimental design techniques for optimization parameters." *Biochemical Engineering Journal* 41, no. 2 (2008): 157–165.

Gredes, Tomasz, Christiane Kunert-Keil, Marzena Dominiak, Tomasz Gedrange, Magdalena Wróbel-Kwiatkowska, and Jan Szopa. "The influence of biocomposites containing genetically modified flax fibers on gene expression in rat skeletal muscle." *Biomedizinische Technik/Biomedical Engineering* 55, no.6 (2010): 323–329.

Gübitz, Georg M., and Artur Cavaco Paulo. "New substrates for reliable enzymes: Enzymatic modification of polymers." *Current Opinion in Biotechnology* 14, no. 6 (2003): 577–582.

Guerriero, Gea, Lauralie Mangeot-Peter, Sylvain Legay, Marc Behr, Stanley Lutts, Khawar Sohail Siddiqui, and Jean-Francois Hausman. "Identification of fasciclin-like arabinogalactan proteins in textile hemp (*Cannabis sativa* L.): In silico analyses and gene expression patterns in different tissues." *BMC Genomics* 18, no. 1 (2017b): 1–13.

Guerriero, Gea, Marc Behr, Jean-Francois Hausman, and Sylvain Legay. "Textile hemp vs. salinity: Insights from a targeted gene expression analysis." *Genes* 8, no. 10 (2017c): 242.

Guerriero, Gea, Marc Behr, Sylvain Legay, Lauralie Mangeot-Peter, Simone Zorzan, Mohammad Ghoniem, and Jean-Francois Hausman. "Transcriptomic profiling of hemp bast fibres at different developmental stages." *Scientific Reports* 7, no. 1 (2017a): 1–11.

Gulrajani, M. L., Ritu Agarwal, Amrit Grover, and Mona Suri. "Degumming of silk with lipase and protease." *Indian Journal of Fibre and Textile Research* 25, no. 1 (2000): 69–74.

Guven, Olgun, Sergio N. Monteiro, Esperidiana AB Moura, and Jaroslaw W. Drelich. "Re-emerging field of lignocellulosic fiber – polymer composites and ionizing radiation technology in their formulation." *Polymer Reviews* 54, no. 4 (2016): 702–736.

Häusler, Rainer E., Thomas Rademacher, Jun Li, Volker Lipka, Katrin L. Fischer, Sabine Schubert, Fritz Kreuzaler, and Heinz-Josef Hirsch. "Single and double overexpression of C4-cycle genes had differential effects on the pattern of endogenous enzymes, attenuation of photorespiration and on contents of UV protectants in transgenic potato and tobacco plants." *Journal of Experimental Botany* 52, no. 362 (2001): 1785–1803.

Ho, Wai Mun, Lai Hoe Ang, and Don Koo Lee. "Assessment of Pb uptake, translocation and immobilization in kenaf (*Hibiscus cannabinus* L.) for phytoremediation of sand tailings." *Journal of Environmental Sciences* 20, no. 11 (2008): 1341–1347.

Hu, Qin, Ling Min, Xiyan Yang, Shuangxia Jin, Lin Zhang, Yaoyao Li, Yizan Ma, et al. "Laccase GhLac1 modulates broad-spectrum biotic stress tolerance via manipulating phenylpropanoid pathway and jasmonic acid synthesis." *Plant Physiology* 176, no. 2 (2018): 1808–1823.

Hu, R., Z. Y. Fan, B. Y. Wang, S. L. Deng, X. S. Zhang, J. L. Zhang, H. B. Han, and Z. X. Lian. "RAPID COMMUNICATION: Generation of FGF5 knockout sheep via the CRISPR/Cas9 system." *Journal of Animal Science* 95, no. 5 (2017): 2019–2024.

Huang, Chengjian, Jinghua Zhou, Yucheng Jie, Hucheng Xing, Yingli Zhong, Weilin Yu, Wei She, Yushen Ma, Zehang Liu, and Ying Zhang. "A ramie bZIP transcription factor *BnbZIP2* is involved in drought, salt, and heavy metal stress response." *DNA and Cell Biology* 35, no. 12 (2016a): 776–786.

Huang, Junfeng, Feng Chen, Siyu Wu, Juan Li, and Wenliang Xu. "Cotton GhMYB7 is predominantly expressed in developing fibers and regulates secondary cell wall biosynthesis in transgenic *Arabidopsis*." *Science China Life Sciences* 59, no. 2 (2016b): 194–205.

Huis, Rudy, Kris Morreel, Ophélie Fliniaux, Anca Lucau-Danila, Stéphane Fénart, Sébastien Grec, Godfrey Neutelings, et al. "Natural hypolignification is associated with extensive oligolignol accumulation in flax stems." *Plant Physiology* 158, no. 4 (2012): 1893–1915.

Ibarra, David, Viviana Köpcke, Per Tomas Larsson, Anna-Stiina Jääskeläinen, and Monica Ek. "Combination of alkaline and enzymatic treatments as a process for upgrading sisal paper-grade pulp to dissolving-grade pulp." *Bioresource Technology* 101, no. 19 (2010): 7416–7423.

Ibrahim, N. A., K. El-Badry, B. M. Eid, and T. M. Hassan. "A new approach for biofinishing of cellulose-containing fabrics using acid cellulases." *Carbohydrate Polymers* 83, no. 1 (2011): 116–121.

International Cotton Advisory Committee. "Global cotton production continues to recover." 2017. https://www.icac.org/Press-Release/2017/PR-24-17-Global-Cotton-Production-Continues-to-Rec (accessed 28 July 2018).

International Energy Agency. "IEA bioenergy task 42 biorefining, country reports." 2017. http://www.iea-bioenergy.task42-biorefineries.com/en/ieabiorefinery/Country-Reports.htm (accessed 02 May 2021).

International Energy Agency. "World energy investment." 2018. https://iea.org/reports/world-energy-investment-2018 (accessed 02 May 2021).

International Sericulture Commission. "Statistics." 2017. https://inserco.org/en/statistics. (accessed 02 May 2021).

International Wool Textile Organization. "Wool production.". 2015. http://www.iwto.org. (accessed 02 May 2021).

Ip, Kenneth, and Andrew Miller. "Life cycle greenhouse gas emissions of hemp – lime wall constructions in the UK." *Resources, Conservation and Recycling* 69, (2012): 1–9.

Islam, M. R., M. D. H. Beg, and M. F. Mina. "Fibre surface modifications through different treatments with the help of design expert software for natural fibre-based biocomposites." *Journal of Composite Materials* 48, no. 15 (2014): 1887–1899.

Islam, Md Shahidul, Jennifer A. Saito, Emdadul Mannan Emdad, Borhan Ahmed, Mohammad Moinul Islam, Abdul Halim, Quazi Md Mosaddeque Hossen, et al. "Comparative genomics of two jute species and insight into fibre biogenesis." *Nature Plants* 3, no. 2 (2017): 1–7.

Jami, Tarun, Deepak Rawtani, and Yadendra K. Agrawal. "Hemp concrete: Carbon-negative construction." *Emerging Materials Research* 5, no. 2 (2016): 240–247.

Jami, Tarun, S. Karade, and L. P. Singh. "Hemp Concrete-A traditional and novel green building material." *Proceedings of the International Conference on Advances in Construction Materials and Structures (ACMS-2018) IIT Roorkee,* Roorkee, Uttarakhand, India, pp. 7–8. 2018.

Jana, K., S. K. Das, D. C. Roy, M. K. Kundu, A. Kundu, and G. Sathish. "Seed yield of linseed varieties grown as â€˜ pairaâ€™ crop as influenced by dates of sowing." *Journal of Applied and Natural Science* 10, no. 1 (2018): 17–23.

Jiang, Jie, Chaohu Pan, Aiping Xiao, Xiai Yang, and Guimin Zhang. "Isolation, identification, and environmental adaptability of heavy-metal-resistant bacteria from ramie rhizosphere soil around mine refinery." *3 Biotech* 7, no. 1 (2017): 5.

Jiang, Xinyuan, Sven G. Sommer, and Knud V. Christensen. "A review of the biogas industry in China." *Energy Policy* 39, no. 10 (2011): 6073–6081.

Jiang, Yanjie, Wangzhen Guo, Huayu Zhu, Yong-Ling Ruan, and Tianzhen Zhang. "Overexpression of GhSusA1 increases plant biomass and improves cotton fiber yield and quality." *Plant Biotechnology Journal* 10, no. 3 (2012): 301–312.

Jung, Sera, Younho Song, Ho Myeong Kim, and Hyeun-Jong Bae. "Enhanced lignocellulosic biomass hydrolysis by oxidative lytic polysaccharide monooxygenases (LPMOs) GH61 from *Gloeophyllum trabeum*." *Enzyme and Microbial Technology* 77, (2015): 38–45.

Kaachra, Anish, Surender Kumar Vats, and Sanjay Kumar. "Heterologous expression of key C and N metabolic enzymes improves re-assimilation of photorespired CO2 and NH3, and growth." *Plant Physiology* 177, no. 4 (2018): 1396–1409.

Kalantzi, Styliani, Diomi Mamma, and Dimitris Kekos. "Physichochemical and low stress mechanical properties of silk fabrics degummed by enzymes." *Eco-Friendly Textile Dyeing and Finishing* (2013): 232–260.

Kalia, Susheel, and Shalu Vashistha. "Surface modification of sisal fibers (*Agave sisalana*) using bacterial cellulase and methyl methacrylate." *Journal of Polymers and the Environment* 20, no. 1 (2012): 142–151.

Karaduman, Y. E. K. T. A., D. Gokcan, and Levent Onal. "Effect of enzymatic pretreatment on the mechanical properties of jute fiber-reinforced polyester composites." *Journal of Composite Materials* 47, no. 10 (2013): 1293–1302.

Kato, Souichiro, Shin Haruta, Zong Jun Cui, Masaharu Ishii, and Yasuo Igarashi. "Stable coexistence of five bacterial strains as a cellulose-degrading community." *Applied and Environmental Microbiology* 71, no. 11 (2005): 7099–7106.

Keller, Greg, Lori Spatola, D. E. N. N. I. S. McCABE, Brian Martinell, William Swain, and Maliyakal E. John. "Transgenic cotton resistant to herbicide bialaphos." *Transgenic Research* 6, no. 6 (1997): 385–392.

Kim, Jonggeun, Bosung Choi, Byoung-Kwan Cho, Hyoun-Sub Lim, Joon Bum Kim, Savithiry Natarajan, Eunjung Kwak, and Hanhong Bae. "Molecular cloning, characterization and expression of the caffeic acid O-methyltransferase (COMT) ortholog from kenaf (*Hibiscus cannabinus*)." *Plant Omics* 6, no. 4 (2013): 246–253.

Kim, Suyeon, C. Lopez, G. Güebitz, and Artur Cavaco-Paulo. "Biological coloration of flax fabrics with flavonoids using laccase from *Trametes hirsuta*." *Engineering in Life Sciences* 8, no. 3 (2008): 324–330.

Kuglarz, Mariusz, Ingólfur B. Gunnarsson, Sven-Erik Svensson, Thomas Prade, Eva Johansson, and Irini Angelidaki. "Ethanol production from industrial hemp: Effect of combined dilute acid/steam pretreatment and economic aspects." *Bioresource Technology* 163, (2014): 236–243.

Kumari, Rajni, and K. Pramanik. "Improved bioethanol production using fusants of Saccharomyces cerevisiae and xylose-fermenting yeasts." *Applied Biochemistry and Biotechnology* 167, no. 4 (2012): 873–884.

Kunert-Keil, Christiane, Tomasz Gredes, Annelie Meyer, Magdalena Wróbel-Kwiatkowska, Marzena Dominiak, and Tomasz Gedrange. "The survival and proliferation of fibroblasts on biocomposites containing genetically modified flax fibers: An in vitro study." *Annals of Anatomy-Anatomischer Anzeiger* 194, no. 6 (2012): 513–517.

Kurek, Itzhak, Thom Kai Chang, Sean M. Bertain, Alfredo Madrigal, Lu Liu, Michael W. Lassner, and Genhai Zhu. "Enhanced thermostability of Arabidopsis Rubisco activase improves photosynthesis and growth rates under moderate heat stress." *The Plant Cell* 19, no. 10 (2007): 3230–3241.

Kurihara, H., M. Watanabe, and T. Hayakawa. "Phytoremediation with kenaf (*Hibiscus cannabinus*) for cadmium-contaminated paddy field in southwest area of Japan." *Japanese Journal of Soil Science and Plant Nutrition (Japan)* (2005).

Lee, Ching Hao, Abdan Khalina, SengHua Lee, and Ming Liu. "A comprehensive review on bast fibre retting process for optimal performance in fibre-reinforced polymer composites." *Advances in Materials Science and Engineering* 2020, (2020): 1–27.

Leitão, Vanessa O., Eliane F. Noronha, Brenda R. Camargo, Pedro R. V. Hamann, Andrei S. Steindorff, Betania F. Quirino, Marcelo Valle de Sousa, Cirano J. Ulhoa, and Carlos R. Felix. "Growth and expression of relevant metabolic genes of *Clostridium thermocellum* cultured on lignocellulosic residues." *Journal of Industrial Microbiology and Biotechnology* 44, no. 6 (2017): 825–834.

Li, Huaying, Yunguo Liu, Guangming Zeng, Lu Zhou, Xin Wang, Yaqin Wang, Chunlin Wang, Xinjiang Hu, and Weihua Xu. "Enhanced efficiency of cadmium removal by *Boehmeria nivea* (L.) Gaud. in the presence of exogenous citric and oxalic acids." *Journal of Environmental Sciences* 26, no. 12 (2014): 2508–2516.

Li, Si-Yu, James D. Stuart, Yi Li, and Richard S. Parnas. "The feasibility of converting *Cannabis sativa* L. oil into biodiesel." *Bioresource technology* 101, no. 21 (2010): 8457–8460.

Li, Wen-Rong, Chen-Xi Liu, Xue-Mei Zhang, Lei Chen, Xin-Rong Peng, San-Gang He, Jia-Peng Lin, et al. "CRISPR/Cas9-mediated loss of FGF5 function increases wool staple length in sheep." *The FEBS Journal* 284, no. 17 (2017): 2764–2773.

Linger, P., J. Müssig, H. Fischer, and J. Kobert. "Industrial hemp (*Cannabis sativa* L.) growing on heavy metal contaminated soil: Fibre quality and phytoremediation potential." *Industrial Crops and Products* 16, no. 1 (2002): 33–42.

Liu, Chenxi, Liqin Wang, Wenrong Li, Xuemei Zhang, Yongzhi Tian, Ning Zhang, Sangang He, Tong Chen, Juncheng Huang, and Mingjun Liu. "Highly efficient generation of transgenic sheep by lentivirus accompanying the alteration of methylation status." *PLoS One* 8, no. 1 (2013): e54614.

Liu, Jiajia, Qin Qiao, Xia Cheng, Guanghui Du, Gang Deng, Mingzhi Zhao, and Feihu Liu. "Transcriptome differences between fiber-type and seed-type *Cannabis sativa* variety exposed to salinity." *Physiology and Molecular Biology of Plants* 22, no. 4 (2016): 429–443.

Liu Laipan, Ruqing Guo, Qin Qin, Jianmei Fu, and Biao Liu. "Expression of Bt protein in transgenic Bt cotton plants and ecological fitness of these plants in different habitats." *Frontiers in Plant Science* 11, (2020):1209

Liu, Ruirui, Aixue Dong, Xuerong Fan, Yuanyuan Yu, Jiugang Yuan, Ping Wang, Qiang Wang, and Artur Cavaco-Paulo. "Enzymatic hydrophobic modification of jute fibers via grafting to reinforce composites." *Applied Biochemistry and Biotechnology* 178, no. 8 (2016): 1612–1629.

Luan, Ming-Bao, Jian-Bo Jian, Ping Chen, Jun-Hui Chen, Jian-Hua Chen, Qiang Gao, Gang Gao, et al. "Draft genome sequence of ramie, *Boehmeria nivea* (L.) Gaudich." *Molecular Ecology Resources* 18, no. 3 (2018): 639–645.

Luthfi, Abdullah Amru Indera, Jamaliah Md Jahim, Shuhaida Harun, Jian Ping Tan, and Abdul Wahab Mohammad. "Biorefinery approach towards greener succinic acid production from oil palm frond bagasse." *Process Biochemistry* 51, no. 10 (2016): 1527–1537.

Luthfi, Abdullah Amru Indera, Shareena Fairuz Abdul Manaf, Rosli Md Illias, Shuhaida Harun, Abdul Wahab Mohammad, and Jamaliah Md Jahim. "Biotechnological route for sustainable succinate production utilizing oil palm frond and kenaf as potential carbon sources." *Applied Microbiology and Biotechnology* 101, no. 8 (2017): 3055–3075.

Luyckx, Marie, Jean-Francois Hausman, Stanley Lutts, and Gea Guerriero. "Impact of silicon in plant biomass production: Focus on bast fibres, hypotheses, and perspectives." *Plants* 6, no. 3 (2017): 37.

Ma, Hongyan, Daniel G. Delafield, Zhe Wang, Jianlan You, and Si Wu. "Finding biomass degrading enzymes through an activity-correlated quantitative proteomics platform (ACPP)." *Journal of the American Society for Mass Spectrometry* 28, no. 4 (2017): 655–663.

Ma, Sanyuan, Jiasong Chang, Xiaogang Wang, Yuanyuan Liu, Jianduo Zhang, Wei Lu, Jie Gao, Run Shi, Ping Zhao, and Qingyou Xia. "CRISPR/Cas9 mediated multiplex genome editing and heritable mutagenesis of *BmKu70* in *Bombyx mori*." *Scientific Reports* 4, no. 1 (2014): 1–6.

Magnuson, Jon K., and Linda L. Lasure. "Organic acid production by filamentous fungi." In *Advances in fungal biotechnology for industry, agriculture, and medicine*, pp. 307–340. Springer, 2004.

Malik, Riffat Naseem, Syed Zahoor Husain, and Ishfaq Nazir. "Heavy metal contamination and accumulation in soil and wild plant species from industrial area of Islamabad, Pakistan." *Pakistan Journal of Botany* 42, no. 1 (2010): 291–301.

Marketsandmarkets.com. "Global succinic acid market set for rapid growth, to reach USD around 1,000.0 Million, by 2020." 2015. https://www.marketresearchstore.com/news/global-succinic-acid-market-set-for-rapid-growth-111. (accessed 02 May 2021).

Martinez, Diego, Randy M. Berka, Bernard Henrissat, Markku Saloheimo, Mikko Arvas, Scott E. Baker, Jarod Chapman, et al. "Genome sequencing and analysis of the biomass-degrading fungus *Trichoderma reesei* (syn. *Hypocrea jecorina*)." *Nature Biotechnology* 26, no. 5 (2008): 553–560.

Maryan, Ali Sadeghian, and Majid Montazer. "A cleaner production of denim garment using one step treatment with amylase/cellulase/laccase." *Journal of Cleaner Production* 57, (2013): 320–326.

Matsuoka, Makoto, Robert T. Furbank, Hiroshi Fukayama, and Mitsue Miyao. "Molecular engineering of C4 photosynthesis." *Annual Review of Plant Biology* 52, no. 1 (2001): 297–314.

Meera, M., and Pariatamby Agamuthu. "Phytoextraction of As and Fe using *Hibiscus cannabinus* L. from soil polluted with landfill leachate." *International Journal of Phytoremediation* 14, no. 2 (2012): 186–199.

Mehmood, Muhammad Aamer, Chen-Guang Liu, and Alissara Reungsang. "Editorial: special issue "Sustainable biorefinery approaches." *Biomass Conversion and Biorefinery* 11, (2021): 741–742.

Melzer, Siegbert, Frederic Lens, Jerôme Gennen, Steffen Vanneste, Antje Rohde, and Tom Beeckman. "Flowering-time genes modulate meristem determinacy and growth form in Arabidopsis thaliana." *Nature Genetics* 40, no. 12 (2008): 1489–1492.

Mohiuddin, Ghulam, Mamunur Rashid, Mizanur Rahman, S. K. Abul Hasib, and Abdur Razzaque. "Biopulping of whole jute plant in soda-anthraquinone (AQ) and kraft processes." *Tappi Journal (2002)* 4, no. 3 (2005): 23–27.

Mokshina, Natalia, Oleg Gorshkov, Nadezda Ibragimova, Tatyana Chernova, and Tatyana Gorshkova. "Cellulosic fibres of flax recruit both primary and secondary cell wall cellulose synthases during deposition of thick tertiary cell walls and in the course of graviresponse." *Functional Plant Biology* 44, no. 8 (2017): 820–831.

Moon, Se-Kwon, Young-Jung Wee, Jong-Sun Yun, and Hwa-Won Ryu. "Production of fumaric acid using rice bran and subsequent conversion to succinic acid through a two-step process." *Applied Biochemistry and Biotechnology* 115, (2004): 843–855.

Ndeh, Didier, Artur Rogowski, Alan Cartmell, Ana S. Luis, Arnaud Baslé, Joseph Gray, Immacolata Venditto, et al. "Complex pectin metabolism by gut bacteria reveals novel catalytic functions." *Nature* 544, no. 7648 (2017): 65–70.

Ni, Xiaoyan, Aixue Dong, Xuerong Fan, Qiang Wang, Yuanyuan Yu, and Artur Cavaco-Paulo. "Jute/polypropylene composites: Effect of enzymatic modification on thermo-mechanical and dynamic mechanical properties." *Fibers and Polymers* 16, no. 10 (2015): 2276–2283.

Ni, Xiaoyan, Aixue Dong, Xuerong Fan, Qiang Wang, Yuanyuan Yu, and Artur Cavaco-Paulo. "Enzyme-mediated surface modification of jute and its influence on the properties of jute/epoxy composites." *Polymer Composites* 38, no. 7 (2017): 1327–1334.

Nie, Shuangxi, Shuangquan Yao, Shuangfei Wang, and Chengrong Qin. "Absorbable organic halide (AOX) reduction in elemental chlorine-free (ECF) bleaching of bagasse pulp from the addition of sodium sulphide." *BioResources* 11, no. 1 (2016): 713–723.

Nithya, E., R. Radhai, R. Rajendran, S. Shalini, V. Rajendran, and S. Jayakumar. "Synergetic effect of DC air plasma and cellulase enzyme treatment on the hydrophilicity of cotton fabric." *Carbohydrate Polymers* 83, no. 4 (2011): 1652–1658.

Niu, Erli, Shuai Fang, Xiaoguang Shang, and Wangzhen Guo. "Ectopic expression of GhCOBL9A, a cotton glycosyl-phosphatidyl inositol-anchored protein encoding gene, promotes cell elongation, thickening and increased plant biomass in transgenic Arabidopsis." *Molecular Genetics and Genomics* 293, no. 5 (2018): 1191–1204.

Nizamuddin, M., M. Mokhlesur Rahman, and Jang-Eok Kim. "Phytoremediation potential of kenaf (*Hibiscus cannabinus* L.), mesta (*Hibiscus sabdariffa* L.), and jute (*Corchorus capsularis* L.) in arsenic-contaminated soil." *Korean Journal of Environmental Agriculture* 35, no. 2 (2016): 111–120.

Palit, P. "Radiation and carbon use efficiency of field grown jute (Corchorus spp.) in relation to potential primary production." *Photosynthetica (Praha)* 28, no. 3 (1993): 369–375.

Pandey, Vimal Chandra, Omesh Bajpai, and Nandita Singh. "Energy crops in sustainable phytoremediation." *Renewable and Sustainable Energy Reviews* 54, (2016): 58–73.

Paraschiv, Dorel, Cristiana Tudor, and Radu Petrariu. "The textile industry and sustainable development: A holt–winters forecasting investigation for the eastern European area." *Sustainability* 7, (2015): 1280–1291.

Paris Agreement. "United nations framework convention on climate change." 2015. https://unfccc.int/process-and-meetings/the-paris-agreement/the-paris-agreement. (accessed 02 May 2021).

Parisi, Maria Laura, Enrico Fatarella, Daniele Spinelli, Rebecca Pogni, and Riccardo Basosi. "Environmental impact assessment of an eco-efficient production for coloured textiles." *Journal of Cleaner Production* 108, (2015): 514–524.

Pazarlioğlu, Nurdan Kaşikara, Merih Sariişik, and Azmi Telefoncu. "Treating denim fabrics with immobilized commercial cellulases." *Process Biochemistry* 40, no. 2 (2005): 767–771.

Pei, Yong, Shuangfei Wang, Chengrong Qin, Jingjing Su, Shuangxi Nie, and Xueping Song. "Optimization of laccase-aided chlorine dioxide bleaching of bagasse pulp." *BioResources* 11, no. 1 (2016): 696–712.

Resch, Michael G., Bryon S. Donohoe, John O. Baker, Stephen R. Decker, Edward A. Bayer, Gregg T. Beckham, and Michael E. Himmel. "Fungal cellulases and complexed cellulosomal enzymes exhibit synergistic mechanisms in cellulose deconstruction." *Energy & Environmental Science* 6, no. 6 (2013): 1858–1867.

Rupp, Hans-Michael, Markus Frank, Tomás Werner, Miroslav Strnad, and Thomas Schmülling. "Increased steady state mRNA levels of the *STM* and *KNAT1* homeobox genes in cytokinin overproducing Arabidopsis thaliana indicate a role for cytokinins in the shoot apical meristem." *The Plant Journal* 18, no. 5 (1999): 557–563.

Sahin, Halil Turgut. "Properties of jute pulps from atmospheric organosolv process." *American Chemical Science Journal* 14, (2016): 1–6.

Salehi, Hassan, Callista B. Ransom, Hesham F. Oraby, Zahra Seddighi, and Mariam B. Sticklen. "Delay in flowering and increase in biomass of transgenic tobacco expressing the Arabidopsis floral repressor gene FLOWERING LOCUS C." *Journal of Plant Physiology* 162, no. 6 (2005): 711–717.

Santoso, Budi, Balai PenelitianTanaman Pemanis, and Serat Indonesian Sweetener. "Manfaat kenaf (*Hibiscus cannabinus*

L.) dalam penyerapan karbondioksida (CO_2)." *Perspektif* 14, no. 2 (2015): 125–133.

Sarkar, Debabrata, Ajay Kumar Mahato, Pratik Satya, Avijit Kundu, Sangeeta Singh, Pawan Kumar Jayaswal, Akshay Singh, et al. "The draft genome of *Corchorus olitorius* cv. JRO-524 (Navin)." *Genomics Data* 12, (2017): 151–154.

Sarkar, Debabrata, Pratik Satya, Nur Alam Mandal, Debajeet Das, Pran Gobinda Karmakar, and Nagendra Kumar Singh. "Jute genomics: Emerging resources and tools for molecular breeding." In *Fiber plants*. Springer, 2016, pp. 155–200.

Satya, Pratik, and Mridul Chakraborti. "Development and utilization of DNA markers for genetic improvement of bast fiber crops." *Applications of Molecular Markers in Plant Genome Analysis and Breeding* (2015): 119–142.

Satya, Pratik, and Ratikanta Maiti. "Bast and leaf fibre crops: Kenaf, hemp, jute, agave, etc." *Biofuel Crops: Production, Physiology and Genetics* (2013): 292–311.

Saviane, Alessio, Ottavia Romoli, Andrea Bozzato, Giuliano Freddi, Chiara Cappelletti, Elena Rosini, Silvia Cappellozza, Gianluca Tettamanti, and Federica Sandrelli. "Intrinsic antimicrobial properties of silk spun by genetically modified silkworm strains." *Transgenic Research* 27, no. 1 (2018): 87–101.

Schroyen, Michel, Han Vervaeren, Hanne Vandepitte, Stijn W. H. Van Hulle, and Katleen Raes. "Effect of enzymatic pretreatment of various lignocellulosic substrates on production of phenolic compounds and biomethane potential." *Bioresource Technology* 192, (2015): 696–702.

Schroyen, Michel, Stijn W. H. Van Hulle, Sander Holemans, Han Vervaeren, and Katleen Raes. "Laccase enzyme detoxifies hydrolysates and improves biogas production from hemp straw and miscanthus." *Bioresource Technology* 244, (2017): 597–604.

Schumacher, Karin, Elisabeth Heine, and Hartwig Höcker. "Extremozymes for improving wool properties." *Journal of Biotechnology* 89, no. 2–3 (2001): 281–288.

Shahid, Mohammad, Faqeer Mohammad, Guoqiang Chen, Ren-Cheng Tang, and Tieling Xing. "Enzymatic processing of natural fibres: White biotechnology for sustainable development." *Green Chemistry* 18, no. 8 (2016): 2256–2281.

Shanthi, R., and G. Krishnabai. "Process optimization for bioscouring of cotton and lycra cotton weft knits by Box and Behnken design." *Carbohydrate Polymers* 96, no. 1 (2013): 291–295.

Shen, Hui, Mitra Mazarei, Hiroshi Hisano, Luis Escamilla-Trevino, Chunxiang Fu, Yunqiao Pu, Mary R. Rudis, et al. "A genomics approach to deciphering lignin biosynthesis in switchgrass." *The Plant Cell* 25, no. 11 (2013): 4342–4361.

Shi, Gangrong, and Qingsheng Cai. "Cadmium tolerance and accumulation in eight potential energy crops." *Biotechnology Advances* 27, no. 5 (2009): 555–561.

Shi, Gangrong, Caifeng Liu, Meicheng Cui, Yuhua Ma, and Qingsheng Cai. "Cadmium tolerance and bioaccumulation of 18 hemp accessions." *Applied Biochemistry and Biotechnology* 168, no. 1 (2012): 163–173.

Shin, Hoonsik, Georg Guebitz, and Artur Cavaco-Paulo. "'In situ' enzymatically prepared polymers for wool coloration." *Macromolecular Materials and Engineering* 286, no. 11 (2001): 691–694.

Silva, Caio O. G, Raissa P. Vaz, and Edivaldo X. F. Filho. "Bringing plant cell walldegrading enzymes into the lignocellulosic biorefinery concept." *Biofuels, Bioproducts and Biorefining* 12, no. 2 (2018): 277–289.

Sindhu, Raveendran, Parameswaran Binod, and Ashok Pandey. "Biological pretreatment of lignocellulosic biomass – an overview." *Bioresource Technology* 199, (2016): 76–82.

Singh, Anita, Somvir Bajar, and Narsi R. Bishnoi. "Physico-chemical pretreatment and enzymatic hydrolysis of cotton stalk for ethanol production by *Saccharomyces cerevisiae*." *Bioresource Technology* 244, (2017): 71–77.

Singh, R. K., Amit Ujwal, and Suresh Chaluvadi. "Study of various pretreatments on waste jute caddies for the production of ethanol." *International Journal of Science, Engineering and Management* 3, (2018): 20–23.

Singh, R., N. K. Bhardwaj, and B. Choudhury. "Cellulase-assisted refining optimization for saving electrical energy demand and pulp quality evaluation." *Journal of Scientific and Industrial Research* 74, (2015): 471–475.

Spina, Federica, Charles Junghanns, Ilaria Donelli, Rakesh Nair, Philippe Demarche, Alice Romagnolo, Giuliano Freddi, Spiros N. Agathos, and Giovanna Cristina Varese. "Stimulation of laccases from Trametes pubescens: Use in dye decolorization and cotton bleaching." *Preparative Biochemistry and Biotechnology* 46, no. 7 (2016): 639–647.

Sugimoto, Hiroki, Satoshi Kondo, Tomoko Tanaka, Chie Imamura, Nobuhiko Muramoto, Etsuko Hattori, Ken'ichi Ogawa, Norihiro Mitsukawa, and Chikara Ohto. "Overexpression of a novel Arabidopsis *PP2C* isoform, *AtPP2CF1*, enhances plant biomass production by increasing inflorescence stem growth." *Journal of Experimental Botany* 65, no. 18 (2014): 5385–5400.

Sun, Sha-Sha, Tieling Xing, and Ren-Cheng Tang. "Enzymatic dyeing and functional finishing of textile fibres with ferulic acid." *Indian Journal of Fibre & Textile Research* 40, no. 1 (2015): 62–69.

Sun, Xiaohuan, James Cahill, Tom Van Hautegem, Kim Feys, Clinton Whipple, Ondrej Novák, Sofie Delbare, et al. "Altered expression of maize *PLASTOCHRON1* enhances biomass and seed yield by extending cell division duration." *Nature Communications* 8, no. 1 (2017): 1–11.

Surendra, K. C., Devin Takara, Andrew G. Hashimoto, and Samir Kumar Khanal. "Biogas as a sustainable energy source for developing countries: Opportunities and challenges." *Renewable and Sustainable Energy Reviews* 31, (2014): 846–859.

Taiwo, A. M., A. M. Gbadebo, J. A. Oyedepo, Z. O. Ojekunle, O. M. Alo, A. A. Oyeniran, O. J. Onalaja, D. Ogunjimi, and O. T. Taiwo. "Bioremediation of industrially contaminated soil using compost and plant technology." *Journal of Hazardous Materials* 304, (2016): 166–172.

Tian, Liqiang, C. Branford-White, Wen Wang, Huali Nie, and Limin Zhu. "Laccase-mediated system pretreatment to enhance the effect of hydrogen peroxide bleaching of cotton fabric." *International Journal of Biological Macromolecules* 50, no. 3 (2012): 782–787.

Trumbo, Jennifer Lynn, Baohong Zhang, and Charles Neal Stewart Jr. "Manipulating micro RNA s for improved biomass and biofuels from plant feedstocks." *Plant Biotechnology Journal* 13, no. 3 (2015): 337–354.

Tzanov, Tzanko, Silgia A. Costa, Georg M. Gübitz, and Artur Cavaco-Paulo. "Hydrogen peroxide generation with immobilized glucose oxidase for textile bleaching." *Journal of Biotechnology* 93, no. 1 (2002): 87–94.

Ul-Haq, Noaman, and Habib Nasir. "Cleaner production technologies in desizing of cotton fabric." *Journal of the Textile Institute* 103, no. 3 (2012): 304–310.

UNFCCC (United Nations Framework Convention on Climate Change). "The Paris Agreement." (2016). https://unfccc.int/files/meetings/paris_nov_2015/application/pdf/paris_agreement_english_.pdf. (accessed 02 May 2021).

Väisänen, Taneli, Antti Haapala, Reijo Lappalainen, and Laura Tomppo. "Utilization of agricultural and forest industry waste and residues in natural fiber-polymer composites: A review." *Waste Management* 54, (2016): 62–73.

Van Acker, Rebecca, Ruben Vanholme, Véronique Storme, Jennifer C. Mortimer, Paul Dupree, and Wout Boerjan. "Lignin biosynthesis perturbations affect secondary cell wall composition and saccharification yield in Arabidopsis thaliana." *Biotechnology for Biofuels* 6, no. 1 (2013): 1–17.

Van Bakel, Harm, Jake M. Stout, Atina G. Cote, Carling M. Tallon, Andrew G. Sharpe, Timothy R. Hughes, and Jonathan E. Page. "The draft genome and transcriptome of *Cannabis sativa*." *Genome Biology* 12, no. 10 (2011): 1–18.

Van den Broeck, Hetty C., Chris Maliepaard, Michel J. M. Ebskamp, Marcel A. J. Toonen, and Andries J. Koops. "Differential expression of genes involved in C1 metabolism and lignin biosynthesis in wooden core and bast tissues of fibre hemp (*Cannabis sativa* L.)." *Plant Science* 174, no. 2 (2008): 205–220.

Van Dyk, J. S., and B. I. Pletschke. "A review of lignocellulose bioconversion using enzymatic hydrolysis and synergistic cooperation between enzymes – factors affecting enzymes, conversion and synergy." *Biotechnology Advances* 30, no. 6 (2012): 1458–1480.

Voorend, Wannes, Hilde Nelissen, Ruben Vanholme, Alex De Vliegher, Frank Van Breusegem, Wout Boerjan, Isabel Roldán-Ruiz, Hilde Muylle, and Dirk Inzé. "Overexpression of GA 20-OXIDASE 1 impacts plant height, biomass allocation and saccharification efficiency in maize." *Plant Biotechnology Journal* 14, no. 3 (2016): 997–1007.

Vosper, J. "The role of industrial hemp in carbon farming." 2011. https://hemp-copenhagen.com/images/Hemp-cph-Carbon-sink.pdf (accessed 02 May 2021).

Walford, Sally Ann, Yingru Wu, Danny J. Llewellyn, and Elizabeth S. Dennis. "GhMYB25-like: a key factor in early cotton fibre development." *The Plant Journal* 65, no. 5 (2011): 785–797.

Wang, Cuiting, Yan Bao, Qiuqing Wang, and Hongxia Zhang. "Introduction of the rice CYP714D1 gene into Populus inhibits expression of its homologous genes and promotes growth, biomass production and xylem fibre length in transgenic trees." *Journal of experimental botany* 64, no. 10 (2013): 2847–2857.

Wang, Peng, Natalia Dudareva, John A. Morgan, and Clint Chapple. "Genetic manipulation of lignocellulosic biomass for bioenergy." *Current Opinion in Chemical Biology* 29, (2015): 32–39.

Wang, Xiaolong, Bei Cai, Jiankui Zhou, Haijing Zhu, Yiyuan Niu, Baohua Ma, Honghao Yu, et al. "Disruption of FGF5 in cashmere goats using CRISPR/Cas9 results in more secondary hair follicles and longer fibers." *PLoS One* 11, no. 10 (2016): e0164640.

Wei, Suzhen. "The application of biotechnology on the enhancing of biogas production from lignocellulosic waste." *Applied Microbiology and Biotechnology* 100, no. 23 (2016): 9821–9836.

Willis, Judith H., Adam S. Wilkins, and Marian R. Goldsmith. "A brief history of Lepidoptera as model systems." *Molecular Model Systems in the Lepidoptera* (1995): 1–20.

Withanage, Samanthi Priyanka, Md Aktar Hossain, Hairul Azman B. Roslan, Mohammad Puad Abdullah, Suhaimi B. Napis, and Nor Aini Ab Shukor. "Overexpression of *Arabidopsis thaliana* gibberellic acid 20 oxidase (AtGA20ox) gene enhances the vegetative growth and fiber quality in kenaf (*Hibiscus cannabinus* L.) plants." *Breeding Science* 65, no. 3 (2015): 177–191.

Wróbel-Kwiatkowska, Magdalena, Jacek Zebrowski, Michał Starzycki, Jan Oszmiański, and Jan Szopa. "Engineering of PHB synthesis causes improved elastic properties of flax fibers." *Biotechnology Progress* 23, no. 1 (2007b): 269–277.

Wróbel-Kwiatkowska, Magdalena, Michał Starzycki, Jacek Zebrowski, Jan Oszmiański, and Jan Szopa. "Lignin deficiency in transgenic flax resulted in plants with improved mechanical properties." *Journal of Biotechnology* 128, no. 4 (2007a): 919–934.

Wróbel-Kwiatkowska, Magdalena, Sławomir Jabłoński, Jakub Szperlik, Lucyna Dymińska, Marcin Łukaszewicz, Waldemar Rymowicz, Jerzy Hanuza, and Jan Szopa. "Impact of CAD-deficiency in flax on biogas production." *Transgenic Research* 24, no. 6 (2015): 971–978.

Wu, Shujing, Libo Shan, and Ping He. "Microbial signature-triggered plant defense responses and early signaling mechanisms." *Plant Science* 228, (2014): 118–126.

Xie, Qi, Giovanna Frugis, Diana Colgan, and Nam-Hai Chua. "Arabidopsis NAC1 transduces auxin signal downstream of TIR1 to promote lateral root development." *Genes & Development* 14, no. 23 (2000): 3024–3036.

Xu, Suyun, Chuanqiu He, Liwen Luo, Fan Lü, Pinjing He, and Lifeng Cui. "Comparing activated carbon of different particle sizes on enhancing methane generation in upflow anaerobic digester." *Bioresource Technology* 196, (2015): 606–612.

Yu, X. S., B. J. Chu, R. E. Liu, J. Sun, Joseph Jones Brian, H. Z. Wang, Zhu Shuijin, and Y. Q. Sun. "Characteristics of fertile somatic hybrids of *G. hirsutum* L. and *G. trilobum* generated via protoplast fusion." *Theoretical and Applied Genetics* 125, no. 7 (2012): 1503–1516.

Yu, Ying, Guangwen Wu, Hongmei Yuan, Lili Cheng, Dongsheng Zhao, Wengong Huang, Shuquan Zhang, et al. "Identification and characterization of miRNAs and targets in flax (*Linum usitatissimum*) under saline, alkaline, and saline-alkaline stresses." *BMC Plant Biology* 16, no. 1 (2016): 1–13.

Yuan, Xufeng, Lei Ma, Boting Wen, Dayun Zhou, Meng Kuang, Weihua Yang, and Zongjun Cui. "Enhancing anaerobic digestion of cotton stalk by pretreatment with a microbial consortium (MC1)." *Bioresource Technology* 207, (2016): 293–301.

Zahid, Kashif Rafiq, Farhan Ali, Farooq Shah, Muhammad Younas, Tariq Shah, Durri Shahwar, Waseem Hassan, et al. "Response and tolerance mechanism of cotton *Gossypium hirsutum* L. to elevated temperature stress: A review." *Frontiers in Plant Science* 7, (2016): 937.

Zhang, Panli, Shumin Liu, Hong-Sheng Song, Guozheng Zhang, Qiangqiang Jia, and Sheng Li. "YorkieCA overexpression in the posterior silk gland improves silk yield in *Bombyx mori*." *Journal of Insect Physiology* 100, (2017): 93–99.

Zhang, Xuewen, Lihua Huang, Yan Zhao, Chao Hu, Qingquan Guo, and Jianrong Chen. "Decrease of auxin binding protein 1 gene expression alters shoot development in Ramie." *Journal of Plant Biochemistry & Physiology* 2, no.2 (2014): e1000125.

Zhang, Zhiyuan, Jun Zhao, Lingyun Ding, Lifang Zou, Yurong Li, Gongyou Chen, and Tianzhen Zhang. "Constitutive expression of a novel antimicrobial protein, Hcm1, confers resistance to both *Verticillium* and *Fusarium* wilts in cotton." *Scientific Reports* 6, no. 1 (2016): 1–13.

Zhao, Shicheng, Xiaohua Li, Dong Ha Cho, Mariadhas Valan Arasu, Naif Abdullah Al-Dhabi, and Sang Un Park. "Accumulation of kaempferitrin and expression of phenylpropanoid biosynthetic genes in kenaf (*Hibiscus cannabinus*)." *Molecules* 19, no. 10 (2014): 16987–16997.

Zhao, Zhiqiang, Yaobin Zhang, T. L. Woodard, K. P. Nevin, and D. R. Lovley. "Enhancing syntrophic metabolism in up-flow anaerobic sludge blanket reactors with conductive carbon materials." *Bioresource Technology* 191, (2015): 140–145.

Zhou, Xiaohong, Thomas B. Jacobs, LiangJiao Xue, Scott A. Harding, and ChungJui Tsai. "Exploiting SNP s for biallelic CRISPR mutations in the outcrossing woody perennial Populus reveals 4-coumarate: CoA ligase specificity and redundancy." *New Phytologist* 208, no. 2 (2015): 298–301.

13

Sustainable Development in Environmental Biotechnology

Mozammel Hoque

CONTENTS

13.1 Introduction

Natural resources are available in the environment for the survival of life. Their judicious use for comfortable living poses no problem. However, reckless human activities and overexploitation of natural resources confer serious consequences on the environment. The activities deplete, contaminate, change or interfere with the equilibrium in environment and thus pose a threat to human existence. Development is a dynamic process in the quest for the needs of an ever-increasing human population. There should be a balance between the development and environmental welfare and what is sustainable development. Sustainable development refers to the development that meets the needs of the present without compromising the ability of future generations to meet their own needs. The International Society for Environmental Biotechnology defines *environmental biotechnology* as "the development, use and regulation of biological systems for remediation of contaminated environments (land, air and water), and for environmentally friendly processes (green manufacturing technologies and sustainable development). Of late, environmental biotechnology has emerged as a great help in achieving the goal of sustainable development. It is high time that we should adopt environmental biotechnology to impart the least damage to the environment and continue with development activities in the quest the requirements of food, water, air and other comforts of life. Environmental well-being is a global concern. The importance of the environment is emphasized by the statement of Dr Guy McPherson: "[I]f you really think the environment is less important than the economy, try holding your breath while you count your money" is self-explanatory to point. To create global awareness, June 5, World Environment Day, is celebrated every year, and India is the host country for this year's celebrations which is an indication of our interest in environmental protection and conservation.

13.2 What Are the Threats to the Environment, and How Can They Be Alleviated for Sustainable Development?

It is a prerequisite to recognize the various components of the environment that are vulnerable to pollution so as to adopt approaches to mitigate them to protect, conserve and maintain the equilibrium of the environment.

Air: Air is a natural resource that is essential for life. Atmospheric air contains nitrogen (78.09%), oxygen (20.95%), argon (0.93%), carbon dioxide (0.04%) and small amounts of other gases. Air also contains a various amount of water vapor, on average about 1% at sea level and 0.4% over the entire atmosphere. Human activities and some natural processes cause deterioration in the quality of air due to the introduction of harmful or excessive quantities of substances, including gases, particulates and biological molecules. Air pollution causes diseases, allergies and human death and may also cause harm to other living organisms, such as animals and food crops, and may damage the natural or built environment. According to the 2014 World Health Organization report, air pollution in 2012 caused the deaths of about 7 million people worldwide. Air pollution costs the world economy $5 trillion per year as a result of productivity losses and degraded quality of life, according to a joint study by the World Bank and the Institute for Health Metrics and Evaluation (IHME) at the University of Washington.

Air Pollution Mitigation Approaches: Various air pollution control technologies and strategies are available to reduce air pollution. Land-use planning is likely to involve zoning and transport infrastructure planning. In most developed countries, land-use planning is an important part of social policy, ensuring that land is used efficiently for the benefit of the wider economy and population, as well as to protect the environment (Fensterstock *et al.*, 1971, 1972). The reduction of combustion

DOI: 10.1201/9781003186304-13

from fossil fuels, such as coal and oil, can reduce air pollution drastically. Most effective is switching to clean power sources, such as wind power, solar power and hydropower, which don't cause air pollution (Jacobson *et al.*, 2015). Efforts to reduce pollution from mobile sources include primary regulation (many developing countries have permissive regulations), expanding regulation to new sources (such as cruise and transport ships, farm equipment and small gas-powered equipment, such as string trimmers, chainsaws and snowmobiles), increased fuel efficiency (such as through the use of hybrid vehicles), conversion to cleaner fuels or conversion to electric vehicles. Titanium dioxide has the ability to reduce air pollution. Ultraviolet light will release free electrons from material, thereby creating free radicals, which break up volatile organic compounds and nitrogen oxide gases. One form is superhydrophilic (Jason, 2011). In 2014, Prof. Tony Ryan and Prof. Simon Armitage of the University of Sheffield prepared a 10-meter by 20-meter-sized poster coated with microscopic pollution-eating nanoparticles of titanium dioxide. Placed on a building, this giant poster can absorb the toxic emission from about 20 cars each day (BBC News, 2014). A very effective means to reduce air pollution is the transition to renewable energy. According to a study published in Energy and Environmental Science in 2015, the switch to 100% renewable energy in the United States would eliminate about 62,000 premature mortalities per year and about 42,000 in 2050 if no biomass was used. This would save about $600 billion in health costs a year due to reduced air pollution in 2050, or about 3.6% of the 2014 U.S. gross domestic product (GDP; Jacobson, *et al.*, 2015).

Water: Water is a renewable natural resource. Water is synonymous with the existence of life. Man, animals and plants depend on this precious natural resource. Moreover, water is an integral part of farming, construction, manufacturing sectors, moving wastes and cleaning and sanitation. Statistically, water is so abundant that it covers more than 70% of the surface of the earth. So how there be water scarcity? Less than 3% of water on earth is fresh water, and the bulk of this is trapped in glaciers and snowfields and not easily accessible. The rest forms the oceans and seas and cannot be used as fresh water. Only a small fraction (0.014%) is surface water in the form of rivers, lakes and swamps. Naturally, the 3% of water should be enough for all humans and animals on earth, but unfortunately, human activities have caused a major upset in the flow and use of fresh water and have posed huge crisis in many parts of the world. Overexploitation and nonjudicious use of water resources are responsible for the quantitative reduction of water. Water, air and land pollution together contribute to the reduction of water quality. Experts say that a person in a developed country having a 5-minute shower uses more water than the average person in a developing country slum uses for an entire day. The daily drinking water requirements per person are 2 to 4 liters. However, it takes 2000 to 5000 liters of water to produce a person's daily food (FAO). Agriculture is by far the biggest user of water, accounting for almost 70% of all withdrawal and up to 95% in developing countries. India is the world's largest user of groundwater, which provides 80% of the country's drinking water needs and supplies nearly two thirds of its irrigation water. Over the last four decades, about 84% of the total addition to irrigation has come from groundwater. Water tables are falling for most part, and there is fluoride, arsenic, mercury and even uranium in groundwater. The quantity and quality of river water are reducing and getting dirty due to disposal of waste material, sewer drainage, industrial waste and blocking natural flow or diverted water in huge quantities. The supply and distribution of river water have been a bone of contention within states and country in many parts of the world. The global water crisis is worsening at such a magnitude that experts are asking: Will the next world war be fought over water? A staggering 844 million people (i.e., roughly 1 in 10 people on earth) live without access to clean water. In developing countries, as much as 80% of illness is linked to poor water and sanitation conditions. The average distance that women in Africa and Asia walk to collect water is 3.7 miles.

Water Crisis Mitigation: The United Nations in 1993 designated 22 March as World Water Day and celebrates every year as an international day of awareness for water crises. The UN Sustainable Development Goal 6 aims to bring access to clean water to every single person on earth by 2030. The UN World Water Report 2018 recommends a nature-based solution (NBS) for water crisis. The NBS mainly addresses the water supply through managing precipitation, humidity, and water storage, infiltration and transmission so that improvements are made in the location, timing and quantity of water available for human needs. Rajendra Singh, known as the "Water Man of India", believes that critically depleted aquifers around the world can be revived with community-based water harvesting and water management initiatives.

Rainwater Harvesting: Rainwater harvesting is one of the simplest and oldest methods of self-supplying water for households and a gold standard toward water conservation. Rainwater harvesting is the collection and storage of rainwater for use instead of allowing it to run off. Rainwater can be collected directly from rain or from rooftops and rivers. It can also be collected from dew or fog with nets or other tools. The water collected is redirected to a deep pit, such as a well, a shaft, a borehole or a reservoir with percolation. The rainwater can be used for gardens, livestock, irrigation, domestic purposes and indoor heating for houses. The harvested water can also be used for drinking purposes with appropriate treatment, longer term storage and for other purposes, such as groundwater recharge. Dams and ponds can be constructed to collect large quantities of rainwater so that even on days when little to no rainfall occurs, enough is available to irrigate crops.

New *Approaches*: Instead of using the roof for catchment, the rain saucer, which looks like an upside-down umbrella, collects rain straight from the sky. This decreases the potential for contamination and makes potable water as a potential source in developing countries.

Rainwater harvesting by solar power panels*:* Solar panels can also be used for harvesting most of the rainwater falling on them and drinking quality water, free from bacteria and suspended matter, can be generated by simple filtration and disinfection processes as rainwater is very low in salinity. The exploitation of rainwater for value-added products such as bottled drinking water makes solar photovoltaic (PV) power plants profitable even in high-rainfall/cloudy areas by the augmented income from value-added drinking water generation.

Groundwater *Overuse* and *Heavy Metal Upsurge*: Groundwater may be contaminated with arsenic, fluoride, lead, iron and poses an endemic public health hazard.

Elevated levels of arsenic in groundwater, or arsenic contamination, are caused largely by natural processes and partly due to anthropogenic activities, such as applying fertilizers, burning of coal, leaching from coal-ash tailings and mining activity, as per a Standing Committee on Water Resources Report 2015–2016.

Innovative technology*:* 'Arsiron Nilogon' (arsiron = arsenic + iron; nilogon = "removal" in Assamese) is a simple, low-cost technology developed at Tezpur University, Assam, India, to remove arsenic and iron from contaminated drinking water at a cost of less than 1 rupee (Re. 0.5) per 100 liters (Rs. 1 for 200 liters). The system comprises two 200-liter or 500-liter plastic drums with taps perched on bamboo stands, placed one below the other. Chemical action takes place in the top drum and filters the water through a sand-gravel bed in the bottom tank. It is a two-stage process based on oxidation–coagulation–adsorption at an optimized pH. It does not require electricity and can be custom-made at home with easily available materials. It is suitable for household- and community-level settings. The method involves the addition of three common chemicals (baking soda, potassium permanganate, ferric chloride) in specific amounts to the water to be treated, which takes about 3 minutes, followed by filtration after 1 hour of resting. The patented technology can bring down arsenic concentration from levels as high as 500 ppb to below 2 ppb, and the iron concentration is reduced to less than 0.1 ppm. No toxic residue is leftover, and the uptick in concentrations of sodium, potassium and chlorine, due to the addition of the chemicals, is insignificant. Furthermore, when compared to other methods, such as nano-based technologies, it is cheaper, and nanoparticles are potentially carcinogenic (Ghosh, 2018).

Wastewater Treatment, Recycling and Desalination: Globally, about 80% of wastewater produced is discharged into the environment, causing widespread pollution (WWAP, 2017). There are several technologies (ozonization, ultra-filtration, aerobic treatment, forward/reverse osmosis, advance oxidation) used to treat wastewater for reuse. Treated wastewater can be reused in industry (e.g., in cooling towers) to recharge aquifers, in farming, in the rehabilitation of the natural ecosystem (e.g., wetlands) and rarely for drinking purposes. Wastewater reclamation provides multiple benefits over the desalination of water. Desalinating saltwater produces water suitable for human consumption or irrigation. Desalinated water and recycled wastewater are regarded as rainfall-independent water sources (Fischetti, 2007).). There are various methods of desalination of water, such as different types of distillation, reverse osmosis, solar evaporation, freeze-thaw and so on, with their advantages and disadvantages. Fresh water from rivers or groundwater, water recycling and water conservation are economical compared to desalinated seawater as the latter process consumes energy. The emissions created by desalination plants also contribute to climate change. Desalination is focused on the cost-effective provision of fresh water for human use, as well as minimizing pollution. Currently approximately 1% of world's population is dependent on desalinated water to meet their daily water requirements. The single largest desalination project is Ras Al-Khair in Saudi Arabia, which produces 1,025,000 cubic meters per day in 2014. Kuwait produces a higher proportion of its water than any other country, totaling 100% of its water use.

Water *Conservation:* Water conservation should be launched as a mass movement to save water, regulation of groundwater utilization, rejuvenation of water bodies and rivers.

Land: Land pollution is a serious global issue today affecting humans, animals and plants. It is caused largely by human activities such as agriculture, mining, industrialization, deforestation, and improper disposal of nonbiodegradable waste. Various measures can be taken to mitigate land pollution, including reducing the use of nonbiodegradable materials, reusing materials and recycling various products.

Agriculture: According to the Environmental Protection Agency (EPA), agricultural soil management practices can lead to the production and emission of nitrous oxide (N_2O), a major greenhouse gas and air pollutant. Activities that can contribute to N_2O emissions include fertilizer usage, irrigation and tillage. The management of soils accounts for over half of emissions from the agriculture sector. Manure management and rice cultivation also produce gaseous emissions. Methods that significantly enhance carbon sequestration in soil include no-till farming, residue mulching, cover cropping and crop rotation, all of which are more widely used in organic farming than in conventional farming (EPA, 2015; Pimentel *et al.*, 2005). Farming can deplete soil carbon and render soil incapable of supporting life; however, conservation farming can protect carbon in soils, and repair damage over time. The farming practice of cover crops has been recognized as climate-smart agriculture Burning of crop residues in north Indian states (Punjab, Haryana, UP and Delhi) causes a serious concern about pollution of environment, and an initiative for public awareness is being promoted to discourage the practice (Wiki- Desalination).

Animal Husbandry: Livestock account for one third of emissions through methane emissions. Livestock production occupies 70% of all land used for agriculture, or 30% of the land surface of the globe. This sector is a major source of greenhouse gases, responsible for 18% of the world's greenhouse gas emissions as measured in carbon dioxide (CO2) equivalents. The UNEP states that methane emissions from global livestock are projected to increase by 60% by 2030 under current practices and consumption patterns. The number of unproductive or poor yielding should be minimized in order to reduce methane production and utilization of feed and fodders. It is argued that the consumption of grain by the livestock acts as competitive inhibition to food security for humans. About 70% of the grains used by developed countries are fed to animals. Livestock consumes an estimated one third or more of the world's cereal grains, with 40% of such feed going to ruminants, mainly cattle. Sustainable approaches are aimed to use alternate feeds for the animals that do not compete with human food. Feeding of silage, hay and high-fiber crop residue unsuitable for human consumption should be considered. With improved breeding and cultivation, ruminant animals can yield food (Eisler, *et al.*, 2014).

Fishery: Fisheries contribute approximately US$274 billion to global GDP. However, they are currently an underperforming asset. The World Bank estimates that if fisheries were managed optimally, they could deliver an additional \$50 billion each year. Fisheries provide employment for hundreds of millions of people in direct and indirect ways. Millions of people in developing countries are involved in fishing activities, and thus, the sector often plays a key role in preventing and reducing poverty. Only sustainably managed fish stocks can ensure the viability of these livelihoods and, following recovery, generate more employment in the long term. It goes beyond numbers as sustainable fisheries often also provide better quality employment with higher safety standards. Fish is a renewable and healthy food source, especially for protein, to 1 billion people. Rebuilding fisheries provides direct economic, social and food security benefits, and the transition to its sustainable management would make marine ecosystems more resilient to the effect of climate change and pollution. Fisheries are vital components of ecosystems, and healthy ecosystems are pivotal to the continued productivity of fisheries. The biggest threat to marine capture fisheries comes from climate change and developing resource-poor countries is the most vulnerable to its adverse effect. The fishery sector has the potential to play a crucial role in achieving food security, offering sustainable livelihoods and providing resilient economies by managing the fishery sector in a responsible manner (ISU Marine Programme, 2012).

Forestry: About 30% of Earth's land surface is covered by forests. More than half of all plant and land animal species in the world live in tropical forests (Rainforest facts). Forests are homes to many florae and faunae, preserve biodiversity and maintain equilibrium in ecosystems. Plants absorb CO_2, a major source of pollution-causing climate change. Deforestation is the removal of a forest cover to convert the land for farming, ranches or urban use. Sometimes, the illegal destruction of forest takes place for wood, timber and other products. Deforestation causes habitat damage, biodiversity loss and aridity. It has adverse impacts on biosequestration of atmospheric carbon dioxide. Deforestation causes extinction, changes to climatic conditions, desertification and the displacement of populations. Tropical rainforests have been the main victim of deforestation. Between 2000 and 2012, 2.3 million square kilometers (890,000 sq mi) of forests around the world were cut down. It is estimated that 6.2 million square kilometers remained of the original 16 million square kilometers of forest that formerly covered the earth due to deforestation (Facts about Rainforest).

Sustainable Forest Management Practices: NBSs in forest management are yielding rich dividends. Legislation empowering local people as stakeholder help in forest management. Social forestry (planting trees on public land, such as road and rail tracts, riversides, barren lands, etc.) is gaining ground to boost the environment. Since 1981, China initiated a mass tree-planting movement, and its forest cover had now reached 16.55% of the landmass, as compared to only 12% two decades ago (Gittings, 2001). Alternate use of bamboo as fuel in place of wood has the advantage of burning cleaner, and bamboo matures much faster than wood; thus, deforestation is reduced as the supply can be replenished faster (Rosenberg, 2012). In many parts of the world, especially in East Asian countries, reforestation and afforestation are increasing the area of forested lands (Foley *et al.*, 2005). The amount of woodland has increased in 22 of the world's 50 most forested nations. Asia as a whole gained 1 million hectares of forest between 2000 and 2005 (Wiki- Deforestation).

Marine Oil Spill: More than 70% of our planet's surface is covered by ocean, contains the largest diversity of life on earth and affects everything from global weather patterns to food systems. Oil spill is a form of pollution due to release of liquid petroleum hydrocarbon into the environment due to human activity. Commonly, oil is released into the ocean or coastal waters and occasionally on land. Sources of oil spills are tanker's crude oil; offshore platforms; drilling rigs and wells; refined petroleum products such as gasoline, diesel and their by-products; heavier fuels used by large ships such as bunker fuel; or the spill of any waste oil or oily refuse. It damages marine life and has disastrous consequences on marine ecosystems. Oil spills penetrate the structure of the plumage of birds and the fur of mammals, reducing their insulating ability and making them more vulnerable to temperature fluctuations and much less buoyant in the water.

Oil Spill Cleanup and Recovery Strategy: Cleanup and recovery from an oil spill is difficult and depends on many factors, including the type of oil spilled, the temperature of the water (affecting evaporation and biodegradation) and the types of shorelines and beaches involved. It may take weeks, months or even years to clean up (Wiki-Oil-Spills).

Bioremediation: It is a process where biological agents like microorganisms (ELC, 2008) are used to break down or remove the spilled oil. Oil-consuming bacteria are classified into three kinds: Sulfate-reducing bacteria (SRB), acid-producing anaerobic bacteria and general aerobic bacteria (GAB). They include *Alcanivorax* (Kasai *et al.*, 2002) or *Methylocella silvestris* (Paddy, 2014). These bacteria are found naturally and act to remove oil from an ecosystem, and their biomass will tend to replace other populations in the food chain. The chemicals from the oil which dissolve in water, and hence are available to bacteria, are those in the water-associated fraction of the oil.

Bioremediation accelerator: These are oleophilic and hydrophobic chemicals that bond chemically and physically to both soluble and insoluble hydrocarbons. They act as a herding agent in water and on the surface, floating molecules to the surface of the water, including soluble such as phenols and BTEX, forming gel-like agglomerations. Undetectable levels of hydrocarbons can be obtained in produced water and manageable water columns. By overspraying a sheen with a bioremediation accelerator, the sheen is eliminated within minutes. Whether applied on land or on water, the nutrient-rich emulsion creates a bloom of local, indigenous, preexisting, hydrocarbon-consuming bacteria. Those specific bacteria break down the hydrocarbons into water and CO_2, with EPA tests showing 98% of alkanes biodegraded in 28 days and aromatics being biodegraded 200 times faster than in nature; they also sometimes use a hydrofire boom to clean up oil by taking away most of it and burning it (EPA, 2010).

Controlled burning can effectively reduce the amount of oil in water, if done properly (NOAA, 2007). But it can only be done in low wind (Mullin and Champ, 2003) and can cause air pollution (library.thinkquest.org). Controlled burning can effectively reduce the amount of oil in water, if done properly

(NOAA, 2007). But it can only be done in low wind (Mullin and Champ, 2003), and can cause air pollution (library.thinkquest.org).

Dispersants: A dispersant is either a non-surface-active polymer or a surface-active substance added to a suspension, usually a colloid, to improve the separation of particles and to prevent settling or clumping. It can be used to dissipate oil slicks (www.itopfl.com). They may rapidly disperse large amounts of certain oil types from the sea surface by transferring it into the water column. They will cause the oil slick to break up and form water-soluble micelles that are rapidly diluted. The oil is then effectively spread throughout a larger volume of water than the surface from where the oil was dispersed. They can also delay the formation of persistent oil-in-water emulsions. However, laboratory experiments showed that dispersants increased toxic hydrocarbon levels in fish by a factor of up to 100 and may kill fish eggs (www. nbcnews. com). Dispersed oil droplets infiltrate into deeper water and can lethally contaminate coral. Research indicates that some dispersants are toxic to corals (Barry, 2007). A 2012 study found that Corexit dispersant had increased the toxicity of oil by up to 52 times (www.msnbc.com)

Watch and wait: In some cases, natural attenuation of oil may be most appropriate due to the invasive nature of facilitated methods of remediation, particularly in ecologically sensitive areas such as wetlands (Pezeshki *et al.*, 2000).

Dredging: This technique is used for oils dispersed with detergents and other oils denser than water.

Skimming: During this process, calm waters are required at all times.

Solidifying: Solidifiers are composed of tiny, floating, dry ice pellets (Gordon, 1994) and hydrophobic polymers that both adsorb and absorb. They clean up oil spills by changing the physical state of spilled oil from liquid to a solid, semi-solid or rubber-like material that floats on the water (Sarbatly *et al.*, 2016). Solidifiers are insoluble in water, therefore, the removal of the solidified oil is easy, and the oil will not leach out. Solidifiers have been proven to be relatively nontoxic to aquatic and wildlife and have been proved to suppress harmful vapors commonly associated with hydrocarbons, such as benzene, xylene and naphtha.

Vacuum and centrifuge: Oil can be sucked up along with the water, and then a centrifuge can be used to separate the oil from the water—allowing a tanker to be filled with near pure oil. Usually, the water is returned to the sea, making the process more efficient, but allowing small amounts of oil to go back as well. This issue has hampered the use of centrifuges due to a United States regulation limiting the amount of oil in water returned to the sea (Fountain, 2010).

Beach raking: Coagulated oil left on the beach can be picked up by machinery.

The use of clean energy in marine transport may considerably reduce contamination.

Energy: Various forms of energy have been utilized in industrial development, transportation, agriculture and civic amenities. Fossil fuels, such as coal and petroleum products, have been the main form of energy utilization and have the dual concerns of depletion of the nonrenewable energy source and the addition of burden in environmental pollution.

Mitigation of Energy Crisis: Various approaches are being adopted to mitigate the energy crisis globally. They include the following.

Use of *a* renewable source of energy (*green energy/clean* energy): This includes exploring the use of renewable energy sources such as solar power, wind power, hydroelectricity, geothermal power, biofuel, biomass and ocean energy. Economic analysts expect market gains for renewable energy and efficient energy use. Globally, there are an estimated 3 million direct jobs in renewable energy industries, with about half of them in the biofuels industry. Some countries with favorable geography, geology and weather well suited to the economic exploitation of renewable energy sources already get most of their electricity from renewable sources, including from geothermal energy in Iceland (100%) and hydroelectric power in Brazil (85%), Austria (62%), New Zealand (65%) and Sweden (54%). Renewable power generators are spread across many countries, with wind power providing a significant share of electricity in some regional areas; for example, 14% in the U.S. state of Iowa, 40% in the northern German state of Schleswig-Holstein and 20% in Denmark. Solar water heating makes an important and growing contribution in many countries, most notably in China, which now has 70% of the global total (180 GWth). Worldwide, total installed solar water heating systems meet a portion of the water heating needs of more than 70 million households. Newer sources of biomass, such as algae, can be transformed into a variety of renewable fuels. The use of biomass for heating continues to grow as well. In Sweden, national use of biomass energy has surpassed that of oil. Direct geothermal heating is also growing rapidly (REN21, 2010).

Biofuels are being developed as a more environmentally friendly alternate energy source to alleviate our dependency on limited fossil fuels. The production of biofuels is being refined so that minimal fossil fuels is used for its manufacturing, thus causing less pollution. Biofuels can be made from cellulose in wood, grasses and nonedible parts of plants. These emit fewer GHG emissions in comparison to fossil fuels. Renewable biofuels for transportation, such as ethanol fuel and biodiesel, have contributed to a significant decline in oil consumption in the United States since 2006. The 93 billion liters of biofuels produced worldwide in 2009 displaced the equivalent of an estimated 68 billion liters of gasoline, equal to about 5% of the world's gasoline production (REN21, 2010). Biogas as a renewable fuel can be harvested from organic waste from industry.

Transportation emissions account for roughly a quarter of emissions worldwide. Energy efficient vehicle, hybrid engine and electric mass transportation may cut down emission.

Use *of energy*-efficient products: Replace traditional light with CFLs and LEDs. They require fewer watts of electricity to illuminate and last longer. Their large-scale use would cut down the demand of energy.

The design of buildings and rescheduling time zones to utilize maximum natural light may cut down energy demand.

Lighting controls: Various innovative technologies for lighting controls, such as preset lighting controls, slide lighting, touch dimmers and integrated lighting control, can help conserve energy and reduce overall lighting costs.

Energy audit: It helps identify the areas of energy leakage, and hence controlling it, energy efficiency can be improved.

Nuclear energy*:* Nuclear power, with a 10.6% share of world electricity production as of 2013, is second only to hydroelectricity as the largest source of low-carbon power. More than 400 reactors generate electricity in 31 countries (WES, 2016). A lot of public concerns, such as those about the disposal of nuclear waste, security, accidents and high cost, are associated with nuclear power.

Climate Change: Climate change poses a great challenge to humans, animals and plants for their survival. More frequent and intense droughts, storms and heat waves; rising sea levels; melting glaciers; and warming oceans can directly harm living organisms, destroy the places they habitat and wreak havoc on people's livelihood and communities.

Climate Change Mitigation: Multidirectional approaches are to be adopted to mitigate climate change: The include a reduction of energy demand by increasing energy efficiency, phasing out fossil fuels by switching to low-carbon energy sources, increasing the capacity of carbon sink (e.g., through reforestation), removing CO_2 from the earth's atmosphere and improved building design. New buildings can be constructed (Climate Engineering) using passive solar building design, low-energy building, or zero-energy building techniques, using renewable heat sources. Existing buildings can be made more efficient through the use of insulation, high-efficiency appliances (particularly hot water heaters and furnaces), double- or triple-glazed gas-filled windows, external window shades, and building orientation and siting (CSEPP, 1992; Rosenfeld, *et al.*, 1997). Carbon capture and storage (CCS) is a method to mitigate climate change by capturing CO_2 from large point sources such as power plants and subsequently storing it away safely instead of releasing it into the atmosphere. The Intergovernmental Panel on Climate Change estimates that the costs of halting global warming would double without CCS (The Guardian, 2016). The International Energy Agency says that CCS is "the most important single new technology for CO_2 savings" in power generation and industry (Robinson, 2010). Although it requires up to 40% more energy to run a CCS coal power plant than a regular coal plant, CCS could potentially capture about 90% of all the carbon emitted by the plant (Robinson, 2010). Norway's Sleipner gas field, since 1996, stores almost a million tons of CO_2 a year to avoid penalties in producing natural gas with unusually high levels of CO_2 (Robinson, 2010). As of late 2011, the total planned CO_2 storage capacity of all 14 projects in operation or under construction was more than 33 million tons a year. This is broadly equivalent to preventing the emissions from more than 6 million cars from entering the atmosphere each year (CCS Report, 2011). On one hand, increasing levels of CO_2 in the atmosphere are responsible for climate change and global warming. On the other hand, a CO_2 shortage is mainly affecting the food and drink industry. The obvious question is, Why we can't just capture the excess CO_2 from the atmosphere and use that? As more of these CO_2 utilization technologies emerge, the demand for the gas will increase, and so will the need for more localized production. The future is about turning waste into a commodity (Styring *et al.*, 2015).

Solar *Radiation Management:* The main purpose of solar radiation management seeks to reflect sunlight and thus reduce global warming. The ability of stratospheric sulfate aerosols to create a global dimming effect has made them a possible candidate for use in climate engineering projects (Launder and Thompson, 2008).

Industry: Industry is a major player in environmental pollution. The various industries that cause pollution are the mining and metallurgical industries, power-generating plants, manufacturing factories and processing industries, among others. The pollution is caused due to burning of coal and other fossil fuels, such as oil, petroleum and natural gas; using chemical solvents in the tanning and dyeing industries and others; releasing untreated wastes into the environment; and having unchecked noise emanating from machines, such as those used in oil exploration and drilling constructions.

Controlling Industrial Pollution: By careful selection of raw material, proper maintenance of machinery, adoption of newer technologies, proper **selection of industry site, treatment of industrial wastes and plantation that** act as buffers and natural purifiers for the environment and stringent law to protect the environment. The Ministry of Environment, Forest and Climate Change (MoEFCC) has developed the criteria of categorization of industrial sectors based on the Pollution Index (PI), which is a function of the emissions (air pollutants), effluents (water pollutants), hazardous wastes generated and consumption of resources. The PI of any industrial sector is a number from 0 to 100, and the increasing value of PI denotes an increasing degree of pollution load from the industrial sector. The following criteria on 'Range of Pollution Index' for the purpose of categorization of industrial sectors are finalized as follows: Industrial Sectors having Pollution Index score of 60 and above (Red category), Industrial Sectors having Pollution Index score of 41 to 59 (Orange category), Industrial Sectors having Pollution Index score of 21 to 40 (Green category) and Industrial Sectors having Pollution Index score incl.& up to 20 (White category). There shall be no necessity of obtaining the Consent to Operate" for White category of industries. Intimation to concerned SPCB/PCC shall suffice. No Red category of industries shall normally be permitted in the ecologically fragile area/ protected area. The purpose of the categorization is to ensure that the industry is established in a manner which is consistent with the environmental objectives. The new criteria will prompt industrial sectors willing to adopt cleaner technologies, ultimately resulting in generation of fewer pollutants. Another feature of the new categorization system lies in facilitating self-assessment by industries as the subjectivity of earlier assessment has been eliminated. This 'Re-categorization' is a part of the efforts, policies and objective of present government to create a clean & transparent working environment in the country and promote the Ease of Doing Business. Other similar efforts include installation of Continuous Online Emissions/ Effluent Monitoring Systems in the polluting industries, Revisiting the CEPI (Comprehensive Environment Pollution Index) concept for assessment of polluted industrial clusters, revision of existing industrial emission/effluent discharge standards, initiation of special drive on pollution control activities in Ganga River basin and many more in the coming days. (Press Information Bureau, Government of India, Ministry of Environment, Forest and Climate Change. 05-March-2016).

Green Industry Initiative: Green *industry* refers to economies striving for growth through a more sustainable pathway by undertaking green public investments and implementing public policy initiatives that encourage environmentally responsible private investments.

Resource Efficient and Cleaner Production (RECP): RECP covers the application of preventive management strategies that increase the productive use of natural resources (materials, energy, water), minimize generation of waste and emissions and foster safe and responsible production.

Cleaner Production (CP): RECP uses CP to accelerate the application of preventive environmental strategies to processes, products and services, to increase efficiency and reduce risks to humans and the environment. It addresses (a) production efficiency, optimization of the productive use of natural resources (materials, energy, water); (b) environmental management, minimization of impacts on environment and nature through reduction of wastes and emissions; and (c) human development, minimization of risks to people and communities and support for their development.

The United Nations Industrial Development Organization (UNIDO) focuses on cost-effective ways to reduce ozone-depleting substances (ODSs), such as freons, halons and chlorofluorocarbons, in the areas of refrigeration, plastic foams, halons, solvents, fumigants and aerosol.

Chemicals Management: The UNIDO works with projects, policies and regulations, institutions and sectoral capacity building, development of preventative approaches and new business models, such as chemical leasing, to assist enterprises in reducing the risks and impacts associated to the use of chemicals.

Chemical Leasing (ChL): ChL is a strategy that creates a business environment to tackle the challenges of the changing global context and offers solutions for sound management of chemicals and reduction of emissions to the environment. The UNIDO plays a leading and coordinating role for the implementation and further development of ChL.

Bioleaching: This is a significant advantageous environmental technical process used for extracting useful elements from their ore by bacteria to dissolve them (Hoque and Philip, 2011). Bioleaching- A significant advantageous environmental technical process used for extraction of useful elements from their ore by bacteria and to dissolve them (Hoque and Philip, 2011).

Waste management: Waste is created in the forms of solid, liquid and gaseous due to human activities. According to the origin of creation, waste may be from household, industrial and biological types. Waste management is a comprehensive effort that starts from its inception to final disposal with a series of activities, such as collection, transport, treatment, disposal, monitoring and regulation. Waste accumulation has an adverse effect on health and the environment, as well as on the aesthetic point of view, and proper waste management is of utmost importance. Waste management practices vary widely in different countries depending on the region and types of waste. Plastic waste management is a great challenge and so is focused on.

Plastic products are widely used and are a serious threat to environment. They are produced on a large scale, and their production crosses the 150 million tons per year globally. In India, approximately 8 million tons of plastic products are consumed every year (2008). The problem with plastics is that they will never degrade and remain on landscape for several years. The recycled plastics are more harmful to the environment than the virgin products due to mixing of color, additives, stabilizers, flame retardants and the like. Furthermore., recycling virgin plastic material can be done two or three times only because, after every recycling, the strength of the plastic material is reduced due to thermal degradation. Although no authentic estimation is available on the total generation of plastic waste in the country, considering, however, that 70% of total plastic consumption is discarded as waste, approximately 5.6 million tons per annum of plastic waste is generated in country, which is about 15,342 tons per day (CPCB Report, 2013).

Plastic *Waste* Disposal*:* The disposal of plastic waste is a serious concern, and to date no full proof technology is available for their disposal. Several approaches are in practice with their merits and demerits.

Incineration*:* It is the globally accepted technology used for plastic disposal. However, due to poor maintenance of incinerators, it releases several harmful gases, such as dioxins and furans in the case of chlorinated and brominated plastic wastes (CPCB Report, 2013).

Landfilling*:* Plastics are commonly disposed of in landfills. However, landfilling is considered highly wasteful as it requires a vast amount of space, and the chemical constituents and energy contained in plastic are wasted in this disposal route. In countries where landfills are poorly managed, plastic waste can be easily blown into waterways or carried out to sea by floodwater. In addition, when plastics decompose in landfills, they may leak pollutants, such as phthalates and bisphenol-A, into the soil and surrounding environment (Hopewell *et al*., 2009).

Recycling*:* Many plastics can be recycled, and the materials recovered can be given a second life. However, this method is not fully utilized, due to difficulties with the collection and sorting of plastic waste. Poor waste management facilities often result in plastics and other waste being recklessly disposed into rivers and water bodies.

Recycling of plastics through environmentally *a* sound manner*:* The extrusion and palletization processes have been redesigned to make the pollution from the process to a minimum level and, as a result, enhance the efficiency of the process

Innovative technical options for safer disposal of plastic waste: Utilization of Plastic Waste in Road Construction The technique is being implemented successfully for the construction of flexible roads at various places in India. There are several advantages of polymer-bitumen roads: they resist stripping and pothole formation, the polymer will not leach out of the bitumen layer, they reduce bleeding of bitumen during the summers and they inhibit leaching of toxic compounds into soil (CPCB Report, 2013).

Coprocessing of plastic waste as *an alternative fuel* and *raw (AFR) material* in cement kilns and power plants: It offers advantages for the cement industry as well as for the Municipal Authorities responsible for waste management. On the other hand, cement producers or power plants can save on fossil fuel and raw material consumption, contributing to a more eco-efficient production. In addition, one of the advantages of the recovery method used in existing facilities is eliminating the need to invest on other plastic waste practices and secure landfilling (CPCB Report, 2013).

Conversion of plastic waste into liquid RDF *(oil)*: Plastic waste is segregated and converted into the more useable, i.e., liquid, RDF through catalytic pyrolysis. The process is termed random depolymerization as the degradation of bonds to break down into monomers occurs randomly. The entire feed material is converted into either of the products: liquid RDF, gases and some sludge. There is no effluent generated in the process and the unused hot air from the reactor is released through a chimney (CPCB Report, 2013).

Plasma pyrolysis technology (PPT) integrates the thermochemical properties of plasma with the pyrolysis process. The intense and versatile heat-generation capabilities of PPT enable it to dispose of all types of plastic waste, including polymeric, biomedical and hazardous waste, in a safe and reliable manner. Pyrolysis is the thermal disintegration of carbonaceous material in an oxygen-starved atmosphere. When optimized, the most likely compounds formed are methane, carbon monoxide, hydrogen carbon dioxide and water molecules. The extreme conditions of plasma kill stable bacteria. Segregation of the waste is not necessary as the very high temperatures ensure treatment of all types of waste without discrimination (CPCB Report, 2013).

Biodegradable Plastics (Bioplastics*):* These are decomposed by the action of living organisms into CO_2 and water and have the potential to solve a number of waste management issues, especially for disposable packaging that cannot be easily separated from organic waste. However, biodegradable plastics are not without controversy as it is found that oxo-biodegradable plastics may release metals into the environment (North and Halden, 2013).

NBS: This is a nature-based, sustainable managemental approach to tackling environmental and societal challenges, such as climate change, water security, water pollution, food security, forest management, human health and disaster risk management. The approach has existed for decades; however, of late, environmentalists have revisited it to explore the potential benefits. A classic example of the NBS is the conservation of mangroves along coastlines. Mangroves moderate the impact of waves and wind on coastal settlements, cause CO_2 sequestration, help prevent soil erosion and provide safe shelter for marine life, which, in turn, provides livelihoods and food security to the people. The NBS appears to be very successful in forest management in which the people involved participate to generate their livelihoods from the forest and, in turn, ensure the maintenance of forest resources. The NBS for water management has yielded tremendous success in many parts of the world, and United Nations World Water Development Report 2018 had the title *Nature Based Solutions for Water*' (Wikipedia; WWDR 2018). The NBS is strongly connected to ideas such as natural-systems agriculture (Jackson, 2002) natural solutions (Dudley, *et al.*, 2010), ecosystem-based approaches (Cowan *et. al.*, 2010), adaptation services (Lavorel, *et al.*, 2015) natural infrastructure (Smith and Barchiesi, 2009; Ozment, *et al.*, 2015), green infrastructure (Benedict and McMahon, 2006) and ecological engineering (Borsje, *et al.*, 2011; Barot, *et al.*, 2012). For instance, ecosystem-based approaches are increasingly promoted for climate change adaptation and mitigation by organizations like United Nations Environment Programme and non-governmental organizations such as the Nature Conservancy. These organizations refer to "policies and measures that take into account the role of ecosystem services in reducing the vulnerability of society to climate change, in a multi-sectoral and multi-scale approach" (Cowan *et. al.*, 2010). *Natural infrastructure* is defined as a "strategically planned and managed network of natural lands, such as forests and wetlands, working landscapes, and other open spaces that conserves or enhances ecosystem values and functions and provides associated benefits to human populations" (Smith and Barchiesi, 2009; Ozment, *et al.*, 2015), and *green infrastructure* refers to an "interconnected network of green spaces that conserves natural systems and provides assorted benefits to human populations" (Benedict and McMahon, 2006).

Ecological engineering generally refers to "protecting, restoring (i.e. ecosystem restoration) or modifying ecological systems to increase the quantity, quality and sustainability of particular services they provide or to build new ecological systems that provide services that would otherwise be provided through more conventional engineering, based on non-renewable resources."

National Green Tribunal (NGT): During the Rio de Janeiro Summit of United Nations Conference on Environment and Development in June 1992, India vowed the participating states to provide judicial and administrative remedies for the victims of the pollutants and other environmental damage. The NGT was established on 18 October 2010 under the National Green Tribunal Act 2010 for effective and expeditious disposal of cases related to environmental protection and conservation of forests and other natural resources, including enforcement of any legal right to the environment and giving relief and compensation for damages to persons and property and for matters connected therewith or incidental thereto. It is a specialized body equipped with the necessary expertise to handle environmental disputes involving multidisciplinary issues. The tribunal shall not be bound by the procedure laid down under the Code of Civil Procedure, 1908, but shall be guided by principles of natural justice. The tribunal dedicated jurisdiction in environmental matters shall provide speedy environmental justice and help reduce the burden of litigation in the higher courts. The tribunal is mandated to make and endeavor for disposal of application or appeals finally within 6 months of filing of the same. This court can rightly be called 'special" because India is only the third country, following Australia and New Zealand, to have such a system and indicates its concern toward conservation of the environment.

Conclusion and future direction: Human activities and overexploitation of natural resources deplete, contaminate, change or interfere with the equilibrium of the environment and thus pose a threat to human existence. To mitigate the changes, future directions should be focused on taking care of the development, use and regulation of biological systems for remediation of contaminated environmental constituents, such as land, air and water, and adopting environmentally friendly processes, such as green manufacturing technologies and sustainable development. The reduction of combustion of fossil fuels, such as coal and oil, can reduce air pollution drastically. Most effective is the switch to clean power sources, such as wind power, solar power, hydropower, which don't cause air pollution. Water conservation should be launched as a mass movement to save water, regulate groundwater utilization and rejuvenate water bodies and rivers. Sustainable agricultural practices and organic farming should be adopted to increase food production without exhausting global resources and to keep the planet cool and habitable. Allied agricultural resources, such as animal husbandry, fishery and forestry, should be explored in a sustainable manner to achieve global food security. Energy-efficient vehicles, hybrid engines and electric mass transportation may cut down emissions. Industrial pollution can be controlled by a careful selection of raw material, the proper maintenance of machinery, the adoption of newer technologies, the proper selection of industry site, the treatment of industrial wastes and the planting of trees and stringent laws to protect the environment. NBS is a nature-based sustainable approach to tackle environmental challenges, such as climate change, water security, water pollution, food security, forest management, human health and disaster risk management.

REFERENCES

Annual Report. (June 2013). *Central Pollution Control Board*. Overview of Plastic Waste Management. New Delhi.

Barot, S., Lata, J.C. and Lacroix, G. (2012). Meeting the Relational Challenge of Ecological Engineering Within Ecological Sciences. *Ecological Engineering* 45: 13–23.

Barry, C. (2007). Slick Death: Oil-Spill Treatment Kills Coral. *Science News* 172: 67.

BBC News. Nanotechnology to Gobble Up Pollution. Retrieved 29 October 2014.

Benedict, M.A. and McMahon, E.T. (2006). *Green Infrastructure: Linking Landscapes and Communities*. Washington, DC.

Borsje, B.W. et al. (2011). How Ecological Engineering Can Serve in Coastal Protection. *Ecological Engineering* 37/2: 113–122.

Bradford, A. (4 March 2015). Deforestation: Facts, Causes & Effects. *Livescience.com*.

CO_2 Turned into Stone in Iceland in Climate Change Breakthrough. (9 June 2016). *The Guardian*, https://www.theguardian.com/environment/201 6/jun/09/co2-turned-into-stone-in-iceland-in-climate-change-breakthrough.

Committee on Science, Engineering (CSE); Public Policy. (1992). *Policy Implications of Greenhouse Warming: Mitigation, Adaptation, and the Science Base*. Washington, DC: National Academy Press. ISBN 0-309-04386-7.

Cowan, C., Epple, C., Korn, H., Schliep, R. and Stadler, J. (Eds.). (2010). Working with Nature to Tackle Climate Change. Report of the ENCA/BfN Workshop on "Developing Ecosystem-Based Approaches to Climate Change – Why, What and How, https://www.bfn.de/fileadmin/MDB/documents/service/Skript264.pdf". Bonn: Bundesamt für Naturschutz.

Deforestation | Threats | WWF. Worldwildlife.org. Retrieved 13 November 2016.

Dudley, N. et al. (2010). *Natural Solutions: Protected Areas Helping People Cope with Climate Change*. Gland: World Wide Fund for Nature.

Eisler, M.C., Lee, R.R.F., Tarlton, J.F., et al. (5 March 2014). Agriculture: Steps to Sustainable Livestock. *Nature*, 507(7490).

Environmental Literacy Council – Oil Spills. (25 June 2008). *Enviroliteracy.org*.

EPA. (2010). S-200 | NCP Product Schedule | Emergency Management | US EPA. *Epa.gov*. Retrieved 16 June 2010.

EPA. (2015). *Agriculture: Sources of Greehouse Gas Emissions*, http://www3.epa.gov/climatechange/ghgemissions/sources/agriculture.html.

Facts About Rainforests. The Nature Conservancy. Retrieved 19 October 2015.

Fahmideh, L., Khodadadi, E. and Khodadadi, E. (2014). A Review of Applications of Biotechnology in the Environment. *International Journal of Farming and Allied Sciences*, 3(12): 1319–1325.

FAO. (2015). *Global Forest Resource Assessment 2015. How Is the World Forest Changing?*

Fensterstock, J.C., Ketcham, B. and Walsh, M. (May 1972). *The Relationship of Land Use and Transportation Planning to Air Quality Management*. Ed. George Hagevik.

Fensterstock, J.C., Kurtzweg, J.A. and Ozolins, G. (1971). Reduction of Air Pollution Potential Through Environmental Planning. *Journal of the Air Pollution Control Association*, 21(7): 395–399.

Fischetti, M. (2007). Fresh from the Sea. *Scientific American*, 297(3): 118–119.

Foley, J.A., Defries, R., Asner, G.P., Barford, C., Bonan, G., Carpenter, S.R., Chapin, F.S., Coe, M.T., Daily, G.C., Gibbs, H.K., Helkowski, J.H., Holloway, T., Howard, E.A., Kucharik, C.J., Monfreda, C., Patz, J.A., Prentice, I.C., Ramankutty, N. and Snyder, P.K. (2005). Global Consequences of Land Use. *Science*, 309(5734): 570–574.

Fountain, H. (24 June 2010). Advances in Oil Spill Cleanup Lag Since Valdez. *New York Times*.

Gavrilescu, M. (2010). Environmental Biotechnology: Achievements, Opportunities and Challenges. *Dynamic Biochemistry, Process Biotechnology and Molecular Biology*. 4(1): 1–36.

Ghosh, S. (9 March 2018). A Simple Low Cost Filter That Removes Arsenic and Iron from Water. Mongabay Series. *Environment and Health*.

Gittings, J. (20 March 2001). Battling China's Deforestation. *The Guardian*.

Global Status of CCS Report:2011, http://www.globalccsinstitute.com/publications/global-status-ccs-2011/online/ 26856.

Gordon, D. (24 February 1994). Zapping Oil Spills with Dry Ice and Ingenuity. *Los Angeles Times*, South Bay Section Page 1.

Hopewell, J., Dvorak, R. and Kosior, E. (2009). Plastics Recycling: Challenges and Opportunities. *Philosophical Transactions of the Royal Society B: Biological Sciences*, 364(1526): 2115–2126.

Hoque, M. (2018). The Way Ahead. In: *Biotechnology for Sustainable Agriculture: Emerging Approaches and Strategies*. Eds. R.L. Singh and S. Mondal. Woodhead Publishing (Elsevier), pp. 375–397.

Hoque, Md E. and Philip, O.J. (2011). Biotechnological Recovery of Heavy Metals from Secondary Sources- An Overview. *Materials Science and Engineering*, 31: 57–66.

https:// en.wikipedia.org/wiki/Oil-spill.

https:// en.wikipedia.org/wiki/Air-pollution.

https:// en.wikipedia.org/wiki/Nature-based solutions.

https:// en.wikipedia.org/wiki/Climate change mitigation.

https:// en.wikipedia.org/wiki/Rain water harvesting

https://en. wikipedia.org/wiki/Desalination.

https://www.unido.org/Green Industry Initiative.

http:// www. Library.thinkquest.org /Oil Spills.

http:// www. Itopfl.com/Spill Response /clean upresponse/ Dispersants.

http:// www.msnbc.com *"Dispersant makes oil 52 times more toxic – Technology & science – Science – LiveScience – NBC News"*.

International Sustainability Unit (ISU) Marine Programme. (February 2012). Towards Global Sustainable Fisheries-The Opportunity for Transition. Report.

Jackson, D.L. (2002). *The Farm as Natural Habitat: Reconnecting Food Systems with Ecosystems*. Washington, DC.

Jacobson, M.Z., et al. (2015).100% Clean and Renewable Wind, Water, and Sunlight (WWS) All-Sector Energy Road Maps for the 50 United States. *Energy and Environmental Science*, doi:10.1039/C5EE01283J.

Jason, P. (12 November 2011). Smog-Eating Material Breaking into the Big Time. *BBC News*.

Kasai, Y., et al. (2002). Predominant Growth of Alcanivorax Strains in Oil-Contaminated and Nutrient-Supplemented Sea Water. *Environmental Microbiology*, 4(3): 141–147.

Launder, B. and Thompson, J.M.T. (2008). Global and Arctic Climate Engineering: Numerical Model Studies. *Philosophical Transactions of the Royal Society A*., 366(1882): 4039–4056.

Lavorel, S., Colloff, M.J., McIntyre, S., Doherty, M.D., Murphy, H.T., Metcalfe, D.J., Dunlop, M., Williams, R.J., Wise, R.M. and Williams, K.J. (2015). Ecological Mechanisms Underpinning Climate Adaptation Services. *Global Change Biology*, 21: 12–31.

Mullin, Joseph V., and Champ, Michael A (1 August 2003). Introduction/Overview to In Situ Burning of Oil Spills. *Spill Science & Technology Bulletin. In-Situ Burning of Spilled Oil*, 8(4): 323–330, doi:10.1016/S1353-2561(03)00076-8.

Nature-Based Solutions' Is the Latest Green Jargon That Means More Than You Might Think. *Nature*, 541(7636): 133–134. 2017-01- Bibcode:2017Natur.541R.133, doi:10.1038/541133b.

NOAA. (20 June 2007). Emergency Response: Responding to Oil Spills. Office of Response and Restoration. National Oceanic and Atmospheric Administration.

North, E.J. and Halden, R.U. (2013). Plastics and Environmental Health: The Road Ahead. *Reviews on Environmental Health*, 28(1), 1–8.

Ozment, S., DiFrancesco, K. and Gartner, T. (2015). *The Role of Natural Infrastructure in the Water, Energy and Food Nexus*. Nexus Dialogue Synthesis Papers. Gland, Switzerland: IUCN.

Paddy, H. (30 April 2014). *Oil and Natural Gas Eating Bacteria to Clear-Up Spills*. www.oilandgastechnology.net.

Pezeshki, S.R., Hester, M.W., Lin, Q. and Nyman, J.A. (2000). The Effects of Oil Spill Clean-Up on Dominant US Gulf Coast Marsh Macrophytes: A Review. *Environmental Pollution*, 108: 129–139.

Pimentel, D., Hepperly, P., Hanson, J., Douds, D. and Seidel, R. (2005). Environmental, Energetic, and Economic Comparisons of Organic and Conventional Farming Systems. *BioSci*, 55(7): 573–582.

Rainforest Facts. Nature.org (1 November 2016). Retrieved 13 November 2016.

REN21. (2010). Renewable Energy Policy Network for the 21st Century- Renewables 2010 Global Status Report.

Robinson, S. (2010). How to Reduce Carbon Emissions: Capture and Store It? *Time.com*, http://www.time.com/ti me/specials/packages/article/0,28804,1954176_1954175,00.html.

Rosenberg, T. (13 March 2012). In Africa's Vanishing Forests, the Benefits of Bamboo. *New York Times*.

Rosenfeld, A.H., Romm, J.J., Akbari, H. and Lloyd, A.C. (1997). *Technology Review. Painting the Town White – and Green*. Massachusetts Institute of Technology, http://eande.lbl.gov/HeatIsland/PUBS/PAINTING/.

Sahney, S., Benton, M.J. and Falcon-Lang, H.J. (2010). Rainforest Collapse Triggered Pennsylvanian Tetrapod Diversification in Euramerica. *Geology*, 38(12): 1079–1082.

Sarbatly, R., Kamin, Z. and Krishnaiah, D. (2016). A Review of Polymer Nanofibres by Electrospinning and Their Application in Oil-Water Separation for Cleaning Up Marine Oil Spills. *Marine Pollution Bulletin*, 106: 8–16.

Scoullors, M. and Ferragina, E. (2010). Environmental and sustainable development in the Mediterranean. 10 papers for Bercelona. European Institute of the Mediterranean (IEMed) and European Union Institute for Social Studies (EUISS), pp. 1–79.

Singh, R.L. (2016). *Principles and Application of Environmental Biotechnology for a Sustainable Future*. Springer, Verlag, pp. 1–487.

Smith, M. and Barchiesi, S. (2009). *Environment as Infrastructure: Resilience to Climate Change Impacts of Water Through Investments in Nature*. Perspectives on Water and Climate Change Adaptation. Gland, Switzerland: IUCN.

Styring, P., Quadrelli, E.A. and Armstrong, K. (2015). *Carbon Dioxide Utilization-Closing the Carbon Cycle*. Elsevier.

UN-Water. (2018). *World Water Development Report 2018: Nature-Based Solutions for Water*. Geneva, Switzerland.

United Nations Industrial Development Organization (UNIDO). (2010). *A Greener Footprint for Industry: Opportunities and Challenges for Sustainable Industrial Development*. Vienna.

United Nations Report of the World Commission on Environment and Development. Our Common Future, 1987.

Waste Water: The Untapped Resources. Paris. ISBN 978-92-3-100201-4.

WHO. (25 March 2014). 7 Million Premature Deaths Annually Linked to Air Pollution.

World Bank; Institute for Health Metrics and Evaluation at University of Washington – Seattle. (2016). *The Cost of Air Pollution: Strengthening the Economic Case for Action*. Washington, DC: The World Bank, p. xii.

World Energy Statistics. (2016), p. 25. http://www.iea.org/publications/freepublications/publication/KeyWorld_Statistics_2015.pdf (PDF).

World Water Assessment Programme (WWAP)-2017. The United Nations World Water Development Report 2017.

www.www. nbcnews.com/id/US Environment/Study chemicals clean oil spill harm fish. 10 April 2009.

Zhu, Qiang, et al. (2015). *Rainwater Harvesting for Agriculture and Water Supply*. Beijing: Springer, p. 20. ISBN 978-981-287-964-6.

14

Policies and Issues in Environmental Biotechnology Research

Ramanuj Banerjee, Mamoni Banerjee and S. Mondal

CONTENTS

DOI: 10.1201/9781003186304-14

14.1 Introduction

In the modern era, biotechnology is an integral part of natural sciences and engineering. As far as application is concerned, it has several verticals: agriculture, environmental, energy, health and industrial technology, referring to genetic technology, medicine and food technology, among others (Gavrilescu, 2010). Traditional biotechnological knowledge is used by the people for centuries to yield bread, wine, beer, alcohol, fermented milk products and medicines. The arenas of today's biotechnology comprise a large spectrum of subjects like genomics, recombinant DNA technology, proteomics and bioinformatics. In the present era, the health care, agriculture, social well-being, and environmental management industries have contributed a lot to the global economy through retaining biotechnological initiatives.

Environmental biotechnology is a branch of biotechnology that is applied to the study of the natural environment. It is the efficient use of environment and natural habitats, such as animals, plants, bacteria, fungi and algae, to produce energy, including renewable energy, nutrients and food. The development process in environmental biotechnology includes integrated and economic profitable pathways of profit making, in which the waste of each process becomes the raw material for another process. As described by the International Society for Environmental Biotechnology, it is the development of environmentally friendly products and processes and its use, control and regulation of biological systems for redressal of polluted environment, including air, water and land or soil. The processes predominantly include green and clean manufacturing technologies and sustainable development. Environmental biotechnology able to provide a wide range of benefits to humanity, such as new sources of renewable energy and more effective treatment of waste. Waste either solid, liquid or gaseous can be converted to less or zero waste through a recycling process and may be reused as new products or less harmful end waste. Replacing chemical materials and processes with biological technologies can reduce environmental damage. Furthermore, the application of environmental biotechnology has led to new products and processes that have a competitive edge over the traditional ones in terms of effectiveness, productivity, cost and safety. It has been noted that environmental biotechnology is the one of most rapidly growing areas in the development of economic and efficient processes with sustainability.

14.2 Objective and Importance of Environmental Biotechnology

14.2.1 Biotechnology vis-à-vis Environmental Biotechnology

If we categorize domains of applications, there are mainly four subfields as accepted in environmental policies (EIBE 2000; OECD 2001). If we categorize domains of applications, there are mainly four sub-fields as accepted in the environmental policies (EIBE 2000; OECD 2001).

1. Green biotechnology, which is the oldest use of biotechnology by humans and deals with plants and its growing technology
2. Red biotechnology, which is meant to create chemical compounds for medicinal use or to help the body in protecting itself from diseases or illnesses
3. White biotechnology, which deals with using biological organisms to produce or manipulate beneficial products for the industry
4. Blue biotechnology, which deals with related biological technology for aquatic use

However, the main action areas for the preceding domains of biotechnology research and development activities can be categorized into three main groups (Kryl 2001; Johnston 2003; Gavrilescu and Chisti 2005):

1. Industrial supplies (enzymes, biochemicals and other relevant reagents for industrial and food processing)
2. Energy (fuels from renewable resources)
3. Environment (pollution diagnostics, different products to prevent pollution, bioremediation)

Of all these, environmental biotechnology and its application for environmental protection and control are very promising fields of research and development. It includes the production of chemicals biosensors for identifying contaminants; the decontamination of environmental components, mainly water, air and soil; and technologies for prevention of pollution and minimization of waste.

The discussion about the environment and the economy can be made conjointly. They are complementary to each other. As an example, if environmental pollution is strictly considered by the industry, economic growth will be decreased to some extent. Moreover, the financial and economic calamity has provided the impetus for policy issues. It aims at encouraging recovery through reusing, reducing and recycling industrial waste for renewable and sustainable growth. To comply with that a strategic vision, it is necessary to authorize suitability for government policies toward economic efficiency, environmental integrity and social equity at both the national and international levels. Innovations in environmental biotechnology could underpin sustained and sustainable growth. It may be fostered by developing and adopting the requisite environmental policy measures on environmental biotechnology research (OECD 2013).

It has also been noted that there are enormous future prospects and opportunities for new developments in the field of environmental biotechnology (Gavrilescu M., 2010). For example, integrated future biorefineries will use a range of sustainable biomass sources, and eventually, they will be expected to produce biofuels, biobased chemicals and bioplastics to have a supportive policy portfolio. When companies are intended to produce products such as biobased plastics, biobased chemicals or biofuel for environmental improvements, a set of policies should be in place to ensure their capacity building for smooth production instead of regulatory compliance. In the same way, the appropriate policy issues should be in practice under the domain of agricultural biotechnology for genetically modified (GM) crops and GM nonfood crops, which may be future green feedstock for industry (OECD 2013).

This chapter contributes new solutions and directions in the remediation, control and monitoring of contaminant environments, minimizing the release of future waste and creating alternatives for preventing environmental pollution. It elaborates on policies and issues in environmental biotechnology research vis-à-vis climate change.

Environmental biotechnology slowly got its importance with the implementation of laws, rules and legislation for environmental protection in a number of countries. It is also synergized through the setting of standards for industry vis-à-vis enforcements of compliance.

Rapid industrialization, urbanization and other societal developments have caused a continuous decrease of clean environment and have depleted natural resources. With the advancement of human civilization, gradual improvement in the standard of living, higher consumer demand vis-à-vis development in the industrial sector, transport, agriculture, health care and advanced domestic deliberations; pollution of air, water and soil have been amplified manifold. The air has been polluted with more carbon dioxide (CO_2), nitrous oxides (NOx), sulfur dioxide (SO_2), greenhouse gasses and particulate matters; water has been polluted with chemical and biological pollutants, leachate, oil spills and the like; and soil is contaminated from the disposal of hazardous waste, the spreading of pesticides and so on. The pollution synergizes with the use of disposable goods or nonbiodegradable materials and the lack of proper facilities for waste and waste disposals (Gavrilescu M., 2010).

Environmental biotechnology is dealt with by implementing emerging technologies for environmental protection. Some of such technologies are very much age-old, such as composting, wastewater treatment and the like. In the early stage, it was also involved with chemical engineering, but gradually it was added to several other disciplines, such as environmental microbiology, molecular biology, ecology, biochemistry, environmental engineering and others (Hashim and Uijang 2004).

14.2.2 Objectives of Environmental Biotechnology

Environmental biotechnology has as its specific aim taking requisite safety measures, such as prevention, protection and reverse degradation of the environmental through the application of biotechnology and other relevant technologies. While supporting safety measures for the environment, a comprehensive plan of action has been taken globally, nationally and locally through the United Nations system, several governments and major social organizations. The specific objectives of such a comprehensive plan are as follow:

- To implement appropriate processes of production, such as green and clean technologies, through the optimal use of natural resources, reusing and recycling biomass, recovering renewable energy and diminishing waste generation
- To endorse the adoption of biotechnological techniques referring to land and water bioremediation, conservation of soil and land rehabilitation, reforestation, solid, gaseous and liquid waste treatment, biomass utilization and the like
- To instrument biotechnological products and processes to defend and sustain environmental integrity with a view to long-standing ecological safety, stability and security

The critical areas of environmental biotechnology include eco-restoration, the conservation of biodiversity, the generation of indigenous eco-friendly technologies, training and research and updating environmental regulation, among others. India is a signatory member of the 'Kyoto Protocol' and the 'Paris Agreement', and thus, the biotechnology policy will encourage industries to take up the 'Clean Development Mechanism' (CDM) to place the industries in the green category. Such industries will then be able to gain 'Certified Emission Reduction' (CER) credit points, which can be utilized in 'carbon trading' on the international market. Such industries will also be able to attract international collaboration, which will boost the economy of the country. Research in environmental biotechnology is an important pathway in developing effective solutions for averting, mitigating, and retrogressive environmental impairment with the help of living organisms, such as whole cells, microorganism, plant cells or enzymes, among others. A range of advanced analytical devices with live indicators have been developed to detect environmental pollution including air, water and land. The adoption of recombinant DNA technology is an important initiative for environmental protection through different processes such as bioremediation, waste utilization, production of renewable energy and waste disposal.

Environmental biotechnology is very much important through converting waste to wealth and making the whole process economically profitable. Otherwise, waste especially industrial residue is habitually referred to as undesirable things with less or no economic value. In addition, the quantum and diversified wastes generated through industries, transport services, cities and municipalities create serious hazards to environmental, plant, animal and human health. Applying different technology, the waste can be used as a resource with certain economic values for another user. The situation is alarming in developing countries, mainly due to inefficient technologies, inappropriate adoption of policies and insensitivity on the part of the industrial sectors. Chemical waste streams often contain either precious or toxic metals. These streams regularly require treatment to become more environmentally acceptable, but treatment may also be due to the need to recover valuable resources. The recovery and recycling of platinum group metals from catalytic, plating and mining streams are an increasingly essential process. With good reason, strict regulation limits the quantities of numerous toxic and heavy metals that may be released into the environment, notably Cd, Cr, Pb and Hg. The removal of such contaminants may greatly reduce waste processing costs. Considerable revenue can be generated from waste through

a. recycling useful materials from municipal solid waste.
b. generating energy from municipal or agricultural wastes (such as palm oil empty fruit bunches and animal waste),
c. making biocomposts or biofertilizer from organic waste collected from agricultural and municipal waste, and
d. using other specific technologies to convert waste to useful materials (such as converting rice husk to charcoal, sludge to bricks, extraction of oil from used plastics, conversion of used tires to carbon black and rubber granulates etc.)

Other than the common biogas, plant energy can also be generated through biomass-hydrolyzing enzymes from the aquatic weed. Water hyacinth (*Eichhornia crassipes*), one of the most productive aquatic plants, is also an aquatic pest and a menace, creating navigational problems and pollution in water bodies. The use of renewable plant biomass for producing cleaner and cheaper fuels has rejuvenated interest in developing cheaper technologies for the large-scale production of biomass-saccharifying enzymes. Apart from anaerobic digestion and biomass production, there are several technologies in practice for converting municipal solid waste (MSW) to wealth in the form of energy. MSW can be directly combusted as mass burn through minimal processing and converting waste to energy facilities as a fuel. It can be commenced as moderate to extensive processing before being directly combusted as refuse-derived fuel. It has been also noted that by means of pyrolysis or thermal gasification technologies, it can be gasified for making waste to wealth.

14.2.3 Environmental Biotechnology and Control of Pollutants

It has been noted that some pollutants can be easily degraded through minimal biotechnological interference and lead to nonaggressive products. The products are referred as compounds mineralization, transformation or immobilization with the involvement of living microbes, plants and animals under certain conditions with the help of abiotic and biotic factors.

There are some advanced techniques or technologies also to treat waste and degrade pollutants. The advanced technologies are assisted by living organisms through the processes like ensuring environmental health through sensor or artificial intelligence–based biomonitoring; applying improved treatment technologies for solid waste and wastewater; implementing bioremediation for cleaning up contamination and following it with phytoremediation; developing cleaner production technologies, that is, manufacturing with less pollution or fewer raw materials; developing of renewable energy from biomass; and implementing genetic engineering tools for environmental protection and control. In the process of advanced biotechnological implementation, the products and processes that develop will generate less waste, change waste to wealth and attempt to preserve the renewable energy and natural nonrenewable resources (Olguin 1999; EIBE 2000; Gavrilescu and Chisti 2005; Chisti 2007).

However, in the process of decontamination, some environmental contaminants can accumulate in the environment, which are noted as refractory and have a certain degree of toxicity. In addition, conventional treatment of some pollutants through incineration, through incineration, chemical degradation and landfilling can generate other contaminants. Actually, a large variety of noxious waste are present or generated through other processes in the environment, which attract the interference of environmental biotechnology for developing combination products, alternatives and economical and reliable biological treatments for changing environmental non-pollutants (OECD 1994; EFB 1995; Krieg 1998; OECD 1998; Futrell 2000; Evans and Furlong 2003; Kuhn *et al.* 2003; Chen *et al.* 2005; Gavrilescu 2005; Betianu and Gavrilescu 2006a, 2006b).; Krieg 1998; OECD 1998; Futrell 2000; Evans and Furlong 2003; Kuhn *et al.* 2003; Chen *et al.* 2005; Gavrilescu 2005; Betianu and Gavrilescu 2006a, 2006b).

14.2.4 Process-Specific Intervention by Environmental Biotechnology

There are major four points to be considered during interventions through environmental biotechnology, which are (1) detection through biosensors and biomonitoring technology; (2) preventions to be applied in the manufacturing process through single-process steps to generate fewer pollutants, substitute traditional processes, create products using modern biotechnology, and use genetic engineering in various food, pharmaceutical, diagnostics and textiles industries; (3) control; and (4) remediation of emissions of pollutants into the environment by decontamination, degradation of harmful substances through treatment of water/wastewater, pollutant oil and management of solid waste (Olguin 1999; Gavrilescu 2005; Chen *et al.* 2005; Das 2005; Gavrilescu and Nicu 2005).

Environmental biotechnology can contribute to the reduction of pollution significantly in several other areas, such as production of biomolecules, namely, proteins, fats, carbohydrates, lipids, vitamins, amino acids, yield improvement in original plant products and others. Using clean technologies and green chemistry, the application of zero-discharge initiatives in production processes can assist in the reduction of waste and minimization of pollution. The minimization of pollutants process is synergized using other biotechnological issues involving reusing or recycling waste streams, generating energy sources or producing new, viable products (Evans and Furlong 2003; Gavrilescu and Chisti 2005; Gavrilescu *et al.* 2008a, 2008b).

After consideration of all the mentioned phenomena, biotechnology and related techniques may be observed as a driving force for cohesive environmental protection by environmental biomonitoring, bioremediation, waste minimization and overall bio-maintenance.

14.3 Issues and Scope in Environmental Biotechnology

14.3.1 Environmental Biotechnology in Issues Related to Pollution Detection and Monitoring

Environmental monitoring is one of the most important policy issues to measure, detect and monitor environmental quality based on a set of selected parameters at a regular interval. Measuring, detecting and monitoring set parameters are mainly based on two methods: physicochemical and biological (Jamil 2001; Lam and Gray 2003; Hagger *et al.* 2006; Hart and Martínez 2006; Conti 2007). In earlier times, physical and chemical (physico-chemical) methods are most prevalent, which need to be associated with using several pieces of analytical equipment and the specific technical expertise to conduct the analysis. The cost of such costly equipment, trained personnel, the complexity of the samples and a lack of information regarding hazard and toxicological element are limitations of these methods (Cannons and Harwood 2004; Gu *et al.* 2004). Due to these limitations with the physicochemical method and, simultaneously, with the advancement of research in environmental biotechnology, biological variables incorporated into policy issues relate to pollution detection and monitoring.

Environmental monitoring has enormous importance in respect of environmental protection. In the natural ecosystems, the harmful effects of toxic chemicals demanded an early detection of toxicants even at very minimum levels of concentrations (Durrieu *et al.* 2006).

Classically, monitoring of environmental contaminants or pollutants includes measuring various toxicants/chemicals in water, soil, sediment and air, frequently at a particular time interval, namely, monthly, quarterly or annually.

Information technology (IT) has been integrated with environmental biotechnology for expanding the dimensions with innovative ways to detect, monitor and control processes at molecular levels. It includes real-time information on data metrics and it's computational analytics and inferences for the detection, monitoring and control in complex environmental systems (Hashim and Uijang 2004).

14.3.2 Indicators to Reveal Issues in Environmental Biotechnology Research

Monitoring different levels of the environment is an important factor to control pollution. Research and development in environmental biotechnology has been able to deliver several bioindicators and biomarkers to identify and monitor pollutants in different spheres of environmental pollution.

14.3.3 Bioindicators and Biomarkers for Environmental Monitoring

In the present era, environmental monitoring programs have been included as determinations of contaminant levels in biota, as well as the assessment of several parameters and responses related to biological/ecological systems. These are in addition to chemical and physical measurements. The temporal and spatial changes in the selected parameters and biological systems are used to reflect changes in environmental quality/conditions through biomonitoring (Markert *et al.* 2003; Conti 2007; Lam 2009). In this context, some organisms or biocommunities may react to an environmental effect, and as a result, there are some measurable changes observed in their chemical composition and/or biological function. The bioindicators/biomarkers possibly refer to such significant environmental changes through their active process pathways (NRC 1987; Jamil 2001; Markert *et al.* 2003; Conti 2007). In biomonitoring programs, biomarkers are thus used to render several biological information and various effects of pollutants on living organisms. Out of those, three main types of indications can be obtained from those bio-indicators, that is, on exposure, effect and susceptibility.

In recent developments of environmental biotechnology, biomarkers that have a potential application in biomonitoring are of different types, namely, biochemical (enzymatic, specific proteins or indicator compounds), histo-cytopathological (cytological, histopathological), molecular (gene expression, DNA integrity), physiological and behavioral and so on.

It has been noted that the field application of biomarkers has various constraints because they are living in nature. The constraints include the availability of living material, which can limit data acquisition and use of multivariate methods through statistical analysis. Apart from these constraints, they need to have other requisite features, such as being sensitive so as to act as an early warning, being very specific as a single compound or a class of compounds, being dependable, being robust, having extensive applicability, being user-friendly, being excellent in controlling quality, being easy to learn by personnel and ultimately delivering the necessary data and information for arriving inference (Beliaeff and Burgeot 2002; Lam 2009).

14.3.4 Biosensors Indicators for Environmental Monitoring

Research on biosensing technology and the development of related devices for environmental monitoring is one of the latest issues in environmental biotechnology. In this biosensing technology, sensors are basically live cells developed through modern biotechnology or genetic engineering technology. The biological sensing element is referred to and whole-cell (genetically engineered [GE] or natural microorganisms) or antibody or antigen (immunosensors) or other metal binding proteins or enzymes or DNA or RNA and the like, and the transducer may be electrochemical, optical or acoustic (Biotech, 2000). A number of examples, such as GM yeast, whole-cell chlorophyll fluorescence or phosphatase and esterase enzyme inhibition-based biosensors, are used for the real-time detection and online monitoring of environment pollution (Tucker and Fields 2001; Evans and Furlong 2003).

Several plants are also used as biosensors, such as white clover (*Trifolium repens*) clones, are used as descriptors of biomass reduction in crop species, the leaves of *Brassica oleracea* variety acephala are used as a biosampler and common species of trees like wild olive, holm oak, white poplar are also used as biosensors (Bargagli 1998; Mertens *et al.* 2005; Madejon *et al.* 2006; Nali *et al.* 2006; Zelano *et al.* 2006).

Target and nontarget insects under invertebrate species, such as crustaceans, are also used for biomonitoring (**Lagadic *et al.* 1994**; Raeymaekers 2006).

The biosensing element comes into close contact with the physical transducer and converts the biosignals to a signal of another form, such as optical, mass or electrochemical. This is possible as the biocatalyst converts the substrate into a product. The conversion of signals into a form of measurable electrical signal is analyzed as per input software, like a dye that changes color, and is reflected in the form of visible parameter to relate environmental pollution. The biosensors are designed innovatively to exploit maximum biological specificity to measure pollution level. The process is designed within an analytical device or embedded product (Reiss and Hartmeier 1999; Rodríguez-Mozaz *et al.* 2004).

It has been noted that the combination of biological sensing element and an electronic signal-transducing element in biosensors reflects high sensitivity, high selectivity, short-response time, portability and low cost. It is perfect for deployment at any place for monitoring environmental pollution (Lam and Gray 2003; Rodríguez-Mozaz *et al.* 2006). Biosensors are more reliable, since a synergistic combination of biotechnology and microelectronics has been noticed in biosensing technology (Verma and Singh 2005). They enable recurrent analysis with the same recognition element and applicable to a wide spectrum of environmental pollutants, including monitoring of ecological toxicity.

The biosensors are developed as "microchips", which are very selective and sensitive sensors, by combining biological activity with nanowire electronics (Cui *et al.* 2001), which is very efficient in detecting an electric current equivalent to the binding of a single molecule (Evans and Furlong 2003). The biosensors are developed as "microchip" which are very selective and sensitive sensors by combining biological activity with nanowire electronics (Cui *et al.* 2001), which is very efficient to detect an electric current equivalent to the binding of a single molecule (Evans and Furlong 2003).

Biosensors can be applied for several types of environmental monitoring, such as toxicity screening of samples using bioluminescence or fluorescence (Rabbow *et al.* 2002; Weitz *et al.* 2002; Gu *et al.* 2004; Rodríguez-Mozaz *et al.* 2004), biomonitoring atmospheric quality (Nali *et al.* 2006; Zelano *et al.* 2006), monitoring water quality (Ramsden 1999; Ashbolt *et al.* 2001; Cannons and Harwood 2004; Starodub *et al.* 2005; Mauriz *et al.* 2006; Mwinyihija *et al.* 2006) and biomonitoring soil contamination (Doran and Parkin 1994; Tom-Petersen *et al.* 2003; Gu *et al.* 2004; Ahn *et al.* 2005; Tarazona *et al.* 2005). Biosensors can be applied for several types of environmental monitoring like toxicity screening of samples using bioluminescence or fluorescence (Rabbow *et al.* 2002; Weitz *et al.* 2002; Gu *et al.* 2004; Rodríguez-Mozaz *et al.* 2004), biomonitoring of atmospheric quality (Nali *et al.* 2006; Zelano *et al.* 2006), monitoring of water quality (Ramsden 1999; Ashbolt *et al.* 2001; Cannons and Harwood 2004; Starodub *et al.* 2005; Mauriz *et al.* 2006; Mwinyihija *et al.* 2006) and biomonitoring of soil-contamination (Doran and Parkin 1994; Tom-Petersen *et al.* 2003; Gu *et al.* 2004; Ahn *et al.* 2005; Tarazona *et al.* 2005).

14.4 Opportunities and Innovations in Environmental Biotechnology

14.4.1 Innovation and Development of Product and Process Innovation in Environmental Biotechnology for Environmental Protection

The innovative modern biotechnological products and processes are applied in the industry and environment to improve efficiency and cost–benefit ratio and diminish the impact on environmental pollution. Innovation in the process development through several biological processes and alteration or replacement of existing processes are predominantly noticed. The introduction of biotechnological steps, such as cellular, plant, microbial or enzyme action, is progressively being used in industrial operations for primary pollution prevention (Olguin 1999; Gavrilescu 2004b; Gavrilescu and Nicu 2005). Equally, new biofuels and biomaterials having little environmental impact is intensifying fast. In microbial metabolism, three processes, like biodegradation, biotransformation and biocatalysis, are mainly prevalent independently or in combination. Biotransformation involves alterations of organic molecules into products of defined structure, in the presence of microbe, plant or animal cells or enzymes. It also includes reactions under mild and measured conditions biosynthesized of new products. The Fraunhofer Institute for Systems and Innovation Research in Karlsruhe surveyed on behalf of the Ministry of the Environment in Stuttgart. In the survey report, it revealed the enormous potential of product-integrated environmental biotechnology, which was claimed to have 70% reduced environmental pollution, 64% reduced process costs and 22% improved product quality.

In the production process of several industrial products, namely, chemicals, textiles, food, detergents, pulp and paper, plastics, mining, energy and pharmaceuticals, biotechnology aids in saving energy and raw materials, minimizing undesired waste through GM enzymes or other modified processes. The biotechnological processes are predominantly implemented under mild conditions, using biodegradable raw materials, intermediates and water as solvent. Due to high enzymatic specificity, the deliverable output is seen as increased yields and decreased by-products, thus saving additional costs for further purification but simultaneously reducing the environmental load.

It is revealed that environmental biotechnology is a strong and powerful tool toward pollution prevention and cleaner production for different sectors, such as industry, transport, agriculture and others. It also renders sustainability in the development of innovative products and processes with measurable gain in cost–benefit ratio, including total economic efficiency.

14.4.2 Environmental Biotechnology for Eco-Efficiency and Environmental Protection

Eco-efficiency is gradually becoming the most important topic in the economic world through maximizing efficiency and minimizing the impact on the environment. It is accomplished through increasing the value of the product or service, optimization of resources, and reducing environmental impact (Gabriel and Braune 2005; Gavrilescu and Chisti 2005; Bidoki *et al.* 2006). Biotechnology, in general, and environmental biotechnology, in particular, have extravagant potential of becoming eco-efficiency as well as economically profitable simultaneously because it offers a number of practical benefits (Wall-Markowski *et al.* 2005; Saling 2005). Environmental encounters progressively improve global competitiveness for gaining value in the marketplace along with minimizing costs for environmental cleanup and pollution control. It has been noted that eco-efficiency inspires productivity and innovation, improves competitiveness and environmental performance creating more value with less impact (Bidoki *et al.* 2006). Environmental biotechnology reflects several hands-on benefits in respect of eco-efficiency and business sustainability for several applications, such as industry, agriculture and others. For example, minimizing pesticide use for sustainable farming is a practical consideration for futuristic economic and eco-efficient agriculture, and at the same time, it reduces toxic dispersion. Effective eco-designing improves eco-efficiency, which is practically implied for preventing pollution. Eco-designing refers to the improvement of several process steps like reduction in the raw material and energy flow intensity, enhanced recyclability, maximum use of renewable resources for making sustainability in production and consumption and others (Olguin 1999; WBCSD 2000; Gavrilescu 2004b; Gavrilescu and Nicu 2005).

A study was executed by Organization for Economic Co-operation and Development (OECD) in which it was recommended that large companies are well aware of the importance and potential of environmental biotechnology vis-à-vis sustainable development. Environmental biotechnology can improve the environmental friendliness of industrial activities along with lowering both capital and recurring expenditures, which a exhibits sound basis for economy and society (OECD 2001).

14.4.3 Role of Environmental Biotechnology and Issues Related to Cohesive Environmental Protection

Environmental biotechnology is observed as influential for preventing pollution and making a cleaner environment. The methods are dependent on a number of practices in industry, agriculture, transport and the market, such as utilizing more efficient raw materials and using green chemistry to replace typical chemical processes in industrial production, replace hazardous materials with less harmful substances in agriculture and transport, eradicate toxic substances from production process and markets and so on.

It has been noted that raw material consumption in industrial processes results in excessive quantities of waste in society through inefficient production processes. However, unsustainable consumption patterns combined with low sustainability of goods make waste in society from the consumer's side (Cheremisinoff 2003; Gavrilescu 2004a; Gavrilescu 2004b; Gavrilescu and Nicu 2005). Gavrilescu and Nicu 2005). The waste discharged in society can be controlled through the implementation of modern environmental protection procedures. The procedures are predominantly with the preclusion of harmful substances in using pre-, during- and postindustrial production processes. The scientists have characterized an ideal industrial process, which is environmentally acceptable, simple and safe; with maximum yield and minimum or zero waste (Doble *et al.* 2004).

14.5 Policy Framework

14.5.1 Environmental Biotechnology and Policy Framework for Environmental Protection

Emissions of environmental pollution are one of the most important policy issues to be considered under a pollution control legislation framework. Specific projects are to be implemented to control environmental pollution from different sources such as industry, agriculture, transport and others. The development and application of green industrial processes, control at resources, following waste disposal norms, special projects for zero discharge or minimum discharge as per permissible limit under the legislations of state or central pollution control boards are important considerations to adhere to effectively.

Environmental biotechnology acts in two ways: either eradicating hazardous pollutants at their source before they are released in the environment or using robust biological processes to make less polluting, industrially useful products. Such products are intended to be more efficient and environmentally friendly. The products also have properties for eliminating solid, liquid or gaseous waste by-products, air emissions, energy consumption and toxic chemicals in several industries (Bull 1995; Olguin 1999; Gavrilescu and Chisti 2005). The primary focus of environmental biotechnology is to develop innovative technologies to treat aqueous, solid and gaseous wastes. But in respect of industrial implementation, the technologies or processes should be more economically viable so as to accept for gaining the industrial performance and cost–benefit ratio. The pivotal point is the application of those processes as Best Available Technology Not Entailing Excessive Costs (BATNEEC) to adhere the framework of strict and transparent environmental legislation (Grommen and Verstraete 2002).

The application of biotechnology in conventional manufacturing process makes an environmentally friendly alternative for preventing pollution through reduction at the source, waste minimization, recycling and reuse. The processes are necessarily considered in lowering production costs, creating less pollution and resource conservation for acceptance by the industries. Therefore, the application of environmental biotechnology is referred to the pollution control through sustainable or gainful industrial development. The main focused areas in which biotechnology can contribute for environmental pollution fall into three broad categories, such as process changes, biological control and bio-substitutions (Evans and Furlong 2003).

The biotechnological processes in industrial production are considered cheaper than traditional methods in addition to environmental protection. The processes are industrially acceptable because they support companies saving on production cost and continuously improving their public image with respect of pollution control (Olguin 1999; Gavrilescu and Nicu 2005; Evans and Furlong 2003; Willke *et al.* 2006).

The application of biotechnology is greatly instrumental in permitting procedural changes so as to meet pollution control norms. It refers to substituting multistep chemical processes with a one-step biological process using GM organisms (GMOs) for preventing environment pollution (Reis *et al.* 2006). The industrial processes are greatly gained profit through cutting down the production costs for land disposal of hazardous solid and liquid waste, by-products, wastewater and hazardous air emissions.

14.5.2 Environmental Policies

Environmental policies may be either enacted as laws by governing bodies or created and enforced by government agencies. The policies are instigated by several layers of governments, namely, national, regional or local. The policies address an array of problems considering air, water, soil, excess withdrawal of fossil fuel, energy conservation, habitat protection, afforestation, restoration of environmental damage, judicial use of pesticides, waste disposal, proper storage and zero waste of hazardous materials, reusing, recycling at different process innovation and trafficking in endangered species, among others. The policies are interdisciplinary in nature covering several aspects like technology, economics and natural and social sciences. In order to develop sustainable policies, therefore, it is necessary to have sound knowledge of the actual and potential environmental impacts of certain activities and some knowledge of the technical characteristics, economic costs, social acceptability and possible side effects of alternative policy options. Environment quality has tremendous impact directly or indirectly on the standard of living. It has been noticed that environmental degradation is not necessarily an unwarranted by-product of economic activities as per people's wish; it sometimes depends on the priorities and consequences set by states in their economic policies.

In India, the National Environmental Policy (NEP) was launched in 2006, and it was revamped on the existing policies (like the National Forest Policy, 1988; Policy Statement on Abatement of Pollution, 1992; National Conservation Strategy and Policy Statement on Environment and Development, 1992; the National Agriculture Policy, 2000; National Population Policy, 2000 and National Water Policy, 2002; etc.). The policy was formulated to stimulate partnerships among different stakeholders, that is, public agencies, private agencies, communities at local level, academician, scientific institutions, investment agencies and international developmental partners, in harnessing their respective resources and strengths for environmental management to protect environment and climate change.

The various environmental policies follow:

1. Economics and environmental policies
2. Industries and environmental policies
3. Agriculture and environmental policies
4. Ecosystem and environmental policies
5. Environmental policy instruments (EPIs)

14.5.2.1 Economics and Environmental Policies

High economic growth sometimes is seen at the costs of environment and health. It is one way to adopt the policy toward achieving a sustainable economy. It is reflected at consumer prices, that is, cost–benefit products, and such economic policy instruments support sustainable development. Environmental policies ought to complement economic instruments. Environmental policies involve certain measures aimed at achieving a sound environment. They are usually developed in the context of public policy, based on economic theory, which focuses more on the level of costs and benefits associated with the implementation of environmental policies than on the quality of the environment.

However, there are certain limitations. The immediate benefits resulting from environmental policies are extremely difficult to assess. Therefore, more importance and attention is given to the costs of environmental measures rather than to the benefits anticipated subsequent to policy implementation. Some detailed understanding of sectorial economic policies that influence the environmental policies directly or indirectly follow.

14.5.2.1.1 ***Agricultural*** *Sector*

The entire agricultural sector, including the food chain, appeals for several direct or indirect subsidies at the cost of consumers and taxpayers. Farmers get influential indications on those subsidies, that is, small grants for soil and water conservation and subsidies on seed, fertilizers, farm equipment and the like. They encourage farmers to occupy marginal land and to clear forests and woodlands, make excessive use of pesticides and fertilizers, and use underground and surface waters in irrigation indiscriminately. It influences environmental policies directly or indirectly.

14.5.2.1.2 ***Forestry*** *Sector*

The pressures on forests throughout the world vary greatly in both developed and developing countries. The pressure is reinforced by government policies. The policy and legal aspects of forestry industry attract a variety of direct and indirect subsidies. The subsidies are in the form of willful incentives that inspire excess harvesting of temperate, as well as tropical, forest commodities, including nonwood forest products. The harvested forest commodities appeal tremendously to the world trade of forest products. It affects environmental policies directly or indirectly.

14.5.2.1.3 ***Transport*** *Sector*

Economic policies benefit the transport sector, especially motor vehicles that are environmentally perverse. It is not possible to distinguish the fuel taxes due to differential environmental effect caused by different types of fuels (petrol or diesel, leaded or unleaded etc.). Direct and indirect subsidies include adopted tax and tariff structures inspire running heavier vehicles, energy-intensive vehicles and road freight. Moreover, there are some income tax exemptions for the expenses for private vehicles in some countries. It distracts from environmental policies directly or indirectly.

14.5.2.1.4 ***Energy*** *Sector*

The major obstacle to energy efficiency is the existing framework of incentives for energy exploration, development and consumption. The incentives endorse several energy components like oil, gas and coal disregarding the costs of land, water and air pollution and increasing unwarranted harmful waste. It is known that several industrialized countries have been spending billions to capture the market and consumer prices in several ways at the cost of promoting acid rain and global warming. However, very rarely, they have been spending only a few million on effective measures to encourage energy efficiency. There is an extreme necessity for creating a stringent national authority to control pollution problems related to energy before the problem is national. However, there are energy-related environmental problems that are becoming progressively international (bilateral, multilateral) or global. This complicates environmental policies considerably. On one hand, international cooperation in the fighting of environmental problems is absolutely necessary. Alternatively, different economic interests are entrusted by different countries. In addition, polluting sectors from energy among countries are not distributed evenly.

Economic-based environment policies have been designed in such a way that they allow simultaneous commercial business, ensuring the sustainability of the environment. They are also designed to facilitate major economic growth as well as achieve better economic efficiency.

14.5.2.2 Industries and Environmental Policies

A country's economic growth is dependent on industry. Therefore, there is a tendency of protecting industries responsible for the extreme growth of the national economy. But industrial waste is responsible for environmental pollution. Environmental policies are adopted for industries for requisite environmental management that are economically feasible and that help restrain environmental degradation.

Environmental management is driven by various factors, such as the following:

- Adoption of stringent environmental regulations
- Pressure from stakeholders, insurers and investors
- Customers' pressure on supply-chain management
- Weak relations between local communities and regulatory bodies

Several industries have recognized environmental management systems (EMSs) to overcome industry-generated threats to ecosystems. The EMS-comprising processes regarding the efficient use of raw materials, the implementation of green and clean industrial processes and the mechanism of waste disposal in the form of minimum or zero waste. It helps accomplish efficient environmental performance by identifying key activities related to environmental measures. It follows the adoption of the requisite management controls to meet its legal and policy requirements as per the rules. Conventionally, industrial sectors, especially the chemical, waste management and oil refinery sectors, having extreme possibilities of affecting the environment, are strictly adopting EMSs. In addition to this basic need, to control pollution, some companies have identified EMSs as an effective way of improving productivity through waste and resource minimization initiatives. There are some mechanisms for increasing sales as customer demand is also for environmentally safe products. Possibly, implementing effective EMSs can be a useful regulatory tool for business. As an example, the business houses achieve major improvements in their air emissions norms reducing the use of certain hazardous substances. As a result, companies are coming out of the strict compliance policy and norms imposed on them by the Environmental Protection Act (EPA), 1990. In addition, such industries will develop improved working environments for their employees and abolish the storage of unwarranted raw material and disposal of hazardous industrial effluents.

14.5.2.3 Agriculture and Environmental Policies

Agriculture has a huge influence and impact on the environment and biodiversity, especially on land, water and air. Applying unwarranted pesticides on agricultural land makes environmental pollution. Similarly, using nitrogen and pesticides in agricultural water has a high risk of soil erosion and depletion of water resources in many regions and countries. In the current scenario, there have been improvements in wildlife habitats, landscapes and greenhouse gases provided by agriculture and farming, but the most significant progress has happened where environmental pressures are extreme.

The environmental impacts due to agriculture are measured through the advantageous or disadvantageous effect of agricultural activities on the points that follow:

- Measuring soil quality (e.g., erosion, nutrient supply, moisture balance, salinity, etc.)
- Based on land quality (e.g., ecological management of agricultural land); water quality (e.g., nutrient, pesticide and sediment runoff and leaching, salinity)
- Based on water quality and quantity (e.g., use efficiency, irrigation consumption, flood prevention, water retention capacity, etc.)
- Measuring air quality (e.g., odor, emissions of dust, absorption of carbon dioxide, ammonia and other greenhouse gases, etc.)
- Affecting biodiversity (e.g., indigenous and farm animals and entire plant diversity)
- Affecting wildlife and semi-natural habitats (e.g., habitats of animal and plant)
- Measuring rural landscape (e.g., biodiversity features designed by farming, accompanying historic buildings and landmarks, etc.)

14.5.2.3.1 Role of Agro-Environmental Policies

To facilitate the enhancement and dissemination of appropriate farming practices, private and public initiatives are designed to meet anticipated environmental practices. However, when market forces are not adequate to meet the demand of environmental services, precise agriculture for taking care of environmental measures at the farm level may be desirable. The relevant measures target either reducing environmental pollution or enhancing environmental benefits for agricultural or farming activities. When formulating and implementing such policy measures, a number of universal principles should be taken into account, namely, incentives such as subsidies, exemptions or disincentives like payment of fines or taxes, among others.

PPP (polluter-pays principle) is recommended to be applied to reduce environmental harm for which both farmers or any other polluter should be accountable. It is also recommended that the guiding principles of PPP identify the opportunity of property rights and reference levels among countries. In this regard, financial incentives may be offered to encourage farmers to implement suitable production practices to improve environmental performance by means of reducing environmental pollution.

14.5.2.3.2 General Policy Principles

When formulating and adopting policy measures, the environmental problem is required to be clearly defined and the following principles are to be considered:

- While implementing an agriculture–environmental policy, the subsequent environmental benefits exceed the costs associated with the policy. The costs include reducing outputs and adopting more environmentally friendly technologies, products, processes and practices. It also includes policy implementation and enforcement costs.
- When farmers or other efficient economic agents are providing a specific environmental service, the level of environmental benefit should be clearly indicated.
- A wide range of policy decisions and tactics may be achieved through possible provisions by an individual farmer when a specific environmental consequence is mutually the effect of agricultural output. It may happen in several ways, that is, either offering incentives, such as subsidies and financial grant, or negative incentives, such as different types of taxes. Individual farmers will be interested in receiving the incentives and accordingly will try their level best to achieve the outcome in respect of agriculture vis-à-vis environmental pollution if the incentives are set appropriately.
- The tax or a penalty will be set in such a way that will confront farmers regarding not only payment amounts but also the degree to which the associated obligations can be enforced for not respecting environmental necessities. It is always expected to be tailored to the local environmental circumstances and demands and hence will vary locally or regionally.

14.5.2.4 Ecosystem and Environmental Policies

The objective related to conservation and management of ecosystem as specified in the Forest Policy is to certify stability of ecosystem through conservation of forest biodiversity, wildlife, water and soil. It is noted that forest biodiversity is gradually decreasing due to unwarranted exploitation of forest and nontimber forest products, illegal logging and poaching, forest encroachment, shifting cultivation, wildfires and others. Wildlife populations are also declining. Exploitation continues because there is a lack of timely organized forest management, integration of wildlife management in forest management plans. A lack of proper synchronization among the government and nongovernment wildlife and forest management institutions has also been noticed. The lack of coordination at various aspects of forest and wildlife management has been noticed due to inadequate infrastructure and staff to manage wide forest areas and prevent excessive resource exportation. There has been rise in the population burden and ineffective forest management or protection initiatives in watershed areas. Due to these reasons, the problems of biodiversity conservation, watershed management and soil conservation have been aggravated. A decline of watershed areas is resulting in water shortages due to poor management and protection. Planting inappropriate species in watershed areas, illegal logging and unsuitable logging processes have further reduced the quantity and quality of water. Due to these reasons, there is worsening, droughts and sedimentation in rivers. Erosion is seen on the riverbanks outside forest reserves due to cultivation, which is also a key apprehension in watershed management.

In order to minimize the damage to the ecosystem due to human activities, the Forest Policy stipulates an environmental impact assessment (EIA) before implementing developmental projects. Rigorous collaboration and association with other sectors and stakeholders are needed to frame the guidelines and scope of an EIA. To accomplish environmental ecosystem, conservation and management in the areas of watershed management, soil conservation, wildlife biodiversity, there is a necessity to consider the following points:

- New reserve forest may be developed for improving biodiversity and conservation.
- To protect forests' biodiversity, conservation and management are required in management plans. Participation in a local eco-development committee (EDC) and a forest protection committee (FPC) and other stakeholders will enhance conservation efforts and forest management through joint management agreements.
- To expand biodiversity conservation and management, research and development on biodiversity and appropriate dissemination of research output should be strengthened.
- In the management plan of natural forests and plantations, biodiversity conservation and necessary managemental guidelines must be incorporated. This minimizes the replacement of natural forests by exotic plantations.
- It is also necessary to include watershed management and soil conservation in the management plan of forest production and protection.
- Requisite research and development and necessary dissemination must be strengthened in order to progress soil conservation and watershed management.
- The environmental impact assessment (EIA) is required to release publicly so that investments for the conversion of forest land to other land that could cause potential damage to the forest environment may be ascertained.

14.5.2.5 EPIs

EPIs interfere and take official actions to curtail and eliminate the harmful environmental influences instigated by society. Sustainable development indicators under EPIs are in place to continuously measure the damage to the environment. EPIs are generally divided into three categories:

1. Regulatory policy instruments that command specific behaviors
2. Market-based instruments that are referred to as incentives for particular activities
3. Information-based instruments that pursue the provision of information

There is a specific strategy to make cleaner and greener production, which is controlled by any of the preceding instruments or in combination as per the following forms.

14.5.2.5.1 ***Product*** *Bans*

Any product susceptible for environmental production may be banned or phased out gradually to promote cleaner production. This implementation technique includes the application of product choice from initial stage or the execution of a substitution principle.

14.5.2.5.2 ***Environmental and*** *Other Audits* ***for*** *Cleaner Production*

To monitor permitting production requirements and admissible processes, it is mandatory to do environmental and other audits as per guidelines. It will be helpful for production industries to carry out cleaner production protocols and implement findings as long as they do not harm the environment.

14.5.2.5.3 ***Financial and*** *Technical Incentives*

Governments may inspire industries to follow cleaner production measures providing specific technical assistance, offering subsidies, exemptions, grants, loans and favorable tax regimes, among others.

14.5.3 Regulatory Instruments

The regulatory instruments are referred to as standards sets by government authorities, which are inspected, monitored and enforced for compliance. There are provisions of punishment with formal legal action for noncompliance. The regulations target environmental benefits, such as having a drop of carbon dioxide emissions by a particular period or specified date. There are provisions of following some mandate toward using a particular technology or process. The regulator will have maximum authority to control over the allocation of resources to accomplish environmental objectives. Regulations regarding quality and quantity of pollution level will be easily understandable by the stakeholders. There are some specified and negotiable compliance in respect of regulatory framework, which are appended below.

The negotiated compliance approach more or less a cooperative-type approach between the government (regulators) and the industry (regulated) in setting and enforcing standards. The joint responsibility between the regulator and regulated agencies allows more open discussions and inspires complying with the standards with greater flexibility. In addition, many countries have initiated regulations targeting a specific attainment, for example, recycling, reuse, zero waste disposal and the like. However, the detailed process flow, methodologies of such targets are left in the hands of industries through nonprescriptive regulations. The end parameters are required to meet with the set standards, which may be inspected by the regulators. It is noted that the followed process improves the economic efficiency of the industry through developing process innovation and making their large quantity of waste wealth.

14.5.3.1 Market-Based Instruments

Markets often fail to produce an environmentally efficient goods, since it is very hard to make a profit from the market after incurring the requisite costs for a particular level of environmental quality. The environmental externalities are related with cost–benefit impact of an industry, as well as cost competitiveness of a product to get desired market volume and profitable price. Market-based instruments address the market failure of environmental externalities either by integrating the external cost of an industry in respect of polluting activities through taxation or by facilitating the establishment of a proxy market, like offering tradable pollution permits. Market based instruments includes some direct or indirect economic-based instruments, such as incentives and subsidies, to make local industries more competitive. On one hand, taxes, fees and charges may be applied to promote cleaner production practices through increasing the costs of undesirable outputs or by providing incentives to promote a more efficient use of natural resources. As an example, energy and water resources, if overused, create both pollution and shortages, so adopting market-based instruments for efficient utilization and controlling environmental pollution.

Sometimes market-based instruments, especially the adoption of high taxes to accomplish desired environmental goals, are not politically feasible. It is also noticed that sometimes, governments face resistance if taxation related to the environment is taken merely as a means for increasing revenues and controlling environmental pollution.

14.5.3.2 Information-Based Strategies

The practices of cleaner production may be disseminated appropriately through information-based strategies. The information includes the right incentive through public disclosure of information and relevant strategies adopted by the firm to achieve environmental performance. To inspire industries to make capacity building inside through the publication and dissemination of relevant successful case studies. Some information-based strategies follow:

- Promoting targeted high-profile demonstration projects in which demonstration includes the deliberations of techniques and cost-saving prospects accompanying with cleaner production
- Encouraging educational institutions to include courses related to the prevention of environmental pollution and environmental management within their curriculum, especially for science, engineering and business courses

- Demanding public disclosure of evidence on environmental performance, namely, introducing a pollutant release and transfer register, encouraging better voluntary corporate reporting and necessitating the provision of information on specific materials related to environmental pollution
- Supporting and promoting the implementation of effective training initiatives on cleaner industrial production and protection of environmental pollution
- Offering high-profile awards for industries that have efficiently executed cleaner and greener production and make permissible or zero environmental pollution

14.5.3.3 EMS

It is a comprehensive regulatory system to review and monitor all environmental aspects of an organization. It gives various degrees of inclusion, regulatory permissible limits and possible strategies of continual improvement. Implementing an EMS covers all aspects of environmental impacts starting with the initial raw materials to the ultimate disposal of the product, covering all intermediate processes and products, including waste management. A series of standards, namely, ISO 14000, are followed in an EMS, which have been established by the International Standards Organization (ISO), Geneva, Switzerland. The ISO 14000 series recommends the necessary elements for an EMS, of which small to large organizations' manufacturing plants be certified under the ISO 14001 for total environmental responsibility. Companies adore regulatory relief and relish continual improvement through compliance with an EMS following ISO 14000 standard **(Waters 1998)**.

14.6 Environmental Laws and Legislations

Environmental law denotes rules and regulations that control the misconduct of humans that affect the environment. Environmental law is not very precise and does not have clear-cut boundaries like other traditional legal subjects. Environment law and regulations are attentive to the rationalization and streamlining of existing measures rather than the formulation of substantial law. The existing measures, such as the public dissemination of environmental law, aid in the study and execution of regulations toward the efficient use of resources, the protection of the biodiversity and environment, mediation, conflict resolution and conciliation and development of stable, unambiguous agreement.

Legislation is always developed in response to a problem. Therefore, there may be some delay between the demand and the execution of a satisfactory law. Without effective legislation, most pitches of human activity are likely to drop into chaos and conflict. Law always takes a lead to inspire satisfactory performance and empower authorities to punish those who infringe legislation regarding biodiversity and environmental management. Legislation also empowers authorities to confiscate faulty equipment or even shut down a company. Depending on the environmental issue, environmental laws can be categorized as public and private laws.

14.6.1 Private and Public Law

Private law comprises the law of property, the law of contracts, the law of torts and the law of obligations. A private bill enacted into law is applicable to an individual or a corporation. Under private law fall several laws, such as labor law, commercial law, corporation law and competition law. The law that shows the relationships between governments and private individuals is referred to as common law. The relationships between the individuals without the intervention of the state or government is referred as law of contract and is governed by private law. Private law can be applied in environmental cases to someone who makes a nuisance.

Public law governs the relationships between individuals and the state. The different types of public law are administrative law, constitutional law and criminal law. This type of law is how public authorities make decisions in cases of infringement of laws. Public law regulates the relationship between the state and individuals and is governed by public bodies, such as a government environment agency. Sometimes, private law plays a negative role in environmental cases. This law protects individual interests, such as challenging uses of land rather than environmental protection.

14.6.2 Principles of International Law

International law governs the relations between states in which there is no direct connection with domestic law or individuals. It depends on voluntary agreements by governments and international bodies. The modern-day international environmental law was initiated during United Nations Conference on Human Environment that was held in Stockholm in 1972. The conference manifested the worldwide acknowledgment of the environment crisis that is a matter of international concern and in need of an integrated international response. After two decades, the second major milestone in the development of international environmental law was constituted during the 1992 UN Conference on Environment and Development (UNCED) held in Rio de Janeiro. The second conference was engrossed with the association between the environment and development, especially the need for sustainable development. The most important outcome of the UNCED was the establishment of the UN Commission on Sustainable Development in 1993. The commission comprises 53 member states that are elected for a period of 3 years. As per the reporting structure, it is stated that it reports to the UN Economic and Social Council. The commission is mandated to review and monitor the implementation of the UNCED's program. The development of international environmental law is encouraged by such international conferences. The basic principles of international environmental law are as stated in the following subsections.

14.6.2.1 Environmental Protection Principles

These comprise the sustainability principle, the precautionary principle and the polluter payment principle:

1. Sustainability principle: It reflects an awareness of the global ecological crisis vis-à-vis traditional nature conservation concerns. It refers to the maintenance of an equilibrium state in development

versus biodiversity and environment management with sustainability.
2. Precautionary principle: This principle is to prevent environmental degradation. The authentic implication of this principle is referred to the measurement taken under the 'Montreal Protocol on Substances that Deplete the Ozone Layer'.
3. Polluter payment principle: In this principle the expenses for anti-pollution measures as decided by the public authorities will be borne by the polluter.

14.6.2.2 Social Principles

These principles comprise the of human rights principle, the equity principle and the participation principle.

14.6.2.3 Other types of Principles

There are other types of principles, such as nondiscrimination, the principle of common differentiated responsibilities, the principle of intergenerational equity and others are noticed.

14.6.2.3.1 Principle of Nondiscrimination

It is equivalent to PPP. In this principle, polluters causing transboundary pollution would be treated the same as if they make similar pollution within their own country.

14.6.2.3.2 Principle of Common Differentiated Responsibilities

As per this principle, nations should split the costs of measures to protect the environment on the basis of different contributions to global environmental degradation. It is seen in the convention on climate change. This convention refers to the fact that developing countries have to comply relatively less stringent standards than do developed countries. In addition, they are entitled to get technological and financial assistance to meet their obligations under the treaty.

14.6.2.3.3 Principle of Intergenerational Equity

Per this principle, nations are obliged to consider the long-term effects of human actions affecting the environment.

14.6.3 Indian Environmental Laws

In general, Indian environmental laws follow the principles of international law. However, few changes are noticed in accordance with the circumstances and environmental issues of the country. In India, there is an urgent need to reform various sectors of law and torts that were ignored earlier. The habit of ignorance is mostly due to the high penalty against offensive litigation. In modern times, common law or the law of torts as advocated in India is the English law customized to Indian conditions and as modified by acts of the Indian legislature time to time. It is noted that the PPP is now being accepted through various judicial declarations in India. It is opined that combined laws with modern legislation may be adopted to combat pollution and degradation of resource effectively. It is important to adopt and implement environmental laws to protect the limited resources as well as its effective utilization in our country. The Ministry of Environment and Forest (MOEF) is engaged in policy formulation in India. The MOEF is also involved in planning, promoting, supporting and coordinating the environmental programs apart from policy issues. There are certain other enforcement parties that assist the MOEF in executing the responsibilities in India.

Industry has a pivotal role in the economic development of a country. However, it generates a substantial amount of pollution to exploit industrial production. Therefore, it is important to manage pollution contributed by industries vis-à-vis economic development of a country. In India, the Central Pollution Control Board at the national level and the State Pollution Control Board at the state level review and monitor industrial pollution. These are notified statutory authorities attached to MOEF. Different laws have evolved from modern legislation in India that illustrates the judicial activism in the field of environmental jurisprudence by the Supreme Court of India.

14.6.4 Objectives and Principles of Legislation

The objectives and principles of legislation follow:

1. To protect, preserve and improve the quality of the environment
2. To protect human health
3. To exploit natural resources in a practical and balanced way
4. To promote initiatives at the international level to deal with environmental problems at the regional level or worldwide.

The environmental legislation comprises rules relating to the promotion of biological diversity, protection of endangered species, conservation of the natural environment, environment management and the protection of forests. The legislation also contains rules related to the pursuit of environmentally friendly agriculture, industrialization, transport and the like. With respect to the human environment, it protects humans with respect to the consumer against contaminated food, defective or dangerous consumer products, economic harm and danger in travel. It also renders protection of historical and cultural environment. It formulates several standards, and enforcement is adopted if the standards are not followed. Environmental legislation is generally formulated based on the principles of quality and quantum of environmental damage. It also includes the concept of rectification from the source, and that cost should be paid by the polluter.

Apart from the preceding objectives, some legal principles formulated by the expert members that are to be followed by the countries follow:

- Conservation and effective use the natural environment, including resources for the advantage of both present and future generations
- Maintaining processes toward ecosystems, biodiversity and essential steps for the functioning of the

biosphere, natural ecosystem and to preserve biological diversity
- Follow the attitude of optimum sustainable yield for effective utilization with respect to natural resources and ecosystems

14.6.5 International Institutions in Implementing Environmental Law and Legislation

The major international institutions and agencies which were influential in formulating and bringing about environmental reforms through constitutions of laws and legislations are stated as follows:

14.6.5.1 The United Nations Environment Programme (UNEP)

It was set up during the Stockholm conference. The. UNEP is a program without executive powers. The budget is mainly dependent on voluntary contributions from states. Its main program is mainly better coordination between existing UN programs in the field of the environment.

14.6.5.2 The Food and Agricultural Organization (FAO)

The organization mainly takes care of deforestation and conservation of fisheries.

14.6.5.3 The World Meteorological Organization (WMO)

The organization has a vital role in the field of climate change and global warming.

14.6.5.4 The International Atomic Energy Agency (IAEA)

This agency was active toward environmental pollution since the Chernobyl disaster in Ukraine.

14.6.5.5 International Bank for Reconstruction and Development/The World Bank

The International Bank for Reconstruction and Development or the World Bank initiated to put their prime attention on the environmental side effects as well as environmental policies.

Apart from United Nations (UN) institutions and organizations, there are certain regional organizations that play active roles in the field of the environment. The regional organizations are as follow.

14.6.5.6 The Organization for Economic Co-operation and Development (OECD)

The organization is involved in implementing environmental policies. It helps countries to design policies followed by effective implementation to address environmental problems, management of natural resources and sustainably. It makes necessary bridges between society and economy and its effect on the environment.

14.6.5.7 The Conference on Security & Cooperation in Europe (CSCE)

The Conference on Security and Cooperation in Europe (CSCE) was unfolded in Helsinki during 1973 by 33 European countries and the United States. It was concerned on the issues related to environment, transportation sustainable development, geotechnical engineering etc.

14.6.5.8 The Organization of African Unity (OAU)

It was established as an intergovernmental organization in Africa during 1963 with an objective of encouraging political and economic growth among 32 signatory governments. It is also trafficking with the improvement of environmental protection, economic empowerment and societal development.

14.6.5.9 The Key International Treaties

The key international environmental treaties deal with air, water, hazardous waste, nuclear wastes and other topics. The treaties are as follow:

1. The 1973 Convention for the prevention of pollution from ships (MARPOL)
2. The 1979 ECE Convention on long-range transboundary air pollution protocols on sulfur dioxide (SO2), nitrogen oxides (NOX) and volatile organic compounds (VOCs)
3. The 1982 UN Convention on the Law of the Sea (UNCLOS)
4. The 1985 Vienna Convention for the protection of the ozone layer (Montreal Protocol and London and Copenhagen Amendments)
5. The 1986 IAEA Convention on early notification of a nuclear accident
6. The 1986 IAEA Convention on assistance in the case of a nuclear accident
7. The 1991 Bamako Convention on the ban of the import into Africa and the control of transboundary movement and management of hazardous wastes within Africa
8. The 1991 ECE Convention on environmental impact assessment in a transboundary context
9. The 1992 UN Convention on biological diversity
10. The 1992 Helsinki Convention on the protection and use of transboundary watercourses and international lakes. (This convention deals with international watercourses.)
11. The 1992 Paris Convention for the protection of the marine environment of the Northeast Atlantic
12. The 1992 UN Convention on climate change
13. 1994 IAEA Convention on nuclear safety

14.6.6 Environmental Policies and Programmes in India

In India, 1972 marks a turning point in environmental management. The then Planning Commission of India prepared three reports about India for the 24th UN General Assembly of 1972. The report emphasized the impact of the population explosion on the natural environment, including the existing state of environmental problems. Subsequently, the then Planning Commission constituted an expert committee to formulate long-term sectoral policies for the environment and forests. Environmental problems were making serious problems for nature, such as the loss of topsoil and vegetative cover, the degradation of forests, enduring pollution by toxic substances, careless industrial and agricultural practices and unplanned urban growth. Environmental degradation was seriously threatening the economic and social growth of the country. On the basis of recommendations of the Union government, the Government of India adopted a National Conservation Strategy (NCS) and Policy Statement on Environment and Development.

The NCS adopts the policy of conservation and sustainable development in harmony with the environmental perspective. The NCS refers to environmental problems of India and announces strategies for action in various domains such as environment, forestry, agriculture, industrial development, mining, transport and tourism, among others. In February 1992, the Union government published a policy for the reduction of pollution. It declared the objective of the government to be integrating environmental considerations into decision-making at all levels. It includes the adoption of fundamental guiding principles, that is, the adoption of the best available technology, the prevention of pollution at the source, public participation in decision-making and the PPP. Therefore, in India, over the years, policies and legislation toward environmental management meeting the need of forest conservation, sustainable development, environmental protection for improvement and management of the environment, such as the quality air, water and soil. In addition to the government, the nongovernment organizations (NGOs), are also instrumental for environment management vis-à-vis the sustainable development of a nation.

14.6.6.1 Forest Conservation Activities

India has a strong forest culture around the hilly regions of the Himalayas, the Northern and North-Eastern hills, the Central Indian plateaus and the Western and Eastern ghats. People around the forest villages are living symbiotically with forest flora and fauna. India's forest tribes, such as the Santhals, worship a number of trees and natural phenomena. The colonial state fundamentally redefined proprietary rights with the introduction of the comprehensive Indian Forest Act in 1878. The act imposed a comprehensive system of management and control on the forests, the priorities of which sharply opposed with earlier systems of local use and control. Subsequently, the Forest Act, 1878 eliminated traditional rights and customary practices governing forest management. It follows the Indian Forest Act of 1927, which offered total control on forest resources to the state. The processes under the acts were continued in independent India, which engaged the forest laws of the colonial government. The revenues were maximized from the sale of common forest resources in the name of national interest and economic growth. The maximum exploitation of the forests' resources, primarily for producing valuable timber for industry and other vital purposes of national interest, was highlighted in the Forest Policy of 1952. The policy also strongly emphasized that forest villages do not have rights over the property. A second policy with respect to commercial exploitation, the viability of forests and the production of wood for industrial purposes was illustrated by the National Commission on Agriculture, 1976. The two policies largely led to a reduction in the natural forest cover. It happened due to the strategy of charging very low or concessional rates toward raw materials for industries. There were some major negative effects noticed, like the nonregeneration of forests due to unsustainable commercial exploitation. The other detrimental effects were, namely, the rejection of peoples' legitimate rights to forest products and disparity in ecologically sensitive regions leading to loss of biodiversity, soil erosion, landslides, air pollution. It also makes ultimate loss, that is, letting down protection for forest flora and fauna.

14.6.6.2 NGO Movements for Environmental Protection in India

In their perseverance of the Industrial Revolution, urban areas often ignored the benefit of the environment. The voices of the local community were drowned out due to the rapid growth of industrialization and in the publicity of large organizations.

14.6.6.3 Silent Valley Struggle

In the Western Ghats, a unique tract of 9,000 hectares of forest is known as the Silent Valley. There was a nationwide movement led by Kerala Sastra Sahitya Parishad, students and a galaxy of intellectuals, academics, artists and naturalists against a hydroelectric project during the 1970s that would have destroyed part of the Silent Valley. None of the government departments/agencies was competent in dealing with such massive public opposition to developmental projects threatening the natural environment. Finally, the decisions were taken from the Prime Minister's Office to stop the construction of the hydroelectric dam project. In the Silent Valley, the discovery of nine plant species and a new genus by the Botanical Survey of India reinforced the activists' willpower to protect the forest.

14.6.6.4 Chipko Movement

Chipko stands for "embrace," "hug" or "to stick". The method of protecting trees from contractors' axes by hugging them was referred as the Chipko movement. It was initiated in the tiny hamlet of Gopeshwar in the Garhwal region during 1973. In 1970s, devastating flooding of the Alaknanda River occurred in Garhwal. The people of the area were directly affected due to disaster and with the ecological change that had taken place in the watershed due to the felling of forests. As a result, the movement called the "Chipko movement" was born in March 1973. The villagers were deprived of access to residue wood for their agricultural implements. The villagers took up a strategy of rushing to protect the trees from contractors' axes by hugging them when the trees were allotted to a sporting goods manufacturing company. The hugging of trees for their protection blew out like wildfire throughout the Garhwal region for the next few years. It was interpreted as a fight for people's rights on common property resources and quickly became a true ecological movement. Henceforth, the importance of forests in maintaining the fragile ecosystem was recognized nationally and internationally. The Garhwal villagers proclaimed that forests are the foundation of their cultural and material life, and therefore, they thought that it is their birthright.

The Chipko movement flourished in altering the official strategy on forest exploitation. The movement became more popular, and by 1977, it became a full-fledged ecological movement and was familiar to the highest level policymakers. As a consequence, in Garhwal, the largest afforestation program, namely, planting millions of trees to restore the barren hillsides was organized voluntarily by the Dashauli Gram Swarajyasangh.

14.6.6.5 Litigation for the Protection of Taj Mahal

The Taj Mahal, one of the seven wonders of the world, is situated in Agra of Uttar Pradesh, India. Mathura Refinery, which is one of the largest refineries, is located in the vicinity of the Taj mahal. A joint-parliamentary committee expressed its serious concerns about the location of the refinery within the Taj Trapezium, a 10,400-square-kilometer area. It has been noted that the Mathura refinery along with more than 2,300 small and medium-sized industries around the refinery, including iron foundries, have continuously been emitting pollutants. Pollutants, such as sulfur dioxide and other poisonous gases, might be changing the color of the Taj Mahal's white stones to yellowish. In this context, a public interest litigation was filed in the Supreme Court by an environmental advocate for the protection of the Taj from serious pollution. The Supreme Court issued a number of orders directing industries to install the necessary devices for controlling pollution or closure. The Mathura Refinery was also asked to minimize sulfur dioxide emissions, and the Union Ministry of Environment and Forests was instructed to take suitable measures to save the Taj Mahal.

14.6.7 Environmental Legislations in India

India has a long history of adopting several acts or codifications for protection of environment. Those are the Indian Forest Act, the Forest (Conservation) Act, the Indian Motor Vehicle Act, the Factories Act, the Indian Penal Code, the Criminal Procedure Code, the Bengal Smoke Nuisance Act of 1905, the Mines and Minerals (Regulation and Development) Act, the Industries (Development and Regulation) Act and the Merchant Shipping Act. The Indian Penal Code was passed in 1860. The Indian Penal Code penalizes persons with imprisonment or fines for causing defilement of water of a public spring or reservoir. Usually, the code has been regarded as a conventional effort of enforcement, although punishment and fines have been considered scant. In addition, most pollution occurs through fouling a "public spring" or "public river. Then the specific language related to the code makes a problem on the proof of the prosecution. In addition, polluters intentionally or knowingly discharge effluent, and for that making successful prosecution is also problematic in a court of law. Subsequently, the Factories Act also addresses public health and public safety issues. Other sections of the act delegate power to the state governments to enact their state rules. As a consequence, a number of states approved versions of the Factory Act, namely, in 1950 in Uttar Pradesh and Tamil Nadu, in 1958 in West Bengal, in 1963 in Maharashtra and in 1969 in Mysore.

14.6.7.1 Evolution of Environmental Legislation in India

There are several environmental legislations are enacted in India. The Indian Forest Act, 1927 combines the law concerning forests and shipment of forest produce, such as timber and nontimber; duties leviable on timber; and so on. To prevent unnecessary suffering and cruelty to animals, the Prevention of Cruelty to Animals Act was enacted in 1960. To promote Cruelty to Animals Act, the Animal Board of India was formed. Subsequently, the Indian Forest Service was created in 1966 by the Government of India under the All-India Services Act, 1951. The prime objective of the Indian Forest Service was to implement the country's National the Forest Policy, scientific management of forest flora and fauna with sustainable exploitation of timber and nontimber products. It follows the implementation of the Wildlife Act in 1972 for the effective protection of wildlife of the country. The other objectives of this Act were to control poaching, smuggling and unlawful trade in wildlife and its by-products. In January 2003, the act was amended with more stringent clauses, namely, the protection of flora and fauna along with ecologically important protected areas. In 1974 Water Act was enacted for the maintenance of water and the prevention and control of water pollution within the country. In 1977, the Water Cess Act was enacted for levy and collection of Cess on water consumption for industrial activities. In 1980, the Forest Conservation Act, which was last amended in 2003, was enacted to protect and conserve the country's forests. The Air Act for the prevention, control and abatement of air pollution in India was enacted in 1981 and amended in 1987. The very familiar EPA, 1986 came into existence in India, 14 years after the UN conference. The objective of the act is for environmental protection and improvement of the country. Amendments on EPA, 1986 were made in 1991.

In 1971, the Man and Biosphere (MAB) Programme of UNESCO was launched. After the formation of the Bioreserve Committee, India joined the MAB Programme in 1988. The main objective of MAB Programme is to develop a base of coherent uses of natural resources and their conservation and to maintain the relationship between humans and the environment.

In India, the coastline is seen at Mandapam (Tamil Nadu), Gangetic Sundarbans (West Bengal), Malvan (Maharashtra), Okha (Gujarat), Lakshadweep and the Andaman group of islands. The length of Indian coastline is more than 7860 km and includes enriched marine flora and fauna. It is our duty to safeguard the coastal natural resources and pass it on to our successors. In addition, India has developed 15 bioreserves, of which the first one is at Nokrek (Meghalaya) in 1988.

To increase environmental awareness among citizens, the Eco-Mark Scheme of India was introduced in 1991. This scheme inspires people to purchase eco-friendly products. In India, the Hazardous Waste Rules, 1989 outlined a list of hazardous chemicals. To afford an accident that occurs to the handling hazardous substances, the Public Liability Insurance Act was enacted in 1991. National Environment Tribunal (1995) has come up for strict liability toward damage or accident due to handling of hazardous waste. Subsequently, Biomedical Waste Rules, 1998 dealing with collection, reception, storage, treatment and disposal of the waste was implemented. The Noise Pollution Rules 2000 was adopted by the state government to categorize industrial, commercial, residential or silence zones. To follow the objectives of UN Convention on Biological Diversity (CBD), the Biodiversity Act 2002, stating that the country should use its own biological resources, was enacted in 1992. The Scheduled Tribes and Other Traditional Forest Dwellers Act, 2006, recognizes the rights and framework of traditional forest dwellers over the forest areas inhabited by them. Ministry of Environment and Forests (MoEF) created the National Environment Appellate Authority (NEAA) to deal with the need of environment clearances in certain restricted areas.

14.6.7.2 Provision of Environmental Legislations in Indian Constitution

India made amendments to its constitution toward allowing states for environmental protection and improvement for safeguarding forests flora and fauna and public health. The government imposes restrictions on potentially harmful entities, such as industries that have polluted the environment. The constitutional article shall not be enforceable by any court, but it shall be the duty of the state to apply the principles of the articles in making laws. The constitution provides rational distribution of legislative powers between the Union and the states to control internal environmental matters. Three jurisdictional listings, namely, Union, state and concurrent, have been developed. In the case of water, at least two or more State legislatures pass resolutions empowering Parliament to pass water-related legislation. It was historically led to cumbersome delays approaching the state to take lead on water-related legislation within its jurisdiction. It was noted that it was almost a decade after the first draft bill was being debated before Parliament was able to pass India's first major water legislation in 1974. The state legislatures voting process to authorize the parliament to pass water-related measures repeated itself during following amendments to the water law in 1978 and 1988.

14.7 Special Initiatives Taken by the Government of India on Environmental Protection

The Union government took several initiatives. The Union Ministry of Health appointed a study committee and began addressing water pollution issues in 1962. The committee recommended for both central- and state-level action; however, questions regarding jurisdiction of the states and the Union government remained unresolved. By 1965, a draft bill was disseminated that expressed that states should pass resolutions authorizing Parliament to enact legislation on their behalf. Accordingly, a bill, that is, the Prevention of Water Pollution, was announced in 1969. Finally, a modified version, that is, the Water (Prevention and Control of Pollution) Act, was passed in 1974. The Pollution Control Boards at the central government and state government levels were established by the Water Act. Later on, then prime minister Indira Gandhi moved to enact a series of environmental measures. In 1980, the Department of Environment (DOE) was created with a mandate of advisory, review and monitoring role for the Union government. The DOE is responsible for environmental appraisals of development projects in detail, monitoring air and water quality, establishing environmental information system, promoting research and development and coordination among Union, state and local governments.

14.7.1 Establishment of Ministry of Environment & Forests

Understanding the need for environmental management and issues, Ministry of Environment and Forests (MOEF) was created in 1985 by the then Indian prime minister. Later, it was changed to the Ministry of Environment and Forests and climate Change (MOEF&CC). In line with the DOE, the MOEF continued with the same functions, such as conducting environmental assessments, management, surveys, reviews, monitoring and necessary enforcement. The MOEF also started doing promotional work related to environmental management. Some umbrella legislation was designed to act on the environment. The legislations are the Air (Prevention and Control of Pollution) Act, which was passed in 1981, and the Environment (Protection) Act, which was passed in the parliament in 1986. The responsibility to MOEF was delegated to manage the new legislations, including Union and State Pollution Control Boards. After taking the responsibility, the MOEF imposed significant increases in its water cess from early 1992 and decided to raise it again since the charges were still lower than the current effluent treatment costs. The MOEF has come up with a new idea of shifting standards from concentration- to load-based. The MOEF targeted small-scale industries that greatly add to the pollution load. It is a very hard task to control small-scale industries, but the ministry provides technical support and small grants to promote central effluent treatment plants as well as regulate them. The ministry created industrial zones to encourage clusters of similar industries to use similar utilities. The cluster concept was introduced to reduce the cost of similar utilities and necessary environmental services. Subsequently, in January 1994, the Union government introduced new restrictions and prohibitions on the extension and/or upgrading of any existing activity or new project unless an environmental clearance is granted.

The Climate Change Division of the MOEF&CC, Government of India, looks after the issues related to climate change and the related domestic policies and actions. The division is also responsible for submitting National Communications (NATCOMs) and the Biennial Update Reports (BURs) to the United Nations Framework Convention on Climate Change (UNFCCC) since India is a party to the

UNFCCC. In order to create and strengthen the scientific and analytical capacity for assessment of climate change in the country, different studies have been initiated under the Climate Change Action Programme, including the National Carbonaceous Aerosols Programme, the Long Term Ecological Observatories Programme and the GHG (Green House Gas) Modelling Studies (MoEF&CC, 2018–2019).

In fact, before introducing new projects, the industry needs to submit an EIA and an EMP for necessary approval. The jurisdiction of the MOEF or the state government in verifying the EIA is designated based on the industrial sector and the size of the project. However, the MOEF usually clears the EIA of those in some highly polluting sectors, such as petrochemicals, power plants and cement, among others. Moreover, state governments may lay down more strict limitations for heavily polluted areas, such as the Tarapur region of Maharashtra, Vapi region in Gujarat and others. Gujarat state has already taken some stringent restrictions and/or bans on new activities for the country's 17 most highly polluting sectors in the Vapi region.

The major Government of India programs/activities related to waste management and Chemicals Safety Rules are as follow (MoEF&CC 2019–20):

1. Solid Waste Management Rules, 2016 includes source segregation, door-to-door collection of waste and channelization of waste for useful purposes such as recovery, reuse and recycle. It has been extended to municipal and urban areas.
2. E-Waste (Management) Rules, 2016 has been extended to manufacturers, dealers, prefabricators, e-retailers and others to address leakage of e-waste to informal sectors at any stage of the chain.
3. Hazardous Waste Management, 2016 has been notified for prevention, minimization, reuse, recycling, recovery, co-processing and safe disposal.
4. Bio-Medical Waste Management Rules, 2016 have been notified with the objectives of improving segregation, collection, processing, treatment and disposal of infectious biomedical waste in an environmentally sound and safe manner. The rules may be applied to any institution, such as hospitals, nursing homes, clinics, dispensaries, veterinary institutions, pathological laboratories, blood banks, health camps and research labs, among others. It is also applicable to all persons who generate, assemble, accept, stock, carriage, treat, dispose or handle biomedical waste.
5. Plastic Waste Management Rules, 2016 have been notified to apply for the manufacturer, import stocking, distribution, sale and use of carry bags, plastic sheets or multilayered packaging, among others.
6. Construction & Demolition (C&D) Waste Management Rules, 2016 has notified for application to building materials, debris and wreckage waste resulting from construction, revamping, restoration and destruction of any civil structures or to an individual who produces such waste.
7. Fly Ash Notifications has been issued by Ministry of Environment, Forest and Climate Change which has subsequently been amended in 2003, 2009 and 2016. Fly ash notification mandates the use of fly ash for the purpose of manufacturing ash-based products such as cement, concrete blocks, bricks, panels or any other material or for construction of roads, embankments, dams, mines backfilling/stowing, agriculture and so on.
8. Chemical Safety: The Ministry of Environment, Forest and Climate Change has notified the Manufacture, Storage and Import of Hazardous Chemicals (MSIHC) 1996 and Ministry of Environment, Forest and Climate Change Rules, 1989 and the Chemical Accidents (Emergency Planning, Preparedness and Response) (CAEPPR) Rules, 1996 for ensuring chemical safety in the Country.
9. International Conventions: MoEF&CC is the nodal ministry of Government of India to participate and take National role in International Conventions. For example, the Basel Convention had the theme "Control of Transboundary Movement of Hazardous Waste and Their Disposal"; the Rotterdam Convention, "Prior Informed Consent Procedure for Certain Chemicals and Pesticides in International Trade"; the Stockholm Convention, "Persistent Organic Pollutants, the Minamata Convention on Mercury and Strategic Approach to International Chemicals Management".
10. The MoEF&CC has implemented a scheme, namely, "Creation of Management Structure for Hazardous Substances". The sub-schemes under this scheme are as follow (MoEF&CC, 2019–20).
    - I. SAMPATTI—'Sustainable Management of Pre-owned Asset through Trade Initiatives', which is basically innovative solutions for municipal solid waste segregation, handling and treatment.
    - II. Capacity building of government agencies/organizations/department/ civil society/institute with respect to environmentally sound management of chemicals and wastes.
    - III. Organizing awareness program with various stakeholders for implementing various waste and chemicals management rules.
    - IV. Innovative technologies for environmentally sound management of chemicals and wastes.
    - V. Setting up facilities for managing biomedical waste (CBMWTF) and treatment, storage and disposal of hazardous waste (TSDF).

In addition to that, the government is also implementing the scheme 'National Adaptation Fund for Climate Change' to support adaptation initiatives at vulnerable areas having adverse impacts of climate change under states/Union territories (MoAFW 2019). The MOEF&CC has been implementing several projects, such as the CDM and the Green Climate, Fund with the assistance of bilateral and multilateral funding agencies, such as Deutsche Gesellschaftfür Internationale Zusammenarbeit GmbH and the United Nations Development Programme **(MoAFW, 2020)**.

14.7.2 Radioactive Waste Management

Radiation and radioactive substances are one of the most hazardous elements for environmental pollution. However, it cannot be avoided because it has many applications, such as in the health care industry, power production, other several industrial applications, agriculture and research and development. The workers and the people involved with radiation work shall remain within the prescribed limit of exposure and also be as low as reasonably achievable. Radiation waste is required to be disposed of in the environment without any adverse effect on the public and the environment since it is a potential hazard. For managing radioactive waste, there are national- and international-level safety codes and guidelines. In India, radioactive waste management activities are regulated by the Atomic Energy Regulatory Board (AERB) and Bhabha Atomic Research Centre (BARC) Safety Council under the Department of Atomic Energy (DAE), Government of India.

14.7.3 Rules and Regulations and Acts for Radiation and Radioactive Substances

Internationally standards, guidelines and recommendations related to radiation and radioactive materials are formulated by several international organizations, such as the International Atomic Energy Agency (IAEA) and the International Commission on Radiological Protection (ICRP). For all activities of radioactive waste management, the IAEA Safety Fundamentals are followed stringently. At the country level, regulations, policies and legislation related to radioactive waste management are developed from these international recommendations, guidelines and standards.

14.7.4 Regulatory Frameworks: Radiation and Radioactive Materials

The Government of India enacted Atomic Energy Act, 1962, which provides the basic regulatory framework regarding Regulatory Bodies, Safety Codes and Guidelines Acts and Rules for activities relating to atomic energy programs in India. The Atomic Energy (Safe Disposal of Radioactive Wastes) Rules, 1987 (also called as GSR-125) govern radioactive waste in the environment. The functions and responsibilities of the employer, the licensee, the workers and radiological safety officers are governed by the Radiation Protection Rules, 2004. In addition, Atomic Energy (Factories) Rules, 1996 allow statutory authorities to impose regulatory control over inspectors, workers, workplace hygiene, safe use of machinery protective equipment and so others.

The AERB and the BARC Safety Council (BSC) was constituted under the Atomic Energy Act, 1962 to control usage and disposal of radiation and radioactive materials. The responsibility of protecting workers, the public and the environment against harmful effects of ionizing radiation was entrusted to the AERB and the BSC. The regulation of nuclear and radiological facilities is done by the AERB, whereas the safety of BARC facilities is done by the BSC in the public domain.

14.7.5 Classification of Radioactive Waste

Classifying radioactive waste helps in segregation, treatment process, selection process selection, storage and disposal. Based on physical characteristics, heat generation, level of radioactivity, activity concentration, type of radiation and half-life, wastes are classified. As an example, based on activity concentration, radioactive liquid waste is categorized as low-level waste, intermediate-level waste and high-level waste. Furthermore, based on its half-lives, it is subcategorized as short-lived and long-lived waste. Most of the radioactive waste has either beta or gamma emitters. Very rarely is it noted to have alpha emitters. To get the necessary approval for transfer and disposal, radioactive wastes are categorized as follows:

1. Based on the radiation dose rate on the surface of the waste package, solid waste having beta–gamma emitters and containing alpha activity less than 4000 Bq/gm are are classified into three (I to III) categories. The highest surface dose rate is Category III. However, solid wastes bearing alpha activity greater the 4000 Bq/gm are classified as Category IV.
2. Based on the activity level, liquid waste with beta–gamma activity is classified into five (I to V) categories.

14.7.6 Effluent Discharge and Dose

Effluent discharge and dose are very much important in respect of environmental pollution. The primary principles of effluent releases are dependent on the control of sources. The application of dose limit and release is permissible as low as achievable reasonably. To monitor and control of release limit and to compliance with the prescribed limit, an Environmental Survey Laboratory (ESL) is set up at every nuclear power plant site. In agreement with the recommendation of ICRP, an effective dose limit of 1 mSv per year at the boundary of a radiological facility to the public during normal operating conditions has been prescribed by AERB and BSC. This distributed dose is further divided among air, water and terrestrial pathways. The dose is further divided among the radionuclides, which are precise to the installation. Suitable environmental modeling and studies are adopted for estimating the dose to the public from released activity. Stringent review in respect of effluent discharge from a radiological facility is done in all aspects, namely, right from the siting stage. After reviewing potential radiological impacts on the environment and detail monitoring of nuclear waste disposal facility, the regulatory authority grants siting clearance. The working organization or the industry is required to specify the limit of levels for gaseous and liquid discharges at all steps of the processes prior to commissioning and regular operation of the facility.

14.7.7 Regulations on Biosafety of Genetic Engineering/Recombinant DNA Research

Genetic engineering and /or recombinant DNA (rDNA) technology is an influential tool in generating environmentally friendly alternatives for products and processes. It may speed up developmental processes as well as add diversity to species, biomolecules and metabolic pathways. Genetically modified biological organisms can be supplemented with additional genetic properties for the biodegradation of specific pollutants that is not possible by naturally occurring organisms. The techniques are used to produce many recombinant proteins for pharmaceuticals and other specialized purposes. A number of GE products, including human growth hormone, alpha interferon, recombinant erythropoietin and tissue plasminogen activator, are approved by the regulatory authority. rDNA technology is also applied for improving the composition and functional characteristics in plant materials. The first commercially available whole-food product is slow-ripening tomato. The gene that expresses the enzyme polygalacturonase is responsible for softening is turned off in this tomato. The rDNA technology is used to develop disease-resistant, pest-resistant and environmental condition–resistant plants.

In 1995, the EPA allowed transgenic corn seed, cottonseed, and seed potatoes that contain recombinant genetic material to challenge certain insects. Such transgenic products are environmentally friendly genetic herbicides and pesticides that are less toxic for nature. In animals, recombinant bovine somatotropin (BST) was the first approved for application of rDNA technology in animal production, which helped increase milk production by 10% to 25% under ideal management conditions. Other applications of recombinant biotechnology in animal production include the development of vaccines, cloning, artificial insemination, the improvement of growth rate and/or feed efficiency and rapid disease diagnostics. The Department of Biotechnology under the Ministry of Science & Technology, Government of India constituted the Review Committee for Genetic Manipulation (RCGM). The RCGM has updated or formulated several rDNA guidelines, such as "Recombinant DNA Safety Guidelines, 1990" and "Revised Guidelines for Research in Transgenic Plants, 1998" and "Regulations and Guidelines on Biosafety of Recombinant DNA Research and Biocontainment, 2017". These guidelines are developed after a detailed review of contemporary scientific information as well as from the factual experience gained while executing biosafety frameworks inside and outside the country. The guideline stipulates the processes for handling hazardous biological material, recombinant nucleic acids, cells, organisms and viruses. The document contains methods and processes for ensuring the protection of the public and occupational health and of the environment from recombinant materials/products. The guideline also includes details about biosafety requirements and recommendations for laboratory facilities. The laboratory facility recommendations include facility design, biosafety equipment, personal protective equipment, laboratory practices and techniques, waste management and others.

The regulations which are applicable in India under the provisions of the Rules 1989 of EPA, 1986 for the storage of hazardous microorganisms, development of GE organisms or cells and manufacture, use, import and export follow:

a. In the document sale, offers for sale, storage for the purpose of sale, offers and any kind of handling over with or without a consideration are mentioned in detail.
b. Risk group microorganisms for making export and import of GE cells or organisms are listed in this document.
c. The document also consists of production, manufacturing, processing, storage, import, drawing off, packaging and repacking of the GE products.
d. The document also includes use of simple microorganisms or GE microorganisms in the production, manufacturing of drugs, pharmaceuticals, food stuffs distilleries and tanneries, among others.

The research and development, as well as large-scale handling, of hazardous microorganisms, GE organisms, cells and products is under the scope of the Rules 1989. In India, to adopt the Rules, six competent authorities and their role have been notified for the following activities:

1. For both GE and non-GE: the regulation and control of contained research activities with hazardous microorganisms, organisms and cells.
2. In production activity, the regulation and control of large-scale use of GE organisms.
3. The action related to import, export and transfer of GE organisms and products.
4. Under statutory provisions, the release of GE organisms in environmental applications.

14.8 Conclusion

In the present scenario, environmental challenges continue to develop emerging issues. Due to such issues, there is a significant climatic change, which has developed several permanent detrimental alterations to the globe. It is a great concern of total flora and fauna family. The overall geological, biological, environmental and ecological damage led to large-scale environmental hazards to biota and abiota, including human health. With the continued increase in the use of chemicals, fertilizers, pesticides, agricultural products, dyes and pigments and nanomaterials and the overconsumption of energy, natural resources and nonrenewable resources to meet the demands of expanding global population, the associated environmental problems are increasing. An unproportioned increase in world population vis-à-vis unsustainable land use, limited potable water and afforestation caused unprecedented exploitation of the environment. In addition, despite innovative initiatives to prevent waste accumulation through reusing and recycling, the amount of overconsumption generates more waste, causing more environmental damage. In the case of waste generation, the degree of waste due to biohazards, rDNA technology hazards and chemical and radioactive hazards appear to be more detrimental for the environment. A deleterious impact on environment is seen through the gradual threatening of anthropogenic activities day by day.

However, perceptions of developing new technologies to combat emerging issues for environmental protection and control are also under development. Innovative technologies are implemented to explore the potential of whole cells, microorganisms, antigen–antibody, enzymes and plants to perform as eco-efficient and robust cleanup agents in a variety of practical situations, such as evolutionary and genomic approaches, enzyme engineering, designing strains and process engineering for upgrading or boosting biodegradation. Other advanced and emerging technologies, such as biomembrane reactor technology, wastewater treatment plant design, the reuse of treated wastewater through decentralized sanitation initiatives, the deployment of anaerobic digestion to treat biowaste, the development of alternative and renewable energy resources from biowaste and applications of soil remediation and cleanup of contaminated sites, are deployed with efficiency (Urbain *et al.* 1996; van Wyk 2001; Grommen and Verstraete 2002; Cicek 2003; Kohli and Martin 2005).; Grommen and Verstraete 2002; Cicek 2003; Kohli and Martin 2005).

In addition to emerging and advanced technologies, a wide group of technologies have pivotal roles in the fields of food production, renewable energy and raw materials, bioremediation and pollution prevention and pollution control.

Environmental biotechnology has extensive potential for issues related to prevention, detection, remediation and control of environmental degradation and pollution. Simultaneously, it speaks about sustainability through the development of innovative products and clean or green processes to be implemented successfully in several sectors, such as industry, agriculture, agro-forestry, food, raw materials and minerals and others. Emerging technologies, such as GMOs and its regulation for release in the environment and safe application for human, animal or plant use, is important.

Environmental biotechnology can offer both economic and environmental profits in the development, manufacturing, monitoring, waste management and pollution control. On one hand, it reduces environmental impact, and on other hand, it enhances sustainability. But there are several emerging challenges that are still required to be solved with the more advanced technologies.

The policy and issues related to environmental biotechnology are already in place nationally and internationally, which were already discussed in detail in this chapter. A detailed appraisal of the challenges, perceptions and issues of modern environmental biotechnology is imperative both for policymakers and for the deployment stakeholders, such as industry, agriculture, transport, agro-forestry, food, raw materials and several related sectors.

The effective adoption of policies related to environmental biotechnology augments the improvement of the whole process rationally and efficiently. However, there is a need to integrate emerging challenges vis-à-vis innovative solutions this decade related to environmental health. An extensive evaluation process for reviewing and monitoring of the consequences, challenges and opportunities of modern environmental biotechnology is important for both policymakers and other stakeholders to uphold a sustainable global environment.

Disclaimer

The paper has been written in the personal capacity of the authors and the views do not necessarily represent those of DSIR or any public, private, society, trust or other organizations nationally/ internationally.

Conflict of Interests

The author declares that there is no conflicts of interest regarding publication of this paper as a book chapter.

ACKNOWLEDGMENTS

The first author acknowledges Department of Scientific & Industrial Research (DSIR) for giving me the opportunity of working on policy and regulatory programs of the Government of India. The first author also acknowledges DSIR, Department of Science & Technology (DST) and Department of Biotechnology (DBT) under Ministry of Science & Technology; Ministry of Environment, Forest & Climate Change (MoEF&CC), Ministry of Health and Family Welfare (MoHFW), Ministry of Agriculture & Farmers Welfare (MoAFW), Government of India for taking initiatives on environmental biotechnology. The authors also acknowledge contribution and initiatives of different scientists, researchers, policymakers, officials and different departments in formulating different rules, acts, guidelines and policies related to environmental biotechnology and environmental protection that help frame this chapter in a structured manner.

REFERENCES

Ahn, Y., Jung, H., Tatavarty, R., Choi, H., Yang, J., and I.S. Kim. 2005. Monitoring of petroleum hydrocarbon degradative potential of indigenous microorganisms in ozonated soil. *Biodegradation* 16:45–56.

Ashbolt, N.J., Grabow, W.O.K., and M. Snozzi. 2001. Indicators of microbial water quality. In: Fewtrell, L., and Bartram, J. (Eds.), *Water Quality: Guidelines, Standards and Health*, IWA Publishing, London, pp. 289–316.

Bargagli, R. 1998. *Trace Elements in Terrestrial Plants: An Ecophysiological Approach to Biomonitoring and Biorecovery*, Springer-Verlag, Berlin, 324 pp.

Beliaeff, B., and T. Burgeot. 2002. Integrated biomarker response: A useful tool for ecological risk assessment. *Environmental Toxicology and Chemistry* 21:1316–1322.

Betianu, C., and M. Gavrilescu. 2006a. Environmental behavior and assessment of persistent organic pollutants. *Environmental Engineering and Management Journal* 5:213–241.

Betianu, C., and M. Gavrilescu. 2006b. Persistent organic pollutants in environment: Inventory procedures and management in the context of the Stockholm Conention. *Environmental Engineering and Management Journal* 5:1011–1028.

Bidoki, S.M., Wittlinger, R., Alamdar, A.A., and J. Burger. 2006. Eco-efficiency analysis of textile coating materials. *Journal of the Iranian Chemical Society* 3:351–359.

Biotech. 2000. *Environmental Biotechnology, The Welcome Trust.* http://www.biochemistry.org/education/pdfs/enviro-card.pdf.

Bull, A.T. 1995. Biotechnology for environmental quality: Closing the circles. *Biodiversity and Conservation* 5:1–25.

Cannons, A.C., and V.J. Harwood. 2004. *Sensor Technology for Water Quality Monitoring: Fiber-Optic Biosensor (WERF Report)*, IWA Publishing, London, 72 pp.

Chen, W., Mulchandani, A., and M.A. Deshusses. 2005. Environmental biotechnology: Challenges and opportunities for chemical engineers. *American Institute of Chemical Engineers* 51:690–695.

Cheremisinoff, N.P. 2003. *Handbook of Solid Waste Management and Waste Minimization Technologies*, Butterworth-Heinemann/Elsevier Science, Amsterdam, 477 pp.

Chisti, Y. 2007. Biodiesel from microalgae. *Biotechnology Advances* 25:294–306.

Cicek, N. 2003. A review of membrane bioreactors and their potential application in the treatment of agricultural wastewater. *Canadian Biosystems Engineering* 45:6.37–6.49.

Conti, M.E. 2007. *Biological Monitoring; Theory and Applications, Bioindicators and Biomarkers for Environmental Quality and Human Exposure Assessment*, WIT Press, Southampton, UK, 228 pp.

Cui, Y.I., Wei, Q., Park, H., and C.M. Lieber. 2001. Nanowire nano sensors for highly sensitive and selective detection of biological and chemical species. *Science* 293 (5533) (17 August):1289–1292.

Das, T.K. 2005. *Toward Zero Discharge. Innovative Methodology and Technologies for Process Pollution Prevention*, John Wiley and Sons, Hoboken, NJ, 744 pp.

Doble, M., Kruthiventi, A.K., and V.G. Gaikar. 2004. *Biotransformations and Bio- Processes*, Marcel Dekker, New York, Basel, 371 pp.

Doran, J.W., and T.B. Parkin. 1994. Defining and assessing soil quality. In: *Defining Soil Quality for a Sustainable Environment*. Soil Science Society of America Special Publication no. 35. SSSA, Madison, Wisconsin, USA.

Durrieu, C., Tran-Minh, C., Chovelon, J.M., Barthet, L., Chouteau, C., and C. Védrine. 2006. Algal biosensors for aquatic ecosystems monitoring. *The European Physical Journal – Applied Physics* 36:205–209.

EFB. 1995. Environmental Biotechnology, EFB Task Group on Public Perceptions of Biotechnology. http://www.bioportfolio.com/efb4.htm

EIBE. 2000. Biotechnology and Environment. European Initiative for Biotechnology Education. http://www.ipn.uni-kiel.de/eibe/UNIT16EN.PDF.

Evans, G.M., and J.C. Furlong. 2003. *Environmental Biotechnology. Theory and Application*, John Wiley and Sons, Chichester, 300 pp.

Futrell, R. 2000. Politics of space and the political economy of toxic waste. *International Journal of Politics, Culture, and Society* 13:447–476.

Gabriel, R., and A. Braune. 2005. Eco-efficiency analysis: Applications and user contacts. *Journal of Industrial Ecology* 9:19–21.

Gavrilescu, M. 2004a. Cleaner production as a tool for sustainable development. *Environmental Engineering and Management Journal* 3:45–70.

Gavrilescu, M. 2004b. Removal of heavy metals from the environment by biosorption. *Engineering in Life Sciences* 4:219–232.

Gavrilescu, M. 2005. Fate of pesticides in the environment and its bioremediation. *Engineering in Life Sciences* 5:497–526.

Gavrilescu, M. 2010. Environmental biotechnology: Achievements, opportunities and challenges. *Dynamic Biochemistry, Process Biotechnology and Molecular Biology* 4 (1):1–36 ©2010 Global Science Books.

Gavrilescu, M., and Y. Chisti. 2005. Biotechnology – a sustainable alternative for chemical industry. *Biotechnology Advances* 23:471–499.

Gavrilescu, M., and M. Nicu. 2005. *Source Reduction and Waste Minimization* (in Romanian) (2ndEdn), EcoZONE Press, Iasi, Romania, 230 pp.

Gavrilescu, M., Pavel, L.V., and I. Cretescu. 2008a. Characterization and remediation of soils polluted with uranium. *Journal of Hazardous Materials* 163:475–510.

Gavrilescu, M., Teodosiu, C., Gavrilescu, D., and L. Lupu. 2008b. Strategies and practices for sustainable use of water in industrial papermaking processes. *Engineering in Life Sciences* 8:99–124.

Grommen, R., and W. Verstraete. 2002. Environmental biotechnology: The ongoing quest. *Journal of Biotechnology* 98:113–123.

Gu, M.B., Mitchell, R.J., and B.C. Kim. 2004. Whole-cell-based biosensors for environmetal biomonitoring and application. In: Zhong, J.J. (Ed.), *Biomanufacturing, Scheper T (Series Ed) Advances in Biochemical Engineering/Biotechnology*, Springer-Verlag, Berlin, Heidelberg, New York, pp. 269–305.

Hagger, J.A., Jones, M.B., Leonard, D.R.P., Owen, R., and T.S. Galloway. 2006. Biomarkers and integrated environmental risk assessment: Are there more questions than answers? *Integrated Environmental Assessment and Management* 2:312–329.

Hart, J.K., and K. Martínez. 2006. Environmental sensor networks: A revolution in the earth system science? *Earth-Science Reviews* 78:177–191.

Hashim, M.A., and Z. Uijang. 2004. Environmental biotechnology: Its relevance and prospects for developing countries. In: Vjang, Z., and Menze, M. (Eds.), *Environmental Biotechnology*, IWA Publishing, London, UK, pp. 7–12.

Jamil, K. 2001. *Bioindicators and Biomarkers of Environmental Pollution and Risk Assessment*, Science Publishers, New Hampshire, 204 pp.

Johnston, D.J. 2003. Biotechnology: The next wave of innovation technologies for sustainable development. In: Seralgedin, I., and Persley, G.J. (Eds.), *Biotechnology and Sustainable Development: Voices of the South and North*, CABI Publishing, England, pp. 67–74.

Kohli, P., and C.R. Martin. 2005. Smart nanotubes for biotechnology. *Current Pharmaceutical Biotechnology* 6:35–47.

Krieg, E.J. 1998. The two faces of toxic waste: Trends in the spread of environmental hazards. *Sociological Forum* 13:3–20.

Kryl, D. 2001. Environmental and industrial biotechnology in developing countries. *Electronic Journal of Biotechnology* 4. http://www.ejbiotechnology.info/content/vol4/issue3/issues/03/index.html.

Kuhn, T., Pittel, K., and T. Schulz. 2003. Recycling for sustainability – a long run perspective? *International Journal of Global Environmental Issues* 3:339–355.

Lagadic, L., Caquet, T., and T. Ramade. 1994. The role of biomarkers in environmental assessment (5). Invertebrate populations and communities. *Ecotoxicology* 3:193–208.

Lam, P.K.S. 2009. Use of biomarkers in environmental monitoring. *Ocean and Coastal Management* 52:348–354.

Lam, P.K.S., and J.S. Gray. 2003. The use of biomarkers in environmental monitoring. *Marine Pollution Bulletin* 46:182–186.

Madejon, P., Maranon, T., and J.M. Murillo. 2006. Biomonitoring of trace elements in the leaves and fruits of wild olive and holm oak trees. *Science of the Total Environment* 355:187–203.

Markert, B.A., Breure, A.M., and H.G. Zechmeister. 2003. Definitions, strategies and principles for bioindication/biomonitoring of the environment. In: Markert, B.A., Breure, A.M., and Zechmeister, H.G. (Eds.), *Bioindicators and Biomonitors*, Elsevier Science, Oxford, pp. 3–39.

Mauriz, E., Calle, A., Montoya, A., and L.M. Lechuga. 2006. Determination of environmental organic pollutants with a portable optical immunosensor. *Talanta* 69:359–364.

Mertens, J., Luyssaert, S., and K. Verheyen. 2005. Use and abuse of trace metal concentrations in plant tissue for biomonitoring and phytoextraction. *Environmental Pollution* 138:1–4.

Ministry of Agriculture & Farmers Welfare (MoAFW) Impact of Climate Change on Farmers. Press Information Bureau, Government of India. 2020 March: 1–2.

Ministry of Agriculture & Farmers Welfare (MoAFW) Studies on Impact of Climate Change. Press Information Bureau, Government of India. 2019 November: 1–6.

Ministry of Environment, Forest & Climate Change (MoEF&CC) Government of India. Climate Change. Annual Report. 2018–2019: 67–95 & 208–220.

Ministry of Environment, Forest & Climate Change (MoEF&CC) Government of India. Climate Change. Annual Report. 2019–2020: 70–100 & 180–196. http://moef.gov.in/wp-content/uploads/2017/06/ENVIRONMENT-AR-ENGLISH-2020.pdf.

Mwinyihija, M., Meharg, A., Dawson, J., Strachan, N.J.C., and K. Killham. 2006. An ecotoxicological approach to assessing the impact of tanning industry effluent on river health. *Archives of Environmental Contamination and Toxicology* 50:316–324.

Nali, C., Francini, A., and G. Lorenzini. 2006. Biological monitoring of ozone: The twenty-year Italian experience. *Journal of Environmental Monitoring* 8:25–32.

NRC. 1987. Biological markers in environmental health research (National Research Council. Committee on biological markers). *Environmental Health Perspectives* 74:3–9.

OECD. 1994. *Biotechnology for a Clean Environment: Prevention, Detection, Remediation*, OECD Publishing, Paris, 204 pp.

OECD. 1998. *Biotechnology for Clean Industrial Products and Processes Towards Industrial Sustainability*, OECD Publishing, Paris, 200 pp.

OECD. 2001. *The Application of Biotechnology to Industrial Sustainability*, OECD Publishing, Paris, 158 pp.

OECD. 2013. *Biotechnology for the Environment in the Future: Science, Technology and Policy*, OECD Science, Technology and Industry Policy Papers, No. 3, OECD Publishing, Paris, 80 pp. http://dx.doi.org/10.1787/5k4840hqhp7j-en.

Olguin, E.J. 1999. Cleaner bioprocesses and sustainable development. In: Olguin, E.J., Sánchez, G., and Hernández, E. (Eds.), *Environmental Biotechnology and Cleaner Bioprocesses*, Taylor and Francis, Boca Raton, pp. 3–18.

Rabbow, E., Rettberg, P., Baumstark-Khan, C., and G. Horneck. 2002. SOS-LUX and LAC-FLUORO-TEST for the quantification of genotoxic and/or cytotoxic effects of heavy metal salts. *Analytical Chimica Acta* 456:31–39.

Raeymaekers, B. 2006. A prospective biomonitoring campaign with honey bees in a district of Upper-Bavaria (Germany). *Environmental Monitoring and Assessment* 116:233–243.

Ramsden, J.J. 1999. A sum parameter sensor for water quality. *Water Research* 33:1147–1150.

Reis, L.F.L., Van Sluys, M.A., Garratt, R.C., Pereira, H.M., and M.M. Teixeira. 2006. GMOs: Building the future on the basis of past experience. *Anais da Aca- demiaBrasileira de Ciencias* 78:667–686.

Reiss, M., and W. Hartmeier. 1999. Monitoring of environmental processes with biosensors. In: Rehm, H-I., and Reed, G. (Eds.), *Biotechnology* (2ndEdn), Wiley-VCH, Weinheim, pp. 125–139.

Rodríguez-Mozaz, S., López de Alda, M.J., and D. Barceló. 2006. Biosensors as useful tools for environmental analysis and monitoring. *Analytical and Bioanalytical Chemistry* 386:1025–1041.

Rodríguez-Mozaz, S., Marco, M.P., López de Alda, M.J., and D. Barceló. 2004. Biosensors for environmental applications: Future development trends. *Pure and Applied Chemistry* 76:723–752.

Saling, P. 2005. Eco-efficiency analysis of biotechnological processes. *Applied Microbiology and Biotechnology* 68:1–8.

Starodub, N.F., Katzev, A.M., Starodub, V.M., Levkovetz, I.A., Goncharuk, V.V., Klimenko, N.A., Shmir'ova, A.N., Piven, N.V., and B.B. Dzantijev. 2005. Biosensors for water quality monitoring. In: Omelchenko, A., Pivovarov, A.A., and Swindall, W.J. (Eds.), *Modern Tools and Methods of Water Treatment for Improving Living Standards*, Proceedings of the NATO Advanced Research Workshop on Modern Tools and Methods of Water Treatment for Improving Living Standards Dnepropetrovsk, Ukraine 19–22 November 2003, pp. 51–70.

Tarazona, J.V., Fernández, M.D., and M.M. Vega. 2005. Regulation of contaminated soils in Spain – a new legal instrument. *Journal of Soils and Sediments* 5:121–124.

Tom-Petersen, A., Leser, T.D., Marsh, T.L., and O. Nybroe. 2003. Effects of copper amendment on the bacterial community in agricultural soil analyzed by the T-RFLP technique. *FEMS Microbiology Ecology* 46:53–62.

Tucker, C.L., and Fields, S. 2001. A yeast sensor of ligand binding. *Nature Biotechnology* 19:1042–1046.

Urbain, V., Benoit, R., and Manem, J. 1996. Membrane bioreactor: A new treatment tool. *Journal American Water Works Association* 88:75–86.

van Wyk, J.P.H. 2001. Biotechnology and the utilization of biowaste as a resource for bioproduct development. *Trends in Biotechnology* 19:172–177.

Verma, N., and M. Singh. 2005. Biosensors for heavy metals. *Biometals* 18:121–129.

Wall-Markowski, C.A., Kicherer, A., and P. Saling. 2005. Using eco-efficiency analysis to assess renewable-resource-based technologies. *Environmental Progress* 23:329–333.

Waters, J.S. 1998. *Environmental Management Systems for the Small Business Leadership Training Program*, Pollution Prevention Institute, University of Kansas, p. 19.

WBCSD. 2000. *Eco-efficiency. Creating More Value with Less Impact*, World Business Council for Sustainable Development, Geneva, Switzerland.

Weitz, H.J., Campbell, C.D., and Killham, K. 2002. Development of a novel, bioluminescence based fungal bioassay for toxicity testing. *Environmental Microbiology* 4:422–429.

Willke, T., Prüsee, U., and K.D. Vorlop. 2006. Biocatalytic and catalytic routes for the production of bulk and fine chemicals from renewable resources. In: Kamm, B., Gruber, P.R., and

Kamm, M. (Eds) *Biorefineries – Industrial Processes and Products. Status Quo and Future Directions* (Vol. 1). Wiley-VCH, Weinheim, Germany, pp. 385–406.

Zelano, V., Torazzo, A., Berto, S., Ginepro, M., Prenesti, E., and A. Ferrari. 2006. Biomonitoring of traffic originated PAHs in the air. *International Journal of Environmental and Analytical Chemistry* 86:527–540.

SUGGESTED READINGS

1. AERB Safety Code for Nuclear Medicine Laboratories (SC/MED 4) Appendix 511.1989. pp. 47–49.
2. Cherry SR, Sorenson JA, Phelps ME. Physics on Nuclear Medicine. 3rd Edition. Saunders; Philadelphia, Pennsylvania: 1980.
3. GovidaRajan KN. BARC; Mumbai: 2002. Basic Safety Standards. Accredition Programme for Nuclear Medicine Technologists in Radiation Safety; pp. 8.1–8.9.
4. ICRP Recommendations of the International Commission on Radiological Protection. Brit J Radiol. 1995;(Suppl 6).
5. International Commission on Radiation Units and Measurement (ICRU) Recommendations. 1980.
6. Kinni KS. Planning of Nuclear Medicine Laboratories for Diagnostic and Therapeutic Procedures. Indian J Nuclear Med. 1998;13(4):165–192.
7. Murthy BKS. BARC; Mumbai: 2000. Operational limits. Training Workshop on Radiation Safety in Nuclear Medicine and RSO Certification Examination; pp. 6.1–6.6.
8. Nagalakshmi B. BARC; Mumbai: 2000. Radioactive Waste Disposal with Special Reference to Nuclear Medicine Laboratories; pp. 16.1–16.4. Training Workshop on Radiation Safety in Nuclear Medicine and RSO Certification Examination.
9. Radiation Protection for Medical and Allied Health Personnel 1989. NCRP Report No 105.
10. Tabish SA. Hospital Infection Control: Conceptual Framework. Academa Publishers; 2005. Ecohealth: Management of Biomedical Waste; pp. 139–145.
11. The Radiation Protection Rules (RPR) 1971, Under Section 30 of The Atomic Energy Act, 1962.

15

Enzymes in Environment Management

Tek Chand Bhalla, Neerja Thakur, Sheetal, Monika, Shashi Kant Bhatia and Jog Raj

CONTENTS

DOI: 10.1201/9781003186304-15

15.1 Introduction

The search for comforts and a utilitarian approach of humans no doubt have brought industrial and agricultural revolutions and transformed the lifestyle of humans vis-à-vis the witnessed economic development in every part of the globe but at the cost of the environment. Human activities in the course of development and modernization have resulted in the release of substances and energies in harmful form or amount (called pollutants), leading to environmental pollution, that is, undesirable changes in physical, chemical or biological characteristics of soil, water and air that harm human, animals, microorganisms, industry, agriculture, living conditions, cultural and heritage assets and natural resources. Environmental pollutants are of three types. The first type are those substances that occur naturally, but due to human activity, their concentrations increase in the atmosphere, for example, carbon dioxide. The higher concentration of carbon dioxide is significantly impacting our climate (Bhatia et al., 2019a).Bhatia et al., 2019a). The second type of pollutants are those toxic substances produced industrially and are not found naturally (also called xenobiotics), for example, pesticides and plastics that often considerably persist in the environment for a long time. The third type of pollutants are substances that are not directly toxic, but their interaction with the environment leads to unfortunate consequences, for example, the effects of chlorofluorocarbons on the ozone layer.

The rate of release of xenobiotics and recalcitrant organic pollutants into the environment is on the rise, and it is becoming increasingly difficult to achieve the required degree of removal of these pollutants using conventional chemical processes; therefore, there is a need for the development of alternative treatment methods that are faster, cheaper, more reliable and simpler to implement than current processes (Song et al., 2021; Gurav et al., 2021; Choi et al., 2021). The rate of release of xenobiotics and recalcitrant organic pollutants into the environment is on the rise, and it is becoming increasingly difficult to achieve the required degree of removal of these pollutants using conventional chemical processes; therefore, there is a need for the development of alternative treatment methods that are faster, cheaper, more reliable, and simpler to implement than current processes (Song et al., 2021; Gurav et al., 2021; Choi et al., 2021; Kim et al., 2020). There are several disadvantages associated with chemical processes and efforts are being made to find ecofriendly and economically viable alternatives (Bhatia et al., 2017; Kim et al., 2020). There are several disadvantages associated with chemical processes and efforts are being made to find ecofriendly and economically viable alternatives (Bhatia et al., 2017; Bhatia et al., 2014; Mehta et al., 2014). Much attention is on the biological approaches based on microorganisms and enzymes as these methods are ecofriendly and generate revenue (Bhatia et al., 2021; Bhatia et al., 2020a). Enzymes are specialized proteins that act as important tools in degrading chemical pollutants and change the characteristics of a given toxic substance into nonhazardous products (Karam and Nicell, 1997). They are increasingly exhibiting their enormous potential in the treatment of industrial effluents (Bhatia et al., 2020b). The enzymes to be used in environmental management can be obtained from a variety of plants and microbial sources. Microorganisms degrading a range of aliphatic, aromatic and heterocyclic compounds have been isolated, identified and their enzymes chemical pollutants degradations pathways have been well characterized. Soil is the biggest repository of microorganisms and exhibits rich microbial diversity (Bhatia et al., 2013; Mehta et al., 2013). There are microorganisms in the soil that harbor enzymes able to catalyze, modify or break down hazardous chemicals and transform them into nontoxic forms. Microorganisms also have a very wide amplitude of adaptation to the harsh environment and so evolve faster than other animals and plants. The microbial evolution in response to the change in physicochemical environmental conditions is proportional to the change in their enzymes that operate under the extremes of the environment (Park et al., 2020). Microbes and enzymes because of their remarkable adaptability to extreme conditions are finding applications not only in processing raw materials to desired products in dairy, paper and pulp, petroleum, textile, chemical and agro-processing industries but also in the management of pollution arising due to agricultural and industrial activities (Bhatia et al., 2019b, 2020c). The present chapter is focused on the use of enzymes in the management of environmental pollution.; Bhatia et al., 2014; Mehta et al., 2014). Much attention is on the biological approaches based on microorganisms and enzymes as these methods are ecofriendly and also generate revenue (Bhatia et al., 2021; Bhatia et al., 2020a). Enzymes are specialized proteins that act as an important tool to degrade chemical pollutants and change the characteristics of a given toxic substance into non-hazardous products (Karam and Nicell, 1997). They are increasingly exhibiting their enormous potential in the treatment of industrial effluents (Bhatia et al., 2020b). The enzymes to be used in environmental management can be obtained from a variety of plants and microbial sources. Microorganisms degrading a range of aliphatic, aromatic, and heterocyclic compounds have been isolated, identified and their enzymes chemical pollutants degradations pathways have been well characterized. Soil is the biggest repository of microorganisms and exhibits rich microbial diversity (Bhatia et al., 2013; Mehta et al., 2013). There are microorganisms in the soil that harbor enzymes able to catalyze, modify, or breakdown hazardous chemicals and transform them into nontoxic forms. Microorganisms also have a very wide amplitude of adaptation to the harsh environment and so evolve faster than other animals and plants. The microbial evolution in response to the change in physicochemical environmental conditions is proportional to the change in their enzymes that operate under the extremes of the environment (Park et al., 2020). Microbes and enzymes because of their remarkable adaptability to extreme conditions are finding applications not only in processing raw materials to desired products in dairy, paper and pulp, petroleum, textile, chemical, and agroprocessing industries but also in the management of pollution arising due to agricultural and industrial activities (Bhatia et al., 2019b, 2020c). The present chapter is focused on the use of enzymes in the management of environmental pollution.

15.2 Types of Pollutants and Their Enzymatic Treatments

The principal source of major contaminants in the environment is industrial sites that manufacture, store and distribute organic chemicals which spill effluents and wastes. Harmful, toxic and undesirable by-products from the factories and manufacturing hubs are released into different sites. These toxic by-products when dumped in the environment such as water bodies (ponds, rivers, and sea), landfills and air result in serious catastrophic effects on the environment, humans and animal health (Commission for Environmental Cooperation, 2004). The waste and pollutants, if not treated properly, make their way from landfills to air, then air to water and again to land and vice versa as shown in Figure 15.1.

15.2.1 Aromatic Contaminants and Related Compounds

Aromatic compounds are organic molecules containing one or more aromatic rings and are considered the most prevalent and persistent pollutants in the environment. There are three major categories of aromatic compounds: polycyclic aromatic hydrocarbons (PAHs), heterocyclics and substituted aromatics (Seo et al., 2009). The various sources of PAHs include the incomplete combustion of fossil fuels, gas production, wood treatment facilities, industrial activities, coal burning and waste incineration (Ahn et al., 1999; Kanaly and Harayama, 2000; Tang et al., 2005; Kim et al., 2003). Thermal geologic reactions, burning of vegetation in forests, grasslands and bush fires are natural sources of PAHs production (Juhasz and Naidu, 2000; Wilson and Jones, 1993). Anthropogenic sources, particularly fuel combustion, automobiles, spillage of petroleum products and waste incinerators are significant sources of phenols and aromatic amines, which constitute one of the major classes of pollutants. They are present in the effluents of a wide variety of industries including petroleum, refining, coal, resins and plastics, wood preservation, metal coating, dyes and other chemicals, textiles, mining and dressing and pulp and paper. Aromatic compounds are organic molecules containing one or more aromatic rings and considered the most prevalent and persistent pollutants in the environment. There are three major categories of aromatic compounds: polycyclic aromatic hydrocarbons (PAHs), heterocyclics, and substituted aromatics (Seo et al., 2009). The various sources of PAHs include the incomplete combustion of fossil fuels, gas production, wood treatment facilities, industrial activities, coal burning, and waste incineration (Ahn et al., 1999; Kanaly and Harayama, 2000; Tang et al., 2005; Kim et al., 2003). Thermal geologic reactions, burning of vegetation in forests, grasslands, and bush fires are natural sources of PAHs production (Juhasz and Naidu, 2000; Wilson and Jones, 1993). Anthropogenic sources, particularly fuel combustion, automobiles, spillage of petroleum products, and waste incinerators are significant sources of phenols and aromatic amines which constitute one of the major classes of pollutants. They are present in the effluents of a wide variety of industries including petroleum, refining, coal, resins and plastics, wood preservation, metal coating, dyes and other chemicals, textiles, mining and dressing, and pulp and paper.

15.2.1.1 PAHs

PAHs are the most commonly found environmental pollutants which belong to hydrophobic organic compounds. The high concentration and frequency of PAHs in the environment impact human health due to their carcinogenic, mutagenic and teratogenic properties (Maigari and Maigari, 2015). During the last three decades, a higher incidence of cancer in urban and industrial areas than in rural areas has been observed due to exposure of humans to PAHs bioaccumulation, metabolism and disposition of these pollutants in aquatic and terrestrial ecosystems. The major sources of PAHs are the combustion of organic materials, such as crude oil, coal tar, petrol, creosote and anthropogenic activities (Abdel – Shafy and Mansour, 2016). The possible fates of PAHs in the environment include volatilization, photooxidation, chemical oxidation, bioaccumulation, adsorption in soil particles or sediments, leaching and microbial degradation and are highly persistent in the environment.

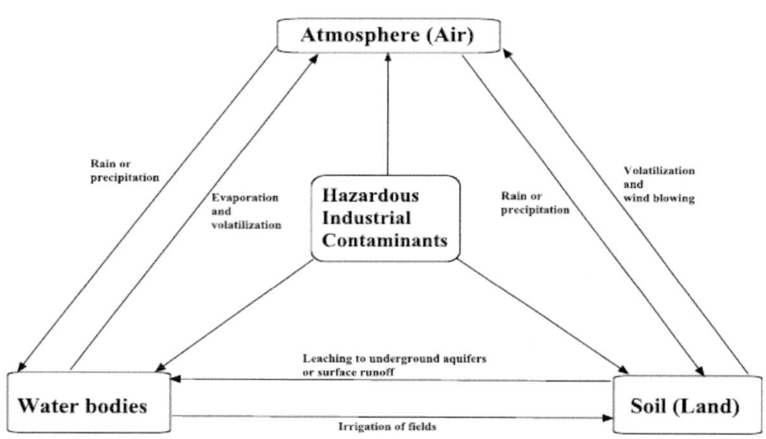

FIGURE 15.1 Cycle of a hazardous industrial contaminant in the environment (modified from original source, Shen, 1995).

15.2.1.2 Heterocyclic Compounds

Heterocyclic compounds, including dibenzothiophene and carbazole, are components of creosote, crude oils and shale oils and often coexist in the environment with PAHs and other aromatic compounds. These heterocyclic compounds are responsible for immense environmental concerns such as they are toxic and carcinogenic in nature and harmful for normal microbial flora of the environment (Padoley et al., 2008).

15.2.2 Agriculture Pollutants

The use of synthetic, highly toxic pesticides in agriculture is increasing globally and has raised many environmental issues related to animals and human health during the last several decades (Singh, 2007; Aktar et al., 2009; Bhat and Padmaja, 2014). The use of insecticides/pesticides not only has increased crop production, but it has also proved hazardous for humans, animals and environmental health (Forget, 1993). The harmful effects of insecticides and pesticides on different aspects of the environment are discussed in Table 15.1. Pesticides, which include herbicides, insecticides and fungicides, are widely used throughout the world for crop protection, and they are being increasingly used in agriculture. The global agricultural sector is the primary user of pesticides, consuming more than 4 million tons of pesticides annually. The main forms of pollution caused by pesticides are direct applications to crops, accidental spills during transport and manufacturing and the release of waste from tanks where cattle are treated to ectoparasites control. Among the microbial communities, bacteria, fungi and actinomycetes are the main transformers and pesticide degraders.

15.2.3 Effluents of Leather Industries

Leather industries play a vital role in the economy of developing countries and contribute to a large part of their annual gross domestic product (GDP; Ghafoor et al., 2012; Siddiqui et al., 2016). At the same time, these industries become big sources of environmental pollution and hazard to humans. These industries are mostly established near crowded cities to ensure raw material and others, and their effluent is mostly discharged in nearby rivers, water bodies and lowland areas without any treatment because these industries usually do not have any pollution control plant (Ferdous et al., 2012; Ghafoor and Zafar, 2014; Swathi et al., 2014). The processing of hides and skins requires a huge amount of water, chemicals and dyes; after processing, this water, along with residual chemicals, biological waste such as fat, lime, alkali, bleaching agents, heavy metals and different obnoxious gases, is the main contaminant in the effluent (Kankaria et al., 2011; Chowdhury et al., 2013). The lather tanning process requires about 170 types of chemicals and dyes, which includes sodium sulfate, sulfuric acid, chromium sulfate, wetting agents, calcium oxide, bactericides, sodium chloride, sodium hypochlorite, sodium bisulfate, formic acid, polyurethane, fat emulsion and formaldehyde (Chowdhury et al., 1996; Kankaria et al.,

TABLE 15.1

Different Pesticides/Insecticides and Their Effects on Human Health

Pesticides/Insecticides	Hazard on Environment/Health	References
Chlorantraniliprole	Less soluble in water, acute toxic for fish (*Channapunctatus*)	Bantu and Vakita, 2013
Indoxacarb	Long persistent in soil, water, and air; proven toxic for birds, aquatic life, bees and humans	Ferraro and McEuen, 1996; US EPA, 2000
Acetamiprid	Contaminate water bodies with agricultural runoff water, toxic to humans	Imamura et al., 2010; Raj and Joseph, 2017
Imidacloprid	Toxic for mammals	Free, 1993; Aitken et al., 2004; Suchail et al., 2000
Fipronil	Bioaccumulative	Holder et al., 2018
Atrazine; 2,4-D	Frequently detected in surface water bodies near agricultural fields, toxic for microalgae (*Chlorella fuscavar-fusca*), having an antagonistic effect on organisms	Kotrikla et al., 1999
Flubendiamide	Toxic against many nontargeted organisms, Neurotoxic potentials Protein metabolism alteration reported in freshwater fish *Labeorohita* and disrupt soil enzyme activity in tropical soil	Sarkar et al., 2014; Wei et al., 2014 Nirmalakallagadda and Rathnamma, 2014; Shrinivas and David, 2015
DDT	Hazardous for birds, mammals, fishes, reptiles, etc., bioaccumulative	US-EPA, 1975; US-EPA, 2000; Peterson and Talcot, 2006
Endosulfan	Toxic against humans and other mammals, affect WBCs and macrophages activity, impair the immune system, many deleterious effects were reported on the male reproductive system, mutagenic and cancerous	Susan and Sania, 1999; Singh et al., 2007
Lindane	Carcinogenic, neurotoxic, induce birth defects, damages kidneys, liver, neural system of human	Sahoo et al., 2008; Padma et al., 2011
Pentachlorophenol	Various human body organs are affected like respiratory tracts, kidney and liver, irritant for eyes, nose and skin	ATSDR, 1999

2011). All these are present as pollutants in the effluents of such industries.

15.2.4 Textile Effluents

Textile industries are mainly responsible for water and air pollution. Every year the textile industry discharges 400 to 500 million toxic chemicals used in the processing of fibers, which includes sulfur, cyanide and other radioactive substances that are directly discharged into water bodies and rivers (UNEP, 2010). Textiles consume vast quantities of natural resources, such as water, oil, and land; in addition, they use toxic chemicals and generate large amounts of carbon dioxide. Millions of tonnes of textiles are discarded every year. In Europe and America, it is estimated that 10 million tonnes of textiles are discarded every year.

In China the total annual production of pre- and postconsumer textile waste is assessed to be more than 20 million tonnes. Not only does this textile waste pollute the environment and clog landfills around the world, but the vulnerable resources that went into making these textiles are also wasted. It generates tons of waste in the form of dumps of clothes. The clothing takes time to break down, and some textiles, such as polyester, even create chemical pollution leading to impact the desired quality of air and water need for human consumption. The dyes used for textiles are used with water and chemicals, which when dumped in soil or water badly affects the groundwater and soil. The burning of all the products inside the mills also pollutes the air with chemicals. Textile industries perform various operations in processing fibers to final products that require a number of chemicals and finally discharged effluent carries harmful residual chemicals and process water (Table 15.2). Different chemicals are used in the treatment of fiber, such as acids, alkali, bleaches, salt, stabilizers, and surfactants, among others.

15.2.5 Petroleum Industries

Hydrocarbons are important for meeting the energy requirements for development, prosperity and growth of the global economy until any new reliable, renewable, more efficient and less polluting source of energy replace them. Petroleum industries have a significant share in the total environmental pollution due to the operations carried out for oil exploration, production, gas flaring, oil spillage, oil refining and so on (Olajire, 2014). Petroleum is a naturally occurring material in underground reservoirs and can exist in the form of crude oil, coal, tar sand, bitumen and natural gas. It has many impurities such as hydrogen sulfides, carbon dioxide, nitrogen metal, water and other material, so its further processing is needed to be used as fuel and to manufacture other products derived from petroleum (Olajire, 2014). Petroleum-contaminated soil and sediment commonly contain a mixture of PAHs and heterocyclic aromatics (Seo et al., 2009). Most of the pollution is caused by oil pipeline leakage, gas flaring, oil waste dumping on land and oil spillage from cargo ships during transportation. Gas flares produce soot, this soot, when it mixes with rainwater, adversely affect the fertility of the surrounding soil. Leakage of oil from oil-transporting pipelines also severely affects soil and water bodies (Olajire, 2014). Petroleum processing uses a large amount of water, which is recycled in the refineries after it is treated, some of which is lost during this process. A large amount of chemicals are also used in petroleum processing, and the remaining oily solid and glycol-based dehydrated waste is discarded out in the environment (API, 2009). Aqueous waste is mostly discharged in water bodies by various sources, such as liquid effluent pipeline systems and others, which makes nonpolar oil films on the surface of the water and prevents the exchange of gases necessary for aquatic life.

15.2.6 Food-Processing Wastes

The food-processing industry is a major industry that produces very large amounts of wastes. Food-processing industries, including wheat mills, paddy mills, milk and dairy, beverage, soft drink bottling, vegetable oil and meat-processing industries, operate on a large scale; wastes produced in the form of gas (smoke); solid and residual water are thrown in open environment sources such as water bodies and land (Newton, 1970; Okereke, 2007). The waste streams and by-products generated from food processing are rich in lipids, carbohydrates and proteins, which contribute to its negative environmental impact (Andler et al., 2017). The dairy, poultry, brewery and food-processing industries produce solid waste, such as vegetable and fruit trimmings, dry brewer's grain, spent grain, plastic packet and bottles, spent hop liquor and liquor from yeast recovery systems and others, and generate large heaps of foul-smelling solid waste (Newton, 1970; Otta and Kendall, 1987). The chocolate industry is among the most water-polluting

TABLE 15.2

Chemicals in the Textile Industry Effluent and Their Hazardous Effects (modified from original source, Jaganathan et al., 2014)

Chemical Effluent	Textile Process	Hazardous Effect
Benzene	Dying Sizing	Carcinogenic, mutagenic and affects central nervous system
Cyanide	Bleaching	Prolonged exposure will affect kidney and liver and lead to death
Sulfate	Dyeing	Eye and respiratory problems
Nitrate	Printing	Methemoglobinemia
Phosphate	Printing	Kidney damage and osteoporosis
Lead	Finishing	Suppression of hematological system

industries in terms of water consumption and discharge of effluents. Many chemicals are also used in the various operations of food treatments like sulfuric acid, acetaldehyde, methanol, hydrochloric acid, ammonia and nitrate compounds, among others. The wastes generated by food industries have the advantage over other wastes of being amenable to conversion into food, feed or nonfood products with value addition. Recently, the potential for using enzymes in food waste handling has been suggested. Enzymes could be used to decrease food wastes via enzymatic processing to yield higher value by-products and to aid in the cleanup of food waste streams.

15.2.7 Paper Wastes

Paper is a major product of the forestry industry and is used widely in our society. Paper is made by pulping wood, bleaching this pulp and then spreading it out into sheets to make it into paper. At various stages of the process, chemicals are used to impart the paper particular properties, such as the bleaching chemicals that make paper white. Effluents from pulp and paper mills are only weakly toxic by conventional measurements, especially if they have been through biological treatment (International Finance Corporation, 2007). However, the effluent from the bleaching process contains a variety of substances, some of which are known or suspected of being toxic, genotoxic or mutagenic (Xiog et al., 2001; Singh et al., 2007). Chlorinated organics generated in the chlorine bleaching processes are of particular concern. The Kraft process is largely used in papermaking these days. The residual process is responsible for the characteristic brown color of the pulp and is commercially removed by the use of bleaching agents, for example, chlorine and chlorine oxides. Bleaching operations generate dark brown–colored effluents that contain environmentally hazardous, toxic and mutagenic chlorinated products (Kringstad and Lindstrom, 1984; Solomon et al., 1993). Dioxin is the family of 210 chlorinated chemicals known as chlorinated dibenzo-para-dioxins and chlorinated dibenzofurans. The most toxic among these compounds is 2,3,7,8-tetrachlo rop-dibenzodioxin (2,3,7,8-TCDD or TCDD) and the closely related compound 2,3,7,8-tetrachlorodibenzofuran (2,3,7,8-TCDF or TCDF) (Martin and Stutt, 2009). These chemicals can be formed during the manufacture of bleached wood pulp if chlorine is used as a bleaching agent. The chlorinated components of the waste stream consist of a variety of compounds, including simple phenols, high- and low-molecular-weight polymers and neutral and acidic materials from the breakdown of the phenolic rings in lignin (Solomon et al., 1993).

15.2.8 Heavy Metal Ions

Inorganic pollutants are composed of heavy metals such as arsenic, cadmium, mercury, lead and chromium. The heavy metals are dangerous contaminants released from different industries as effluent and mining waste streams, as well as in solid wastes, municipal sewage sludge, nuclear radiation and releases of heavy metals and landfill leachate (Akhtar et al., 2013).

Moreover, metals, such as mercury, lead and arsenic, have exhibited their toxic effect on the kidneys and the nervous system, which may lead to the symptoms of mental disorder and cause weakness, headaches, abdominal cramps, diarrhea and anemia. The sustained exposure to these metals may lead to permanent damage of organelles. Cadmium, for instance, is present in wastes from the mixing, smelting, electroplating and pigment manufacture while other metals are used in the nuclear power, defense and fuel-reprocessing industries. In view of the high-toxicity risk of these heavy metals, they must be prevented from being discharged or dumped into receiving water bodies or soil. Low concentrations of heavy metals lead to a decrease in metabolic activity while high concentrations could act in a more deleterious way by blocking essential functional groups, displacing other metal ions or modifying the active conformation of biological macro- and other molecules. Besides, they are toxic to both higher organisms and microorganisms.

15.3 Chemical Treatment Methods of Pollutants

The aromatic contaminants, such as pesticides, volatile organic compounds, polychlorinated biphenyls (PCBs), chloroderivatives and PAHs, are treated by various chemical methods such as the Fenton system, photooxidation, ozonation, dechlorination, chemical oxidation and reductions (Fox, 1996).

The commonly used oxidants in the chemical oxidation process are potassium permanganate, chlorine, ozone, H_2O_2, Fenton's reagent and chlorine dioxides (Gogate and Pandit, 2004). In chemical oxidation, aromatic contaminants present in the effluent are easily oxidized to biodegradable products such as water, carbon dioxide, alcohols, aldehydes, ketones and carboxylic acids (Gupta et al., 2012). A single oxidation system is not sufficient for the complete decomposition of aromatic pollutants present in effluents from textile (halogenated anilines, benzenes, anthraquinones, etc.), tannery (phthalic acid, phenol, cresol, polyethylene glycol, benzoic acid, cyclohexane carboxylic acid, phenylacetic acid, methoxycinnamic acid, etc.), petrochemical (xylene, styrene, methylstyrene, indane, etc.), paper and pulp (doline, toluamide, hydrogenated quinolone derivatives, etc.) and other industries (Dsikowitzky and Schwarzbauer, 2013).

Advanced oxidation processes (AOPs) involves the simultaneous use of more than one oxidation process for the production of the highly reactive hydroxyl free radical (Yoon et al., 2001; Comninellis et al., 2008; Chong et al., 2010). AOPs use strong oxidizing agents and high-energy radiation that nonselectively oxidize a range of recalcitrant organic pollutants, such as benzene, benzoquinone, phenols, dyes, formaldehyde (Lofrano et al., 2013; Dixit et al., 2015) detergents, pesticides and others, and can also be applied for the oxidation of inorganic contaminants, such as cyanides, sulfides, and nitrites (Munter, 2001), into stable inorganic compounds, such as water, carbon dioxide and salts (Rameshraja and Suresh, 2011). There are various types of AOPs such as chemical oxidation processes using ozone (Wu et al., 2004), combined ozone and peroxide, ultraviolet-enhanced oxidation with Fenton (Ghaly et al., 2001), hydrogen peroxide and ozone (Esplugas et al.,

TABLE 15.3

Chemical Methods in the Treatment of Various Types of Environmental Contaminants

Type of Contaminants	Sources of Effluent	Type of Chemical Method
Phenols and aromatic amines, volatile organic compounds, polychlorinated biphenyls (PCBs), polycyclic aromatic hydrocarbons (PAHs)	Petroleum, refining, coal, resins and plastics, wood preservation, metal coating, textiles, mining and dressing, and pulp and paper industries	Coagulation, flocculation, chemical oxidation, and reduction, ozonation
DDT, endosulfane, lindane, and pentachlorophenol	Agriculture pollutants	Advanced oxidation process (AOP), ozonation, sonolysis, and photocatalysis
Fats, lime, alkali, bleaching agents, heavy metals	Leather industries	Ion exchange method, AOPs, solvent extraction, chemical precipitation, coagulation, flocculation
Benzene, cyanide, sulfate Nitrate, phosphate, lead	Textile industries	Chemical redox, ion exchange AOPs, coagulation, flocculation
Oily solid and glycol-based dehydrated waste		AOPs
Chlorinated organics	Pulp and paper industries	AOPs, coagulation, flocculation
Heavy metals (arsenic, copper, cadmium, lead, chromium)	Electroplating, fertilizer, nuclear fission plant, metal smelter, pesticides, herbicides, automobile	Ion exchange, AOPs, solvent extraction, chemical precipitation

2007), catalytic wet air oxidation, sonolysis and photocatalysis (Li et al., 2003). The Fenton system and ozonation are used for the remediation of more recalcitrant contaminants, such as PAHs and PCBs from contaminated soil (Awaleh and Soubaneh, 2014; Das et al., 2014). In the Fenton reaction, a high concentration of Fe^{2+} and low pH values are required; this necessitates an additional acidification step with an undesirable precipitate of ferric hydroxide. An excess of ozone must be destroyed because it is a potentially hazardous, highly irritating and toxic gas, and its oxidizing action is nonselective, with a high cost of generation and transportation (Rodríguez, 2003; Awaleh and Soubaneh, 2014).

In chemical reduction, aromatic contaminants are transformed into less toxic compounds by a reducing chemical reaction, such as dehalogenation or dechlorination. A potential limitation of these processes is the formation of by products, such as nitrosamines, bromate or formaldehyde, due to partial oxidation of the compounds and reaction with matrix components (Von Gunten, 2003; Wert et al., 2007).

Phenol is used as a commodity chemical in manufacturing processes, such as synthetic resins, plastics, pharmaceuticals and intermediate of dyes (Liu et al., 2013). Solvent extraction is an effective method to remove phenolic compounds from industrial effluent. Many extractants have been applied for the extraction of phenol, such as octanol (Jiang et al., 2003), diethyl carbonate (Olejniczak et al., 2005), amines (Xu et al., 2006) and cumene (Liu et al., 2013). Chemical precipitation is applied for the removal of heavy metals (arsenic, cadmium, chromium, copper, nickel, zinc, lead) and organic contaminants from industrial effluent but affected by the oily and greasy substances. The most commonly used chemicals for this purpose are alum, sodium bicarbonates, ferric chloride, ferrous sulfate and lime (Bhargava, 2016). The pH and temperature are the primary controlling factors for the precipitation process.

Coagulation and flocculation are used for the removal of suspended and dissolved solids, colloids and organic matter present in various types of wastewater, such as effluent of palm oil mills, pulp and paper mills (Wong et al., 2006; Razali et al., 2011), sanitary landfill leachates, coal waste slurry (Sabah and Erkan, 2006) and the tannery and textile industries (Renault et al., 2009; Lee et al., 2014). The commonly used chemical coagulants are aluminum coagulants (alum, aluminum chloride, sodium aluminate) and iron coagulants, such as ferric chloride, ferric sulfate, ferrous sulfate and ferric chloride sulfate and lime (Bhargava, 2016). These coagulants are removing pollutants but are usually more expensive than natural coagulants. The disposal of the resulting sludge, the application of chemicals and the transfer of toxic compounds into the solid phase are major limitations of this process. Commercially used polymeric flocculants include polyacrylamide, polyacrylic acid, polydiallyl dimethyl ammonium chloride and polyamine (Singh et al., 2000). Although water-soluble synthetic polymers find wide applications as flocculants, however, their usage is debatable because of their market cost, environmental consequences, and health hazards (Lee et al., 2014). The monomer units and reaction by-products of the polymers in water are toxic to aquatic organisms (Bolto and Gregory, 2007; Wu et al., 2012).

The treatments of food wastes included oxidation, hydrolysis, acylation, phosphorylation of carbohydrates, deamination, glycosylation of proteins and hydrogenation and esterification of lipids. The esterification reactions are widely employed to manufacture several value-added food and agricultural products. Oils can be esterified with alcohol to produce biodiesel, sugars can be esterified for use as surfactants, starches can be esterified for use as biodegradable plastics, coatings and hot-melt adhesives and esterification of flavonoids can improve their bioavailability and efficacy in promoting health and wellness. Traditional approaches to these modifications require a chemical catalyst and significant energy input, have limited reaction specificity and result in the formation of by-products, particularly when performed in complex matrices, such as food waste streams. While the enzymes offer benefits over traditional chemical catalysts in the valorization of waste streams.

The common procedures for heavy metal removal are chemical precipitation, dialysis, ion exchange, reverse osmosis, and solvent extraction. Lime or caustic soda is used for precipitation of heavy metals in hydroxides, sulfides, carbonates and phosphates (Gunatilake, 2015). Chemical precipitation requires a large amount of chemicals to reduce metals to an acceptable level for discharge. These are very expensive and low-efficiency methods. Sometimes, it may change the soil properties and spread contaminants from one to other places after remediation (Barakat, 2011). Ion exchange is a reversible reaction in which a toxic metal ion such as arsenic, cadmium, chromium, copper, lead, nickel and zinc in effluent from leather, tannery and electroplating industries are exchanged with a similarly charged ion, which is electrostatically attached to an immobile solid particle (Mier et al., 2001; Dabrowski et al., 2004). The drawback of this method is that it cannot handle concentrated metal solutions as the matrix gets easily fouled by organics and other solids in the effluent. Moreover, ion exchange is nonselective, is highly sensitive to the pH of the aqueous phase and produces waste by-products (Barakat, 2011).

15.4 Advantages of Enzymes over Chemical Methods for Treatment of Pollutants

The previous section has shown that chemical treatment methods of pollutants are used as an essential pretreatment, post-treatment and main treatment process for various industrial effluents. However, there are many drawbacks accompanying these chemical treatments, such as the toxicity and health hazards possessed by inorganic coagulants, the production of secondary pollutants from chemical treatment processes that required extra cost and separate handling and their being less effective in the complete removal of heavy metals or emerging contaminants. There is the complexity of scaling up the chemical-based treatment process that requires precise control of various parameters to ensure the efficiency of the process. Furthermore, each method is also found to be specific on the type of wastewater. Oxidation is the nonselective process and decreases their effectiveness due to reaction with free radicals. Oxidizing agents promote strong interaction with surrounding materials, frequently leading to degradation required in high concentrations; overdosing is a serious cause of health problems and must be removed. The cost of chemicals and instruments is relatively high, and transportation is potentially hazardous. Some chemicals are explosive when concentrated and can be fatal if swallowed. For these reasons, at present, there is a great interest in improving and implementing new safe, cost-effective and eco-friendly wastewater treatment methods. The introduction and implementation of stringent standards for waste discharge into the environment have necessitated the need for the development of enzymatic processes.

Enzymatic treatment has been proposed by many researchers as a potential alternative to conventional methods as enzymes are highly selective and can effectively treat even dilute wastes and are less likely to be inhibited by substances that may be toxic to living organisms and their cost could eventually be less than that of other methods if commercially available enzymes are produced in bulk quantities. Moreover, enzymes operate over a broad substrate concentration range and require low retention times with respect to other treatment methods.

The major category of enzymes involved in chemical pollutant degradation reactions includes hydrolases, oxidoreductases and halogenates. The potential advantages of enzymatic treatment as compared with conventional treatment include application to biorefractory compounds; operation at high- and low-contaminant concentrations; operation over a wide range of pH, temperature, and salinity; absence of shock loading effects; absence of delays associated with the acclimatization of biomass; reduction in sludge volume (no biomass generated); and the ease and simplicity of controlling the process (Gianfreda and Bollag, 2002). In recognition of these potential advantages, recent research has focused on the development of enzyme processes for the treatment of wastewaters and solid wastes.

15.5 Enzymes in Treatment of Pollutants

15.5.1 Treatment of Aromatic Contaminants

The representative classes of enzymes in pesticide remediation involved hydrolases, dehalogenases, transferases, oxidoreductases and mixed-function oxidases (Li et al., 2007). Hydrolases are phosphotriesterases, amidases, proteases, carbohydrases (cellulases and amylases) and depolymerases. Oxidoreductases are mono or di-oxygenases, reductases, dehalogenases, cytochrome P450 monooxygenases, phenoloxidases (laccases, tyrosinases) and peroxidases (lignin and manganese peroxidases). The breakdown of esteric, amidic and peptidic bonds by esterases, amidases and proteases lead to less toxic products. For instance, bacterial hydrolases, such as carbamate or parathion hydrolases, have been successfully used in the transformation of pesticide pollutants, such as carbofuran and carbaryl or parathion, diazinon and coumaphos. The desulfurization, dehalogenation, denitrification, ammonification, hydroxylation, biotransformation and biodegradation of various aromatic and aliphatic compounds are catalyzed by monooxygenases. These properties have been explored in recent years for important application in biodegradation and biotransformation of aromatic compounds. Methane monooxygenase enzyme is the best characterized one, involved in the degradation of substituted methanes, alkanes, cycloalkanes, alkenes, haloalkenes, ethers and aromatic and heterocyclic hydrocarbons (Figure 15.2). Under oxygen-rich conditions, monooxygenase catalyzes oxidative dehalogenation reactions, whereas, under low oxygen levels, reductive dechlorination takes place.

15.5.2 PAH Degradation

PAHs in the environment are degraded via volatilization, photo-oxidation, chemical oxidation, bioaccumulation and adsorption to soil particles, with the principal role of microbial transformations and degradation reactions (Tian et al., 2008; Baboshin and Golovleva, 2012). The microbial degradation of alkyl PAHs involves a wide variety of enzymes from aerobic bacteria (Mahajan et al., 1994). The reactions catalyzed by

TABLE 15.4

Different Types of Contaminants and Their Enzymatic Treatment and Application

Type of Contaminant	Sources	Enzymes for Treatments	Application of Enzymes
Phenols and aromatic compounds	Coal conversion, petroleum refining, resins and plastics, metal coating, textiles, mining and dressing, pulp and paper industries		Removal of phenols and oxidation of ortho and paradiphenols, aminophenols, polyphenols, polyamines, lignins, and aryl diamines Removal of aromatic compounds Oxidation of monoaromatic phenols and aromatic dyes Removal of phenols and aromatic amine
Pesticides	Agriculture waste	Parathion hydrolase	Hydrolyzation of organophosphate pesticides
Refractory organics	Paper and pulp mill wastewater	Alkylsulfatase	Surfactant degradation
Heavy metals	Industrial and mining waste,	Phosphatase	Removal of heavy metals
Cellulosic sludge	Pulp and paper industry	**Cellulolytic enzymes:** cellulase, cellobio-hydrolase, cellobiase	Cellulose degradation
Textile waste	Cloth mills,	Laccase, peroxidases, cellulase xylanase	Decolorization of Kraft bleaching effluents Removal of cellulose microfibril
Paper and pulp waste	Pulp and paper industry	Cellulase	Removal of ink during paper recycling
Dairy waste	Dairy industry	Lactases	Dairy waste processing and production of value-added products
Cyanide	Chemical, synthetic fibers and rubber synthesis, pharmaceuticals, ore leaching, coal processing and metal plating	Cyanide hydratase Cyanidase	Conversion to cyanide to ammonia and formic acid/formamide

TABLE 15.5

Example of Aromatic Contaminants and Enzymes Involved in Their Degradation

Aromatic Compounds	Potential Source	Enzymes in Degradation	Mechanism of Enzyme Action
Pesticides: Organochlorines (derivatives of diphenyl ethane, cyclodienes and chlorinated hydrocarbons) Organophosphates (aliphatic, phenyl and heterocyclic derivatives) Carbamates (esters of carbamates and organic compounds derived from carbamic acid Glyphosates	Pesticide manufacture industries, pesticide applications in agriculture	Dehalogenases Phosphotriesterase, Organophosphorus hydrolase (OPH) and Organophosphorus acid anhydrolase (OPAA) Carboxyl esterases Peroxidases and Laccases	Reductive dehalogenation, dehydrochlorination Hydrolysis of the P-O alkyl or P-O aryl bonds Acylation, hydrolysis Oxidative dehalogenation
Volatile organic compounds: Benzene, Toluene, Xylenes, Chloromethane, Trichloroethanes	Industrial and commercial waste, chemical factories, dumping sites	Monoxygenases, Dioxygenases, Halogenases	Oxidation, Reductive dechlorination
Polychlorinated biphenyls (PCBs)	Electrical equipment	Mono- and dioxygenases, Laccase (LAC), Mn-dependent peroxidase (MnP), and Mn-independent peroxidase (MiP)	Oxidation
Chloroderivatives: Chlorophenols, Chlorobenzene	Paper mill effluents, wood industries	Chlorocatechol dioxygenases, cycloisomerases, chloromuconolactone isomerase and dienelactone hydrolases	Hydrolysis, dehalogenation and cycloisomerization
Polycyclic aromatic hydrocarbons (PAHs): Naphthalene, Anthracene, Fluorine, Phenanthrene	Fossil fuel wastes, creosote waste sites, crude oil, coal tar, and wastes from the combustion of fossil fuel, coal gasification and liquefaction and incineration of industrial wastes	Laccase, decarboxylase	Oxidation

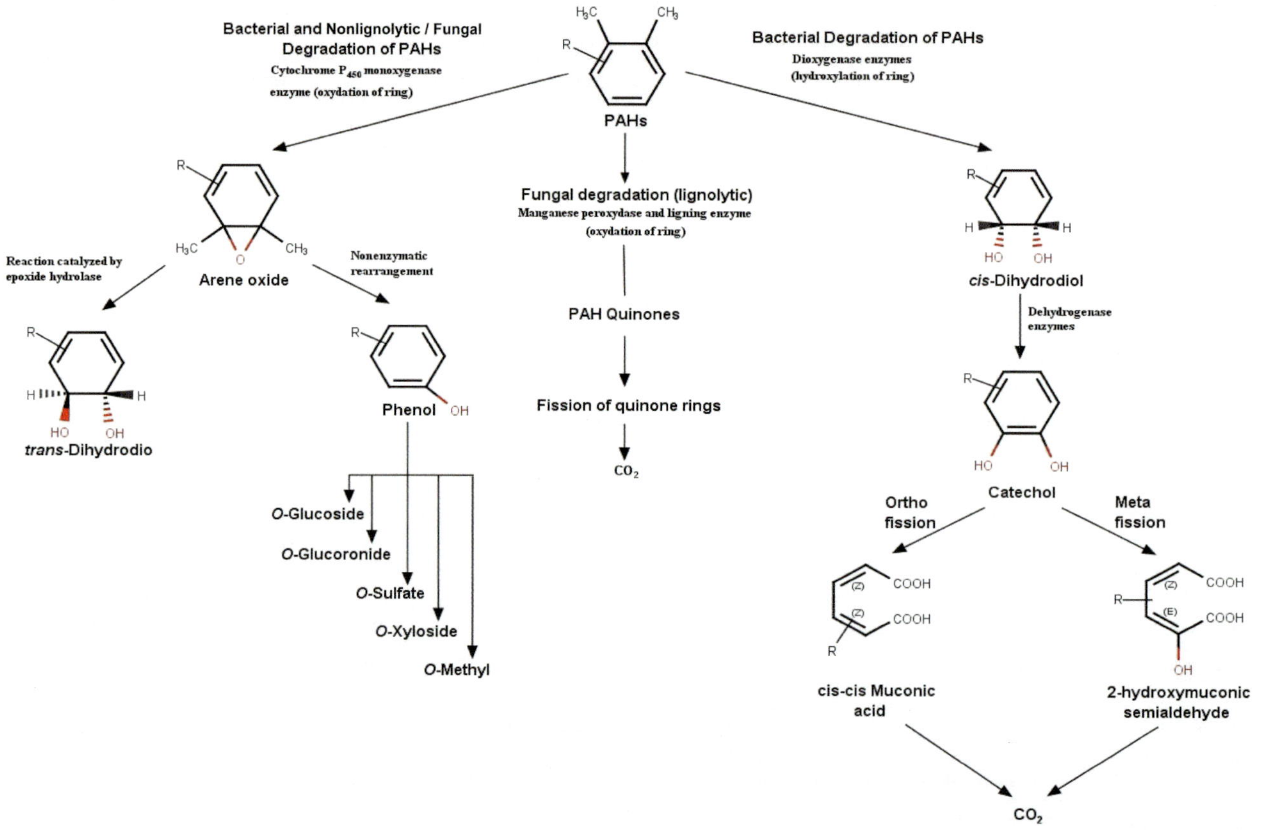

FIGURE 15.2 Pathways for the degradation of polycylic aromatic hydrocarbons (PAHs; modified from original source, Bamforth and Singleton, 2005).

such enzymes involve initial activation and oxidation reactions, such as oxidation of methyl group to alcohol, aldehyde or carboxylic acid, decarboxylation, demethylation and dioxygenation. Oxygenases are the important enzymes produced by microorganisms that catalyze oxygen-fixing reactions. The two groups of oxygenases, that is, monooxygenases and dioxygenases, can be used to degrade different types of PAHs (Maigari and Maigari, 2015). The other two proposed mechanisms for PAH degradation by bacteria are mineralization and co-metabolism. In mineralization, bacteria can use some PAHs as their sole carbon and energy source. The proposed mechanism pathways involved in the degradation of PAHs are presented in Figure 15.2.

15.5.3 Enzymes in Pesticide Degradation

Applications of enzymes in waste treatment arise from the disposal of wastes formed during the production and formulation of pesticides, the detoxification of pesticide containers and spray tanks and the pollution of surface and groundwater by pesticide runoff. There are several enzymes, such as esterases, phosphotriesterases, parathion hydrolases and diisopropylfluorophosphatases, that have been studied for the transformation reactions of such compounds (Kumar et al., 2018). The main enzymes involved are discussed in the following subsections.

15.5.3.1 Hydrolases

Hydrolases are a broad group of enzymes involved in pesticide biodegradation that catalyze the hydrolysis of several major biochemical classes of pesticide and generally operate in the absence of redox cofactors, making them ideal candidates for all of the current bioremediation strategies (Scott et al., 2008). Carbofuran pesticides can be transformed in the environment and different metabolites are generated and accumulated in potentially contaminated sites (soil, water, and sediments, mainly).

15.5.3.2 Phosphotriesterases

Phosphotriesterases (PTEs) are one of the most important studied groups of enzymes to degrade pesticides (Chino-Flores et al., 2012). These enzymes have been isolated from different microorganisms that hydrolyze and detoxify organophosphate pesticides and reduce their toxicity (Worrall et al., 2001; Ghanem and Raushel, 2005). The phosphotriesterases are encoded by a gene called opd (organophosphate-degrading). These enzymes specifically hydrolyze phosphoester bonds, such as P–O, P–F, P–NC and P–S, and the hydrolysis mechanism involves a water molecule at the phosphorus center (Ortiz-Hernandez et al., 2003). In the PTE organophosphorous pesticide hydrolysis mechanism, the enzymatic active site removes a proton from water, activating this molecule; the activated water directly attacks the central phosphorus of the pesticide molecule, producing an inversion in its configuration. The oxygen is polarized by the active site, with the participation of a zinc atom (Ortiz-Hernandez et al., 2003; Kapoor and Rajagopal, 2011).

15.5.3.3 Esterases

Esterases can be categorized into three groups according to the nature of their interactions with organophosphorus insecticides. They have the capability to hydrolyze organophosphate pesticides. In the esterase-catalyze reactions, a wide range of substrates are converted to respective alcohol and acid products. One category of esterases is carboxylesterases, which are capable of hydrolyzing ester bonds of many insecticides, such as organophosphates, carbamates and pyrethroids, having carboxylic ester (Karigar and Rao, 2011). These are the enzymes that can hydrolyze ester bond–containing xenobiotic compounds (Hatfield et al., 2016).

15.5.3.4 Oxidoreductases

Oxidoreductases are the enzymes that carry out oxidation–reduction reactions by using O_2 as an electron acceptor, with its reduction to water or hydrogen peroxide. Additional cofactors are also required by many of these enzymes to meet the requirement of either an electron acceptor or an electron donor. Several of these have applications in bioremediation as they catalyze an oxidation/reduction reaction by including the molecular oxygen as an electron acceptor, which is further reduced to water or hydrogen peroxide. One of the applications of such enzymes is in the degradation of pesticides, such as endosulfan, an organochlorine compound of the cyclodiene family of pesticides that is highly toxic and an endocrine disruptor. The endosulfan degradation pathway involves either oxidation or hydrolysis reactions via oxidoreductase enzymes. Several intensive studies on the degradation of endosulfan have been conducted showing the primary metabolites to normally be endosulfan sulfate and endosulfan diol (Ortiz-Hernandez et al., 2013).

15.5.3.5 Mixed-Function Oxidases

Mixed-function oxidase (MFO) (EC 1.14.14.1) is an enzyme system comprising two enzymes, cytochrome P450 and nicotinamide-adenine dinucleotide phosphate (NADPH)–cytochrome P450 reductase. It requires NADPH and O_2 for its activity in which an atom of one molecule of oxygen is incorporated into the substrate and the other is reduced to water (Ortiz-Hernandez et al., 2013). They are also known as cytochrome P-450 monooxygenases systems and are active in the metabolism of a wide variety of xenobiotics. The cytochrome P450 family is a large, well-characterized group of monooxygenase enzymes with the potential to metabolize both endogenous and exogenous substances. Due to this capability, these enzymes can degrade a wide range of compounds such as organophosphates, carbamates, pyrethroids, DDT and others (Alzahrani, 2009).

15.5.3.6 Parathion hydrolase

It is also known as phosphotriesterase produced by a number of bacteria and has a role in pesticide detoxification; thus, it is used in more common treatment methods. It has been shown to hydrolyze a number of the most widely used organophosphate pesticides, such as methyl and ethyl parathion, diazinon, fensulfothion, dursban and coumaphos. Coumaphos is an insecticide used to kill cattle disease ticks by dipping the cattle in large vats of this chemical. Parathion hydrolase has been shown to successfully hydrolyze coumaphos much more rapidly than chemical hydrolysis.

15.5.4 Enzymes in Textile Effluent Treatment

The effluents of the major operations in the textile industry that heavily pollute water are dyeing, desizing and scouring. Dyeing effluents generating colored compounds and toxic heavy metals are the serious problems associated with textile industries. There is preferred use of synthetic dyes over natural dyes in dyeing industries, and such dyeing effluents are nondegradable by conventional treatment methods due to their persistence and stable nature as well as disadvantages of the used methods (Maddhinni et al., 2006). Several physico-chemical methods such as ozonation, coagulation or flocculation, adsorption and membrane filtration have been used for

TABLE 15.6
Enzymes Involved in the Degradation of Pesticides

Enzyme	Pesticide/Herbicide
Oxidoreductases	Glyphosate
	Endosulfan and endosulfato Atrazine, norflurazon and isoproturon Hexachlorobenzene and pentachlorobenzene
Dioxygenases	Trifural
Phosphotriesterases: Parathion hydrolase Organophosphorous hydrolase	Ethyl parathion, diazinon, fensulfothion, dursban and coumaphos Insecticides phosphotrieste
Haloalkane dehalogenases	Dicamba

decolorization of azo dyes; however, these methods are expensive and generate large amounts of sludge after treatment that requires safe disposal (Uday et al., 2016). The use of enzymes in the textile industry has proved to be eco-friendly, nontoxic and environmentally friendly over the conventional treatment methods. A majority of the enzymes being investigated for their dye degradation potential belong to the enzyme class oxidoreductases. These enzymes are involved in electron transfer reactions. Ligninolytic enzymes, such as lignin peroxidase, manganese peroxidase and laccase, are produced by white rot fungi and have emerged as a reliable alternative to the conventional effluent treatments (Mojsov et al., 2016).

15.5.4.1 Lignin Peroxidase

One of the largest classes of dyes used in the printing and textile processing industries involves azo dyes, characterized by the functional azo group (-N=N-). Enzymes produced by white rot fungi have shown great potential in degrading azo dyes and related effluents. Lignin peroxidases are one of such group of enzymes that oxidize a large number of different compounds and, therefore, can be used for textile wastewater treatment. They can degrade azo dyes by oxidation of the phenolic group to produce a free radical at the carbon bearing the azo linkage. The phenolic carbon is then attacked by a water molecule, resulting in its breakdown to produce phenyldiazene, which is easily oxidized by a one-electron reaction to generate nitrogen (Rodriguez et al., 1999).

15.5.4.2 Laccase and Azo Reductase

Azo reductases and laccases are the promising enzymes in the remediation of dyes. Laccases find applications in the treatment of industrial effluents, such as textile, and have been extensively studied for their degradation of azo dyes (Chivukula et al., 1995; Kirby et al., 2000; Peralta-Zamora et al., 2003; Blanquez et al., 2004; Novotny et al., 2004). These enzymes are oxidases, and their mechanism involves the removal of H^+ ion from hydroxyl and amino groups of the ortho- and para-substituted mono- and polyphenolic substrates and aromatic amines. Laccases decolorize azodyes through a highly nonspecific free radical mechanism forming phenolic compounds, thereby avoiding the formation of toxic aromatic amines (Chivukula et al., 1995; Wong and Yu, 1999). Azo reductase catalyzes the reductive cleavage of azo groups in the azo dyes, thus converting these groups to amines (Zimmermann et al., 1982).

15.5.5 Food Waste Treatment

Enzymes could be used to decrease food wastes via enzymatic processing to yield higher value by-products and to aid in the cleanup of food waste streams. The most commonly used enzymes in food waste treatment are discussed in the following subsections.

15.5.5.1 Proteases

Proteases are a group of hydrolases that are widely used in the food industry in processing fish and meat waste. Proteases can solubilize proteins in waste streams, resulting in recoverable liquid concentrates or dry solids of nutritional value for fish or livestock. Proteases hydrolyze insoluble proteins through a multistep process whereby the enzyme, which is initially adsorbed on the solid substrate, cleaves of polypeptide chains that are loosely bound to the surface. Then, the solubilization of the more compact core occurs at a slower rate dependent on the diffusion of the enzyme to surface active sites and core particles. Feathers constitute 5% of the body weight of poultry and can be considered as a high protein source for food and feed provided their rigid keratin structure is completely destroyed. Total solubilization of the feathers was achieved after pretreatment with NaOH, mechanical disintegration and enzyme hydrolysis. The end product was a heavy, grayish powder with a very high protein content that could be used mainly as a feed supplement.

15.5.5.2 Amylases

Amylases belong to hydrolases that have been used in the saccharification and fermentation of starch and can be used in the treatment of starch-based food waste streams (Blasheck, 1992). Such waste streams arise from the processing of starch-rich food contents, such as corn, rice, potato and sweet potato (Andler and Goddard, 2018). Amylases can be used to produce alcohol from rice-processing wastewater. It was also reported that amylases can be used to break down longer starch molecules into smaller fragments and enhance activated sludge wastewater treatment, followed by the attack of glucoamylase to produce glucose through saccharification. The glucose is fermented to lactic acid strains of lactic acid bacteria. Lactic acid is finally recovered, purified and used in the production of photo and biodegradable plastics. Thus, this enzyme can be used to form useful products from the waste (Karam and Nicell, 1997).

15.5.5.3 Lactases

Lactose is present at a concentration of about 4.7% (w/v) in milk and the whey (supernatant) left after the coagulation stage of cheese-making. Lactose has the problem of its low solubility resulting in crystal formation at higher concentrations. The disposal of lactose waste in whey is expensive due to its high biological oxygen demand. These problems can be overcome by hydrolysis of the lactose by lactase enzyme, which makes it more soluble and capable of forming concentrated, microbiologically secure syrups.

15.5.5.4 Lipases

These are the enzymes that hydrolyze triacylglycerols to glycerol and free fatty acids and can be extracted from bacteria, plant, actinomycetes and animal cell. Lipases can be used to treat contaminated soil containing hydrocarbons and oil spills (Karigar and Rao, 2011). They can carry out hydrolysis, esterification, alcoholysis and ammonolysis reactions (Prasad and Manjunath, 2011).

TABLE 15.7

Heavy Metal Ions from Different Sources and Their Enzymatic Remediation

Heavy Metal Ion	Source of Release	Enzyme Remediation	Mechanism of Action
Mercury	Pesticides, batteries, paper industry	Mercuric reductase	Reduction
Chromium	Chemical factories	Chromate reductase	Reduction
Cadmium	Welding, electroplating, pesticides, fertilizers, batteries, nuclear fission plant	Cadmium containing carbonic anhydrase	Chelation
Arsenic	Anthropogenic discharge, pesticides, fungicides, metal smelting	Oxidoreductase	Oxidation, methylation, reductive detoxification
Selenium	Agriculture drainage, organic Se-containing compounds	Selenate reductase, S-adenosyl-methionine dependent selenide methyltransferase	Reduction, methylation

15.5.6 Paper Wastes

Chlorinated phenolic compounds are among the most abundant recalcitrant wastes found in the effluents generated by the paper and pulp industry. These compounds are produced on the partial degradation of lignin during the pulp-bleaching process. Many fungal species are considered suitable for the removal of chlorinated phenolic compounds from contaminated environments. The activity of fungi is mainly due to the action of extracellular oxidoreductase enzymes, such as laccase, manganese peroxidase and lignin peroxidase, which are released from fungal mycelium into their nearby environment (Karigar and Rao, 2011).

15.5.6.1 Laccase and Peroxidases

Laccases are multicopper enzymes that belong to the family of blue-oxidase enzymes with the capability to degrade a wide variety of aromatic compounds and, therefore, are very useful in their application in bioremediation. Laccases are useful in the treatment of wastewater from textile industries due to their capability to decolorize textile dyes and have a broad substrate specificity. Peroxidases and laccases are the enzymes that can perform the task to treat the bleaching effluents. Kraft mill effluents colors can be removed by the use of horseradish peroxidase and lignin peroxidase enzymes.

15.5.6.2 Cellulolytic Enzymes

Cellulosic materials are polysaccharides that can be hydrolyzed to monosaccharides. Cellulase enzymes are capable of degrading cellulose to glucose and can be employed to remove cellulose microfibrils, formed during washing and the use of cotton-based cloths (Karigar and Rao, 2011). The cellulolytic enzymes can be used in the treatment of sludge from pulping and ink removal operations, thereby converting sludge into fermentable sugars (Duff et al., 1994). The cellulolytic enzymes include a blend of cellobiohydrolase, cellulase and β-glucosidase enzymes to treat cellulose waste (Andler and Goddard, 2018; Duff et al., 1994).

15.5.7 Heavy Metal Ions Treatment

The heavy metal–contaminated soil, sediment, and water sources through industrial activities, such as mining, refining and electroplating. Mercury, arsenic, lead and chromium are often prevalent at contaminated sites. Bacterial remediation is the process of using metal-reducing bacteria to break down the contaminants (White et al., 1997). There are many biotransformation reactions carried out by bacteria to reduce toxic metal to less toxic forms. Macaskie et al., (2000) found that sulfate-reducing bacteria successfully treat the metal leachates generated by sulfuric acid–producing bacteria. Baheri and Meysami (2002) reported that white rot fungi have the ability to transform pollutants from contaminated soil through ligninolytic enzymes. Moreover, Lang et al. (1998) have successfully used the production and activity of lignolytic enzymes in the bioremediation of contaminated soil under field conditions. There are several enzymes listed for the removal or treatment of heavy metal compounds in Table 15.7.

15.6 Conclusion

Industries process various kinds of raw materials to manufacture desired products, and as a result, a large quantity of by-products, toxic chemicals, are released into the air or dumped into land and aquatic bodies. The dumping of waste imposes an acute problem of pollution to the environment, thereby requiring different treatment methods to convert them into less toxic forms. The nature of waste can be organic or inorganic and both forms are toxic and require cost-effective, eco-friendly methods of treatment to convert them into less toxic forms. There are various conventional physio-chemical methods used for the degradation of toxic recalcitrant chemicals for wastewater or solid waste treatment. Although the chemical treatment processes are effective for a large amount of waste, yet they are laborious, expensive and less efficient and adversely affect the environment. Bioremediation is an alternative to chemical methods that involve microorganism-mediated transformations of hazardous contaminants to non-hazardous compounds. Microorganisms bear enzymes that can catalyze such reactions and can be used in remediation processes that are less hazardous to the environment than conventional techniques. Moreover, enzymes have several advantages such as versatility of substrates, specificity, non-toxicity and reusability. They operate in mild, as well as in harsh, reaction conditions and can convert toxic compounds into innocuous products. They can also be applied directly to the contaminated sites. New enzymatic processes are being developed with innovative applications that can replace the use of chemicals in conventional techniques of remediation in different industries. Research and development techniques are focusing on enzyme engineering to impart desirable properties to the enzymes. The use of enzymes as a replacement for chemical treatments in various industries has resulted in reduced waste chemical discharge into the environment. Thus, the enzymes are promising candidates for the detoxification of pollutants due to their immense catalytic potential, eco-friendly transformations, specificity for desired substrates, and activity under a wide range of environmental conditions.

REFERENCES

Abdel-Shafy, H.I., and M.S.M. Mansour. 2016. A review on polycyclic aromatic hydrocarbons: Source, environmental impact, effect on human health and remediation. *Egyptian Journal of Petroleum* 25:107–123.

Agency for Toxic Substances and Disease Registry (ATSDR). 1999. *Toxicological Profile for Pentachlorophenol.* Public Health Service, U.S. Department of Health and Human Services, Atlanta, GA.

Ahn, Y.H., J. Sanseverino., and G.S. Sayler. 1999. Analyses of polycyclic aromatic hydrocarbon – degrading bacteria isolated from contaminated soils. *Biodegradation* 101:149–157.

Aitken, R.J., P. Koopman., and S.E.M. Lewis. 2004. Seeds of concern. *Nature* 432:48–52.

Akhtar, M.K., N.J. Turner, and P.R. Jones. 2013. Carboxylic acid reductase is a versatile enzyme for the conversion of fatty acids into fuels and chemical commodities. *Proceedings of the National Academy of Science of the United States of America* 110:87–92.

Aktar, M.W., D. Sengupta, and A. Chowdhury. 2009. Impact of pesticides use in agriculture: Their benefits and hazards. *Interdisciplinary Toxicology* 2:1–12.

Alzahrani, A.M. 2009. Insects cytochrome P450 enzymes: Evolution, functions and methods.

Andler, R., and A. Steinbuchel. 2017. A simple, rapid and cost – effective process for production of latex clearing protein to produce oligopolyisoprene molecules. *Journal of Biotechnology* 241:184–192.

Andler, S.M., and J.M. Goddard. 2018. Transforming food waste: How immobilized enzymes can valorize waste streams into revenue streams. *NPJ Science of Food*:2–19.

API. 2009. *Overview of Exploration and Production Waste Volumes and Waste Management Practices in the United States.*

Awaleh, M.O., and Y.D. Soubaneh. 2014. Waste water treatment in chemical industries: The concept and current technologies. *Hydrology Current Research* 5:164–176.

Baboshin, M.A., and L.A. Golovleva. 2012. Aerobic bacterial degradation of polycyclic aromatic hydrocarbons (PAHs) and its kinetic aspects. *Microbiology* 6:639–650.

Baheri, H., and P. Meysami. 2002. Feasibility of fungi bioaugmentation in composting a flare pit soil. *Journal of Hazardous Materials* 89:279–286.

Bamforth, S.M., and I. Singleton. 2005. Bioremediation of polycyclic aromatic hydrocarbons. *Journal of Chemical Technology and Biotechnology* 80:723–736.

Bantu, N., and V.R. Vakita. 2013. Acute toxicity of chlorantraniliprole to freshwater fish *Channapunctatus* (Bloch). *Advances in Zoology and Botany* 1:78–82.

Barakat, M.A. 2011. New trends in removing heavy metals from industrial wastewater. *Arabian Journal of Chemistry* 4:361–377.

Bhargava, A. 2016. Physico – chemical waste water treatment technologies: An overview. *International Journal of Scientific Research and Education* 4:5308–5319.

Bhat, D., and P. Padmaja. 2014. Assessment of organic pesticides in ground and surface water in Bhopal India. *IOSR Journal of Environmental Science, Toxicology and Food Technology* 8:51–52.

Bhatia, S.K., R.K. Bhatia, J-M. Jeon, G. Kumar, and Y-H. Yang. 2019a. Carbon dioxide capture and bioenergy production using biological system – A review. *Renewable and Sustainable Energy Reviews* 110:143–158.

Bhatia, S.K., R.K. Bhatia, Y.H. Yang. 2016. Biosynthesis of polyesters and polyamide building blocks using microbial fermentation and biotransformation. *Reviews in Environmental Science and Bio/Technology* 15:639–63.

Bhatia, S.K., R. Gurav, T-R. Choi, H-R. Jung, S-Y. Yang, H-S. Song, J-M. Jeon, J-S. Kim, Y-K. Lee, and Y.-H. Yang. 2019b. Poly (3-hydroxybutyrate-co-3-hydroxyhexanoate) production from engineered *Ralstoniaeutropha* using synthetic and anaerobically digested food waste derived volatile fatty acids. *International Journal of Biological Macromolecule* 133:1–10.

Bhatia, S.K., R. Gurav, T-R. Choi, H.J. Kim, S-Y. Yang, H-S. Song, J. Young Park, Y-L. Park, Y-H. Han, Y-K. Choi, S-H. Kim, J-J. Yoon, and Y-H. Yang. 2020a. Conversion of waste cooking oil into biodiesel using heterogenous catalyst derived from cork biochar. *Bioresource Technology*:122872.

Bhatia, S.K., S. Mehariya, R. Bhatia, M. Kumar, A. Pugazhendhi, M. Awasthi, A. Atabani, G. Kumar, W. Kim, S-O. Seo, and Y.H. Yang. 2020c. Wastewater based microalgal biorefinery for bioenergy production: progress and challenges. *Science of the Total Environment* 751:141599.

Bhatia, S.K., P.K. Mehta, R.K. Bhatia, and T.C. Bhalla. 2013. An isobutyronitrile-induced bienzymatic system of *Alcaligenes* sp. MTCC 10674 and its application in the synthesis of α-hydroxyisobutyric acid. *Bioprocess and Biosystems Engineering* 36:613–625.

Bhatia, S.K., P.K., Mehta, R.K., Bhatia, and T.C. Bhalla. 2014. Optimization of arylacetonitrilase production from *Alcaligenes* sp. MTCC 10675 and its application in mandelic acid synthesis. *Applied Microbiology and Biotechnology* 98:83–94.

Bhatia, S.K., S.V. Otari, J-M. Jeon, R. Gurav, Y-K. Choi, R.K. Bhatia, A. Pugazhendhi, V. Kumar, J. Rajesh Banu, J-J. Yoon, K-Y. Choi, and Y-H. Yang. 2021. Biowaste-to-bioplastic (polyhydroxyalkanoates): Conversion technologies, strategies, challenges, and perspective. *Bioresource Technology* 326:124733.

Bhatia, S.K., N. Vivek, V. Kumar, N. Chandel, M. Thakur, D. Kumar, Y-H. Yang, A. Pugazendhi, and G. Kumar. 2020b. Molecular biology interventions for activity improvement and production of industrial enzymes. *Bioresource Technology*:124596.

Blanquez, P., N. Casas, X. Font, X. Gabarrell, M. Sarra, G. Caminal, and T. Vicent. 2004. Mechanism of textile metal dye biotransformation by Trametesversicolor. *Journal of Water Research* 38:2166–2172.

Blasheck, H.P. 1992. Approaches to making the food processing industry more environmentally friendly. *Trends in Food Science and Technology* 3:107–110.

Bolto, B., and J. Gregory. 2007. Organic polyelectrolytes in water treatment. *Water Research* 41:2301–2324.

Chino-Flores, C., E. Dantan-Gonzalez, A. Vazquez-Ramos, R. Tinoco-Valencia, R. Diaz-Mendez, E. Sanchez-Salinas, M.L. Castrejon-Godinez, F. Ramos-Quintana, and M.L. Ortiz-Hernandez. 2012. Isolation of the opdE gene that encodes for a new hydrolase of *Enterobacter* sp. capable of degrading organophosphorus pesticides. *Biodegradation* 23:387–397.

Chivukula, M., and V. Renganathan. 1995. Phenolic azo dye oxidation by laccase from Pyriculariaoryzae. *Applied and Environmental Microbiology* 61:4374–4377.

Choi, Y-K., T-R. Choi, R. Gurav, S.K. Bhatia, Y-L. Park, H.J. Kim, E. Kan, and Y-H. Yang. 2019. Adsorption behavior of tetracycline onto Spirulina sp. (microalgae)-derived biochars produced at different temperatures. *Science of The Total Environment*:136282.

Chong, M.N., B. Jin, C.W. Chow, and C. Saint. 2010. Recent developments in photocatalytic water treatment technology: A review. *Water Research* 44:2997–3027.

Chowdhury, F.J., S.M. InamulHaq, and M. Aminul Islam. 1996. Accumulation of various pollutants by some aquatic macrophytes found in the Bariganga River. Proceedings of the 25th Bangladesh Science Conference.

Chowdhury, M., M.G. Mostafa, B.T. Kumar, and S.A. Kumar. 2013. Treatment of leather industrial effluent by filtration and coagulation processes. *Water Resource and Industry* 3:11–22.

Commission for Environmental Cooperation. 2004. The North American Mosaic: An overview of key environmental issues. Retrieved from industrial pollution and waste. http://www3.cec.org/islandora/en/item/987 – north – american – mosaic – overview – key – environmental – issues – en.pdf.

Comninellis, C., A. Kapalka, S. Malato, S.A. Parsons, I. Poulios, and D. Mantzavinos. 2008. Advanced oxidation processes for water treatment: Advances and trends for R&D. *Journal of Chemical Technology and Biotechnology* 83:769–776.

Dabrowski, A., Z. Hubicki, P. Podkoscielny, and E. Robens. 2004. Selective removal of the heavy metal ions from waters and industrial wastewaters by ion – exchange method. *Chemosphere* 56:91–106.

Das, N., and D. Das. 2014. Strategies for remediation of polycyclic aromatic hydrocarbons from contaminated soil – an overview. *Journal of Critical Reviews* 2:20–25.

Dixit, S., A. Yadav, P.D. Dwivedi, and M. Das. 2015. Toxic hazards of leather industry and technologies to combat threat: A review. *Journal of Cleaner Production* 87:39–49.

Dsikowitzky, L., and J. Schwarzbauer. 2013. Organic contaminants from industrial waste waters: Identification, toxicity and fate in the environment. In: *Environmental Chemistry for a Sustainable World 4*, eds. E. Lichtfouse, J. Schwarzbauer, and D. Robert, 45–101. Springer International Publishing, Switzerland.

Duff, S.J., J.W. Moritz, and K.L. Andersen. 1994. Simultaneous hydrolysis and fermentation of pulp mill primary clarifier sludge. *Canadian Journal of Chemical Engineering* 72:1013–1020.

Esplugas, S., D.M. Bila, L.G. Krause, and M. Dezotti. 2007. Ozonation and advanced oxidation technologies to remove endocrine disrupting chemicals (EDCs) and pharmaceuticals and personal care products (PPCPs) in water effluents. *Journal of Hazardous Materials* 149:631–642.

Ferdous, Z., S. Akter, M. Hasan, R.A. Begum, and R.M. Shahajahan. 2012. Phytoplankton diversity and abundance in relation to pollution levels in the Hazaribagh tannery effluent sewage water of the river Buriganga. *Bangladesh Journal of Zoology* 40:121–128.

Ferraro, P., and S.F. McEuen. 1996. Hydrolysis of dpx – jw062 (a racemic mixture of dpx – kn128 and dpx – kn127) in buffer solutions of pH 5, 7, and 9. *DuPon. Agricultural Products Document No. AMR* 2789–2793.

Forget, G. 1993. Balancing the need for pesticides with the risk to human health. In: *Impact of Pesticide Use on Health in Developing Countries*, eds. G. Forget, T. Goodman, and A. de Villiers, 2–16. International Development Research Centre, Ottawa, 2.

Fox, R.D. 1996. Physical/chemical treatment of organically contaminated soils and sediments. *Journal of the Air and Waste Management Association* 46:391–413.

Free, J.B. 1993. *Insect Pollination of Crop*. Academic Press, London.

Ghafoor, A., M. Aslam., and S. Rasool. 2012. Determinants of leather goods exports: A case of Pakistan. *Journal of Business and Economics* 4:256–269.

Ghafoor, A., and M. Zafar. 2014. Determinants of leather exports from Pakistan: A time series analysis. *VFAST Transactions on Education and Social Science* 5:77–86.

Ghaly, M.Y., G. Hartel, R. Mayer, and R. Haseneder. 2001. Photochemical oxidation of p – chlorophenol by UV/H2O2 and photo – Fenton process: A comparative study. *Waste Management* 21:41–47.

Ghanem, E., and F.M. Raushel. 2005. Detoxification of organophosphate nerve agents by bacterial phosphotriesterase. *Toxicology and Applied Pharmacology* 207:459–470.

Gogate, P.R., and A.B. Pandit. 2004. A review of imperative technologies for waste water treatment I: Oxidation technologies at ambient conditions. *Advances in Environmental Research* 8:501–551.

Gunatilake, S.K. 2015. Methods of removing heavy metals from industrial waste water. *Journal of Multidisciplinary Engineering Science Studies* 1:12–18.

Gupta, V.K., I. Ali, T.A. Saleh, A. Nayaka, and S. Agarwal. 2012. Chemical treatment technologies for waste – water recycling – an overview. *Royal Society of Chemistry Advances* 2:6380–6388.

Gurav R., S.K. Bhatia, T-R. Choi, Y-K. Choi, H.J. Kim, H-S. Song, S.L. Park, H.S. Lee, S.M. Lee, K-Y. Choi, and Y-H. Yang. 2021. Adsorptive removal of crude petroleum oil from water using floating pinewood biochar decorated with coconut oil-derived fatty acids. *Science of The Total Environment* 78:146636.

Hatfield, J.M., A.R. Umans, L.J. Hyatt, C.C. Edwards, M. Wierdl, L. Tsurkan, R.M. Taylor, and M.P. Potte. 2016. Carboxylesterases: General detoxifying enzymes. *Chemico – Biological Interactions* 259:327–331.

Holder, P.J., A. Jones, C.R. Tyler, and J.E. Cresswell. 2018. Fipronil pesticide as a suspect in historical mass mortalities of honey bees. *Proceedings of the National Academy of*

Science of the United States of America 115:13033–13038. https://doi.org/10.1038/s41538-018-0028-2.

Imamura, T., Y. Yanagawa, K. Nishikawa, N. Matsumoto, and T. Sakamoto. 2010. Two cases of acute poisoning with acetamiprid in humans. *Clinical Toxicology* 48:851–853.

International Finance Corporation. 2007. Environmental, health, and safety guidelines pulp and paper mills. *World Bank Group*:1–31.

Jaganathan, V., P. Cherurveettil, A. Chellasamy, and M.S. Premapriya. 2014. Environmental pollution risk analysis and management in textile industry: A preventive mechanism. *European Scientific Journal* 2:323–329.

Jiang, H., Y. Fang, and Y. Fu. 2003. Studies on the extraction of phenol in wastewater. *Journal of Hazardous Materials* 101:179–190.

Juhasz, A.L., and R. Naidu. 2000. Bioremediation of high molecular weight polycyclic aromatic hydrocarbons: A review of the microbial degradation of benzo[a]pyrene. *International Biodeterioration and Biodegradation* 45:57–88.

Kanaly, R.A., and S. Harayama. 2000. Biodegradation of high – molecular – weight polycyclic aromatic hydrocarbons by bacteria. *Journal of Bacteriology* 182:2059–2067.

Kankaria, S., A. Andukur, C.G. Hemamailin, and M. Kisishnveni. 2011. Impact of tannery effluents on ground water and agriculture with a remedial measure. A case study. International Conference on Chemical Biological and Environmental Science, Bangkok, 383–388.

Kapoor, M., and R. Rajagopal. 2011. Enzymatic bioremediation of organophosphorus insecticides by recombinant organophosphorous hydrolase. International Biodeterioration and *Biodegradation* 65:896–901.

Karam, J., and J.A. Nicell. 1997. Potential applications of enzymes in waste treatment. *Journal of Chemical Technology and Biotechnology* 69:141–153.

Karigar, S.C., and S.S. Rao. 2011. Role of microbial enzymes in the bioremediation of pollutants: A review. *Enzyme Research*:1–11.

Kim, J.E., S.K. Bhatia, H.J. Song, E. Yoo, H.J. Jeon, J-Y. Yoon, Y. Yang, R. Gurav, Y-H. Yang, H.J. Kim, and Y-K. Choi. 2020. Adsorptive removal of tetracycline from aqueous solution by maple leaf-derived biochar. *Bioresource Technology* 306:123092.

Kim, T.J., E.Y. Lee, Y.J. Kim, K.S. Cho, and H.W. Ryu. 2003. Degradation of polyaromatic hydrocarbons by *Burkholderiacepacia* 2A – 12. *World Journal of Microbiology and Biotechnology* 19:411–417.

Kirby, N., R. Marchant, and G. McMullan. 2000. Decolourisation of synthetic textile dyes by *Phlebiatremellosa*. *FEMS Microbiology Letters* 188:93–96.

Kotrikla, A., G. Gatidou, and T.D. Lekkas. 1999. Toxic effects of atrazine, deethyl – atrazine, deisopropyl – atrazine and metolachlor on *chlorella fuscavar – fusca*. *Global Nest. The International Journal* 1:39–45.

Kringstad, K.P., and K. Lindstrom. 1984. Spent liquors from pulp bleaching. *Environmental Science and Technology* 18:236A–247A.

Kumar, S., G. Kaushik, A.M. Dar, S. Nimesh, J.U. Lopez-Chuken, and F.J. Villarreal-Chiu. 2018. Microbial degradation of organophosphate pesticides: A review. *Pedosphere* 28:190–208.

Lang, E., F. Nerud, and F. Zadrazil. 1998. Production of ligninolytic enzymes by *Pleurotus*sp., and *Dichomitussqualens* in soil and lignocellulose substrate as influence by soil microorganisms. *Microbiology* 167:239–244.

Lee, C.S., J. Robinson, and M.F. Chong. 2014. A review on application of flocculants in waste water treatment. *Process Safety and Environmental Protection* 437:489–508.

Li, X.Z., H. Liu, L.F. Cheng, and H.J. Tong. 2003. Photocatalytic oxidation using a new catalyst – TiO2 Microsphere for water and waste water treatment. *Environmental Science and Technology* 37:3989–3994.

Li, X.Z., M.A. Schuler, and M.R. Berenbaum. 2007. Molecular mechanisms of metabolic resistance to synthetic and natural xenobiotics. *Annual Review of Entomology* 52:231–253.

Liu, J., J. Xie, Z. Ren, and W. Zhang. 2013. Solvent extraction of phenol with cumene from wastewater. *Desalination and Water Treatment* 51:3826–383.

Lofrano, G., S. Meric, G.E. Zengin, and D. Orhon. 2013. Chemical and biological treatment technologies for leather tannery chemicals and wastewaters: A review. *Science of the Total Environment* 461:265–281.

Macaskie, L.E., K.M. Bonthrone, P. Yong, and D.T. Goddard. 2000. Enzymically mediated bioprecipitation of uranium by a *Citrobacter* sp.: A concerted role for exocellular lipopolysaccharide and associated phosphatase in biomineral formation. *Microbiology* 146:1855–1867.

Maddhinni, V. L., H.B. Vurimindi, and A. Yerramilli. 2006. Degradation of azo dye with horseradish peroxidase (HRP). *Journal of the Indian Institute of Science* 86:507–514.

Mahajan, M.C., P.S. Phale, and C.S. Vaidyanathan. 1994. Evidence for the involvement of multiple pathways in the biodegradation of 1- and 2-methylnaphthalene by Pseudomonas putida CSV86. *Archives of Microbiology* 161:425–433.

Martin, I., and E. Stutt. 2009. Soil guideline values for dioxins, furans and dioxin-like PCBs in soil. Science Report SC050021 / Dioxins SGV. Environment Agency, Rio House, Waterside Drive, Aztec West, Almondsbury, *Bristol BS32 4UD*, 1–22.

Mehta, P.K., S.K. Bhatia, R.K. Bhatia, and Bhalla, T.C. 2013. Purification and characterization of a novel thermo-active amidase from *Geobacillussubterraneus* RL-2a. *Extremophiles* 17:637–648.

Mehta, P.K., S.K. Bhatia, R.K. Bhatia, and T.C. Bhalla. 2014. Bench scale production of nicotinic acid using a versatile amide-hydrolysing *Geobacillus subterraneus* RL-2a isolated from thermal spring of manikaran, india. *Journal of Molecular Catalysis B Enzyme* 105:58–65.

Mier, M.V., R. L. Callejasa, R. Gehr, B.E.J. Cisnerosc, and P.J.J. Alvarezd. 2001. Heavy metal removal with mexicanclinoptilolite: Multi-component ionic exchange. *Water Research* 35:373–378.

Mojsov, D.K., D. Andronikov, A. Janevski, A. Kuzelov, and S. Gaber. 2016. The application of enzymes for the removal of dyes from textile effluents. *Advanced Technologies* 5:36–41.

Munter, R. 2001. Advanced oxidation processes-current status and prospects. *Proceeding of the Estonian Academy of Sciences Chemistry* 50:59–80.

Newton, D.F. 1970. *Role of the Dairy and Food Industry in Environmental Pollution Control*. Environmental Health Technology dej:jaitmentbroome Technical Community College Binghamton, New York, 13902.

Nirmalakallagadda, and V. Rathnamma. 2014. Flubendiamide-a Phthalic acid diamide, effect on protein metabolism of freshwater fish Labeorohita (Hamilton). *International Journal of Recent Scientific Research* 5:1554–1557.

Novotny, C., K. Svobodova, A. Kasinath, and P. Erbanova. 2004. Biodegradation of synthetic dyes by Irpexlacteus under various growth conditions. *International Biodeterioration and Biodegradation* 54:215–223.

Of analysis. *Global Journal of Molecular Sciences* 4:167–179.

Okereke, C.D. 2007. *Environmental Pollution Control*. 1st Edition. Barloz Publication, Owerri, Nigeria.

Olajire, A.A. 2014. The petroleum industry and environmental challenges. *Journal of Petroleum and Environmental Biotechnology* 5:1–19.

Olejniczak, J., J. Staniewski, and J. Szymanowski 2005. Extraction of phenols and phenyl acetates with diethyl carbonate. *AnalaticaChimica Acta* 53:5251–257.

Ortiz-Hernandez, M.L., R. Quintero-Ramirez, A.A. Nava-Ocampo, and A.M. Bello-Ramirez. 2003. Study of the mechanism of *Flavobacterium* sp. for hydrolyzing organophosphate pesticides. *Fundamental and Clinical Pharmacology* 17:717–723.

Otta, W.J., and C.J. Kendall. 1987. Land application of waste from brewery and distilling plants. Proceedings of the 1987 Food Processing Waste Conference, Sept 1–2, 1987. Georgia tech Research Institute, Atlanta, Georgia, 129–144.

Padma, V.V., P. Sowmya, T.A. Felix, R. Baskaran, and P. Poornima. 2011. Protective effect of gallic acid against lindane induced toxicity in experimental rats. *Food and Chemical Toxicology* 49:991–998.

Padoley, K.V., S.N. Mudliar, and R.A. Pandey. 2008. Heterocyclic nitrogenous pollutants in the environment and their treatment options – An overview. *Bioresource Technology* 99:4029–4043.

Park, Y-L., Y-H. Han, H-S. Song, J-Y. Park, S.K. Bhatia, R. Gurav, K-Y. Choi, Y-G. Kim, and Y-H. Yang. 2020. Effects of osmolytes on salt resistance of *Halomonas socia* CKY01 and identification of osmolytes-related genes by genome sequencing. *Journal of Biotechnology* 10:21–28.

Peralta-Zamora, P., C.M. Pereira, E.R.L. Tiburtius, S.G. Moraes, M.A. Rosa, R.C. Minussi, and N. Duran. 2003. Decolorization of reactive dyes by immobilized laccase. *Applied catalysis B: Environmental* 42:131–144.

Peterson, M.E., and P.A. Talcot. 2006. *Small Animal Toxicology*, 2nd ed. Elsevier, St. Louis, 526–527, 941–949.

Prasad, M.P., and K. Manjunath. 2011. Comparative study on biodegradation of lipid-rich wastewater using lipase producing bacterial species. *Indian Journal of Biotechnology* 10:121–124.

Raj, J., and B. Joseph. 2017. Impact of acetamiprid toxicity on electrophoretic patterns in liver, brain and gill tissues of the fish *Oreochromismossambicus*. *International Journal of Zoological Research* 13:120–124.

Rameshraja, D., and S. Suresh. 2011. Treatment of tannery wastewater by various oxidation and combined processes. *International Journal of Environment Research* 5:349–360.

Razali, M.A.A., Z. Ahmad, M.S.B. Ahmad, and A. Ariffin. 2011. Treatment of pulp and paper mill wastewater with various molecular weight of polyDADMAC induced flocculation. *Chemical Engineering Journal* 166:529–535.

Renault, F., B. Sancey, J. Charles, N. Morin-Crini, P.M. Badot, P. Winterton, and G. Crini. 2009. Chitosan flocculation of cardboard-mill secondary biological wastewater. *Chemical Engineering Journal* 155:775–783.

Rodriguez, E., M.A. Pickard, and R. Vazquez-Duhalt. 1999. Industrial dye decolorization by laccases from ligninolytic fungi. *Current Microbiology* 38:27–32.

Rodríguez, M. 2003. *Fenton and UV-vis Based Advanced Oxidation Processes in Wastewater Treatment: Degradation, Mineralization and Biodegradability Enhancement*. Universitat de Barcelona, Spain.

Sabah, E., and Z.E. Erkan. 2006. Interaction mechanism of flocculants with coal waste slurry. *Fuel* 85:350–359.

Sahoo, D.K., A. Roy and G.B. Chainy. 2008. Protective effect of vitamin E and curcumin on L- thyroxine-induced rat testicular oxidative stress. *Chemical and Biological Interactions* 176:121–128.

Sarkar, S., M. Dutta, and S. Roy. 2014. Potential toxicity of flubendiamide in Drosophila melanogaster and associated structural alterations of its compound eye. *Toxicology and Environmental Chemistry* 96:1075–1087.

Scott, C., G Pandey, C.J. Hartley, C.J. Jackson, M.J. Cheesman, M.C. Taylor, R. Pandey, J. L. Khurana, M. Teese, C.W. Coppin, K.M. Weir, R.K. Jain, R. Lal, R.J. Russell, and J.G. Oakeshott. 2008. The enzymatic basis for pesticide bioremediation. *Indian Journal of Microbiology* 48:65–79.

Seo, S.J., S.Y. Keum, and X.Q. Li. 2009. Bacterial degradation of aromatic compounds. *International Journal of Environment Research and Public Health* 6:278–309.

Shen, T.T. 1995. Industrial pollution prevention. In: *Industrial Pollution Prevention. Environmental Engineering*, 15–35. Springer, Berlin, Heidelberg.

Shrinivas, S.J., and M. David. 2015. Modulatory impact of flubendiamide on enzyme activities in tropical black and red agriculture soils of Dharwad (North Karnataka), India. *International Journal of Agricultural and Food Science* 5:43–49.

Siddiqui, A., A. Uroos, and F. Nadeem. 2016. Sectoral competitiveness and value chain analysis: leather Gloves value chain analysis in Pakistan. Trade Development Authority of Pakistan.

Singh, D.K. 2007. Insecticidal methods of pest control. In: *Applied Entomology*, eds. D.K. Singh and T.P. Trivedi. http://nsdl.niscair.res.in/jspui/handle/123456789/416.

Singh, N., A. Sharma, P. Dwivedi, R. Patil, and M. Kumar. 2007. Citrinin and endosulfan induced teratogeniceffects in Wistar rats. *Journal of Applied Toxicology* 27:143–151.

Singh, R.P., G.P. Karmakar, S.K. Rath, N.C. Karmakar, S.R. Pandey, T. Tripathy, J. Panda, K. Kanan, S.K. Jain, and N.T. Lan. 2000. Biodegradable drag reducing agents and flocculant based on polysaccharides: Materials and applications. *Polymer Engineering and Science* 40:46–60.

Singh, S.K. 2007. Effect of irrigation with paper mill effluent on the nutrient status of soil. *International Journal of Soil Science* 2:74–77.

Solomon, K., H. Bergman, R. Huggett, D. Mackay, and B. McKague. 1993. A review and assessment of the ecological risks associated with the use of chlorine dioxide for the bleaching of pulp. *Association for Environmental Technology Erin ON*:73.

Song, H.J., R. Gurav, S.K. Bhatia, E.B. Lee, H.J. Kim, Y.-H. Yang, E. Kan, H.H. Kim, S.H. Lee, and Y.K. Choi. 2021. Treatment of microcystin-LR cyanotoxin contaminated water using Kentucky bluegrass-derived biochar. *Journal of Water Process Engineering* 41:102054.

Suchail, Se V., D. Guez, and L.P. Belzunces. 2000. Characteristics of imidacloprid toxicity in two *Apismellifera* subspecies. *Environmental Toxicology and Chemistry* 7:1901–1905.

Susan, S., and P. Sania. 1999. Endosulfan-A review of its toxicity and its effects on the endocrine system. *WWF (World Wild Life Fund – Canada)* 11:123.

Swathi, M., A.S. Singh, S.A. Aravind, P.K. Ashisudhakar, R. Gobinath, and S. Devi. 2014. Adsorption studies on tannery wastewater using bamboo dust as absorbent. *World Journal of Environmental Biosciences* 3:56–60.

Tang, L., X.Y. Tang, Y.G. Zhu, M.H. Zheng, and Q.L. Miao. 2005. Contamination of polycyclic aromatic hydrocarbons (PAHs) in urban soils in Beijing, China. *Environment International* 31: 822–828.

Tian, Y, Y.R. Luo, T.L. Zheng, L.Z. Cai, X.X. Cao, and C.L. Yan. 2008. Contamination and protential biodegradation of polycyclic aromatic hydrocarbons in mangrove sediments of Xiamen, China. *Marine Pollution Bulletin* 6:1184–1191.

Uday, U.S.P., T.K. Bandyopadhyay, and B. Bhunia. 2016. Bioremediation and detoxification technology for treatment of dye(s) from textile effluent. In: *Textile Wastewater Treatment*, eds. E.P.K. Kumbasar and A.E. Korlu, 75–92.

United Nations Environment Programme (UNEP). 2010. Environmental policy implementation. www.unep.org/eou.

US-EPA. 1975. DDT- A review of scientific and economic aspects of the decision to ban its use as a pesticide; EPA 540/1–75–022; U.S. Environmental Protection Agency, Office of Pesticide Programs, U.S. Government Printing Office: Washington, DC.

US-EPA. 2000. Environmental risk assessment for diazinon; U.S. Environmental protection agency, office of prevention, pesticides and toxic substances, office of pesticide programs, U.S. Government Printing Office: Washington, DC.

Von Gunten, U. 2003. Ozonation of drinking water: Part II disinfection and by-product formation in presence of bromide, iodide or chlorine. *Water Research* 37:1469–1487.

Wei, L., W.W. Shao, G.H. Ding, X.L. Fan, M.L. Yu, and Z.H. Lin. 2014. Acute and joint toxicity of three agrochemicals to Chinese tiger frog (Hoplobatrachuschinensis) tadpoles. *Dong wuxue Yan jiu* 35:272–279.

Wert, E.C., F.L. Rosario-Ortiza, D.D. Drury and S.A. Snyder. 2007. Formation of oxidation byproducts from ozonation of wastewater. *Water Research* 41:1481–1490.

White, I., M.D. Melville, B.P. Wilson, and J. Sammut. 1997. Reducing acidic discharges from coastal wetlands in eastern Australia. *Wetlands Ecology and Management* 5:55–72.

Wilson, S.C., and K.C. Jones. 1993. Bioremediation of soil contaminated with polynucleararomatichydrocarbons (PAHs): A review. *Environmental Pollution* 81:229–249.

Wong, S.S., T.T. Teng, A.L. Ahmad, A. Zuhairi, and G. Najafpour. 2006. Treatment of pulp and paper mill wastewater by polyacrylamide (PAM) in polymer induced flocculation. *Journal of Hazardous Materials* 135:378–388.

Wong, Y., and J. Yu. 1999. Laccase-catalyzed decolorization of synthetic dyes. *Water Research* 33:3512–3520.

Worrall, F., M. Fernandez-Perez, A. Johnson, F. Flores-Cesperedes, and E. Gonzalez-Pradas. 2001. Limitations on the role of incorporated organic matter in reducing pesticide leaching. *Journal of Contaminant Hydrology* 49: 241–262.

Wu, C., Y. Wang, B. Gao, Y. Zhao, and Q. Yue. 2012. Coagulation performance and floc characteristics of aluminum sulfate using sodium alginate as coagulant aid for synthetic dying wastewater treatment. *Separation and Purification Technology* 95:180–187.

Wu, J.J., C.C. Wu, H.W. Ma, and C.C. Chang. 2004. Treatment of landfill leachate by ozone-based advanced oxidation processes. *Chemosphere* 54: 997–1003.

Xiog, X., F. Stagnitti, J. Peterson, G. Allinson, and N. Turoczy. 2001. Heavy metal to contamination of pasture soils by irrigated municipal sewage. *Bulletin of Environmental Contamination and Toxicology* 67: 535–540.

Xu, J.Q., W.H. Duan, and X.Z. Zhou. 2006. Extraction of phenol in wastewater with annular centrifugal contactors. *Journal of Hazardous Materials* 131: 98–102.

Yoon, J., Y. Lee, and S. Kim. 2001. Investigation of the reaction pathway of OH radicals produced by Fenton oxidation in the conditions of wastewater treatment. *Water Science Technology* 44:15–21.

Zimmermann, T., H.G. Kulla, and T. Leisinger. 1982. Properties of purified orange II azoreductase, the enzyme initiating azo dye degradation by Pseudomonas KF46. *European Journal of Biochemistry* 129:197–203.

16

Health Issues and Environmental Biotechnology

Anish Singhal, Prathamesh H. Kamble and Prasunpriya Nayak

CONTENTS

16.1 Introduction: Environmental Biotechnology and Health

Integration of technological improvisation in the biological process and application of that for the benefit of society is the focal theme of biotechnology. Over the past decade, practicability in biotechnology has accelerated rapidly. We are actually heading towards a remarkable modification in the scientific arena, where efficiency, economicalness and pervasiveness of a process/system/product should be the key feature for its appropriateness as a biotechnological tool.

Biotechnology has already brought a variety of reforms in human behavior, along with edible vegetation and animal life (Health Policy Institute, 2017). These changes have also impacted the environment, directly or indirectly. The growth in biotechnology is vast and rapid and often is based on immediate needs; there is hardly any scope to evaluate the distributive and long-standing consequences of biotechnological

DOI: 10.1201/9781003186304-16

applications. To avoid undue alterations in the environment, as some of biotechnological processes are deemed to do, special emphasis has to be given to the protection of the environment without compromising biotechnological advancements. The identification of the biotechnology-associated sources of threat toward environmental upsets, as well as their possible remedial measures, should be considered under the umbrella of 'environmental biotechnology'. Apart from these, preventing contamination in the natural environment should also be considered. There are ways in which (a) wastes or by-products can be treated biologically with the help of microbes before they could turn into harmful contaminants, (b) biological redressal of already polluted environment could be carried out by wild or biotechnologically modified microbugs and (c) customized instant monitoring of the level of identified contamination or pollution in the environment could be possible. One important threat of massive biotechnological progress could be defalcated use of its benefits concealing unintended implications. Other threats may include inadvertent deviations from the extrapolated interactions of the new application with the existing physico-chemical, biological and even social environs.

When the environment is essentially threatened with pollution and ecology is facing a crisis due to shortage of raw materials and usable water supply, environmental biotechnology can come into rescue by uncovering the possibilities for the prevention of pollution, management of wastes and improvising the manufacturing process for reduced consumption of resources and controlled pollution. In other words, the health of the environment can be protected by the proper use of environmental biotechnology and biomonitoring-cum-bioengineering tools. With the advent of newer and finer molecular tools, options for customized biomolecules and organisms are on the rise. It is believed that application of these tools and concepts of biotechnology; environment can have a face change while simultaneously achieving integrated and sustainable development. This may involve the health, industrial, agroforestry, food, energy fuel, waste management, raw material and, obviously, environment management sectors and can contribute to clean technology (Singh, 2017). However, it is worth mentioning that the impact of these applications on health can be direct or indirect.

Although the primary goal of environmental biotechnology is the protection of the environment, the possibility of escaped long-term or improvident consequences cannot be undermined. To ensure the sustainable development intended to continue for generations, a framework of collaborating environmental scientists and engineers with integrating environmentally friendly policies and developmental strategies having wide and changing coverage is required. In-depth understanding of environmental physiology and its possible interactions with physical and chemical associates are implied requirements of environmental health and so to so for environmental biotechnology. Sometimes, biotechnology is considered just the widened next step of plant and animal breeding that has been used for generations in farming. However, modern biotechnology is distinctly moving away from breeding by manipulating chosen genes without considering the natural selection. While the genes are chosen with a specific goal, unlike the natural selection procedure, the untoward events are not likely to be considered or not possible to evaluate in its totality with other genetic or epigenetic counterparts. Deliberating the interactions with other nontargeted components of biota, environmental biotechnology came into the existence and promoted the concept of sustainable development with environmental protection. When the unlikely outcomes of biotechnological processes become a possibility, even to a minuscule extent, the undesirable changes may come into existence. This, in turn, may initiate some unsuitable environmental changes.

However, with market/economical pressure, evaluations of temporal and spatial spans of concerning health impacts often remain incomplete. Hence, the chapter is targeted to provide a priming about the direct and indirect health issues related to several aspects of environmental biotechnology. Before portraying the health issues of environmental biotechnology, a glance through the spectrum of biotechnology would be helpful in understanding the health implication in the current context.

16.2 Scope of Biotechnology and Its Relation to Health

The health sector is one important recipient of biotechnological applications. Even though other sectors, as mentioned already, use a good deal of biotechnology, this chapter is mainly enumerating the probable, maybe even conjectured, problems of biotechnology in health issues.

These branches/subspecialities of biotechnology and their applications are theoretically presented in color codes. As Kafarski (2012) mentioned, the colors of biotechnology are not hard-and-fast rules of coding the biotechnology field; rather, they indicate the wide areas of biotechnology applications, and the addition of more colors are also possible and open. Roughly, 10 branches of biotechnology are coded with colors (Figure 16.1).

Red biotechnology encompasses the health and medical sector. However, in a true sense, most of inventions of biotechnology are for the benefit of physical or psychological health of individuals or public at large. Natural and biotechnologically modified microbes are used in the process of biodegradation or detoxification of non–environmentally friendly substances. Biobased processes include biogas and biofuel production, biosensing and biomonitoring, bioreactors and biofabrication, biofouling, bioremediation, bioelectronics and some other new prospects of human well-being. Bioprocess engineering is also

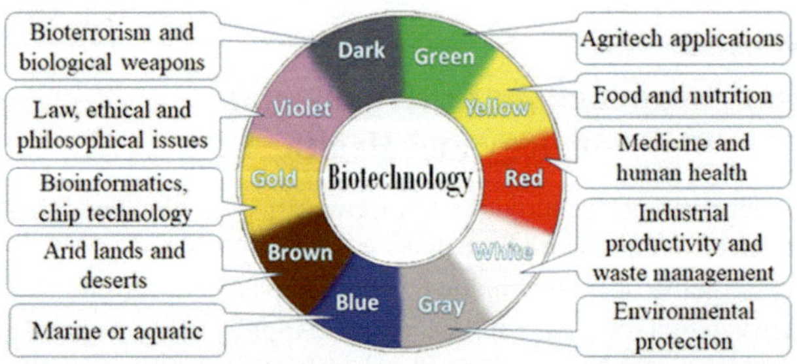

FIGURE 16.1 Colors of biotechnology—the wide horizon.

growingly applied in food production and food processing, like the 'golden rice', which is a product of *green biotechnology*, and was useful in improving the vision of good number of Asian children as their traditional food lacked vitamin A (Kafarski, 2012).

Additionally, environmental biotechnology became an inevitable part of the waste treatment and management, cosmetics and pharmaceuticals, leather and textile technology, plastic and pulp and paper industry, as well as environmental management. All these and pharmaceutical industries are part of *white biotechnology*; nevertheless, their products are used for health benefits. Even if the *white biotechnology* uses any process of biocatalysis or biofueling or biorefinaries, the control in environmental pollution is providing health benefits indirectly. On the other hand, ethical and/or lawful protection of health and/or environment comes under the *violet biotechnology*. Nutrition-related *yellow biotechnology* or aquatic *blue biotechnology* are also involved in the development of functional foods and, in turn, are related to the *red biotechnology*. The use of bioinformatics (*golden biotechnology*), an interdisciplinary field of science that combines biology, technology and statistics to analyze and interpret biological data, is of high importance for public health, indirectly forming a part of *red biotechnology*. Other arenas of biotechnology, for example, *brown biotechnology*, also contribute to *red biotechnology* in special cases. Interestingly, there is no color identified to represent environmental biotechnology; nonetheless, *gray biotechnology* is used for problems of environmental protection.

16.3 Biotechnology in Medicine

There is no doubt that biotechnology has contributed significantly to the growing needs of individual and global health. All the areas of health sector, be it diagnosis, medicine, intervention or maintenance, have been influenced greatly by the biotechnological approaches in contending them (Afzal et al., 2016). Continuous advancements in molecular biology and its corresponding techniques have allowed the manipulation of biological systems toward the targeted approach of human well-being (Mohammed et al., 2016).

Red biotechnology is very useful in identifying and managing different types of disease processes, in addition to safeguarding the population from harmful ailments. With the advent of this area of biotechnology, genetic engineering has come up with procedures, such as genetics treatment, recombinant DNA technology and polymerase domino effect, which utilize genetics and DNA particles to detect diseases and place brand-new and healthy and balanced genes in the body that substitute the wrecked cells. *Red biotechnology* encompasses and embraces other areas of biotechnology and has flourished extensively in recent decades. Ideally, *green*, *yellow*, *red*, *white* and *gray biotechnology* should consort together for the betterment of health in totality. Unfortunately, a failure of *white biotechnology*, especially waste management, is a major concern for individual, societal and environmental health, while other areas of biotechnology may also lead to some unfortunate issues.

16.3.1 Diagnostic Biotechnology

Diagnosis, particularly if it is early, counts maximum in the health issue. Biotechnological progresses on many occasions have provided health advisers tools for the early and accurate detection of the disease and provided a handsome window to manage the ailments. On the other hand, the inclusion of biotechnology has enabled the health care consumers to fight with bottlenecks of lengthy and costly diagnostic processes (Wu et al., 2018).

In fact, it is only the biotechnology that led the possibility of molecular diagnosis to a level of precision, which was beyond the imagination even few years earlier and surpassed the possibilities of clinical chemistry, microbiological, hematological or immunological diagnostic procedures. Different polymerase chain reaction (PCR; including multiplex, reverse transcriptase, real-time), solid phase–based techniques/hybridization or blotting techniques, microarrays and sequencing are a few of the biotechnological processes that have revolutionized molecular diagnosis processes (Dwivedi et al., 2017). Automated diagnostic test kits and autoanalyzer machines with robotics have made the diagnosis not only quick but dependable and reproducible also. Because of biotechnology only, even a minute volume of body fluids or biopsy sample could be analyzed with a high level of precision and contribute authoritatively to the process of diagnosis. Chip-based detection procedures have been developed and specific diagnosis with the help of lab-on-chip technology became handy in providing support to the effective clinical management of many debilitating diseases (Wu et al., 2018).

While the two-dimensional imaging by X-ray and ultrasound and three-dimensional imaging by magnetic resonance imaging, computed tomography, positron emission tomography, single-photon emission computed tomography and optical imaging have improved at the level of instrumentation, the analysis by these modalities are also improved a lot. Side by side, an analysis of the images and their application in the diagnosis have improved diligently (i.e., quietly and steadily continuing a task despite any difficulties). Particularly in cancer diagnosis, three-dimensional contrast-enhanced ultrasound, positron emission mammography of the breast and spectroscopy-enhanced colonoscopy for cancer screening, hyperpolarization magnetic resonance imaging, dynamic contrast-enhanced magnetic resonance imaging, time-of-flight positron emission tomography for staging and ion beam-induced positron emission tomography scanning are playing important roles (Frangioni, 2008).

Recombinant DNA technology has also come in handy in detecting some infectious diseases, such as measles, hepatitis, tuberculosis and others, while it is useful in identifying faulty proteins, especially in prion diseases. The application of techniques has not only made the identification of the disease or its cause easier and quick; it has also broadened the area of possible inclusion of genetic and epigenetic interactions into the diagnosis process. Genetic testing is commonly being done in establishing the nature of a tumor, the likelihood of some familial disease and the detection the presence of some dominant disease-causing genes. In some cases, close members of the family or family tree of the person diagnosed with some debilitating conditions are advised to undergo genetic testing. In this way, diseases can be diagnosed in at a very early stage or individuals with a high risk can be identified. This genetic screening, for some specific debilitating diseases, is also possible even in fetal life. Recently, the use of microRNAs (miRNAs) and their expressions in the diagnosis process are also suggested (D'Angelo et al., 2016; Huang, 2017).

Nanotechnology is also contributing to the field of diagnostic biotechnology. In fact, this is an alternative technique, which can detect a minute amount of the molecular targets, and, thus, spares the steps of molecular amplification during diagnosis. With the help of nanotechnology, the diagnosis process became quicker and more flexible along with enhancement of their sensitivity (Nikalje, 2015).

Even in health, as in the case of pregnancy, detection by kit method at home is an excellent example of diagnostic ease served by the biotechnological progress. The same is also true for glucometers, remote electrocardiogram (ECG) monitoring and mobile apps for monitoring calorie intake and expenditure, in which technological progress has eased monitoring health without much of a requirement of sophisticated instruments and skilled personnel. These simplified instruments also act as data providers for health surveys, and this, in turn, can contribute to the national health policy.

Each part of the supply and use chain, including production, packaging, utilization and waste disposal, may become a threat to environment if not handled properly.

16.3.2 Medical and Drug Biotechnology

The uses of biotechnology for medical purposes are quite obvious. The invention of newer molecular techniques, the completion of the Human Genome Project and ever-expanding information about the genetic basis of diseases, whether those are heritable or not, have made the progress possible in the field of medical and drug biotechnology and allowed humans to enjoy the benefit of it. While the essentiality of knowledge cannot be ignored, the most important determinant for the flourishing of this field of biotechnology is its market potential. Investors would like to show their interest more on the products, which are likely to be used regularly by the patients than the products that are used one time. Hence, the growth of medical and drug biotechnology is market-oriented rather than knowledge-based. The developmental phase of any biotechnology product demands a lot of financial commitments to evaluate the overall impact of the product on the environment and ecology. However, there is always a chance of laxity in this part, particularly in absence of strict vigil.

Biopharmaceuticals have taken up a good share in the current pharma market. A growing number of companies are coming up with multiple options for the production of biopharmaceuticals, such as antibiotics, factors of blood origin, hormones, cyto(chemo)kines, enzymes, vaccines and monoclonal antibodies (Almeida et al., 2011). One of the major achievements of biotechnology is the expression of desired proteins in a large but controllable scale. These desired proteins are often recombinant product of genetic engineering and transgenic organisms, such as bacteria, yeasts, plants, insects and mammals, have served as biofactories for these recombinant therapeutic proteins (Almeida et al., 2011; Afzal et al., 2016; Lumencandela, 2018). The uses of recombinantly produced insulin, erythropoietin, parathormone, tissue plasminogen activators and blood-clotting factors are well-known. Most important, these biotechnological industries are environmentally friendly with efficient production processes having low-energy consumption and the reduced emission of greenhouse gases (Almeida et al., 2011). The reduced consumption of energy, along with the reduction in greenhouse gas emissions, provides these industries a double-benefit support for the environment. These industries first produce the master cell line containing the gene that makes the desired protein. In the next step, the industry grows them in large numbers and maintains them in such a condition so that these cells can produce the desired protein in large quantities. Isolating and purifying the desired proteins are carried out before these proteins are packaged and marketed for patient/hospital use. Each of these steps needs thorough monitoring for volume, pH, temperature and ingredients of the maintenance media. Even though the manufacturers maintain the conditions of bioreactors for the best production outcome, subtle changes in them may lead to changes in protein functions and properties. These products are to be evaluated for their safety in every lot of productions before it could reach for the use in patients. Therefore, these biopharmaceutical industries are capable of producing biological or near-biological wastes. As normally the productions are of large quantity, strict management procedures are to be in place for handling these large quantities of biological wastes.

Highly purified amino acids are produced by biotechnological means. These amino acids are used as specialty nutrients (in depression, premenstrual dysphoric disorder, smoking cessation, bruxism, attention deficit-hyperactivity disorder etc.; Ivanov et al., 2013). By means of genetic engineering or metabolic engineering, microbes are modified to produce high-yield species. However, the problems of contamination and back mutation are always there in the process of developing and maintaining these genetically modified organisms. Therefore, the threat to the environment and the ecology can originate from these metabolically engineered organisms.

miRNA and small-interfering RNA (siRNA) have recently been gaining their space in the field of commercial therapeutic biopharmaceuticals and are likely to contribute significantly in the future of medicine (Chakraborty et al., 2017). Gene therapy, another promising tool of biotechnology, is also being used for treating diseases such as AIDS, cystic fibrosis, cancer and high cholesterol. However, preventing the destruction of modified cells by a person's immune system and the induction of a desired level of therapeutic impact are the two major bottlenecks in this process (Lumencandela, 2018). These newer techniques and procedures have not passed the level of initiation yet; hence, the environmental threats from them are limited to the imaginary consequences. However, a real test of environmental impact is only possible when these health benefits come to the market with bounteous uses.

16.3.3 Interventional Biotechnology

Biotechnology has also contributed to improving the efficacy of drug delivery systems (Afzal et al., 2016). In this context, the blooming of nanotechnology in recent years is most noteworthy. In addition, the health sector is also benefited from the newer approaches of therapeutic and surgical intervention like near-infrared fluorescence-guided surgery for cancer treatment (Frangioni, 2008). In the same line, a number of grafts and transplant materials have been successfully engineered, namely the bone, collagen, heart valves and others. With the progress of tissue engineering technology, substitutes for organ transplantation come into the field. Functional organs made in laboratory with the help of a three-dimensional arrangement of cells also added to the benefits of modern health biotechnology.

Biotechnology participated a lot in the discovery of various types of vaccines or diagnostic antigens (Balamurugan et al., 2006). Vaccination is an important preventive and interventional tool for number of infections (Lauer et al., 2017). Importantly, the stability of utilized whole proteins or their parts of viral or microbial origin was the mainstay of those to be used as vaccination means (Balamurugan et al., 2006). This 'stability' is also a cause of concern for the nontarget organisms of linked species. Similarly, the beginning of the 21st century witnessed a wide variety of gene or gene compositions (naked DNA vaccine) as vaccine substrates for immune betterment. Recently, vaccines made up of recombinant viral vectors are in use for their stability, easy insertion and inductibility (Lauer et al., 2017). In some cases, plant-origin vaccines are also found worthy. These vaccines are on the market because they are environmentally safe. Recent thrives (vigorous growth) are toward the multipotent and multivariable vaccine development. However, restrictions implied in the process for the multitude of targets are crucial along with the robustness of the evaluation processes so that when these vaccines are in use, environmental safety is taken care of.

Like the viral vectors, genetic engineering techniques are also employed for the purpose of gene therapy, in which an external gene could be introduced as an appurtenant gene for a particular function or to compensate for the absence of the same gene. Therapeutic uses of engineered proteins, such as muteins, are also of great use in the current health setup. They not only provide better suitable pharmacodynamic characters by means of their stability, solubility, specificity, substrate affinity and the like, but they can also compensate the deficiency created by nonfunctional or less functional proteins. While vaccines are great tools against communicable disease, therapeutic proteins can be helpful in noncommunicable diseases.

Nanoparticles, such as dendrimers, nano-porous materials, nanocrystals and solid lipid nanoparticles, and nanotechnologies, such as nanotubes, 'minicells', nanoparticle chains, nano-sponges and liposomes, are used in highly selective drug delivery (Nikalje, 2015; Zdrojewicz et al., 2015). Nanocomposites, nano-knives and nanorobots are recent improvisations that have taken interventional biotechnology to an all-time high. These nanotechnology tools are likely to be very useful in minimally invasive surgeries and monitoring the patients prior to surgeries or during the surgery and postsurgery developments (Vijayakumar et al., 2013); these can be misused in fields other than interventional medicine. Apart from these, the toxicity of these nanoparticles is a cause of concern, and worldwide efforts in identifying the problems have been noted (Andronescu et al., 2016). In fact, when some of these nanomaterials have already shown their adverse effects on the biological system even with short-term exposure; the long-term effects are simply unknown. Interestingly, the quest for environmentally friendly nanomaterials is still on (Andronescu et al., 2016).

Stenting is another area that revolutionized the scope of health benefits by applying biotechnology for the intervention of disorders. Then come the varied versions of stents—drug-coated, drug-eluting, biodegradable and, finally, bioabsorbable; however, the long-term problems of stenting are not yet fully overcome.

16.3.4 Health Sustainment Biotechnology

Keeping long-term health concerns in mind, there are some biotechnological developments that allow the body to get equipped in such ways that it can take care of the problem on its own. There are vaccinations that assist in protecting against, as well as reducing, the impact from ailments that, otherwise, are really complicated to handle. Using the same concept, communication among intestinal epithelial cells, enteric commensals, pathogens and their interaction with the immune system in humans and animals, taking the concept of probiotic and immunobiotic to new heights (Hosoya et al., 2011; Villena et al., 2014). Along with maintaining good health, the beneficial effect of probiotics and 'enhanced' probiotics are being suggested (Pandey et al., 2015). In this line, the achievements of biotechnology are supporting the replenishment or rejuvenation of gut microflora with the help of probiotic, prebiotic and synbiotic support (Markowiak and Śliżewska, 2017). The long-term impact of them on human health is unknown, especially when the environment is facing continuous change, with new pollutants being added regularly. Similarly, the nanoparticles, which are so active in body cells, will also likely be active in microbes, and possible changes in these modified microbes cannot be predicted with a guarantee.

16.4 Health Concerns due to Biotechnology

Advancements in biotechnology have revolutionized the field of medicine. It has led to the development of highly sensitive and specific disease diagnosis methods, novel and safer vaccines, highly efficacious medicines and other advanced treatment modalities, such as gene therapies, tissue engineering, nanomedicine, nanoscale surgery and certain types of targeted drug delivery systems (Ebbesen et al., 2006). Also, biotechnology has improved health indirectly by improving the environment, food quality and lifestyle (WHO, 1997). Nevertheless, at the same time, it has certainly raised health-related concerns in humans. These health concerns can be divided into the following categories.

16.4.1 Direct Health Concerns

Although biotechnology is considered safe and clean, it is associated with many health hazards mainly through the biotechnology-derived products (biologicals). The health hazards can be further classified into unintentional health hazards and intentional health hazards.

16.4.1.1 Unintentional Health Hazards

In a wide range of disorders, biologicals have shown their potential as therapeutic measures. Disorders originating from cancerous emergence, inflammatory imbalances and atherosclerotic changes are included in this category. Biologics are different from traditional drugs both structurally and functionally. Despite the differences, these biologicals can initiate immune responses. Interestingly, many times, the immune responses include both humoral and cell-mediated immune systems. Health hazards associated with the exposure to biologicals can be separated into five categories.

16.4.1.1.1 Infection by Biologicals

The biotechnology personnel are constantly exposed to the risks of infections during handling biologicals or microorganisms in their routine laboratory work. The United Kingdom, Austria and Switzerland have seen higher incidences of infectious diseases in medical laboratory technicians in comparison to control population (May et al., 1967; Cox, 1966; Benbough and Hambleton, 1973). The transmission of such infections to others have been reported. Accidental or even intentional involvements of medical laboratory workers in such transmission are also reported (Benbough et al., 1972; Hambleton, 1970). Apart from human errors, flawed processing techniques used in the biotechnology industry may lead to outbreaks of pathogens in the population. A brucellosis outbreak was experienced (1939) when, in a process to separate *Brucella abortus* in a stairwell in the veterinary college of Michigan State University, a tubular bowl centrifuge was used and an infectious aerosol was formed, affecting 93 people, with one death (May et al., 1957). However, this type of incident is preventable; authorities should observe strict compliance for preparing and practicing standard operating protocols for the setup and for individuals dealing with potential pathogens.

Genetically modified pathogens in the biotechnology industry may change its virulence or toxicity. In 2001, the researchers working on sterilization of rodents by biological means using mousepox virus witnessed a genetically modified mousepox virus that became highly lethal vaccinated and naturally resistant species (Jackson et al., 2001). Similarly, in 2012, a group of scientists were working on a project to improve surveillance of H5N1 influenza virus, and they created a strain that was capable of airborne spread and overcoming the obstacle that limits the global impact of natural H5N1 (Herfst et al., 2012). Such highly virulent pathogens produced in the laboratory pose a threat of outbreak. Regulatory authorities should make policies and implement them to prevent such infections.

16.4.1.1.2 Allergic Reactions to Biologicals

Biological agents are immune modulators. Many biologicals can cause grievous allergic reactions, and sometimes these allergic responses can be fatal too. The biotechnology workers are exposed to allergens of biological or occupational origin, and patients treated with biological agents are also prone to these allergic reactions (Table 16.1).

16.4.1.1.3 Reaction to Biotechnology-Derived Endotoxin

Endotoxins are the cell membrane–associated toxins present in gram-negative bacteria whether the organisms are pathogenic or not. They are capable of interfering with the structure and function of other cell membranes and, thus, hamper normal cell growth and their functions. Varied responses of cells toward endotoxins kept the critical dose of endotoxin undecisive for the initiation of damage at the level of membrane or cell as a whole. Similarly, the form of endotoxin and the mechanism of these damages are also not understood clearly. Endotoxin inhalation has been assigned to number of diseases, and interestingly, several biotechnology industries are associated with these diseases. In this context, the example of chronic respiratory disease is citable. Textile workers were exposed to endotoxin derived from *Enterobacter agglomerans* in dirty cotton (Ferguson et al., 1993). The use of microorganisms such as *Methylophilus methylotrophus* and *Methylomonas methanolica* for single-cell protein production in biotechnology was shown to be associated with an influenza-like illness that was due to the endotoxins transmitted through the inhalational route (Mayes, 1982).

Palchak et al., (1988, 1990) felt the need to establish an action threshold value to identify the critical concentration of airborne endotoxin of clinical importance. Based on the available literature on the clinical impacts of different levels of inhaled endotoxin and a monitoring campaign for airborne endotoxin carried out in a biotechnology plant, they have proposed that the airborne endotoxin level of 30 ng/m^3 should be accepted as an action threshold value (Palchak et al., 1988, 1990).

16.4.1.1.4 Toxic Reactions to Biologicals

The American Chemical Society has identified that the byproducts discharged during the production of carbon nanotubes have cancer-causing compounds, air pollutants and other substances of concern (American Chemical Society, 2007).

In workers of an industry involved in oral contraceptive formulation, a number of problems related to the reproductive system have been reported. Menstrual problems, loss of libido, postmenopausal vaginal bleeding and gynecomastia topped the list of reported problems (Harrington et al., 1978).

The pharmaceutical ind`ustry's handling of antibiotics can be a source of airborne antibiotic substances. Workers of these industries have had multidrug-resistant species isolated from nasal swabs, mucus and stool samples (Sarker et al., 2014). Eventually, this type of exposure to can give rise to a varied range of symptoms.

In addition to biomolecules, organisms used in the biotechnology industry can also be a threat to workers or even nearby inhabitants. Sometimes, normally innocuous organisms can produce harmful biomolecules when exposed to certain environmental conditions. As an example, lethal toxins can originate from *Aspergillus flavus* and *Aspergillus oryzae* when exposed to conditions of altered temperature and pH. These toxins can also be carcinogenic upon inhalation exposure (Naguib and El-Khadem, 1976). Therefore, as emphasized earlier, ensuring the standard operating conditions for the targeted fermentation procedure is very essential for processes

TABLE 16.1

Types of Allergic Reactions Because of Biologicals and Their Intended Uses

Type of Allergic reaction	Biologicals	Its Use
Cytokine release syndrome (also known as *cytokine storm*) (Suntharalingam et al., 2006)		Novel monoclonal antibody designed as a humanized super-agonist anti-CD28 monoclonal antibody
Hypersensitivity reaction, Delayed injection reactions (Bavbek et al., 2015)	Etanercept, Adalimumab	Biologic medicine used to treat rheumatoid arthritis, plaque psoriasis, psoriatic arthritis, polyarticular juvenile idiopathic arthritis and ankylosing spondylitis
Anaphylaxis reactions, Serum Sickness-Like Reactions (Galvao and Castells, 2015)	Omalizumab	To reduce the sensitivity to allergens
Secondary Immunodeficiency (Murdaca et al., 2015)	Infliximab, Etanercept	Used against autoimmune diseases
Autoimmune disorder: Systemic lupus erythromatosus (Niewold and Swedler, 2005)	Interferon-alpha	Used as anticancer drug
Atopic dermatitis (Chan et al., 2004)	Infliximab	Used against autoimmune diseases
Immunological cross-reactivity to anti-epidermal growth factor receptor producing acne (Perez-Soler and Saltz, 2005)	Gefitinib	Used as anticancer drug
Allergic asthma (Bennett, 1994)	Ampicillin Benzyl Penicillin Penicillin Penicillamine Tetracycline	Used in production of antibiotics
	Amylase	Used in industrial baking and handling of bulk quantity enzymes
	Cellulase Esperase	Used in production of enzymes
	Cephalosporin Lysozyme Tuberculin	Used in production of pharmaceuticals
	Protein Dust	Used in protein production
	Subtilisin	Used in preparation of washing powder
Microbial-induced asthma outbreaks (Lehrer and Bannon, 2005)(Lehrer and Bannon, 2005)		Used in farming, especially in composting for mushrooms
		Used in fermentation and production of citric acid
	Aureobasidium pullulans Paecilomyces variotii	Used in processing of woodworks
	Bacillus subtilis	Used as probiotic for prevention of ventilation-associated pneumonia
	Baculoviruses	Used in production of pesticide.
	Candida tropicalis	Used in production of protein
	Penicillium sp.	Used in production of citric acid, ATP and making of cheese.
Food allergies	Biotech proteins in foods	To improve the quality of food

that involve these types of microorganisms. Proper control and monitoring measures must be there to check any violation of the standard operating procedure.

16.4.1.1.5 Hazards Posed by Genetic Modification

Gene therapy is another area in which precautionary measures are of utmost importance. The first trial of clinical gene therapy was approved in 1988. Gene therapy has been employed in more than 3000 patients since its commencement (Selkirk, 2004). The hazards associated with gene therapy can be related either to the delivery vector or to the expression of transferred genes.

The most common delivery vectors are viruses. Primarily, there are reports that these vectors have produced inflammations. Vogel (2000), for the first time, reported the death of 18-year-old participant due to viral vector. Second is the malignant transformation of the incorporated retrovirus (Selkirk, 2004). In 2003, patients treated for severe combined immunodeficiency with gene therapy reported T-cell leukemia (Hacein-Bey-Abina et al., 2003a, 2003b).

Apart from the theoretical risk of expression of the transferred gene, clinical trials of gene therapy have reported aberrant inflammatory responses. This type of response can be seen with any other therapeutic agent, but in the case of gene therapy, it is not possible to undo the process (Selkirk, 2004).

16.4.1.2 Intentional Health Hazards

This involves the intentional release or dissemination of hazardous biological agents. These agents can be pathological bacteria, viruses, fungi, or toxins. These agents may be in a naturally occurring or a human-modified form. These are released to cause illness or death in people, animals or plants. Some of the examples of such pathogenic agents are smallpox, Ebola and Marburg virus. These agents may be preferred as biowarfare weapons because of their atrocious capacity to induce malady. Biowarfare attacks are now possibilities with an intention of not only causing sickness and death in a large number of victims but also of creating fear and panic in society.

In 1972, the Biological Weapon Convention (BWC) was drafted and named "*Convention on the Prohibition of the Development, Production, and Stockpiling of Bacteriological (Biological) and Toxin Weapons and on Their Destruction*". This treaty bans the development, production, and storage of pathogens or toxins in for unfair purposes. As of July 2018, the BWC has 181 states as members.

To fight against biological warfare, the health care providers, as well as the public, should be familiar with epidemiology and control measures of likely used pathogenic agents. Strategies should be developed for timely diagnosis and management of such diseases. Further education focusing on recognizing this threat is necessary.

16.4.2 Indirect Health Hazards

Indirect health hazards are the biotechnology-triggered human health hazards due to negative environmental and ecosystem functioning, including harm to plants, animals, soil and, ultimately, biodiversity.

16.4.2.1 Hazards due to Agricultural Biotechnology

With the increase in population load worldwide, there is a need for improving food yield—both in quantity and in quality. This demand of the food supply can be met by biotechnological improvement of crops, poultry, farming and aquaculture. Biotechnological principles are being utilized to increase the resistance to disease, pests and environmental stress in crops so that the target for "Good Health for All" (WHO, 2018) can be achieved. However, there are some associated risks in the process of increased food production using biotechnology. These risks include increased food allergens, the chance of cross-pollination and unwanted changes in the existing food or the biotechnology food so that they become inedible. In addition, the increased use of herbicides and insecticides by industrial agriculture could also be a source of threat.

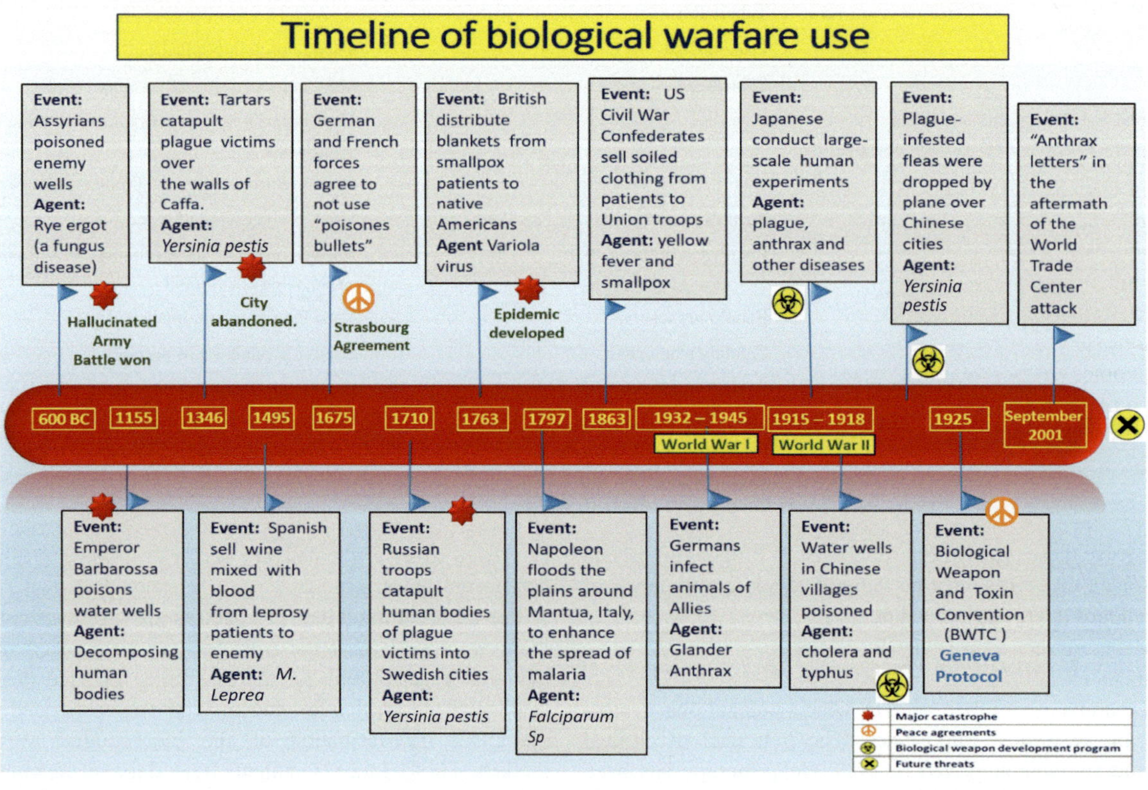

FIGURE 16.2 Timeline of biological warfare use.

Of late, the harm done to human and environmental health by the overuse of genetically modified plants and organisms is drawing attention. Genetically modified plants and organisms are unique crops/organisms created by mixing genes of two different species that have never shared genes naturally. Thus, there are many ambiguities in these crops and organisms. In the United States, a majority of soybean is genetically engineered so that it withstands Monsato's Roundup herbicide. It has been shown that usage of this herbicide is associated with various health risks not limited to non-Hodgkin's lymphoma (2008 Sweden), other cancers, skull and brain deformities (2009 Paraguay) and birth defects due to DNA damage (2007 Ecuador). DNA damage is a major population health concern. The effects of these genetically engineered products can not only affect an individual health unenviably, but its extensive usage can also result in the weakening and destruction of the whole population and environment. A study found that tadpoles in Argentina that were exposed to the Roundup herbicide had skull, eye and tail deformities (Samal and Bhattacharya, 2017).

16.4.2.2 Hazards due to Gene-Manipulated Microbe

In the genetic engineering processes, virus vectors are used. There are chances of combining genes of vector viruses with genes from infecting viruses, resulting in a virus that may be more infectious and can cross species borders. One of the important examples is cauliflower mosaic virus (CaMV). It was shown that CaMV can reactivate dormant or create new pathogenic viruses (Martinière et al., 2009). CaMV has been reported to produce cancers in stomach and colon (Yasaka et al., 2014)

16.4.2.3 Hazards due to Transgenic Animals

The major uses of transgenic animals are for the generation of animal models of basic science studies for the production of pharmaceutical products to enhance breeding in farm animals and improve the quality of animal-originated food. Although there are no serious health hazards reported due to the use of these transgenic animals, there are few reports of allergic reactions due to the expression of novel proteins in susceptible people. The production of pharmaceuticals in the milk of transgenic farm animals has also raised some biosafety concerns. The meat contaminated with potentially harmful pharmaceuticals might enter the human food supply and may pose serious health risks.

16.4.2.4 Hazards due to the Impact on Biodiversity

Biodiversity is the variety of plants and organisms living on Earth. The observed diversity in biota is a compounded one that includes noticeable heterogenicity in terms of genetic, species and ecosystem, as well. This multifariousness is a source of food, shelter and clothing for humans. Due to the advent of biotechnology, biodiversity is interfered and harmed, which is indirectly affecting human health. In Europe, herbicide-resistant transgenic crops have disturbed the ecological equilibrium and inherent interactions among the organisms. These transgenic crops have wiped out the weeds, affecting the food chain starting from the weeds, insects and so on, as organisms at every level are dependent on the energy source for the bottom one. One of the commonly noticed changes is that previously common farmland birds are now becoming rare, and it is actually suspected that changes in biodiversity could be the reason of these changes. Similarly, the decline in the number of native ladybirds could also be assigned to man-made alteration in biodiversity. Seven-spot ladybirds were introduced with an intention to contend the Russian wheat aphid; however, these new introductions competed with the native ladybirds in reality. These seven-spot ladybirds turned into predators of the native ladybirds and have made them disappear from the ecosystem (Amini et al., 2014).

16.4.3 Conjecturable Concerns

Because of nonuniformity in raw material and production of nanoparticles, as well as a lack of standard practice of characterization of nanomaterials used in medicine, they possess a wide variety of structures, physical and chemical properties and, accordingly, cytotoxicity too (Zdrojewicz et al., 2015). Kobayashi et al. (2017) reviewed the respiratory toxicity of carbon nanotubes, showing inflammation, injury, and fibrosis in the respiratory system. In addition, the biopersistence of carbon nanotubes in lung and gene damage are also discussed (Kobayashi et al., 2017). The use of carbon nanotubes can induce tissue-level oxidative stress and inflammatory reactions. These altered tissue responses led to malignant transformations, interstitial fibrosis and the formation of granulomas in many organs. In addition, reproductive and developmental toxicity are also reported with the use of carbon nanotubes (Kolosnjaj et al., 2007; Liu et al., 2013; Kobayashi et al., 2017). Similarly, Emma et al. (2016) reviewed the reproductive and developmental toxicity caused by carbon-based nanomaterials in animal models of rat, mice, chicken, zebrafish and others and suggested an appropriate risk-assessment study for these nanomaterials. Based on the potential to induce harmful and fibrotic reactions by the raw materials if reached on the organ, which could be possible in some conditions, Kolosnjaj et al. (2007) suggested a thorough evaluation of the safety of carbon nanoparticles and their environmental impacts. In 2013, a special issue of Toxicology Mechanisms and Methods was published with guest editor Balam Muñoz that encompassed the toxicity of silver nanoparticles on human tumor cells, human normal cells and mice intestinal mucosa in addition to the mechanisms of toxicity and immunomodulation by the nanomaterials (Munoz, 2013). In the case of glycol-coated gold nanoparticles, its high toxicity has been noticed with higher sizes while it is found to be safe with lower sizes (Zhang et al., 2011). These reports clearly indicate that there is toxicity caused by nanotechnology, however, not well defined. Therefore, the environmental impact or the impact on the ecology is not well estimated, hence, no measure to tackle them is formulated yet.

Biotechnologically produced organic insecticide (gene cloned from Bacillus thuringiensis and introduced in plants) proved to be useful until now; however, there is every chance that insects may develop resistance to the protein produced by this gene. On the other hand, plants producing this protein is environmentally safe as it is innocuous for the plants themselves, animals and humans; however, as it is an external protein, there is a possibility that plants may start metabolizing it into products that could be toxic.

16.5 Possible Impacts of Environmental Biotechnology on Health

With the help of genetic modifications, it is actually achievable to make meals with higher amount of minerals, vitamins and antioxidants. Also, through increasing crop yields, deforestation is avoided and, one of the most crucial for the cultivating countries, economic growth is sped up. Biotechnology has actually changed plants to possess their own bug-insusceptible traits, making the use of dangerous pesticides much less than usual on the farms. Through biotechnology, we have been able to make better sampling fruits as well as vegetables. Biotechnology has likewise offered numerous diagnostic sets that make it less complicated for human beings to identify pest problems in vegetations as well as animals. The most effective part is actually that these diagnostic kits are very accurate, and for this reason, growers can rightly forecast the amount needed for pesticides as well as other procedures for optimum plant growth and farming. Vegetation has ended up being extra-resistant to viruses and other infectious insects.

Industrial pollution and their bioremedial measures are the two opposing forces in the issues of health concerns related to environmental biotechnology. On one side, technological progress is producing environmental pollutants while, on the other side, bioremediation is trying to minimize their impacts on the environment. Environmental biotechnology has helped in both reducing the level of pollution and controlling the waste output

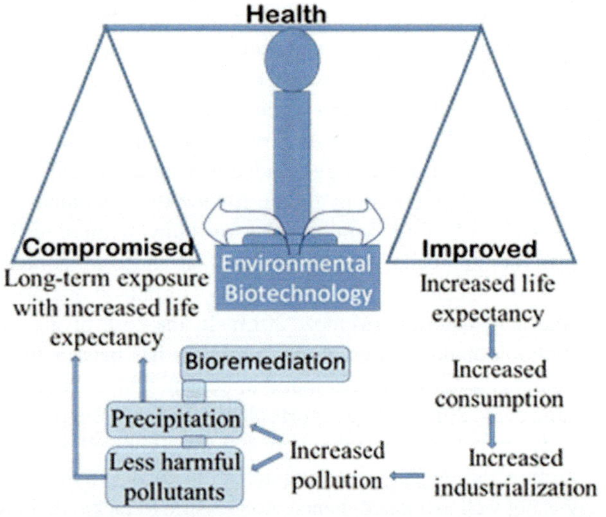

FIGURE 16.3 The other side of biotechnology: unwitting balance between the benefits and risk.

in the environment. Improved health and life expectancy by the application of biotechnology, in fact, increased the demand for consumables and required greater production. Thus, a vicious cycle has formed that was supposed to be in control using environmental biotechnology. In reality, the growth of environmental biotechnology failed to cope with the rising load of pollution and management responsibilities.

The bioremediation depends on biostimulation of microorganisms to convert pollutants into less harmful products and demobilize toxic pollutants. Over a long duration and with excess pollutant generation, the effectiveness of these bioremedial processes becomes questionable. In addition, there are some options available for handling soil and water contaminations; however, developments for handling air pollution are still in their infancy. Recent developments in the field includes phytobioremediation and phyllobioremediation. In this regard, biotreatment and biodegradation options were reviewed by Afzal et al. (2016), who suggested that the major concerns and limitations are cost-effectiveness, the volume to dispose of and the bioremedial capability of the microorganisms.

Deinococcus radiodurans is being used to bioprecipitate heavy metals, including nuclear wastes, due to its high radiation-resistant property. It has been labeled as the toughest bacteria by the Guinness *Book* of *World Records* (DeWeerdt, 2002). In 1956, *Deinococcus radiodurans* was first isolated from canned meat that spoiled even after being exposed to high doses of gamma radiation. It was then realized that this bacterium is resistant to radiation and slowly was brought to use by genetic modifications. About 70 million cubic meters of ground and 3 trillion liters of groundwater have been contaminated due to leaking of the radioactive waste in the United States during the Cold War, and *Deinococcus radiodurans* is being used as the bioremediating cleanup technology (Daly, 2000). This continuous exposure to radiation can make the bacteria resistant to subsequent higher doses of radiation (Mortazavi et al., 2014). Thus, the success of this engineered bioremediating bacteria that once-contaminated human food and is now being used for cleaning up water is itself worrying as it may become a bridge for bioterrorism, for radio resistance and resistance to antibiotics and for genotoxic agents.

The genetically engineered seeds are normally resistant to a single bug, and to provide higher yields, they require more and scheduled supply of fertilizers and water. The dependence on natural rains and the unavailability of costly fertilizers are major reasons that poor Indian farmers failing to achieve a higher yield despite using genetically engineered cotton seeds. It was found that genetically engineered cotton seeds, despite being resistant to its primary pest, the bollworm, were destroyed by mealybug. This is because of a weakening of the crops and the environment. The natural and organic farmers are also paying a toll for these genetically engineered seeds because of contamination, seed mixing and pollen drift from neighboring genetically engineered seed fields that affect not only their seed/crop quality but also yield. This fault of the genetically engineered cotton seeds thus could not help reduce the suicide rate among debt-burdened cotton farmers in India. These issues of quality of produce and its safety for human consumption are still questionable (Samal and Bhattacharya, 2017).

16.6 Policies to Protect Environment

16.6.1 Existing Policies

In our research, we came across the biotechnology policies of various states in India and found that only three states, namely, Andhra Pradesh, Jammu and Kashmir and Madhya Pradesh, have notably mentioned "conservation of biodiversity" as the major objective of the policy. Most of the states considered biotechnology as a means to develop and promote high-quality infrastructure and facilitate the flow of venture capital funds to biotech companies. It appears that the only aim behind the blindfolded (i.e., without noticing the effects) promotion of biotechnology across the country was to achieve high productivity and economic development in various fields by utilizing bioresources, and the effects of this pursuit on health and environment found no place in any state's policy (Samal and Bhattacharya, 2017).

In 1998, due to the fear of toxicity and allergenicity of the transgenic seeds, plants and plant parts the Department of Biotechnology, Ministry of Science and Technology, Government of India, issued guidelines for research in transgenic plants and the evaluation of its health-related side effects. Current guidelines also cover recombinant DNA research on plants and its evaluation (Rastogi, 2016). There is a provision to establish a separate tribunal to explicitly try cases related to biotechnology. This is a cause for great concern as there is no legal case pertaining to biotechnology that would come under the purview of local judiciaries, and the fairness in dispensing off the cases would entirely depend on the special tribunal (Ray, 2012).

Unlike most countries, Canada has been able to formulate a successful environmental health policy and framework. Canadian Environmental Protection Act, 1999 (CEPA, 1999; a joint venture by Health Canada and Environment Canada) controls new products in which biotechnology might have been used. In addition, if any substance poses a threat of human health risk or problem to the environment, the CEPA 1999 also deals with those. Guidelines to evaluate the safety of food and feed products from genetically modified animals have been published both by U.S. Food and Drug Administration (FDA) and European Food Safety Authority (EFSA) in 2012 (USA FDA, 2015). The European guidelines include recommendations on preserving the welfare of genetically modified animals. Other reports from academic groups have defined rules to assure animal welfare. Other institutions, namely, the Codex Alimentarius, have published their own guidelines which are essentially similar to those by the FDA and EFSA. A separate EFSA document takes into account the environmental impact of genetically modified animals (Assuring Animal Welfare, 2007).

16.6.2 Policies Needed

When any biotechnological product is found to have a negative impact on the environment, the product is withdrawn from the market. There is no provision of finding the stakeholders who were involved in the process of development. Similarly, there is no formulation of policies for how environmental damage is to be corrected or how the responsibility for such damage should be identified—discoverer, producer, marketing agency or the sanctioning authority (competent authority to release the product into the market).

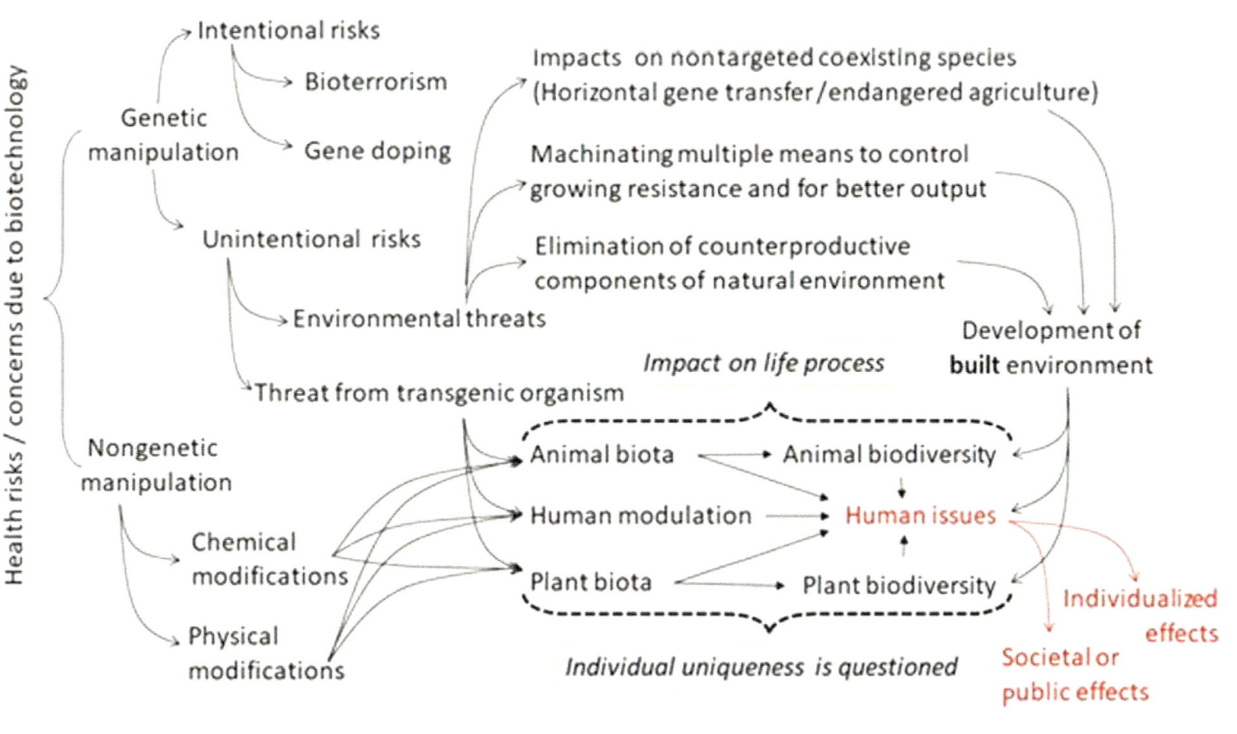

FIGURE 16.4 Conglomeration of environmental biotechnological issues in health perspectives.

There is also need for policies to take care of animal rights. The manipulation of animals for the cause of human benefit through genetic engineering and/or selective breeding is unethical. Not only for the sake of finding new health compound, but also for the evaluation of environment-associated risks, knockout animals are being used. While these steps need to be regulated for the benefit of humans, the rights of animals also need to be considered.

16.6.3 Anticipated Challenges

Major hurdles toward making these policies possible are (1) the identification of the problem, as in most cases, the negative results remained unpublished, and (2) the involvement of the public sector in the process of development with unidentified onus of the responsibility.

Because of the evolving prospects in the field of environmental biotechnology, the investors may like to conceal the novel negative findings to gain economic benefits without staking their public impression. Similarly, unnecessary replications of unsuccessful protocols may generate dubious benefits. These are known but not regulated by any policies.

16.7 Future of Environmental Biotechnology in Relation to Health

Plants are able to yield far better fruit produce and animals yield more dairy products with the help of biotechnology. Strongly reliable, regular and rapid chemicals have been produced with help from biotechnology. These enzymes participate in a major way in food handling as well as deterrence. Knowingly or not, we are habituated to using quite a good number of biotechnology products. They have already become part of our lifestyle routine.

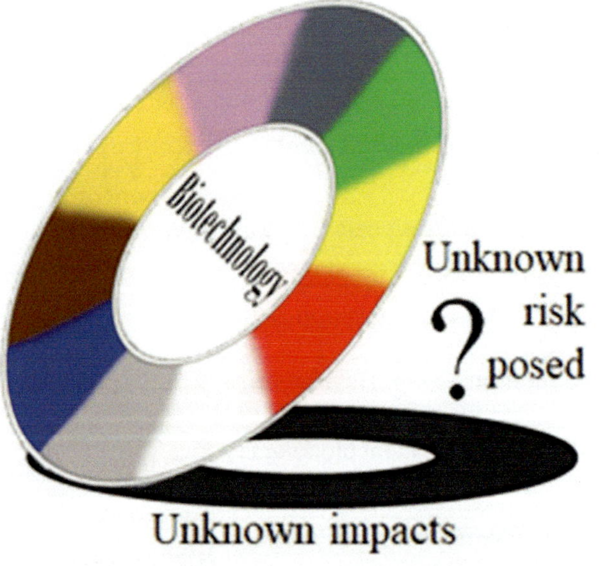

FIGURE 16.5 Black biotechnology hideouts (unidentified but harmful arena of biotechnology).

As discussed already, there are many unexplored areas where the intended beneficial changes in the environment come back to us through environmental modification as unsound alterations. In this context, mobile radiation is great example. With the rampant use of mobile technology, mobile towers and receiver handsets are covering the immediate environment on the earth's surface. Nobody knows how genetically modified organisms will respond to these low-dose long-term radiation exposures. In spite of 'statutory warnings' inscribed on every not-to-be-thrown-away accessory of electronic gadgets, a lack of awareness and the easy availability of low-cost use-and-throw consumables are being disposed of carelessly. This modern technology junk causes additional risk of genetic or epigenetic modulators of so-called harmless tools of environmental biotechnology.

16.8 Conclusion

The biotechnological progress is great, and its importance in protecting the environment is immense. The benefits of environmental biotechnology may be associated with a risk of far-reaching impact on human health, directly or indirectly. Sometimes, the risk is not identifiable immediately, especially when it may not be evaluated properly.

Considering the discussed points, robust regulatory mechanisms are the need of the hour, and environmental life-cycle assessments or impact assessments are necessary. Objectives that need to be added in biotechnology policy include the following (especially in the context of India and its states):

1. Regulation of global impact of novel biotechnological products—with inputs from all possible corners
2. For biotechnological processes—assessment of environmental and human health impacts
3. Campaigning about biotechnology products to raise public awareness in India
4. Raising public concern about the regulation of productions, commercialization and use of biotechnology in all spheres of life in India

REFERENCES

Afzal, H., K. Zahid, Q. Ali, et al. 2016. Role of biotechnology in improving human health. *Journal of Molecular Biomarkers and Diagnosis* 8:1000309.

Almeida, H., M. H. Amaral, and P. Lobão. 2011. Drugs obtained by biotechnology processing. *Brazilian Journal of Pharmaceutical Sciences* 47:199–207.

American Chemical Society. 2007. Toxic byproducts of carbon nanotube manufacturing: Are there green alternatives? www.sciencedaily.com/releases/2007/08/070821081446.htm (accessed 18 September 2018).

Amini, S., S. Sharafi, H. R. Komeili, and N. A. Tabaee. 2014. Effect of biotechnology on biodiversity. *International Journal of Farming and Allied Sciences* 3:910–915.

Andronescu, E., J. M. Brown, F. N. Oktar, S. Agathopoulos, J. Chou, and A. Obata. 2016. Nanomaterials for medical applications: Benefits and risks. *Journal of Nanomaterials*. Doi:10.1155/2016/8284319.

Assuring Animal Welfare. 2007. From societal concerns to implementation. In *Second Welfare Quality® Stakeholder Conference*, ed. I. Veissier, B. Frokman, and B. Jones, 1–93. Project Office Welfare Quality®: Animal Sciences Group of Wageningen-UR, Berlin Germany.

Balamurugan, V., A. Sen, P. Saravanan, and R. K. Singh. 2006. Biotechnology in the production of recombinant vaccine or antigen for animal health. *Journal of Animal and Veterinary Advances* 5:487–495.

Bavbek, S., S. Ataman, A. Akinci, et al. 2015. Rapid subcutaneous desensitization for the management of local and systemic hypersensitivity reactions to etanercept and adalimumab in 12 patients. *The Journal of Allergy and Clinical Immunology: In Practice* 3:629–632.

Benbough, J. E., and P. Hambleton. 1973. Structural, organisational and functional changes associated with envelopes of bacteria sampled from aerosols. In *Fourth International Symposium on Aerobiology*, ed. J. F. P. Hers, and K. E. Winkler, 135–137. Oosthoeck, Ultrecht, The Netherlands.

Benbough, J. E., P. Hambleton, K. L. Martin, and R. E. Strange. 1972. The effect of aerosolization in the transport of a-methyl glucoside and galactosides into Escherichia coli. *Journal of General Microbiology* 72:511–520.

Bennett, A.M. 1994. Health hazards in biotechnology. In *Biosafety in Industrial Biotechnology,* ed. P. Hambleton, J. Melling, and T. T. Salusbury, 109–128. Springer-Science+Business Media, BV, Dordrecht.

CEPA [The Canadian Environmental Protection Act] 1999. https://www.canada.ca/en/environment-climate-change/services/canadian-environmental-protection-act-registry (accessed 20 November 2018).

Chakraborty, C., A. R. Sharma, G. Sharma, C. G. P. Doss, and S. S. Lee. 2017. Therapeutic miRNA and siRNA: Moving from bench to clinic as next generation medicine. *Molecular Therapy – Nucleic Acids* 8:132–143.

Chan, J. L., L. Davis-Reed, and A. B. Kimball. 2004. Counter-regulatory balance: Atopic dermatitis in patients undergoing infliximab infusion therapy. *Journal of Drugs in Dermatology* 3:315–318.

Cox, C. S. 1966. The survival of Escherichia coli in nitrogen atmosphere under changing conditions of relative humidity. *Journal of General Microbiology* 45:283–288.

Daly, M.J. 2000. Engineering radiation-resistant bacteria for environmental biotechnology. *Current Opinion in Biotechnology* 11:280–285.

D'Angelo, B., E. Benedetti, A. Cimini, and A. Giordano. 2016. MicroRNAs: A puzzling tool in cancer diagnostics and therapy. *Anticancer Research* 36:5571–5575.

DeWeerdt, S. E. 2002. The world's toughest bacterium. http://www.genomenewsnetwork.org/articles/07_02/deinococcus.shtml (accessed 9 May 2021).

Dwivedi, S., P. Purohit, R. Misra, et al. 2017. Diseases and molecular diagnostics: A step closer to precision medicine. *Indian Journal of Clinical Biochemistry* 32:374–398.

Ebbesen, M., and T. G. Jensen. 2006. Nanomedicine: Techniques, potentials, and ethical implications. *Journal of Biomedicine and Biotechnology* 5:515–516.

Emma, M., K. S. Hougaard, A. Kishimoto, and K. Honda. 2016. Reproductive and developmental toxicity of carbon-based nanomaterials: A literature review. *Nanotoxicology* 10:391–412.

Ferguson, R., C. Feeney, and V. A. Chirurgi. 1993. Enterobacter agglomerans-associated cotton fever. *Archives of Internal Medicine* 153:2381–2382.

Frangioni, J. V. 2008. New technologies for human cancer imaging. *Journal of Clinical Oncology* 26:4012–4021.

Galvao, V. R. and M. C. Castells. 2015. Hypersensitivity to biological agents-updated diagnosis, management, and treatment. *Journal of Allergy and Clinical Immunology Practice* 3:175–185.

Hacein-Bey-Abina, S., C. Von Kalle, M. Schmidt, et al. 2003a. A serious adverse event after successful gene therapy for X-linked severe combined immunodeficiency. *New England Journal of Medicine* 348:255–256.

Hacein-Bey-Abina, S., C. von Kalle, M. Schmidt, et al. 2003b. LMO2-associated clonal T cell proliferation in two patients after gene therapy for SCID-X1. *Science* 302:415–419.

Hambleton, P. 1970. The sensitivity of Gram-positive bacteria recovered from aerosols to lysozymes and other hydrolytic enzymes. *Journal of General Microbiology* 61:197–204.

Harrington, J. M., G. F. Stein, R. O. Rivera, and A. V. de Morales. 1978. The occupational hazards of formulating oral contraceptives – a survey of plant employees. *Archives of Environmental Health* 33:12–15.

Health Policy Institute. 2017. Effects of biotechnology on human health. http://healthpolicy-institute.org/ (accessed 19 November 2018).

Herfst, S., E. J. Schrauwen, M. Linster, et al. 2012. Airborne transmission of influenza A/H5N1 virus between ferrets. *Science* 336:1534–1541.

Hosoya, S., J. Villena, T. Shimazu, et al. 2011. Immunobiotic lactic acid bacteria beneficially regulate immune response triggered by poly(I:C) in porcine intestinal epithelial cells. *Veterinary Research* 42:111.

Huang, W. 2017. MicroRNAs: Biomarkers, diagnostics, and therapeutics. *Methods in Molecular Biology* 1617:57–67.

Ivanov, K., A. Stoimenova, D. Obreshkova, and L. Saso. 2013. Biotechnology in the production of pharmaceutical industry ingredients: amino acids. *Biotechnology & Biotechnological Equipment*, 27:3620–3626.

Jackson, R. J., A. J. Ramsay, C. D. Christensen, S. Beaton, D. F. Hall, and I. A. Ramshaw. 2001. Expression of mouse interleukin-4 by a recombinant ectromelia virus suppresses cytolytic lymphocyte responses and overcomes genetic resistance to mousepox. *Journal of Virology* 75:1205–1210.

Kafarski, P. 2012. Rainbow code of biotechnology. *Chemik International* 66:811–816.

Kobayashi, N., H. Izumi, and Y. Morimoto. 2017. Review of toxicity studies of carbon nanotubes. *Journal of Occupational Health* 59:394–407.

Kolosnjaj, J., H. Szwarc, and F. Moussa. 2007. Toxicity studies of carbon nanotubes. *Advances in Experimental Medicine and Biology* 620:181–204.

Lauer, K. B., R. Borrow, and T. J. Blanchard. 2017. Multivalent and multipathogen viral vector vaccines. *Clinical Vaccine Immunology*, 24: e00298–16.

Lehrer, S. B. and G. A. Bannon. 2005. Risks of allergic reactions to biotech proteins in foods: Perception and reality. *Allergy* 60:559–564.

Liu, Y., Y. Zhao, B. Sun, and C. Chen. 2013. Understanding the toxicity of carbon nanotubes. *Accounts of Chemical Research* 46:702–713.

Lumencandela. 2018. Genetic engineering products. https://courses.lumenlearning.com/boundless-microbiology/chapter/genetic-engineering-products/ (accessed 9 May 2021).

Markowiak, P., and K. Śliżewska. 2017. Effects of probiotics, prebiotics and synbiotics on human health. *Nutrients* 9:1021.

Martinière, A., A. Zancarini, and M. Drucker. 2009. Aphid transmission of cauliflower mosaic virus: The role of the host plant. *Plant Signaling & Behavior* 4:548–550.

May, K. R. 1967. Physical aspects of sampling airborne microbes, in airborne microbes. In *17th Symposium of General Microbiology*, ed. P. H. Gregory and J. L. Monteith, 6–80. Cambridge University Press, Cambridge.

May, K. R., and G. J. Harper. 1957. The efficiency of various liquid impingers in bacterial aerosols. *British Journal of Industrial Medicine* 14:187–197.

Mayes, R. W. 1982. Lack of allergic reactions in workers exposed to Pruteen (bacterial single-cell protein). *British Journal of Industrial Medicine* 39:183–186.

Mohammed, B. R., S. K. Malang, S. Mailafia, and R. I. S. Agbede. 2016. Application of biotechnology towards diagnosis and treatment in veterinary medicine in Africa: Potentials and future developments. *Journal of Biotechnology and Biomaterials* 6:245.

Mortazavi, S. A., F. Khademi, M. Motamedifar, M. Haghani, and S. M. J. Mortazavi. 2014. Human-induced radioresistance as a possible mechanism for producing biological weapons: A feasible bridge between radioresistance and resistance to antibiotics and genotoxic agents. *Iranian Journal of Public Health* 43:247–248.

Munoz, B. 2013. Perspectives and approaches in nanotoxicology research. *Toxicology Mechanisms and Methods* 23:151–152.

Murdaca, G., F. Spano, M. Contatore, et al. 2015. Infection risk associated with antiTNF-alpha agents: A review. *Expert Opinions in Drug Safety* 14:571–582.

Naguib, M. M., and M. El-Khadem. 1976. Toxicity of Aspergillus flavus cultures isolated from Egypt. *Zentralblatt fur Bakteriologie, Parasitenkunde, Infektionskrankheiten und Hygiene. Zweitenaturwissenschaftliche Abt.: Allgemeine, landwirtschaftliche und technischeMikrobiologie* 131:506–509.

Niewold, T. B., and W. I. Swedler. 2005. Systemic lupus erythematosus arising during interferon-alpha therapy for cryoglobulinemic vasculitis associated with hepatitis C. *Clinical Rheumatology* 24:178–181.

Nikalje, A. P. 2015. Nanotechnology and its applications in medicine. *Medicinal Chemistry* 5:81–89.

Palchak, R. B., R. Cohen, M. Ainslie, and C. L. Hoerner. 1988. Airborne endotoxin associated with industrial-scale production of protein products in gram-negative bacteria. *American Industrial Hygiene Association Journal* 49:420–421.

Palchak, R. B., R. Cohen, and J. Jaugltetter. 1990. A threshold for airborne endotoxins associated with industrial scale production of proteins in Gram-negative bacteria. *Developments in Industrial Microbiology* 31:199–203.

Pandey, K. R., S. R. Naik, and B. V. Vakil. 2015. Probiotics, prebioitics and synbiotics – a review. *Journal of Food Science and Technology* 52:7577–7587.

Perez-Soler, R., and L. Saltz. 2005. Cutaneous adverse effects with HER1/EGFR-targeted agents: Is there a silver lining? *Journal of Clinical Oncology* 23:5235–5246.

Rastogi, P. 2016. India: Protection of biotechnology under Indian laws. https://www.mondaq.com/india/patent/506468/protection-of-biotechnology-under-indian-laws (accessed 9 May 2021).

Ray, P. S. 2012. Biotechnology regulatory authority of India act: The new weapon of the second green revolution. http://sanhati.com/excerpted/4523/ (accessed 9 May 2021).

Samal, A., and P. Bhattacharya. 2017. Biotechnology policy in India. In *Progress of Biotechnology in India*, ed. S. C. Santra and A. Mallick, 190–215. ENVIS Centre on Environmental Biotechnology, University of Kalyani, West Bengal, India.

Sarker, M. M., K. N. Islam, H. Z. Huri, et al. 2014. Studies of the impact of occupational exposure of pharmaceutical workers on the development of antimicrobial drug resistance. *Journal of Occupational Health* 56:260–270.

Selkirk, S. 2004. Gene therapy in clinical medicine. *Postgraduate Medical Journal* 80:560–570.

Singh, R. L. 2017. Introduction to Environmental Biotechnology. In: *Principles and Applications of Environmental Biotechnology for a Sustainable Future*, 1–12. Springer, Singapore.

Suntharalingam, G., M. R. Perry, S. Ward, et al. 2006. Cytokine storm in a phase 1 trial of the anti-CD28 monoclonal antibody TGN1412. *New England Journal of Medicine* 355:1018–1028.

USA FDA. 2015. Guidance for industry: Regulation of genetically engineered animals containing heritable recombinant DNA constructs. Food and Drug Administration (USA FDA), Maryland, USA. https://www.fda.gov/media/135115/download (accessed 9 May 2021).

Vijayakumar, R., R. Agarwal, S. Jagannathan, R. P. Gandhi, and R. Ranjithkumar. 2013. Applications of nanotechnology in medicine: A view. In *Nanotechnology Vol. 7: Diagnostics and Therapeutics*, ed. N. K. Navani, S. Sinha, and J. N. Govil, 89–100. Studium Press LLC., USA.

Villena, J., E. Chiba, M. G. Vizoso-Pinto, et al. 2014. Immunobiotic Lactobacillus rhamnosus strains differentially modulate antiviral immune response in porcine intestinal epithelial and antigen presenting cells. *BMC Microbiology* 14:126.

Vogel, G. 2000. Gene therapy. FDA moves against Penn scientist. *Science* 290:2049–2051.

WHO. 1997. Global programme for vaccines and immunization. Vaccine research and development. Biotechnology and world health: Risks and benefits of vaccines and other medical products produced by genetic engineering: proceedings of a WHO meeting. WHO/VRD/BLG/97.01. http://www.who.int/iris/handle/10665/63726 (accessed 20 November 2018).

WHO. 2018. Good Health for All: Vital for reducing poverty and sustaining development. http://www.who.int/mediacentre/events/goodhealth/en/ (accessed 20 November 2018).

Wu, J., M. Dong, C. Rigato, Y. Liu, and F. Lin. 2018. Lab-on-chip technology for chronic disease diagnosis. *Nature Partner Journal Digital Medicine* 1:7.

Yasaka, R., H. D. Nguyen, S. Y. Ho, et al. 2014. The temporal evolution and global spread of Cauliflower mosaic virus, a plant pararetrovirus. *PloS One* 9:e85641.

Zdrojewicz, Z., M. Waracki, B. Bugaj, D. Pypno, and K. Cabała. 2015. Medical applications of nanotechnology. *Postępy Higienyi Medycyny Doświadczalnej (Advances in Hygiene and Experimental Medicine)* 69:1196–204.

Zhang, X. D., D. Wu, X. Shen, et al. 2011. Size-dependent in vivo toxicity of PEG-coated gold nanoparticles. *International Journal of Nanomedicine* 6:2071–2081.

17

Socioeconomic Status and Its Relation to Pollution

Jalaj Tripathi Devgan

CONTENTS

17.1 Introduction

In the modern world, pollution is a synonym of environmental pollution and a major hurdle in growth of healthy population of a country. History says the environment was a vital issue even in ancient world, and the available literature from various parts of world confirm this. In India, Vedic literature (some of the oldest literature available in the world) uses more than 200 words for air and water. In Vedic literature, *Vayu*, generally used in the sense of physical air or wind, is used as a synonym for Paraná vayu (breathing life). It shows the concern of our ancestors about clean and pure air for everyone. In Vedic literature, the word poison has been used for any substance that can harm the environment. In *Rigveda*, the oldest Vedic literature in Sanskrit, all natural resources including air have been given status of god (Sharma and Sharma 1994) as mark of respect. Because ancient texts do not reveal any evidence regarding pollution and pollution sources, it can be assumed that the ancient Indian natural resources were pure, contaminant-free or had no contamination. There is no reference to polluted air in literature (Sarma, 2015).

DOI: 10.1201/9781003186304-17

Although almost no evidence is available about polluted air or any reference related to it in ancient Indian literature, which indicates a nonpolluted environment, some reports from the Western world suggest the presence of polluted air during the Roman Empire.

It says the residents of Rome referred to their city's smoke cloud as 'gravioris caeli' (heavy heaven) and 'infamis aer' (infamous air) (Jacobson 2002; Mosley 2014). According to Mosley, in 535 AD, Roman emperor Justinian supported the importance of clean air and even formulated some rules that can now be considered as early versions of a clean air act. The story doesn't end here. With time, various pollution levels increased all over the world, and the situation became very serious even before industrialization in various parts of the world. No systematic analysis of pollution is available in literature for the ancient period, so it's difficult to draw any conclusions about when and how human beings first noticed toxic pollutants and their harmful effects. Nothing significant has been discussed for other types of pollutions also in ancient literature.

Environmental pollution probably started with use of metals and other related industrial and commercial aspects of human life and increased with the use of coal and wood as fuel or sources of energy. The extensive use of sea coal around 1200 filled London's air with smoke. This continued through the 17th and 18th centuries when the situation worsened and it became a matter of serious concern in England, the United States and other parts of Western world engrossed with industrial progress. Great Britain and the United States were the worst hit during the 20th century. One aspect is crystal clear: Until then, pollution had affected human population almost uniformly without any social gradient, and probably, populations with higher and lower 'socioeconomic status' (SES) were not aware of any scientific mode to gauge and protect themselves from the adverse impacts of pollution.

Thick fog and smoke continued in various parts of the world until 1963. Countermeasures were taken only when U.S. Congress enacted the first Clean Air Act. In 1970, the U.S. Congress proposed the specific norms for air pollution regulation, and these were enforced. In due course, this growing awareness about environmental pollution induced the North Carolina movement in 1982. This was followed by the adoption of terms such as environmental injustice and *SES* and these being into practice.

In 1972, efforts at the global level were initiated and implemented to curb pollution levels. By this time, various programs related to clean and sustainable environment and efforts were initiated in different parts of world. As a result, pollution levels declined. The high levels of pollution were the result of a lack of awareness, education, poor financial status and restricted access to health facilities, specifically in case of populations with lower SES. These clean environmental programs and efforts brought awareness to the population belonging to lower SES. This sector of the population enjoyed benefits after the implementation of environmental awareness programs.

The parameters related to SES play a major role in explaining the impacts of exposure of all types of pollutions independently and accumulated exposure from multiple sources, such as air, water, various forms of wastes, noise, nutrition, housing quality and neighborhood conditions, among others. The outcome of implementing these parameters is also related to the mode of their applications and the existing functional conditions. SES can be measured at the individual and community levels. The outcomes of these investigations are likely to differ from each other. The SES phenomenon is a dynamic phenomenon; hence, it is very rarely stable. This aspect plays a functional role in investigations related to SES.

Of the five major types of pollution, waste pollution seems to affect individuals of the lower SES population the most. The relationship between pollution and SES has been discussed in detail by other groups also. The triple jeopardy hypothesis, which elaborates the degree of impacts on the population with low SES, throws light on this issue (O'Neill et al. 2003). These individuals are affected relatively more by (1) environmental pollution, (2) a higher degree of susceptibility to air pollution specifically related to health consequences because of financial constraints and (3) the genotypic traits and health status of the individuals of this sector of the population that also add to the degree of susceptibility to air pollution and water pollution. All these parameters are strongly correlated with their SES and hence have impacts on such individuals and communities. It has been shown that individuals and communities with lower SES tend to benefit the most from the implementation of programs concerned with improving air and water quality (Bae 1997).

There are differences in the sources of pollution and its relation to SES in developing and developed countries. There is a strong and urgent need to find effective solutions for providing a clean, safe and sustainable environment not only for individuals of lower SES but others also. These efforts may include providing alternative modes of burning of biomass fuel for cooking, getting to maximum environmental illiteracy (possibly 100%) and finding feasible alternative modes specifically for malnourished children and other workforces, that is, women, pregnant mothers, mothers who feed their young ones and others. In addition to these, there should be some constructive efforts to completely switch over from unauthorized, unhygienic living conditions by providing proper housing and properly planned cities in developing and underdeveloped countries at a global level to curb pollution and reduce the gap between high- and low-SES levels in communities. For this, the role of SES functions as an indicator that can be used to minimize the deleterious impacts of pollution and to make the environment clean, safe and able to sustain the current industrial, agricultural, transportation and food technological innovatory activities occurring globally.

In the recent past, pollution has grown into a huge and drastic silent epidemic that poses health risks to all the biota contents of environment. Even the abiota of the environment are equally affected, and it is difficult to answer which are affected the most. In a particular, in geographic and ecological regions having same environment and same densities of pollutants, the impact is different. This detrimental impact depends on the underlying health of individuals, their economic status, their education with respect to environmental awareness level and their lifestyle, including a wide spectrum from living style, behavior, religion, faith and personal attitude, among others.

17.2 Historical Background

It is worth reviewing the work of earlier workers to understand the relationship between pollution and socioeconomic status. This also elucidates what prompted them to study the relevance of pollution and socioeconomic status. Some of them are mentioned in the following discussion.

Freeman was first to study three U.S. cities—Kansas city, St. Louis and Washington, DC, for pollutants such as sulfur dioxide (SO_2). This study included black and low-income groups that faced environmental inequality. His analysis was based on 1960 census data for Kansas City. SO_2 and other particulate pollutants were the main environmental pollutants. These were found to be present at highest levels compared to the localities in which populations with better environmental awareness and better SES resided (Freeman 1972).

Similar observations were reported by other groups. Zupan and coworkers reported similar observations to those of Freeman, specifically for SO_2 and other particulate pollutants, in the localities for low-income households in New York City (Zupan 1973). An investigation of the non-point-source emission of carbon monoxide (CO) and hydrocarbons in Washington, DC, area led to the conclusion that low incomes, low rents, low proportions of professional and managerial workers and high proportions of black residents characterized the most highly polluted zones in and around the Washington, DC, area (Kruvant 1975). This supported the proposed relationship between pollution and the socioeconomic status in the region of investigation.

An investigation comprising 284 cities and 23 states covering large population was also carried out that supported and affirmed the suggested correlation between pollutants and densely populated cities with lower incomes groups, comprising mainly populations of nonwhite residents, deprived of educational facilities and lacking of environmental awareness (Asch and Joseph 1978). In 1982, U.S. residents successfully launched a legal campaign against environmental inequality existing specifically among people of low-income groups or financially weaker sections. These population groups had no 'say' in the policymaking process. This campaign supported individuals with low SES to ensure a reduction in their suffering due to environmental issues (Chavis 1987; Cleere 2016). Chavis acted as an executive director of United Church of Christ Commission for Racial Justice (UCC) and played key role in drawing the public attention to the locations of 'hazardous waste dumping sites' and poor minority neighborhoods in a town in Warren County, North Carolina, into the limelight. He identified and elucidated the related environmental disparity among the individuals of higher and lower SES levels. He referred it as 'environmental racism' (Chavis 1993). It was also argued by some that 'institutionalized racism' existed within the governmental and corporate policies in practice during that period. The poor and financially weak had less political representation and faced 'environmental racism' (Bullard 2001). Because of the untiring efforts and social awakening, the much-needed attention and support from all corners of world were ensured. This later became the focal point for starting decisive results on the association between SES and pollution. In spite of 'environmental racism' being identified and initiated in the United States, it also prevailed in the other regions of the globe. One of the global reasons for this 'environmental racism' was related to transportation of 'hazardous waste' either as commercial by-products or as scrape materials from developed countries to either developing or underdeveloped countries. In most of these countries, environmental protection and safety policies were not being implemented and practiced strictly. Furthermore, in such countries, the marginalized communities were devoid of equal social status and effective political means that could enable them to oppose the unwarranted environmental practices in that region. Finally, a voice against the transboundary movement of waste material culminated in form of Basel Convention (1989).

Although an enormous amount of data support the relationship among SES, pollution and human health are available from various countries, they are not conclusive. This makes it difficult to formulate a generalized theory that can govern environmental racism and the world population. It becomes equally tedious to seek the relevant answers pertaining to various environmental issues. In spite of such impediments, the concerned data can facilitate the formulation of some sort of general classification (Table 17.1).

TABLE 17.1

Flow Chart Depicting Relationship between Socioeconomic Status (SES) and Pollution

FLOW CHART DEPICTING RELATIONSHIP BETWEEN SOCIOECONOMIC STATUS (SES) AND POLLUTION

POLPULATION WITH UPPER LEVEL OF SES	POPULATION WITH MIDDLE LEVEL OF SES	POPULATION WITH LOWEST LEVEL OF SES
LEAST AFFECTED	LESS AFFECTED	MOST AFFECTED
BECAUSE THEY HAVE POLITICAL INFLUENCE, ENVIRONMENTAL AWARENESS, EDUCATION AND EASY ACCESS TO HEALTH FACILITIES I	BECAUSE THEY HAVE POLITICAL INFLUENCE, ENVIRONMENTAL AWARENESS AND MEDIUM ACCESS TO HEALTH FACILITIES	BECAUSE OF OF LEAST POLITICAL INFLUENCE, ENVIRONENTAL AWARENESS AND CAN'T AFFORD HEALTH FACILITIES

TABLE 17.2

A Comparison between Developed and Developing Countries

Developed Countries	Developing Countries
Majority of population uses gas stove/piped gas connection due to which public health is improved and they have fewer respiratory problems.	Still a major portion of the population use biomass, charcoal, agriculture, waste and animal dung (Gobar in India) for cooking, although governments are putting effort to switch over to alternatives
Strict rules and laws of environmental issues have been implemented keeping socioeconomic issues in mind.	A large population lives in unauthorized colonies. There are no strict rules and laws regarding socioeconomic status and pollution, and if they exist, they are not being implemented.

Another classification can be done on the basis of developed and developing countries showing association of pollution to SES (Table 17.2).

17.3 Various Views on Defining *SES*

Readers come across various terms, that is, *SES*, environment equality/inequality and environmental justice, all interrelated terms, used by different groups/researchers while studying the relationship between pollution and human health, but all focus on one point only: "who gets affected most by pollution". Different approaches and indicators have been used by various researchers and scientific groups to define *SES*. The main components of socioeconomic are social and economic issues varying from income, financial status, poverty, education, residential location, age, immunity level, the ambient environment, social and political awareness and others. It is imperative to review the history of the subject and understand the various definitions proposed by different groups in detail. This is because each definition is correlated with indicators that suit most to the population being studied. For the readers' reference, many of the definitions are presented in tabulated form in Table 17.1

17.3.1 According to Elizabeth Baker (2014)

SES is defined as a measure of one's combined economic and social status that naturally control health and education; occupation and income are other strong indicators for measuring health condition. Differences in SES are responsible for important disparities in the nutrition, housing, safety and the health of large groups of people. In general, the lower one's SES, the greater one's risk of malnutrition, heart disease, infectious diseases and early mortality from all causes. Income, education, occupation, vocation and wealth all contribute to SES (Baker 2014).

17.3.2 According to the National Center for Educational Studies (United States, NCES 2012)

SES considers the individual and the family as two major components in society. It is an economic and sociological combined sum of a person's work experience and of an individual's or family's economic and social position in relation to others, based on income, education and occupation. When analyzing a family's SES, the household income and earners' education and occupations are examined, as well as combined income, whereas for an individual's SES, only their own attributes are assessed. However, SES is more commonly depicted an economic difference in society as a whole (NCES 2012).

The SES of an individual or a family belongs to any one of three statuses in society: high, middle or low. When considering any one of these categories, any one or all three variables (income, education, occupation) can be assessed.

17.4 Some of the Most Probable Parameters That Influence SES

According to the World Health Organization, well-defined parameters should be used as base line to investigate correlative association between SES and pollution. There are two prime parameters: socioeconomic indicators and determinants of health of individual and the ambient environment. Both of these have been well defined and reflect SES as well as health aspects of individuals and the environmental habitat.

17.4.1 Socioeconomic Indicators

According to the proposal of World Health Organization, the basis of investigating socioeconomic conditions are the social and cultural environments in which an individual of a population resides, their economic conditions, the gross national income, multidimensional poverty, deprivation by all dimensions of poverty and adult literacy; these are the parameters for health actions, personal attitudes and lifestyles of the individuals of a population, among others (Blakely et al. 2004).

17.4.2 Determinants of Health

A range of parameters that influence health status of an individual and the concerned population can be considered as determinants of health. These include the social, personal habits and lifestyles of individuals in society, social attitudes and norms, along with economic conditions. Other parameters, for example, the physical environment and individual characteristics, all interact with each other and may not contribute as isolated parameters when considering the determination of health aspects of a population.

According to a recent proposal by World Health Organization, the social, economic and physical environments and an individual's personal characteristics and behaviors also contribute to the determination of health in a major way. Individual characteristics include several factors, such as genetics, social support network, gender issues of the individual and the demography of the population. These also influence the determination of health aspects of the population and the individuals constituting that population. Access and the use of health care services are often considered very important, generally do not play important role in such studies (WHO 2017, 3 February).

Most researchers have considered only those indicators that have a strong correlation with health. For a correct statistical answer, there should be no confusion about the terms *determinant* and *effect*. According to Barnes and coworkers, care should be taken that the indicators used in any analysis as the determinants of SES are not the effects of SES. Therefore, according to these authors, indicators should be used in such a way that the effect of determinants should not overlap and autocorrelation of effects should be avoided (Barnes et al. 2018).

It is very difficult to formulate any functional rule in this regard because different countries have used different indicators to study the association between pollution and SES. These indicators vary according with their respective country of investigation underway and vary according to the prevailing environment, existing weather conditions and the status of the country, that is, whether the country of investigation is underdeveloped, developing or developed. In this presentation, an attempt has been made to provide summary definitions of *SES* given by various researchers and groups. These primarily reflect the concept of SES (Table 17.3).

All data mentioned here and elsewhere in literature and information regarding the effects of pollution on the environment lead to the universally accepted fact that the '[q]uality of environment differs amongst different population for different components of the environment including air, water, land, access to green space, drinking water and waste management etc.' The discussions and examples/models from various ecological and geographical regions of the world should be focused on finding an amicable solution to this problem and proving a safe and sustainable environment for all the related biotic components. George Orwell's concept related to 'animal form' depicts the unsafe environment for some and not for more. This appears to be one of the prevailing current concepts.

17.5 SES and Its Association with Pollution in Developed Countries

17.5.1 Global Report with an Emphasis on the United States

The United States, the United Kingdom and other Western countries were the first to take initiatives for clean and safe environment. Most of the data available are from these countries. Environmental issues related to research and SES appear to be relatively limited as, these countries have taken effective measures to keep the environment clean. The debate/issues of SES and pollution started in the United States in the 1980s, and the latest reports suggest that the struggle continues for finding a solution for environmental equity and not to deprive and discriminate on the basis of race and ethnic groups.

In the United States, a majority of population reaping benefits of living better environmental conditions exclude those in minority groups and those belonging to lower SES groups.

Data available from the United States include almost every parameter from the correlation of SES to the criteria of air

TABLE 17.3
Probable Parameters (defined by organizations/groups) That Influence SES

Study Group/Organization	Summary of Work Done	Definition of *SES*
WHO (2017, 3 February)	Potential health effects on a population (health impact assessment)	Income and social status define SES. Other components such as education access to health services and job conditions also equally important. Determinates of health also include individuals' behaviors and characteristics (genetics and social support).
	Study of links between noise and air pollution	'SES includes income, education, job, housing category and any other measure of deprivation, and most reports have used a combination of them'.
Department of Communities and Local Government (2015) UK. The English Index of Multiple Deprivation (IMD) (2015) (Smith et al. 2015)	Deprivation study: Main: main focus on small areas in England	It is based on several domains e.g., deprivation, income, employment, education, health, disability, crime, housing, access to services and living environment.
National Center for Educational Studies (NCES, USA, November 2012)	SES and three levels of society	Socioeconomic background is based on individuals work experience, individual or joint family income and social status. An individual or family may fall into a high-, middle- or low-income groups and any or combined income, joy and education can be considered for SES study.
American Psychological Association (APA 2007)	Definition of *SES*	APA has considered social and economic factors to define SES, covering almost every parameter: (1) access to education and health care, (2) type and prestige of occupation and (3) income and place of residence or sometimes ethnic origin also.

TABLE 17.4
A Brief Summary of Studies of Air Pollution and Its Relation to SES Inequalities Carried Out by Various Groups

Author/Year	Location	Analysis/Brief Description	Conclusion	
Hajat, A et al. (2015)	USA/European/ developing countries		(1) North American cities from the U.S./Canada (2) Asia, Africa and other developing countries (3) European cities/countries	(1) Populations with lower SES are more affected by high levels of pollution (2) Same as above (3) Not conclusive, mixed results
Kunzmann (2018)	U.S.		SES and racial status both play major roles in defining the effect of pollution	
Tessum W Christopher et al. (2019)	U.S.	Racial/ethnic status also play major role in studying effect of pollutants (specially $PM_{2.5}$) on population	Populations belonging to low SES/class provide all services to higher income groups, but they themselves get affected by pollution because they can't afford what higher income groups can manage for themselves.	

pollutants and racial and ethnic disparities. A study by Hajat and coworkers that reviewed the situation based on criteria air pollutants summarized published research work related to 'Distribution of criteria air pollutants (study of six most common air pollutants, carbon monoxide, lead, ground-level ozone, particulate matter, nitrogen dioxide and sulfur dioxide) across different socioeconomic groups of the world shows that the United States and developing countries exhibit lower SES areas/groups/individuals that are associated withhigher pollutant concentration' (Hajat et al. 2015). Data of such reports make it easier to provide environmental equity to all. They have concluded that a direct relationship exists between CAP (criteria air pollutants) and SES. Some exceptions present are probably due to geographical location and developmental activities for the planning of cities. Some groups do not support above theory and correlate other factors with socioeconomic status and pollution (Kunzmann 2018; Tessum et al. 2019) (Table 17.4).

17.6 Reports from Canada

Some of the research reports available from Canada are summarized in Table 17.2 (Hajat et al.), and specifically two recent studies have been discussed later. These distinctly differ from previous studies because the focus is on ambient air quality near schools (Cakmak et al. 2016).

A few exceptions were noticed in the study zone of North American countries while some reports based on SES census tracts/ indicators showed the opposite association. Higher proportions of students belonging to higher SES levels attended schools located in regions with higher concentrations of pollutants.

17.6.1 Report from Quebec Public Schools

The impact of proximity of public school near the industrial zone and the pollutants like particulate matter, SO_2 and NO_2 were investigated (Batisse et al. 2017). Quebec City was selected as the zone of investigation because of the presence of many industries such as metal smelters, pulp and paper mills industry, refineries and several other industries spread all over city, including both urban and rural areas (Brand et al. 2016). Based on an analysis of the 2006 Canadian Census, four indicators, namely, the low-income threshold indicator, neighborhood SES indicator, and the social and material deprivation indicators (social and material deprivation indicators were developed by Robert Pampalon et al. commonly known as Pampalon indicator/index for public health in Canada (Pampalon et al. 2009)) were choosen for investigation. The Canadian National Pollutant Release inventory data of 2006 was used as standard to calculate industrial air emissions. The investigation of Quebec public schools in Canada showed that students with higher rate of socioeconomic deprivation may be likely to be exposed to higher emissions of industrial air pollutants (Batisse et al. 2017).

17.6.2 Report from Elementary Schools

Similar results were also found from Montreal where ambient air quality was checked near primary schools for pollutant NO_2 because children spend a large part of the day in school. The study was done keeping two objectives in mind: first, the proportion of major roads in city blocks where elementary schools are situated and NO_2 concentrations near schools and, second, to find out any significant differences in the pollution indicators in the selected study zone. Regarding levels of socioeconomic deprivation at these schools, students from socioeconomically disadvantaged background prefer to tend to attend elementary schools located in more polluted environments (Carrier et al. 2014).

17.7 Reports from Australia

The effects of ambient air pollution on fetal development and weight of offspring at the time of birth have attracted researchers worldwide. Studies from North America, the United Kingdom, China, Brazil and other countries have shown contradictory reports related to relationship between pollution and 'intrauterine growth retardation'. Few reports are available that investigate the relationship among pollution, weight of newborns and SES. A study from Australia has reported the impact of five air pollutants: particulate matter (PM_{10} and $PM_{2.5Units}$), carbon monoxide (CO), NO_2 and ozone (O_3) on the weight of children at the time of birth (Mannes et al. 2005). Data related to all births in Sydney from January 1998 to December 31, 2000, was obtained from the Midwives Data Collection at the New South Wales Department of Health, in the metropolitan city of Sydney. Covariates in study (regression models) included sex of child, maternal age, gestational age, gestational age at first antenatal visit, maternal smoking, maternal indigenous status, weather first pregnancy and season of birth. SES (a group-level variable that was significant) was measured using the Index of Relative Socio Economic Disadvantage based on maternal residence (residing within 5 km of the monitoring station).

A variable, 'small for gestational age,' was calculated. The authors conclude that "all SES variables exhibit a statistically significant association with 'small gestational age.'" Compared to the high SES category, there was increasing risk of SGA with decreasing SES (Mannes et al. 2005). A reduction in birth weight has affected lower income group in comparison to the group with a higher SES level.

17.7.1 Pollution and Emissions from Locations Having Coal-Fired Power Stations, Mining, Refineries and Manufacturing Units

Although no concrete data have been reported, the latest research by the Australian Conservation Foundation suggests that pollutants are emitted mostly from coal-fired power stations, mining refineries and manufacturing units. These are located in low-income areas, such as Hunter Valley in New South Wales, Mount Isa in Queensland, New Man and Collie in Western Australia, Latrobe Valley in Victoria and others. Coal-fired power stations and mining operations created the most emissions in these areas (Lisa 2018).

17.8 Reports from Other Developed Countries

Similar conditions were reported from other developed countries. Reports from New Zealand indicate variations in the air pollutants in locations of different social groups. Pollution levels were calculated based on an atmospheric dispersion model (Pearce et al. 2006). Parameters such as the complex interaction of emissions, topography and meteorology (that determinates pollution level in many areas) appear to be relatively more accurate compared to previous analysis done by other researchers. High levels of particulate pollutants have been reported in the city of Christchurch, New Zealand, at least during winter months. The primary source of varied in different social groups; hence, the pollution also varies due to the type of domestic heating used. A range of demographic and socioeconomic indicators, including age, ethnicity, income and deprivation, were used in analysis. The researchers also reported that different social groups are exposed to different levels of both ambient air pollution and extreme air pollution episodes. This condition is higher among most disadvantaged communities and exhibit inverse relationship between populations responsible for pollution and exposed to pollution. Those responsible for producing higher proportion of pollution are less exposed to it (Pearce et al. 2006).

17.9 Air Pollution, Schoolchildren and SES Levels

Because SES is a very sensitive issue for kids and affect their upbringing and development, its association with pollution must be discussed separately. According to the World Health Organization, any type of pollution is a major threat to children. In one of its key messages, it says that pollutant affects the children the most in unique ways. Children are especially vulnerable to pollution during fetal development and the early stages of childhood when their organs are under developing and maturing. Both 'ambient air pollution' and house air pollution contributed to respiratory tract infections that resulted in 543,000 deaths in children younger than 5 years of age in 2016. The World Health Organization admits that "pollution has affected most countries with 'low and middle income' (LMICs)." According to this report, regions in Africa have the highest levels of exposure to 'household air pollution'. It is the single largest environmental risk factor worldwide. The primary cause of this widespread pollution is the fuels and technologies for basic daily needs, such as cooking, heating, lighting and other domestic needs. According to the World Health Organization, premature birth is the only factor that kills more children younger than age 5 than accurate respiratory infection (WHO 2018, October 29). It further reports that poverty is strongly correlated with exposure to air pollutants. Children in LMICS and in low-income communities within high-income countries disproportionally suffering from the effects of air pollution (social determinants of pollution). Schoolchildren are sensitive to environmental pollution due to their increased breathing rates and developing breathing systems, low immunity levels, deficiencies of vitamin D and long durations in outdoor activities, among others. But surprisingly, very few reports are available related to pollution and its association to SES of schoolchildren belonging to lower age groups and juvenile groups.

17.9.1 Report from Windsor, Ontario, Canada, on Pollution, SES and Schoolchildren

There are two reports have focused on SES and pollution among schoolchildren from Canada. Cakmak et al. (2016) used a land-use regression study on schoolchildren in Windsor, Canada. This location is one of the busiest crossings between the United States and Canada. It may be considered as an examples of a site where average lower income groups reside and attend educational institutes (Cakmak et al. 2016). At such sites, the probability of respiratory disorders due to traffic or air pollution exposure is higher. These socioeconomic indicators can be readily correlated with many of the factors such as diet, smoking and indoor/outdoor air quality. Other highlights of their reports are as follows:

a. Low-income groups are at higher health risk due to traffic and air pollutant present in the area.
b. Lung function declined in low-income groups.
c. The risk of developing respiratory conditions was increased in low- and less educated groups.

Other reports are from Canada on existing social gradient in schoolchildren of Quebec public schools to industrial air pollution and concentration of NO_2 near elementary schools.

17.9.2 Correlation between Pollution and SES in Relation to Developing Countries

Developing countries are understudied as far as the correlation between SES and pollution is concerned. Countries with rapidly growing economies, such as India and China, need more data and research by only one competent authority in their respective countries. Latin American countries have made good progress while countries in Africa need more attention compared to the rest of the world. The fast urbanization in developing countries has affected populations by all types of pollution components: air, water, waste, noise and nutrition. But most researchers have focused mainly on air pollution in developing countries. Due to high population density and a lack of well-connected networks for studying pollution across the countries, China and India present almost similar pictures of SES and pollution, that is, having inadequate data for the analysis of effects of pollution on their population.

Another important discriminating factor between developed and developing countries is choice of pollutants and their size for analysis. Unlike developed countries studies from developing countries are based on air pollutants and not specific to size of the particulate matters.

17.10 Report from China

A recently published analysis based on nationwide survey data of China in 2014 applied the multilevel mixed effects model to evaluate SES (using education and income as indicators) and its contribution to the relationship between self-rated air pollution to health status at the community and individual levels (Jiao et al. 2018). Present work from China is an example of logical work based on ground realities with wider practical approach. These researchers have emphasized on the theoretical data based on the conventional parameters. The authors followed a more practical approach by considering self-rated analysis along with a community-level analysis to explain the relationship among health status, pollution and socioeconomic status. Self-rating seems to be more logical provided individuals are well educated and able to explain and provide correct information about themselves. Important finding is a nonlinear relationship between community-level SES and community air pollution in urban China, contrary to the results of similar analyses carried out in developed countries that concluded that the highest level of air pollution is in communities with moderate SES levels. However, findings were not same for self-analysis. Self-rated air pollution has the greatest impact on self-rated health of the lower SES groups.

17.11 Reports from Latin America and the Caribbean

Developing countries are experiencing a problem with a lack of city planning compared to developed countries. Latin America and the Caribbean are also facing the same problem of migration of the population to urban areas, where more than 75% of the population lives. Unplanned growth of cities is directly proportional to the increase in pollution level and low-SES-level localities. It has been reported that groups with low-SES levels live in areas having higher concentrations of air pollutants while groups with high-SES levels live in areas with a low concentration of air pollutants (Naess et al. 2007).

An analysis shows the effect of particulate matter (PM_{10}) concentration in Bogota, capital district of Colombia, on the daily mortality from all causes, including specifically from cardiopulmonary diseases (Blanco-Becerra et al. 2014). The study is based on the methodology applied in the 'Multi-City Study of Air Pollution and Health Effects in Latin America'. A PM_{10} concentration map shows the greatest concentration of pollutant in the south and southwest portions of Bogota, which is divided in 20 administrative units. The result suggests that SES levels significantly modify the environmental exposure to particulate matter.

17.12 Reports Related to SES Levels and Pollution from India

Most studies from India have reported strong correlations between health issues and pollution. There are relatively fewer reports on the relationship between SES levels and pollution. More data are available on an individual basis that is worth analyzing. In contrast to developed countries, highlighting the distributional aspect of human population with respect to socioeconomic status is almost negligible in India.

Few attempts have been reported from India on the relationship between socioeconomic status characteristic and air pollution (Kathuria and Khan 2007). This investigation involves exposure of household to household and a specific air pollution exposure index for 347 houses. Seven pollution monitoring stations in Delhi were selected. Observations were made considering socioeconomic data, and the primary survey was based on a questionnaire related to the study the environmental equity. Data and air pollution (suspended particulate matter [SPM]) was from the Central Pollution Control Board New Delhi for the period from 1999 to 2003. The observations were the product of an analysis involving economically challenged communities. This group is the most affected by exposure to air pollution. Education, status of household members and house ownership are other important factors influencing exposure. But a separate analysis of residential and industrial areas suggests that exposure to air pollution also depends on the location of residences besides their SES because residents near the industrial areas are exposed to outdoor pollution more than others.

This report was published in 2007, and since then, the situation in Indian cities has changed a lot with large number of people migrating from rural to urban areas. Now almost all cities do have mix areas of high- and low-SES population. The theory that in developed countries, higher pollution concentration areas are dominated by population belonging to low-SES levels is not applicable. Here in India, and other developing countries too, although there is a difference in various pollutants concentration in various parts of cities, the difference is minimal.

One of the latest report from Bangalore, the information technology (IT) hub of India, has shown that a linear relationship between SES and air pollution does not hold true for India, unlike developed countries (Sabapathy et al. 2015). The objective of their study was to determine whether air pollution exposure while commuting for workplace relates to commuters' income characteristics. The study measured exposure to CO and PM_{10} using personal samplers for a sample of employees in the public sector and IT sector. They found that the longer commuting times of higher income groups offset the benefits of exposure to lower pollutant concentrations.

17.12.1 Futuristic Approach in the Case of India

Unplanned cities and higher growth rate of population of Indian cities have made it almost impossible to define the areas in most of cities as low- and high-income residential areas with low- and high-SES levels, barring very few cities and industrial areas. In Indian cities, levels of particulate matter (PM_1, $PM_{2.5}$ and PM_{10}) are much higher than the guidelines set by World Health Organization. Biomass, such as stubble burning, dust from on road vehicles and construction sites, is a major contributor, and it is also one of the major cause of smog formation, which engulfs Indian cities and affects population belonging to lower and higher SES levels. Moreover, Delhi, the capital of India, presents a most complicated situation with A-grade (affluent residential areas) colonies surrounded by unauthorized colonies, less affluent areas and industrial zones in the heart of city. To carry out studies of such areas and in selection of SES indicators, one has to be very careful, and the best option to analyze SES is on one-to-one basis. Due to lack of air quality monitoring systems in Indian cities, especially in rural areas, it is difficult to draw a correlative association that involve models, pollution and SES level. Air pollution policies need to focus on rural areas also along with urban areas. According to the Global Burden of Disease report, air pollution levels in rural and urban India are unsafe, with 75% of indoor air pollution–related deaths being reported from villages (Brauer 2016; Arku et al. 2018).

Even though children are most susceptible to pollution, hardly any data are available on the Indian scenario that can be feasibly used to find out about the daily exposure of every student and child.

Although the CPCB and the NGT (National Green Tribunal, New Delhi) have taken several measures to control pollution in various part of country, still there seems to be a large scope of do something for children as growing children need special attention. Top priority must be given to collecting data from each corner of the country, and online monitoring units must be installed in urban and rural areas, especially near all academic institutes. Then there are some chances to check the level of pollution near academic institutes and exposure to every student (if possible using personal sampler). Simultaneously, we must adopt measures to also bring down pollution levels near academic institutes by all natural means. All roadside dividers and planters near academic institutes must be planted with small hedge plants that don't need much water and are able to absorb roadside dust. Such efforts are likely to provide protection to all students irrespective of their SES level. Simultaneously, there must be some efforts to provide dry or suitable nutritious, healthy snack meal packets to all students in the country, irrespective of school category (snack packet must contain nutrients that are uniformly lacking among students of all categories, for example, vitamin D, iron, etc.) to bring equality to the health status of all students belonging to low- and high-SES levels. Care also must be taken not to have any industrial units and mobile towers near schools. Recently, the Karnataka state government has banned mobile towers within 50 meters of schools and hospitals.

Women also spend most of their time in kitchen and are exposed to indoor pollution both in rural and urban areas. The situation is better in urban areas because of the availability of various kitchen appliances and cooking gas, but with the ongoing Ujjwala scheme (Ujjwala 2016) providing cooking stove and gas connection to rural women, hopefully this situation will change. Along with this there must be awareness campaign for women so that spending extra energy and time in kitchen ***(time management to run a kitchen)*** can be avoided, and an emphasis must also be given on ***equal participation of male and female in the kitchen and cooking.***

17.13 Noise Pollution and SES

Most researchers have analyzed the combined effect of air and noise pollution, citing the reason that both have a common source. But readers must give weight to separate analyses of noise and air pollution in which they are not combined as both may affect health differently. Other researchers have also supported this view. There are chances that simultaneously analyzing both noise and air pollution may suppress the effect of the latter, considering that noise is more apparent (Dzhambov 2018). It is true that urbanization leads to more traffic and noise; fast-reducing green spaces, that is, the lungs of rural and urban areas; and less peaceful ambient environments, specifically in unplanned cities. Theoretically speaking its quite tedious to investigate the impacts of noise pollution on socioeconomic status especially on lower and higher income groups in isolation. Noise generated outside and inside residences by the traffic, DJs, loudspeakers in social and religious functions affect all members belonging to all the groups of SES populations. As far as the relationship among noise pollution, SES and poverty is concerned, no conclusion can be made because of a lack of sufficient and significant data. These data should be qualitatively and quantitatively analyzed, as this aspect is relatively tedious. Very few systematic reports on noise pollution are available from different parts of the world, but one report (Barnes et al. 2018), based on findings of European Commission of Science for Environment Policy, covers all the latest evidence explaining the role of SES and the relationship between noise and air pollution of the European region.

There seems to be some indecisive discussions about the relationship between noise pollution and SES and real estate prices. There is a greater need to affirm this. In some cases, studies have found that higher noise levels have been associated with lower housing values, which can be suggestive of a lower SES population staying in the area. Although it is not clear whether housing prices are devalued because of their proximity to noise pollution or vice versa probably noise polluting activities, such as markets, industrial units, and traffic, come up on land where property prices are low resulting the area into noise zones.

Boennec and Salladare studied Nantes, France green capital city of 2013 and have explained the situation that if buyers are willingly pay and prefer the advantages of living closer to facilities such as markets, medical facilities/hospitals, shopping, schooling/education and others, over noise pollution, then there might be no reason for the price to be affected (Boennec and Salladare 2017). Residents belonging to upper SES groups can afford to spend and protect themselves from exposure. No other evidence was observed in support of this aspect in most of the literature. Hence, it may not be considered as supporting the statement 'Higher SES is affected by higher noise pollution'. The preceding example of those with a higher SES being affected by a higher level of noise pollution is mainly due to location of their houses or residential area near shopping area or markets by choice or compulsion. Studies from other developed European countries also draw a conclusion along similar lines: that those with socioeconomic deprivation amass the negative health impacts of pollution (Barnes et al. 2018).

17.14 Waste Pollution and SES

The concept of correlation between SES and waste originated in the 1980s when waste used to be dumped in ground in Warren County in North Carolina. It catalyzed the 'environmental justice movement' in the United States during the 1980s, and since then, the situation has not changed much. It is a known fact that populations with a lower SES level are affected most by such activities. This aspect is not applicable in countries such as Sweden, where only 1% of waste goes to landfill sites. Everywhere, especially in developing countries, waste segregation and management are being practiced by people belonging to low-SES levels, and it affects their lives (Tomaszewski 2017). There is a term, *LULU* (locally undesirable land use), for waste dumping sites in the United States, but it is always a cause of conflict between minority groups and policymakers in the United States because individuals of any SES level would not like to reside near such sites (Kevin 1997). This situation exhibits the linear association between waste and low SES.

17.15 Pollution and SES Status of an Individual: Education Is an Important and Functional SES-Level Indicator

The study of pollution and its relation to SES at an individual level using education as an indicator may provide a decisive view and can play significant role in bridging the gap between populations of high and low SES levels. It may facilitate the influence and derogative effects of pollution on an individual. Effects of pollution on mortality has been discussed but very few studies have analyzed the role of education as SES indicator at the individual level and its relationship with pollution. Education as a determinant deserves to be a key priority when analyzing such interrelationships, especially in developing countries. A study of 20 cities in the United States using education as one of the indicators has shown an inverse relationship between level of education and mortality rate associated with PM_{10}. The results indicate that when an individual has a higher education level, the effect of particulate matter is less when mortality is considered (Zeka 2006). Another example is from Canada. An investigation by Cakmak and colleagues in one analysis clearly shows the risk of developing respiratory conditions as increased in the individuals in low-income and low-education groups (Cakmak et al. 2016).

17.16 Conclusion

The purpose of this chapter was to present a qualitative picture of association between SES levels and pollution. The data available in the literature have been reviewed to understand the overall trend regarding this correlation in a very simple and lucid manner. It is a fact the developing countries seem to have more problems related to pollution and inadequate data. There are two reasons. The first reason is that there are no adequate pollution recording monitors/stations available in country, and the second is that there are insufficient regulatory bodies to manage all aspects of pollution-related issues. Socioeconomic indicators also vary across different countries, and there is a remarkable difference between developed and developing countries. The income of individuals and families, types of housing, access to potable water and energy, nutritional facilities and the availability of appropriate food and education are the essentials. These parameters have been considered as indicators by different groups, depending on the source of pollution and SES levels of their study population samples. Although sometimes using only one indicator in study may not be sufficient to get the correct data, at the same time, one must acknowledge that indicators of SES level are highly correlated, and thus, using multiple indicators within a single model is not acceptable.

To understand the relation between SES level and pollution in a better way role of both individual and community level SES is essential. In fact a typical analysis and data collection on pollution and SES levels needs revolutionary change and focus should be more on individual rather than community. The approach should be highly individually oriented. It also has been observed that a linear relationship between SES and pollution is correct to some extent for developed countries, but the same is not true for developing countries. This theory is not applicable for fast-growing economies, such as India and China, with high population densities and almost homogeneous levels of air pollution density in most cities affecting the populations of both lower and higher SES levels. It is also difficult to the draw conclusion about which out of five major types of pollution has the strongest and most direct association with SES levels. Waste and its management need to be much discussed more compared to other types of pollution because this job in most of the countries is carried out by people of

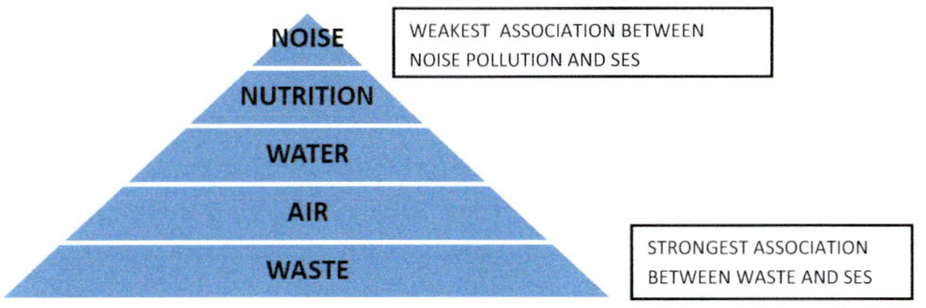

FIGURE 17.1 Association between waste and SES

low-income groups. An association between various type of pollution and SES can be represented by pyramid (Figure 17.1).

Based on available reports from various parts of world, a strong linear association can be drawn between the SES of a population and waste. All over the world, waste dumping sites are in areas where people of lower income groups reside and are affected, but with the coming advanced techniques of waste management in developed countries, this linear relationship seems to be changing very fast without showing indications for much change for developing countries in near future. A 'personalized index of exposure' is essential for every individual, and it should record health, time spent by the individual near pollution sources, personal activities in and outside of the house and distance from the source of pollution during activity to find out the exact association among individuals, SES and pollution. All of the previously mentioned parameters play crucial role in such studies. Changing the SES levels of a population may take time, so it is better to control pollution levels, which can be done faster just by creating more awareness and implementing already framed pollution policies in countries suffering the most, to provide a safe and clean environment for all.

Acknowledgment

The author would like to express gratitude toward Dr. Ambuj Tripathi (IUAC, New Delhi), Dr. Anjum Usmani (Venkateshwara University, UP), and employees of Department of Science, MMH College for their assistance.

REFERENCES

APA (2007). Definition of SES: in APA dictionary of Psychology. https://dictionary.apa.org › socioeconomic-status (accessed 27 September 2021).

Arku R E, Brauer M, Hystad P, Yusf S, Shupler M, and Birch A. Characterizing exposure to household air pollution within the Prospective Urban Rural Epidemiology (PURE) study. *Environment International*, 114 (2018), 307–317. doi: 10.1016/j.envint.2018.02.033

Asch P and Joseph J S. Some evidence on the distribution of air quality. *Land Economics* 54(3) (1978), 279–297. https://doi.org/10.2307/3145999. URL: http://www.jstor.org/stable/3145999sd

Bae C H C. The equity impacts of Los Angles air quality policies. *Environment and Planning A: Economy and Space*, 29(9) (1997), 1563–1584. doi: 10.1068/a291563.

Baker E H. Socioeconomic Status, Definition. *The Wiley Blakcwell Encyclopedia of Health, Illness, Behavior, and Society*, 2210–2214. https://doi.org/10.1002/9781118410868.wbehibs395.

Barnes Jo, Laura De Vito, Enda Hayes et al. Qualitative assessment of inks between exposure to Noise and air pollution and socioeconomic status in conference, air pollution XXVI. *WIT Transaction on Ecology and the Environment*, 230 (2018), 15–25, doi:10.2495/AIR 180021.

Basel Convention (1989). UNEP report in Basel convention on the control of transboundary movement of hazardous waste and their disposal. https://www.unep.org › resources › report › basel-conv (accessed 26 September 2021).

Batisse E, Goudreau S, Baumgartner J, and Audrey S. Socio-economic inequalities in exposure to industrial air pollution emission in Quebec public schools. *Canadian Journal of Public Health*, 108(5–6) (2017), e 503–e509, doi:10.17269/CjPH.108.6166.

Blakely T, Hales S, Woodward A. Poverty: Assessing the distribution of health risks by socioeconomic position at national and local levels. World Health Organization (WHO), Protection of human environment, Geneva, 2004. https://www.who.int/quantifying_ehimpacts/publications/ebd10.pdf (accessed on 28 September 2021).

Blanco-Becerra L C, Miranda-Soberanis V, Villarreal A B, Diaz M H, Junger W, Hurtado-Díaz M, Romieu I. Effect of socioeconomic status on the association between air pollution and mortality in Bogota, Colombia. *Salud Publica de Mexico* 56(4) (2014), 371–378.

Boennec L R, and Salladare F. The impact of air pollution and noise on the real estate marketmarket, the case of the 2013 European green capital. *Nantes, France, Ecological Economics*, 138 (August 2017), 82–89. https://doi.org/10.1016/j.ecolecon.2017.03.030

Brand A, McLean K E, Henderson S B, et al. Respiratory hospital admission in young children living near metal smelters, pulp mills and oil refineries in two Canadian provinces. *Environment International*, 94 (2016), 24–32; PMID: 27203781, doi:10.1016/j.envint.2016.05.002

Brauer M. Current and future burden of disease from major air pollution. HEI Annual Conference, Denver CO, May 2, 2016. https://www.healtheffects.org/sites/default/files/Brauer.pdf

Bullard D R. Environmental justice in the 21st century: Race still matters. *Phylon*, 49(3–4) (2001), 151–171. doi: 10.2307/3132626 (Phylon journal is published from Clark Atlanta university, US).

Cakmak S, Hebbern C, Cakmak D J and Vanos J. The modifying effect of socioeconomic status on the relation ship between traffic, air pollution and respiratory health in elementary school children. *Journal of Environmental Management*, 177 (2016), 1–8. https://doi.org/10.1016/j.jenvman.2016.03.051

Carrier M, Apparicio P, Seguin A M, and Crouse D. Ambient air pollution concentration in Montreal and environmental equity: Are children at risk at school? *Case Studies on Transport Policy*, 2(2) (2014), 61–69, doi:10.1016/j.cstp.2014.06.003

Chavis, Benjamin F., Jr. Toxic waste and race in the United States: A national report on the racial and socioeconomic characteristics of communities with hazardous waste sites by the united church of Christ, commission for racial justice, 1987. https://www.nrc.gov/docs/ML1310/ML13109A339.pdf (accessed 25 September 2021).

Chavis, Benjamin F., Jr. Foreword. In *Confronting Environmental Racism Voices from the Grassroots*, edited by R D Bullard. Boston, MA: South End Press, 1993.

Cleere, Rickie. Environmental racism and the movement for Black lives: Grassroots power in the 21st Century. Pomona Senior Theses, Paper 140. 2016. https://scholarship.claremont.edu/pomona_theses/140

Dzhambov A, Hartig T, Markevych L, Tilov B, and Dimitrova D. Urban residential green space and mental health in youth: Different approaches to testing multiple pathways yield different conclusions. *Environmental Research*, 160 (2018), 47–59. doi:10.1016/j.envres.2017.09.015. Epub 2017 Oct 3. PMID: 28961469.

Freeman A M. The Distribution of Environmental Quality. In *Environmental Quality Analysis, Theory and Methods in Social Sciences*, edited by Kneese A V and Bower B T. Baltimore, MD: John Hopkins University Press, Baltimore, 1972.

Hajat A, Hsia Charlene, and O'Neill Marie S. Socioeconomic disparities and air pollution exposure: A global review. *Current Environmental Health Reports*, 2(4) (2015), 440–445. doi: 10.1007/s40572-015-0069-5. PMID: 26381684; PMCID: PMC4626327.

Jacobson, Mark Z. Urban air pollution. In *Atmospheric Pollution: History, Science, andRegulation*. Cambridge University Press, Cambridge, 2002, 81–114. doi:10.1017/CBO97805118022 87.005.

Jiao K, Xu M, and Liu M. Health status and air pollution related socioeconomic concerns in urban China. *International Journal for Equity in Health* 17, (2018), 18. https://doi.org/10.1186/s12939-018-0719-y

Kathuria V, and Khan N. Vulnerability to air pollution: Is there any inequity in exposure? *Economic and Political Weekly*, 42(30) (2007), 3158–3165. Retrieved September 6, 2021, from http://www.jstor.org/stable/4419846

Kevin, D. Environmental racism and locally undesirable land uses: A critique of environmental justice theories and remedies. *Villanova Environmental Law Journal*, 8(121) (1997). http://digitalcommons.law.villanova.edu/elj/vol8/iss1/3.

Kruvant W J. People, energy and pollution. In *The American Energy Consumer*, D K Newman and D Day (Eds.). Ballinger, Cambridge, MA, 1975.

Kunzmann, Kevin. Socioeconomic status, Racial disparity at the heart of US pollution. Lecture by Balmes John R in American Thoracic Society, International Conference, May 22, 2018. https://www.hcplive.com › view › socioeconomic-status. . .

Lisa C. Australia's most polluting industries are almost all in poor areas – report. *The Guardian*, November 16, 2018. https://www.theguardian.com/environment/2018/nov/16/australias-most-polluting-industries-are-almost-all-in-poor-areas-report

Mannes T, Jalaludin B, MorGan G, Lincoln D, Sheppeard V, and Corbett S. Impact of ambient air pollution on birth weight in Sydney, Australia. *Occupational and Environmental Medicine*, 62 (2005), 524–530, doi:10.1136/oem.2004.014282.

Mosley S. Environment history of air pollution and protection. *The Basic Environmental History*, 4 (2014), 143–169. Springer International Publishing, Switzerland, M. Agnoletti and S. Neri Serneri (eds.), doi: 10.1007/978-3-319-09180-8_5

Naess O, Piro F N, Nafstad P, Smith G D and Leyland A H. Social deprivation and mortality: A multilevel cohort study. *Epidemiology*, 18(6) (2007), 686–694; doi: 10.1097/EDE.0b0 13e3181567d14, PMID18049185.

NCES, Expert Panel (Cowan D C et al.). A Definition of SES 'socioeconomic status' in report on Improving the Measurement of Socioeconomic Status for the National Assessment of Educational Progress: A Theoretical foundation, November 2012. https://nces.ed.gov/nationsreportcard/pdf/researchcenter/socioeconomic_factors.pdf (accessed 26 September 2021).

O'Neill M S, Jerrett M, Kowachi I, Levy J I, Cohen A J, Gouvwia N, Wilkinson P, et al. Health, Wealth and air pollution: Advancing theory and methods. *Environmental Health Perspectives*, 111(16) (2003), 1861–1870. https://doi.org/10.1289/ehp.6334.

Pampalon R, Hamel D, Gamache P, Raymond G. A deprivation index for health planning in Canada. *Chronic Diseases in Canada*, 29(4) (2009), 178–191. PMID: 19804682.

Pearce J, Kingham S and Reza P Z. Every breath you take? Environmental justice and air pollution Christchurch, New Zeland. *Environment and Planning A: Economy and Space*, 38(5) (2006), 919–938. http//doi.org/10.1068/a37446

Sabapathy, A., Saksena, S. & Flachsbart, P. Environmental justice in the context of commuters' exposure to CO and PM_{10} in Bangalore, India. *Journal of Exposure Science & Environmental Epidemiology* 25, 200–207 (2015). https://doi.org/10.1038/jes.2014.34

Sarma R. Environment assessment in Vedic literature. *International Journal of Sanskrit Research*, 1(4) (2015), 5–8.

Science for Environment Policy (SEP). Links between noise and air pollution and socioeconomic status. In-depth Report 13 produced for the European Commission, DG Environment by the Science Communication Unit, UWE, Bristol, 2016. http://ec.europa.eu/science-environment-policy (accessed 05 January 2021).

Sharma R and Sharma B. *Rig-Veda Sanhita/mandala6//hymn 48, mandala 1//hyman10*. Brahma Varchas publisher, Shantikunj Haridwar (UP), India, 1994.

Smith T, Noble M, Noble S, Wright G, McLennan D, and Plunkett E. In technical report on English indices of deprivation 2015, Dept. of Communities and Local Government, September 2015. http://www.gov.uk/dclg, https://assets.publishing.service.gov.uk/government/uploads/system/uploads/attachment_data/file/464485/English_Indices_of_Deprivation_2015_-_Technical-Report.pdf

Tessum C W, Apte J S, Goodkind A L, Muller N Z, Mullins K A, Paolella D A, et al. Inequity in consumption of goods and services adds to racial–ethnic disparities in air pollution. *PNAS*, 116(13) (March 2019), 6001–6006. https://doi.org/10.1073/pnas.1818859116

Tomaszewski T Z. Environmental justice and waste management: Location of landfills and and socioeconomic conditions in Sao Paulo (SP, Brazil). Debora, in Master's Thesis series in Environmental Studies and Sustainability Science MESM02(20171) LUCSUS (Lund University University Center for Sustainability Studies); LupLup.lub.lu.se/student paper /record /891459 0 /file/8914602.pdf.

Ujjwala (2016). PMUY, https://pmuy.gov.in/ (accessed 29 September 2021).

WHO. Determinants of health, February 3 2017. https://www.who.int/news-room/questions-and-answers/item/determinants-of-health (accessed 30 December 2021).

WHO. WHO Social media toolkit for *Air pollution and child health: Prescribing Clean Air* launch, 29 October 2018. https://www.who.int/ceh/publications/Social_Media_Toolkit_AirPollution_Chidren_Health_Report_Launch.pdf (accessed 26September 2021).

Zeka A, Zanobetti A, and Schwartz J. Individual level modifiers of the effects of particulate matter on daily mortality. *American Journal of Epidemiology*, 163(9) (2006, May 1), 849–859. http:doi.org/10.1093/aje/kwj116

Zupan J M. *The Distribution of Air Quality in the New York Region*. (John Hopkins University Press, Baltimore for resources for the future, 1973).

18

Gene–Environment Interaction and the Risk of Chronic Kidney Disease of Uncertain Etiology

Manushi Siddarth, Basu Dev Banerjee, Rishila Ghosh, Neeru Singh and Om Prakash Kalra

CONTENTS

DOI: 10.1201/9781003186304-18

18.1 Chronic Kidney Disease of Unknown Etiology

18.1.1 Overview

The prevalent character of chronic kidney disease of uncertain etiology (CKDu) was documented first in 1999. However, the occurrence of this disease has grown substantially in many geographical regions of the world over the last 15 years, especially in Mexico and Central America (Wesseling et al., 2013a), Tunisia and Morocco (Barsoum, 2013), Egypt (EmadG and Osama, 2010), Tunisia and Morocco (Barsoum, 2013), Egypt (EmadG and Osama, 2010), Saudi Arabia (El Minshawy et al., 2014), Sri Lanka's North-Central Province (Chandrajith et al., 2011) and India (Ganguli, 2016). The distinctive features of CKDu are that it has no connection with the distinguished risk factors, that is, diabetes, hypertension or chronic nephropathies. Singh et al. (2013) illustrated that CKDu occurs more commonly in men than women and frequently in poor adult agricultural farmers of hot tropical areas. Numerous authors showed that rice paddy farmers from Sri Lanka and cashew nut, coconut, and rice farmers from India have a more significant affected population (Wesseling et al., 2013a, Wegman et al., 2016), Saudi Arabia (El Minshawy et al., 2014), Sri Lanka's North-Central Province (Chandrajith et al., 2011), and in India (Ganguli, 2016). CKDu progresses silently to end-stage renal failure, causing thousands of fieldworkers to die prematurely (Ramirez-Rubio et al., 2013). Consulting to India's CKD registry, CKDu accounts for 16%, which is the most common CKD cause next to diabetes. These patients belong to rural low-income groups and young people (Rajapurkar et al., 2012). CKDu affects 8% to 12% of the population in villages across India, including Cuttack, Bhubneshwar, Kalahandi, Koraput, Jaipur, Balangir, Maharashtra and Delhi, according to many records. Odisha and Andhra Pradesh have been named two CKDu hotspots in India by the Indian Council of Medical Sciences (ICMR).

18.1.2 Definition

CKDu has been described in many ways by numerous authors. Although there is no universally agreed-on concept of CKDu, the following characteristics are consistent across studies:

a. Asymptomatic and progressive CKD
b. Absent or sub-nephrotic proteinuria
c. Absence of hematuria
d. Absence of diabetes, chronic or severe arterial hypertension, HIV, snake bite, glomerulonephritis, or other urinary tract diseases
e. Normal glycosylated hemoglobin
f. Blood pressure <160/100 mmHg among untreated patients <140/90 mmHg in patients receiving up to two antihypertensive drugs

18.1.3 Histopathology

CKDu is distinguished by a primary alarm in the proximal epithelium and interstitium, resulting in distinct clinical and histopathological characteristics. The core pathology of CKDu involved damage to proximal tubular epithelium and interstitium. Clinically it is distinguished by tubular proteinuria, typically β_2-microglobulinuria and the absence of edema and hypertension. The histological features of CKDu indicate tubulointerstitial pathology, usually seen in toxic nephropathies (Athuraliya et al., 2003). Recently, a histological study was conducted on renal biopsies of CKDu patients, and the researchers observed the pathology of asymptomatic CKDu, which was characterized by consistent interstitial fibrosis with or without interstitial inflammation tubular atrophy and glomerular sclerosis; viable glomeruli were remarkable. According to their analysis, they divided the patients into four stages: Stage I: Patients have mild interstitial fibrosis; Stage II: Patients have mild to moderate interstitial fibrosis; Stage III: Patients have interstitial inflammation, fibrosis, glomerular sclerosis and tubular atrophy of moderate severity; and Stage IV: Patients have severe tubular atrophy and interstitial fibrosis-related with intermediate to severe glomerular sclerosis and severe interstitial inflammation (Wijetunge et al., 2015).

18.1.4 Risk Factors of CKDu

CKDu develops slowly, most likely beginning in the second decade of adulthood, and is mostly asymptomatic before advanced stages. The unusual characteristic of CKDu is that its causative cause does not include typical CKD risk factors like diabetes, hypertension, or glomerulonephritis. The cause of the increased prevalence of CKDu in many parts of the world is thought to be toxin-mediated renal failure. To explore the causes of CKDu, both occupational and environmental factors have been studied and suggested the involvement of pesticides, heavy metals, nonsteroidal anti-inflammatory drugs (NSAIDs), infection, alcohol, and recurrent dehydration due to occupational heat stress, intake of fructose-rich soft drinks, hyperuricemia and hyperuricosuria. The search for the root of the pandemic focused mostly on pesticides as CKDu was mainly found in men in agricultural areas where pesticides were widely used.

18.2 Organochlorine Pesticides

Organochlorine pesticides (OCPs) such as dichlorodiphenyltrichloroethane (DDT), hexachlorocyclohexane (HCH), endosulfan, aldrin, and dieldrin are among the most frequently used pesticides in Asia's developing countries because of their low price and resourcefulness against numerous pests (Abhilash and Singh, 2009; Gupta, 2004; Lallas, 2001). However, these compounds were debarred in established nations two and half decades ago because of their possible bioaccumulation and biological effects. Their resistance to decay has resulted in pollution throughout the world and has been discovered in several environmental compartments. Such residues may contain various compounds, including some toxicologically important derivatives, such as degradation products, metabolites and congeners. Referring to the Food and Agriculture Organization (FAO) inventory, more than 500,000 tons of idle and outdated pesticides affect the environment and public health in countless countries. Pesticide residues have lately been a significant cause of concern all over the world. Although India is emerging from the euphoria of the green movement, it is also grappling with the long-term effects of commonly used chemical fertilizers and pesticides such as HCH, DDT, and endosuflan (Abhilash and Singh, 2009).

18.2.1 HCH

18.2.1.1 Structure and Formulation

HCH, which is previously known as benzene hexachloride (BHC), an artificial chemical that presents in 8 chemical forms called isomers (Figure 18.1). Technical-grade HCH, a mixture of several chemical forms of HCH, typically contains about 10% to 15% of γ-HCH, as well as the alpha (α), beta (β), delta (δ), and epsilon (ε) forms of HCH. Almost all the insecticidal properties reside in the γ-HCH. It is present in the white solid and can be evaporated into colorless vapor with some musty odor.

18.2.1.2 Presence in Environment

Even though HCH is no longer used as an insecticide in India, it has been detected in surface water and soil at hazardous waste sites because it remains in the environment. HCH isomer vapors bind to soil particles and ashes, where they can be washed away by rain or degraded by other substances in the atmosphere. Depending on the weather conditions, HCH can stay in the air for long periods and fly great distances (Li, 1999). They are metabolized by algae, fungi and bacteria in soil, water and sediments into less toxic compounds, but this process will take a long time.

18.2.1.3 Exposure to Humans

Humans can be exposed to HCH and their isomers by eating foods that may be contaminated drinking water, breathing contaminated air or having contact with soil or water at hazardous waste sites that may contain these compounds. Exposure to α-, β- and δ-HCH is less frequent than exposure to γ-HCH because these compounds are no longer used in India. However, γ-HCH and the other isomers of HCH can enter the human body via food or drinking water contaminated with HCH.

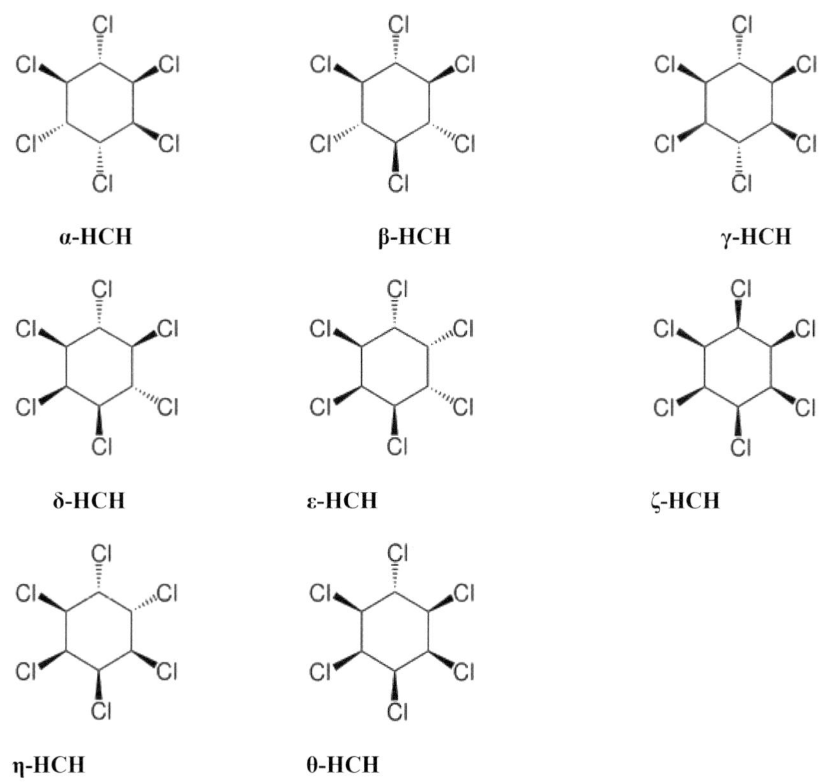

FIGURE 18.1 Different isomers of HCH.

18.2.2 Aldrin and Dieldrin

18.2.2.1 Structure and Formulation

The generic names for two structurally related compounds used as insecticides are aldrin and dieldrin (Figure 18.2). 1,2 ,3,4,10,10-hexachloro-1,4,4,5,8,8-hexahydro-1,4-endo, exo-5,8-dimethanonapthalene is the chemical term for aldrin (HHDN). Aldrin in technical grade is 85.5 percent aldrin. 1,2,3,4,10,10 -hexachloro-6,7-epoxy 1,4,4-,5,6,7,8,8-octahedron-1,4-endo, exo-5,8-di-methanonaphthalene is the chemical term for dieldrin. HEOD is the abbreviation for dieldrin's scientific name. Dieldrin in professional grade is 85% dieldrin. Technical-grade aldrin and dieldrin are tan powders that steadily evaporate in the sunlight, while pure aldrin and dieldrin are white powders. Aldrin dissipates faster than dieldrin. Both have. a slight chemical odor found in dirt, water, and homes where termiticides were used to combat termites. They can also be found in plants and animals living close to hazardous waste sites.

18.2.2.2 Presence in Environment

Accidental spills or leaks from waste storage containers will release aldrin and dieldrin into the atmosphere. Farmers used aldrin and dieldrin to destroy pests on crops, and exterminators used these compounds to kill termites in the past. Aldrin is rapidly converted to dieldrin in the presence of sunlight and bacteria. Dieldrin degrades very slowly in soil or liquids. It adheres to the soil and stays that way for several years. It is difficult to wash off the soil during rainstorms due to its insolubility in water. As a result, it is not present in large amounts in liquids. Dieldrin is mostly found in soil and sediments at the bottom of lakes, wetlands and streams. Dieldrin can fly long distances by binding to dust particles carried by the wind over long distances. Within a few days in the air, dieldrin turns into photo-dieldrin. Dieldrin can be taken up by plants and preserved in their leaves and roots. Dieldrin is stored in vast amounts in the fat of fish and animals that consume dieldrin-contaminated products. Dieldrin levels in the fat of cattle or fish who feed other animals are several times higher than in the fat of those that consume vegetables.

18.2.2.3 Exposure to Humans

Humans are susceptible to aldrin and dieldrin by the ingestion of infected foods. Fish or shellfish from polluted lakes or ponds, root crops, dairy products and meats are among the foods. Aldrin and dieldrin may also be swallowed, inhaled, or come into contact with polluted soil at toxic waste facilities. Workers who used aldrin and dieldrin to kill insects were exposed to these contaminants on their skin and inhaled them. Long after aldrin and dieldrin were used as insecticides in homes, people were introduced to them. Because aldrin easily degrades into dieldrin in the atmosphere, exposure to it is usually minimal. However, dieldrin has a long half-life in the ecosystem and can be present in the soil, sediments and animal fat.

18.2.3 Endosulfan

18.2.3.1 Structure and Formulation

Endosulfan is marketed as a combination of two related chemical forms known as α- and β-endosulfan (Figure 18.3). It can take the shape of a crystalline brown solid powder with a distinct turpentine-like odor. Endosulfan is a chemical used to control insects in food crops like rice, tea, vegetables and fruits and nonfood crops, such as tobacco and cotton. It is also used to prevent the wood from rotting.

18.2.3.2 Presence in Environment

When endosulfan is processed or used as a pesticide, it enters the air, water and soil. Sprayers are commonly used to add endosulfan to crops. Any endosulfan in the air can fly great distances before landing on crops, soil or water (Silva and Gammon, 2009). Endosulfan on crops degrades in a matter of weeks and is released into the soil. Any endosulfan in the soil evaporates into the air, while other endosulfan in the soil degrades. They can, however, persist in the soil for many years before fully decomposing. Endosulfan attached to soil particles may be washed away by rainwater and dissolved in seawater, where it may bind to soil particles and float on the water surface before binding to the soil at the bottom. The tiny amount of endosulfan that dissolves in water takes a long time to degrade. Endosulfan that dissolves in water can break down in as little as one day or as long as several months, depending on the conditions in the water. Any endosulfan in surface water

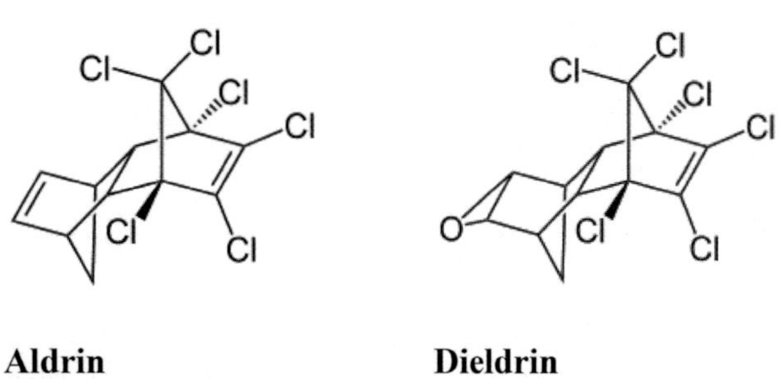

FIGURE 18.2 Structure of aldrin and dieldrin.

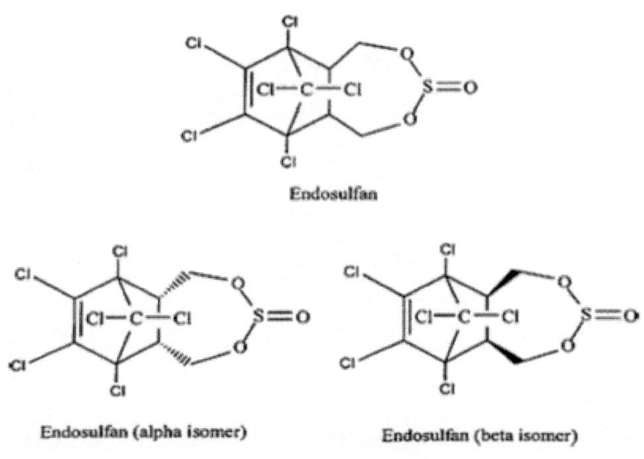

FIGURE 18.3 Different isomers of endosulfan.

evaporates and breaks down in the air. Endosulfan is present in relatively small concentrations in groundwater because it does not dissolve readily in water (water below the soil surface, e.g., well water). As a result, endosulfan can build up in the bodies of animals that live in endosulfan-contaminated waters. Endosulfan levels in their bodies can be many times higher than in the ambient water.

18.2.3.3 Exposure to Humans

Endosulfan has been discovered in various foods, including fruits, grains, oils, and fats. In addition, humans may be exposed to low amounts of endosulfan by skin contact with infected soil or through consuming cigarettes containing endosulfan residues (Papadopoulou-Mourkidou and Milothridou, 1990). It has been proposed that endosulfan exposure is higher for people who live near hazardous waste sites when they come into contact with endosulfan-contaminated soil. Apart from workplace contamination, which has resulted in several instances of poisoning, residues of food and drinking water are found at sufficiently high levels worldwide.

18.2.4 DDT and Its Metabolite DDE

18.2.4.1 Structure and Formulation

DDT is a pesticide that was once commonly used to monitor mosquitoes on farm crops and insects that bear diseases such as malaria and typhus. It was extensively used in India to combat malaria for several years. Technical grade DDT is made up of three types: p,p′-DDT (85%), o,p′-DDT (15%) and o,o′ -DDT (trace amounts; Figure 18.4). These are all solids that are white, crystalline, tasteless and nearly odorless. DDE (1,1-dichloro-2, 2-bis(p-chlorophenyl)ethylene) and DDD (1,1-dichloro-2,2-bis (p-chlorophenyl) ethane) are potential pollutants of technological category DDT. DDE is a DDT-breakdown product.

18.2.4.2 Presence in Environment

DDT was released into the air, water and soil after its manufacture and use as an insecticide until it was banned (Abhilash and Singh, 2009; Gupta, 2004. The bulk of DDT in the atmosphere is a result of its extensive usage in previous years. Because of its current usage in other parts of the world, DDT also makes its way into the environment. DDE is a degradation product of DDT that is only present in the atmosphere due to pollution or DDT breakdown. It was sprayed on crops and trees on a wide scale as pest control in previous years,

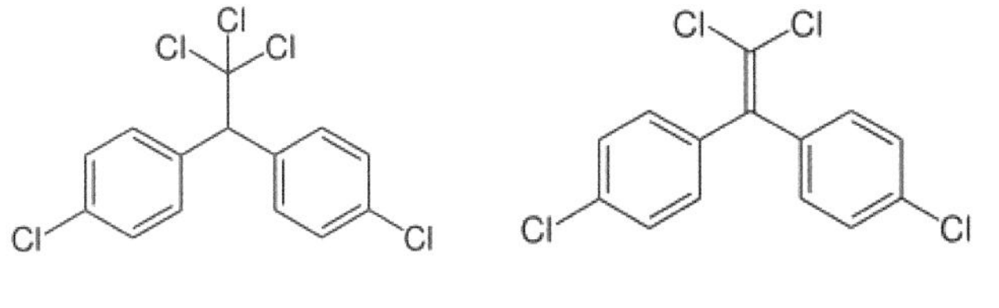

DDT **DDE**

FIGURE 18.4 Structure of DDT and its metabolite DDE.

contaminating the air, soil and water. When DDT and DDE evaporate from polluted water and soil, they may penetrate the air. DDT and DDE can be accumulated on the ground or in seawater if present in the air. This evaporation and deposition period can be replicated forever. As a result, DDT and DDE will fly great distances across the atmosphere. DDT has a half-life in the atmosphere of 2 to 15 years (Ngowi et al., 2007), and the primary breakdown product in the environment and living organisms is p,p′ -DDE. In comparison to the parent compound, p,p′ -DDE lasts much longer and is used as a marker for previous DDT exposure. DDT and DDE are also persistent organic contaminants (POPs), which are industrial compounds with high lipophilicity.

DDT and DDE may occur as a vapor in the atmosphere or bound to solids in the air. Due to a reaction caused by the sun, vapor-phase DDT and DDE can degrade in the atmosphere. The half-life of these compounds as vapor in the atmosphere has been measured to be 1.5 to 3 days. In fact, this half-life calculation is too short to account for DDT's and DDE's ability to migrate long distances through the atmosphere. Therefore, DDT and DDE remain in the soil for a long time, probably hundreds of years.

18.2.4.3 Exposure to Humans

Humans are more susceptible to DDT toxicity when they are at the top of the food chain. People in India are primarily susceptible to DDT, DDE and DDD by consuming foods containing trace quantities of these chemicals. Meat, poultry, dairy products and fish are the primary sources of DDT in the diet. DDT is used in more significant quantities in leafy vegetables than in other vegetables, presumably because DDT in the air is accumulated on the leaves. DDT damage from inhalation or skin absorption during handling or application is infrequent. Breast milk consumption may expose infants. DDT or its breakdown components are still detected in air, water and soil samples.

18.3 OCPs and Disease (International Scenario)

Several long-term surveys from around the globe have revealed the prevalence of OCPs in the general population. In the serum samples of young males in southern Spain, Carreño et al. (2007) reported higher concentrations of p,p'p'-DDE (96%), o,p'p'-DDD (65%) and endosulfan (92%). Higher levels of HCB (0.178 mg/kg), HCH (0.504 mg/kg), pp'pp'-DDE (2.789 mg/kg), o,p'p'-DDT (0.130 mg/kg), p,p'p'-DDT (0.340 mg/kg) and t-DDT (3.258 mg/kg) were observed in blood and abdominal tissues of 126 adult cadavers at the Institute of Forensic Medicine of the University of Veracruz, Mexico (Waliszewski et al., 2004). The impact of OCPs on human health is a major concern, as these pesticides have been linked to a variety of diseases, including diabetes, poor reproductive outcomes, and cardiovascular disease (CVD). Dioxin and non-dioxin-like PCBs, as well as other organochlorine compounds, were shown to be positively correlated with the lowest quartile of serum concentrations; the odds ratio (OR) for CVD across rising quartiles was 0.9, 2.0 and 5.0 for dioxin-like PCBs; 1.2, 1.2 and 3.8 for non-dioxin-like PCBs; and 1.9, 1.7 and 4 for organochlorine compounds (Ha et al., 2007). Mirex, 101 PCB congeners, DDE and HCB were found in blood samples from Akwesasne Mohawks (Goncharov et al., 2008). The researchers (Goncharov et al., 2008). They concluded that PCBs were specifically responsible for enhanced cholesterol synthesis and triglycerides, which are considered significant risk factors for CVD. In a cross-sectional sample of an adult Native American population, Codru et al. (2007) discovered a prospective correlation between diabetes and serum levels of PCBs, DDE, HCB, and mirex. An early study by Rutten et al. (1988) has shown the significantly higher levels of HCB and p,p'p'-DDE in serum of dialyzed (before dialysis, the level of HCB-15.5 nmol/l and pp'pp'-DDE-27.0 nmol/l; after dialysis, the level of HCB-17.2 nmol/l and pp'pp'-DDE-28.0 nmol/l) and non-dialyzeduremic patients (the level of HCH and DDE were 16.2nmol/l and 26.4nmol/l, respectively) than in controls (the level of HCB and pp'pp'-DDE were 7.7nmol/l and 20.1 nmol/l, respectively).

18.4 OCPs and Disease (National Scenario)

Chlorinated and other POCs were contained in the blood and tissue samples of the general population in many studies on the Indian population. Different reported studies have revealed the presence of a variety of OCPs in blood samples obtained from the general population in various parts of India, with amounts ranging from 2.92 to 4.52 parts per billion (ppb) for α-HCH, 1.93 to 10.05 ppb for β-HCH, 1.69 to 5.33 ppb for γ-HCH, 0.03 to 3.32 ppb for aldrin, 1.91–2.77 ppb for dieldrin and 0.01 and 2.21 ppb (Kumar et al., 2010; Mustafa et al., 2010; Pathak et al., 2008; Bhatnagar et al., 2004). Bhatnagar et al. (2004) examined the blood samples of 18 male volunteers from the Ahmedabad (urban) district for residue analysis. They detected p,p'-DDE, o,p'-DDT, p,p'-DDD, p,p'-DDT and DDT in serum samples with average values 20.85, 1.15, 2.03, 9.28 and 32.61g/l, respectively. HCB was present at an average concentration of 0.2 g/l, and α-HCH, β-HCH, γ-HCH, and t-HCH were found at 4.49, 35.06, 1.69, and 41.23 g/l, respectively, in serum samples. BHC and DDE levels were observed higher in the serum samples of explicitly exposed people. Subramaniam and Solomon (2006) reported the BHC residue concentrations ranged from 0.006 to 0.130 ppm, and DDE residue concentrations ranged from 0.02 to 0.033 ppm in serum samples. Chlorinated and other POCs were contained in the blood and tissue samples of the general population in many studies on the Indian population. Different reported studies have revealed the presence of a variety of OCPs in blood samples obtained from the general population in various parts of India, with amounts ranging from 2.92–4.52 parts per billion (ppb) for α-HCH; 1.93–10.05 ppb for β-HCH; 1.69–5.33 ppb for γ-HCH; 0.03–3.32 ppb for aldrin; 1.91–2.77 ppb for dieldrin; 0.01–2.21 ppb (Kumar et al., 2010; Mustafa et al., 2010; Pathak et al., 2008; Bhatnagar et al., 2004). Bhatnagar et al. (2004) examined the blood samples of 18 male volunteers from the Ahmedabad (urban) district for residue analysis. They detected p,p'-DDE, o,p'-DDT, p,p'-DDD, p,p'-DDT, and DDT in serum samples with average values 20.85, 1.15, 2.03, 9.28, and 32.61 g/l, respectively. HCB was present at an average concentration of 0.2 g/l, and α-HCH, β-HCH, γ-HCH, and t-HCH were found at 4.49, 35.06, 1.69, and 41.23 g/l, respectively in serum samples. BHC and DDE levels were observed higher in the serum samples of explicitly exposed people. Subramaniam and Solomon (2006) reported the BHC residue concentrations ranged from 0.006 to 0.130 ppm, and DDE residue concentrations ranged from 0.02 to 0.033 ppm in serum samples.

18.5 Association of OCPs with CKDu

Middle-aged people (40–60 years) living in rural areas have a higher prevalence of late-onset nephropathies such as CKDu, with men having a higher prevalence. According to the results of a survey conducted in Sri Lanka's North Central Province, there is a substantial association between CKDu and environmental factors in farming areas (Chandrajith et al., 2011). Furthermore, a higher level of OCPs was detected in CKDu patients along with a reduced glomerular filtration rate and increased oxidative stress (Siddharth et al., 2012). In this case-control study, we detected the nine OCPs, namely, α-HCH, β-HCH, γ-HCH, aldrin, dieldrin, α-endosulfan, β-endosulfan, DDT and DDE, in the blood sample of healthy people and CKDu patients in variable quantities. In CKDu patients, the level of several pesticides, namely, α-HCH, γ-HCH, total-HCH, aldrin, α-endosulfan, β-endosulfan and p,p' -DDE were found to be increased by 30% to 40% and by 50% in case of total pesticides, which represent the sum of all pesticides and possibly accounts for the cumulative effect of all the pesticides.

Many of the OCPs found in higher concentrations in CKDu patients had a direct link to the disease. Significant association of α-HCH (OR = 2.52), γ-HCH (OR = 2.70), total HCH (OR = 3.18), aldrin (OR = 3.07), α-endosulfan (OR = 3.11), β-endosulfan (OR = 3.19), and *p,p'p'*-DDE (OR = 2.45) with CKDu were observed after the adjustment of age, sex, body mass index, total lipid content and concurrent adjustment of each analyte. A highly significant association of total pesticides was observed with CKDu (OR = 3.52, $p < 0.01$). It is possible that total pesticides, which represent the sum of all the pesticides detected and are a mixture of various OCPs, could have an additive and synergistic effect on kidney function. Having observed that blood OCPs in higher amounts are associated with CKDu, the correlation between OCP levels and GFR of CKD patients was carried out. Total HCH, dieldrin, *p,p*-DDE, β-endosulfan and total pesticide load showed a significant negative correlation with eGFR, indicating that the presence of excess pesticides reduces GFR in CKDu patients. It's unknown if the aggregation of OCPs impacts kidney function. Animals' kidneys have been shown to develop glomerular, tubular, and interstitial fibrosis, as well as acute necrosis, after being exposed to OCPs in the environment (Uboh et al., 2011; Khan and Kumari, 2012). Blood OCP levels have been related to acute renal disease and renal cancer in the workplace (Karami et al., 2008). Several of these OCPs, such as endosulfan, HCH, and DDT, have also been confirmed to be nephrotoxic and cause severe renal toxicity based on laboratory or occupational exposure findings (Karami et al., 2008; Khan and Kumari, 2011; Uboh et al., 2011). It has also been documented that laboratory animals exposed to doses of γ-HCH and endosulfan developed kidney damage (Kramer et al., 2004; Reigart, 2009). Consequently, it has been suggested that elevated levels of OCPs exacerbate kidney failure, which contributes to CKDu.

18.6 Genetic Polymorphism

Polymorphism is defined as two or more distinct phenotypes that coexist in the same population. Genetic polymorphism is a difference in a person's DNA sequence in more than 1% of the population. They are heritable and therefore can be changed by population selection and breeding patterns. Five molecular mechanisms have been suggested to trigger genetic polymorphism in different enzymes. First, the gene sequence that codes for the basic enzymes may be absent. Second, expression can be missing or reduced due to changes in the regulatory portion of the gene. Third, the pre-mRNA can be wrongly spliced as a result of mutations at intron–exon boundaries. Fourth, there could be a decrease in enzyme function due to mutated noncritical amino acids in the proteins. Fifth, essential amino acids in the genes have been altered, resulting in an enzyme no longer involved (Wormhoudt et al., 1999).

If SNPs are present in noncoding regions of the genome, they have little effect on enzyme function. When SNPs are found in promoter regions, they can influence the affinity of a specific DNA sequence for the gene's transcriptional ability (Miao et al., 2003). While most SNPs included in an intron of the primary transcript would be spliced out and do not influence that gene's expression/protein profile, this may not rule out the possibility that they will affect a neighboring gene. Provided that some gene sequences are known to overlap, they can influence gene transcription or have a more direct effect (Makalowska et al., 2005). Some nonsynonymous SNPs in exon regions can cause changes in the amino acid sequence of a protein, affecting its secondary/tertiary structure, which is responsible for receptor interaction and catalytic properties. In addition, these polymorphisms can sometimes add a stop codon instead of coding for an amino acid, resulting in the production of defective or inactive proteins (Hu et al., 1999). Both of these factors could cause an up- or downregulation of a particular pathway, which could affect the host's susceptibility to an unexpected physiological response.

18.7 Xenobiotic-Metabolizing Enzymes Polymorphism

18.7.1 Phase I Metabolizing Enzymes (Cytochrome P 450 Enzymes)

The human cytochrome P450 enzyme family (abbreviated CYP, P450, CYP450, or CYPs) is a large group of enzymes. They can act as a terminal oxidase in the metabolism of xenobiotics. There are 57 functioning human P450 genes and 29 nonfunctional pseudogenes identified. Cytochromes P450 are enzymes that metabolize a wide range of endogenous and exogenous substances. There are heme-containing enzymes in which the ferric (Fe3+) condition of the heme iron in cytochrome P450 is present, decreasing to the ferrous (Fe2+) state throughout ligand binding. These enzymes operate in the same manner as a monooxygenation reaction does, in which one atom of oxygen is introduced into a substrate (Ercal et al., 2001). The general reaction catalyzed by cytochrome P450 enzymes is

$$O_2 + 2e- + RH + 2H+ \rightarrow H_2O + ROH.$$

The CYP450 enzymes are mainly classified into P450 families, namely, CYP1, CYP2 and CYP3, CYP17 and CYP19 (a new family is also reported) as CYP4, and it is encoded in liver microsome), the substrates of which include several fatty acids, eicosanoids, and xenobiotics. Due to the role of CYP450 enzymes in the metabolism of several xenobiotics, carcinogens and endogenous hormones, any variation in these genes might be associated with health risks. It has been well known that genetic polymorphisms in these enzymes cause changes in enzyme function. Furthermore, polymorphisms in these have been related to an elevated risk of most human diseases, including CKD. Despite substantial studies in the fields of toxicology and pharmacology, the significance of this community of xenobiotic-metabolizing enzymes has been ignored.

18.7.1.1 Cytochrome P450 1A1

Cytochrome P450 (CYP) 1A1 (CYP1A1; EC 1.14.14.1) is a phase I enzyme involved in xenobiotic bioactivation. CYP1A1 mostly metabolized organochlorine pesticides such as PCBs and other large flat-structured aromatic hydrocarbons. The human *CYP1A1* isoenzyme gene is found on human chromosome 15q22-q24 and has seven exons and six introns, the first of which is noncoding. The full gene span sequence is 6311 base pairs long. CYP1A1 enzyme exists in several polymorphic forms, namely, (1) *CYP1A1*2A* (T3801C), which generates a MspI restriction site in the 30 noncoding regions of the gene and is believed to result in increased inducibility of the gene following exposure to Ah receptor ligands (Kiyohara et al., 1996); (2) *CYP1A1*2B* (A2455G), which leads to an Ile462→Val substitution in the heme-binding domain of exon 7, with a concurrent increase in the catalytic activity of the protein (Crofts et al., 1993); (3) *CYP1A1*3,* 3205T→C located on 3' noncoding region, that is, upstream of polyadenylation site; and (4) *CYP1A1*4,* 2453C→A, in exon 7 causing Thr461Asn (rs1799814).

The CYP1A1 *2A mutant genotype is common in Asian populations, with frequencies ranging from 2% to 18% for the homozygous mutant CC genotype and 32% to 55% heterozygous mutant TC genotype (Srivastava et al., 2008; Suryanarayana et al., 2004). In European and White Americans, 0% to 5% of populations are homozygous mutant CC genotype, and 9% and 28% are heterozygous TC genotype (Bailey et al., 1998; Li et al., 2004; Longuemaux et al., 1999). Frequencies in African Americans are intermediates between White Americans and Asian populations (4%–6% CC, 35%–39% TC; Bailey et al., 1998; Garte et al., 2001; Li et al., 2004). Similarly, Asians are the most common carriers of *2C, with frequencies ranging from 1% to 10% for homozygous mutant GG genotype and 15% to 50% for heterozygous AG genotype (Mittal and Srivastava, 2007; Sobti et al., 2004). Just about 3% of Europeans and U.S. Whites have the homozygous mutant GG genotype, while up to 15% have the heterozygous AG genotype. African Americans are less likely than Whites to have the homozygous mutation GG genotype (Garte et al., 2001; Masson et al., 2005).

18.7.2 Phase II Metabolizing Enzymes (Glutathione S-Transferases)

Glutathione S-transferases (GSTs; EC 2.5.1.18) are phase II enzymes formed alongside GSH and are abundant in all living organisms. In 1961, GSTs became popular for the first time (Booth et al., 1961). It is a multifunctional protein family that catalyzes the conjugation of electrophilic compounds to the tripeptide glutathione (γ-glutamylcysteinyl glycine) to detoxify a variety of exogenous and endogenous compounds (Mannervik et al., 1985). The bound microsomal and cytosolic family members of human GSTs are classified into two distinct superfamilies. GSTs play a significant part in the endogenous synthesis of prostaglandins and leukotrienes in microsomes. GSTs in the cytosol are categorized into six groups: α, μ, π, θ, ξ, ώ, κ, χ (Table 1). Among the various groups, μ is found in the body, mostly in the liver, θ is expressed most prominently in the renal tubules, and is the primary isoenzyme in RBCs.

18.7.1.2 GSTμ Gene Family

The GST class genes are found on chromosome 1p13.3 in a 20-kb cluster in the following order: *5′5′-GSTM4-GSTM2-GSTM 1-GSTM5-GSTM3–3′3′*. The *GSTM1* gene, as well as the therapeutic importance of its polymorphisms, has been extensively studied. Board (1981) used starch gel electrophoresis to demonstrate polymorphism at the *GSTM1* locus and allelic variants. He discovered a sequence for GST-class enzyme from the liver typical of a polymorphic dimeric protein encoded by three alleles with a single null allele. *GSTM1*0*, *GSTM1*A* and *GSTM1*B* are the three alleles found in the *GSTM1* locus. The *GSTM1*0* allele induces a gene deletion, and homozygotes do not contain any protein. In exon 7, *GSTM1*A* and *GSTM1*B* varied by just one base. The enzymes encoded by these alleles were shown to have identical catalytic properties. The three positive GST M1 phenotypes (*GSTM1*A*, *GSTM1*B*, *GSTM1*A*B*) resulted from homozygotic and heterozygotic variants of the *GSTM1*0*, *GSTM1*A* and *GSTM1*B* alleles Board (1981). *GSTM1* has been of great interest to researchers in molecular epidemiological studies because, in Caucasian populations, about half of those studied have homozygous *GSTM1* gene deletion and, as a result, are believed to be at higher risk when exposed to carcinogens due to decreased detoxification potential than those who are *GSTM1(+)*.

18.7.1.3 GST θ Gene Family

Peter et al. (1989) are the first to note a polymorphic GST θ. They discovered that only 60% of people's erythrocytes catalyzed the conjugation of methyl chloride with glutathione, while the other 40% lack their activity. *GSTT1* and *GSTT2*, two genes of the GST family, are found on chromosome 22 and are separated by about 50 kb. The *GSTT1* locus, like *GSTM1*, has a deleted allele, *GSTT1*0*, and people who have it on both chromosomes don't contain any protein. *GSTT1 +/+*, *GSTT1 +/–*, and *GSTT1 –/–* are three phenotypes thought to result from the merger of functional (*GSTT1*1*) and nonfunctional (*GSTT1*0*) alleles (*GSTT1null*). About 20% of Caucasians are homozygotes for *GSTT1* gene deletion, making them more vulnerable to environmental carcinogens. *GSTT1* null genotype is not rare in the population, according to numerous studies from various geographies, and it has been the focus of extensive study.

18.7.1.4 Other Classes of GSTs

Other types of GSTs can be included as well. The GST π gene family is one of the most prominent. At nucleotides 313 and 341, *GSTP1* contains three allelic polymorphisms. The first *GSTP1*A* has AG transformations from ATC to GTC at codon 104 (Xi et al., 2011). Second, *GSTP1*B*, where C→T transforms from GCG (Ala) to GTG, occurs at codon 113 (Mannervik et al., 1985). Third, there is no improvement in the protein *GSTP1*C*. Both mutations occur in the electrophile-binding active sites, resulting in functionally distinct *GSTP1* proteins. Another kind of *GSTα* is the gene family, which consists of five genes: *GSTA1*, *GSTA2*, *GSTA3*, *GSTA4* and *GSTA5*. There are two alleles of the *GSTA1* gene: *GSTA1*A* and *GSTA1*B*. *GSTA1*A* expression is higher than *GSTA1*B* expression.

18.8 Polymorphism of Xenobiotic Metabolizing Genes (*CYPs* and *GSTs*)

CYPP450 and *GSTs* are xenobiotic-metabolizing enzymes and play a main role in the vital detoxification process. These

TABLE 18.1

Classes of Glutathione S-Transferase

Class	Chromosome	Class	Chromosome
α (Alpha)	6	ξ (zeta)	14
μ (Mu)	1	ώ (omega)	10
π (Smits et al.)	11	κ (Kappa)	7
θ (Theta)	22	χ (Chi)	10

genes are polymorphically distributed among different human ethnic populations. Genetic polymorphism in these genes affects their enzyme activity, resulting in alteration in their substrate's metabolism (which is a harmful substance), leading to accumulation induces the high risk of certain human disorders. These polymorphisms show ethnicity-dependent frequency distribution, and therefore, the risk of developing a disorder in a population might depend on the frequencies of some of these high-risk polymorphisms.

Xenobiotic-metabolizing enzymes CYPP450 and GSTs play a critical role in the detoxification process. Different human racial groups have polymorphic distributions of these genes. Genetic polymorphisms in these genes affect enzyme function, causing changes in the synthesis of their substrate (a toxic substance) and causing aggregation, which raises the risk of many human diseases. Because the frequency distribution of these polymorphisms is ethnicity-based, the probability of having a disorder in a population may be influenced by the occurrence of any of these high-risk polymorphisms. Siddarth et al. (2013) found that both **2A* and **2C* polymorphisms of the CYP1A1 gene are substantially correlated with 1.5- to 2.2-fold increased risk of CKDu in the North Indian population in a previous analysis. For the first time, we found an important connection between CYP1A1 polymorphism and CKDu. However, previous research on diabetes (Wang et al., 2002), coronary artery disease (Achour et al., 2011) and cancers, such as lung cancer, breast cancer and others (Masson et al., 2005; Sergentanis and Economopoulos, 2010; Wright et al., 2010), were accessible. Some researchers have also found a connection between *CYP1A1* gene polymorphism and renal cell cancer, with an unusual ratio varying from 1.5- to 2.0-fold 9 (Chen et al., 2011; Ryan et al., 1994). According to Wang et al. (2002), people with the *CYP1A1*2A* heterozygous genotype have a 2.5-fold increased chance of RCC in the Chinese population. According to Ahmad et al. (2013), mutated alleles of the *CYP1A1*2C* genotype (AG, GG) were related to a two-fold rise in the risk of RCC.

When *CYP1A1*2A* and **2C* polymorphisms were linked to homozygous mutants, Taioli et al. (1998) found a twofold increase in CYP1A1 expression. In a previous study, we used multinomial logistic regression to determine the relationship of various genotype combinations with CKDu using genotype mixture analysis of *2A and *2C variants. We observed that hetero-/ homozygous combinations of both**2A* and **2C* genotypes, that is, TC/AG (OR = 1.86, 95% CI = 1.25–2.79, $p < 0.01$); TC/GG (OR = 3.36, 95% CI = 1.01–11.17; $p < 0.05$); CC/AG (OR = 2.25, 95% CI = 1.04–4.90, $p < 0.05$) and CC/GG (OR = 2.45, 95% CI = 1.10–5.43; $p < 0.01$) were extremely associated with CKD_u. Because each heterozygous and homozygous mutant mixtures *of CYP1A1 *2A (T/C)* and **2C (G/A)* polymorphism enhanced the expression of the protein, the extent of reactive metabolites generated by the phase I reaction mediate by the augmentation of this protein could be through some unknown mechanism(s) react with the urinary organ cells like kidney and thereby directly disturb its functional activity or may occur through enhanced OS probably generated by the enlarged reactive metabolites.

GSTs are phase II detoxification enzymes present in several tissues, in which they control the conversion of toxic compounds to hydrophilic metabolites (Hayes and Strange, 2000). It has been discovered that differential expression influences the anticarcinogenic ability of tissues. It was first proposed as a possible cancer susceptibility marker in 1986 (Seidegård et al., 1986). GSTs are being studied as possible risk biomarkers for diabetes, atherosclerosis and cancer, including RCC (Sui et al., 2011). In a previous analysis, we found that the frequency distribution of *GSTM1(−)* and *GSTT1(−)* genotypes in controls were 33.0% and 21%, respectively, while it was 39% ($p > 0.05$) and 32% (p 0.05) in CKDu patients (Siddarth et al., 2014). The rates of homozygous *GSTM1* deletions in Caucasian, Nigerian, French and Chinese populations have been reported to be about 50%, 21.7%, 43%, and 58.3%, respectively (Sprenger et al., 2000). The *GSTT1* null polymorphism is found in 13% to 26% of Asians, while it is found in 35% to 52% of Caucasians (Raimondi et al., 2006). In North America, the composite deletion level of both genes, *GSTM1* and *GSTT1*, is 6%, 8% in Malaysia and 8% in Egypt (Alshagga et al., 2011). In a previous study, we observed that the prevalence of both null genotypes was 1.8 times greater in CKDu patients than in stable controls. *GSTM1(+)/GSTT1(−)* and *GSTM1(−)/GSTT1(−)* genotypes were shown to be significantly associated with nondiabetic CKD (Datta et al., 2010). In a previous analysis, we found that the frequency distribution of *GSTM1(-)* and *GSTT1(-)* genotypes in controls were 33.0 percent and 21%, respectively, while it was 39 percent ($p > 0.05$) and 32 percent (p 0.05) in CKDu patients (Siddarth et al., 2014). The rates of homozygous *GSTM1* deletions in Caucasian, Nigerian, French and Chinese populations have been reported to be about 50%, 21.7 percent, 43 percent, and 58.3 percent, respectively (Sprenger et al., 2000). The *GSTT1* null polymorphism is found in 13–26 percent of Asians, while it is found in 35 percent to 52 percent of Caucasians (Raimondi et al., 2006). In North America, the composite deletion level of both genes, GSTM1 and GSTT1, is 6%, 8% in Malaysia, and 8% in Egypt (Alshagga et al., 2011). In a previous study, we observed that the prevalence of both null genotypes was 1.8 times greater in CKDu patients than in stable controls. *GSTM1(+)/GSTT1(-)* and *GSTM1(-)/GSTT1(-)* genotypes were shown to be significantly associated with non-diabetic CKD (Datta et al., 2010). Agrawal et al. (2007) found a connection between the *GSTM1(−)* (p = 0.0386; OR = 1.445, 95% CI = 1.033–2.021) and *GSTT1(−)* (p0.0001; OR = 4.568, 95% CI = 3.215–6.492) genotypes and an increased risk of CKD in a north Indian population.

18.9 Gene–Environment Interaction

The word gene–environment interactions is frequently used to describe a particular relationship between a gene and its surroundings. We needed information on all the elements of the relationship to describe gene–environment interactions. The genetic polymorphism and environmental factors were both listed separately in the preceding section. Several collaborators have identified the idea of genetic vulnerability and environmental exposures. The suitable model includes consistency heterogeneity, mathematical models of interaction and the challenge of determining biological plausibility. These models are described in the following subsections.

18.9.1 Qualitative Models

In the dichotomous genotype and environmental exposure analysis, four variations will exist, shown in a 2 × 4 table (Botto and Khoury, 2001), and the relative risk seen, as shown in Figure 18.5. In the dichotomous genotype and environmental exposure analysis, four variations will exist, shown in a 2 X 4 table (Botto and Khoury, 2001) and the relative risk seen in the graph Figure 5.

A | In the most simplified example of a dichotomous genotype (e.g., carriers versus noncarriers of an allele corresponding to a dominant trait) and dichotomous exposure (e.g., 'exposed' versus 'nonexposed'), three categories of joint exposure can be compared with a reference category (for which the relative risk is, by definition, 1.0). In the example shown here, the relative risk of developing a disease is much greater in individuals who are both genetically susceptible to the condition and have been exposed to the environmental variable (cell **d**) than in individuals who carry the wild-type genotype and are not exposed to the environmental variable (cell **a**), or who are either only exposed to the environment or genetically susceptible (cells **b** and **c**, respectively). **B** | In the slightly more complex situation in which there are three categories of exposure, it has been proposed that genetically susceptible individuals could be at risk of disease at lower levels of exposure; in this model, the difference in risk between genotypes among individuals at the medium level of exposure is the only indication of an interaction. Adopted by Hunter (2005).

18.9.2 Statistical Models

The appropriate statistical method is very important to check whether the interaction is significant or not. There are several methods to assess the statistical significance of interaction. For example, in the dichotomous study model of genotype and environmental exposure relation (Rothman and Greenland, 1998), the generally used procedure is to test departure from the multiplicative model of interaction.

18.9.3 Biological Plausibility

Figure 18.6 illustrates five biologically realistic models of associations between a genotype and an environmental response regarding disease risk (Ottman, 1990). If risks are calculated on an additive or multiplicative scale, the relationship among models B, C, and D should always be present. Model E, on the other hand, covers all conditions with and without contact. Alternatively, the measuring scale would be tested to see whether there is any connection. Models B, C, D, and E, to some extent, form an exhaustive collection of alternative gene-environmental interaction models. In terms of their individual effects on disease incidence, there are four potential genotype and exposure combinations: (a) an effect of genotype but not of exposure, (b) an effect of genotype but not of exposure, (c) an effect of neither genotype nor exposure and (d) an effect of both genotype and exposure

A. The effect of the genotype is to produce or increase expression of a "risk factor" that can also be produced non-genetically. B. The genotype exacerbates the effect of the risk factor, but there is no effect of the genotype in unexposed persons. C. The exposure exacerbates the effect of the genotype, but there is no effect of the exposure in persons with the low-risk genotype. D. Both the exposure and the genotype are required to increase risk. E. the exposure and the genotype each affect disease risk, and when they occur together, the risk is higher or lower than when they occur alone. Adopted by Hunter (2005).

The current scientific consensus is that neither genes nor environment are solely responsible for influencing human well-being (Han et al., 2008). Gene-environment association is a useful method for determining the cause of complex multifactorial diseases, including diabetes and CKD. The relationship of genes and environments can help distinguish the danger of disease by vector form, such as (a) differentiating the negotiation for obtaining population health gains and stratifying the probability of epidemic incidence (b) may help in the discovery of new environmental risk factors for disease development or the confirmation of suspected environmental risk factors and (c) may help recognize the occurrence of illness in terms of natural history, duration, etiologic variability and intervention goals at the population level. Advances in genomic technologies and analytical techniques have enabled breakthroughs in our understanding of the molecular details of normal biology. They hold the promise of providing new insights into the molecular mechanisms of a variety of diseases.

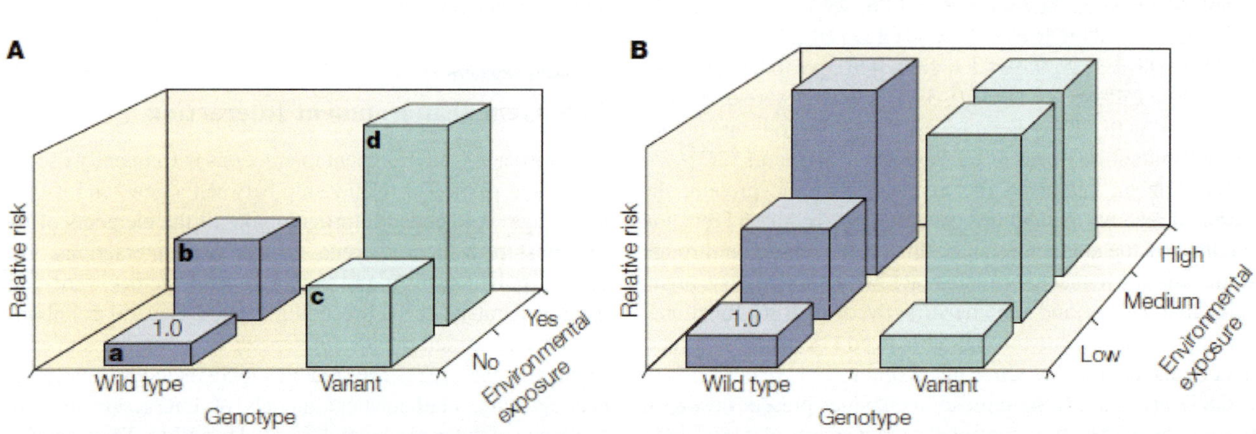

FIGURE 18.5 Models of gene–environment interactions.

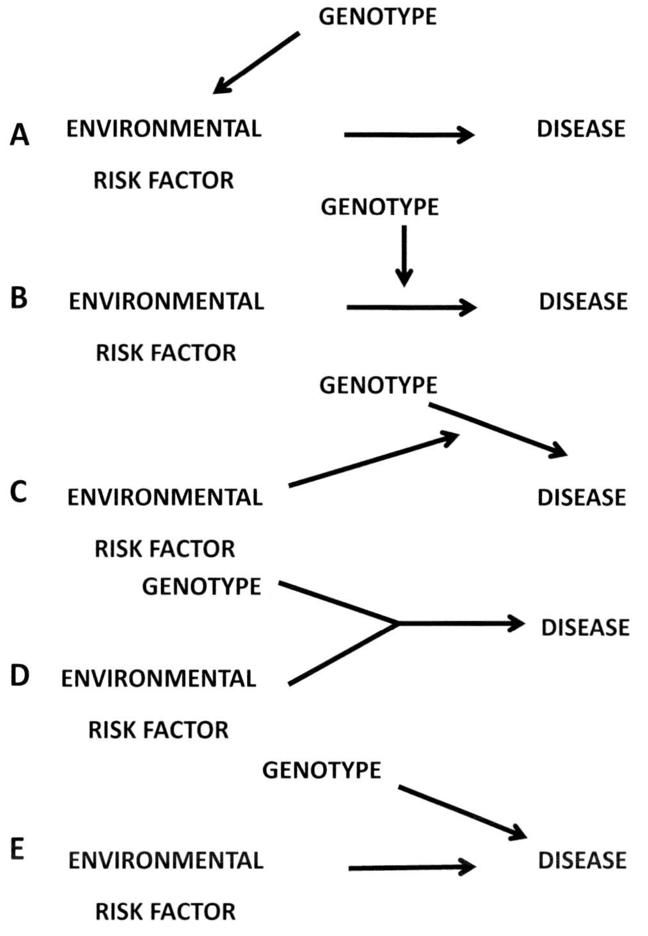

FIGURE 18.6 Models of the relationship between a high-risk genotype and an environmental exposure, in terms of their effect on disease risk.

18.10 Gene–Environment Interaction and CKDu

In CKDu, both genes and environmental factors play a part in determining complex traits. In a lab study, we discovered a close correlation between the genotypes of the xenobiotic-metabolic enzyme gene, *CYP1A1*, and the blood levels of multiple organochlorine pesticides in CKDu patients. The existence of heterozygous genotypes of *CYP1A1 *2A* and **2C (TC/AG)* was shown to be substantially ($p > 0.05$) correlated with the levels of α-endosulfan, γ-HCH, p,p′-DDE, p,p′ -DDT and total pesticides, suggesting that heterozygous genotypes raise the risk of augmentation of these pesticides by 2.5 times. Furthermore, α-endosulfan, γ-HCH, p,p′ -DDE, p,p′ -DDT and complete pesticides were also substantially ($p > 0.05$) correlated with a mutant genotype combination of CC/GG, suggesting that the mutant genotype combination raises the likelihood of increased prevalence of these associated pesticides by two-fold (data not published).

Again, to investigate the association of *GSTM1* and *GSTT1* genotypes with blood OCP levels of CKDu patients, seven pesticides, namely, aldrin (OR = 2.84, 95% CI: 1.29–6.23; OR = 3.29; 95% CI: 1.59–6.79), γ-HCH (OR = 3.13, 95% CI: 1.41–6.94; OR = 2.36, 95% CI: 1.16–4.83), α-HCH (OR = 2.32, 95% CI: 1.06–5.03; OR = 2.35, 95% CI: 1.15–4.78), total-HCH (OR − 3.45; 95% CI: 1.51–7.84; OR = 3.88, 95% CI: 1.80–8033), p,p'-DDE (OR = 3.23, 95% CI: 1.41–7.39), p,p'-DDT (OR = 3.84, 95% CI: 1.71–8.64; OR = 3.55, 95% CI: 1.71–7.38; OR = 3.74, 95% CI: 1.74–8.03) and total pesticides (OR = 5.08, 95% CI: 2.16–11.93; OR = 6.24, 95% CI: 2.86–14.42) showed significant ($p < 0.05$) association with *GSTM1(+)/GSTT1(−)* and *GSTM1(−)/GSTT1(−)* genotypes. γ-HCH (OR = 2.03, 95% CI: 1.05–3.90), p,p'-DDT (OR = 3.64, 95% CI: 1.38–5.16) and total pesticides (OR = 2.27; 95% CI: 1.15–4.47) showed a significant ($p < 0.05$) association with *GSTM1(−)/GSTT1(+)* genotype, indicating that patients with either null or both null genotypes tend to build up a majority of pesticides in CKDu patients (Siddarth et al., 2014). There is no other report available to refute these findings. It was hypothesized that weaker OCP detoxification caused by GST polymorphism leads to higher levels of these pesticides in CKDu patients with the *GSTM1(−)/GSTT1(−)* genotype. According to several studies, if inactive variants of enzyme (null genotypes of *GSTM1* and *GSTT1*) are present, detoxification of oxidative metabolites of xenobiotics formed in phase I reactions is decreased, which may contribute to CKD progression (Agrawal et al., 2007; Jain et al., 2006).

The relationships between genetic and environmental CKD risk factors are currently unknown to us. We need relevant patient subgroups to help understand these experiences and promote activities directed at the classification of tailored treatments and prevention. Only a few studies have looked into the role of genetic predisposition in CKD hotspots, which could contribute to discovering new gene–environment interactions. For example, it has been hypothesized that the familial nature of Balkan endemic nephropathy can be explained by a gene found on 3q25 and the instability of chromosome 3. The disease, however, affects people of various ethnic backgrounds, including newcomers to the region, but it is rare in people who have migrated away from endemic areas. In addition, even in places where it used to be more normal, the prevalence is said to be declining. Environmental factors, or a G×E relationship, may explain these findings. For example, a single-nucleotide polymorphism in sodium-dependent dicarboxylate transporter member 3 (SCL13A3) predisposes CKDu in Sri Lankan nephropathy. However, further research is required to validate this connection (Nanayakkara et al., 2014). In addition to genetic analysis, epigenetic studies are needed to fully comprehend the impact of genes and climate on CKD hotspots. Longitudinal studies of the relationship between genetic risk factors and prenatal, childhood, and adult exposures, including diet, would be essential for understanding the progression of CKD, particularly in LMICs (Stanifer et al., 2016; Wesseling et al., 2013b). Environmental factors interfere with genetic factors to alter disease incidence, but they also affect disease progression, as shown by certain renal diseases linked to the immune response in infection, for example, APOL1-associated kidney disease and trypanosomiasis (Thomson et al., 2014) and IgA nephropathy risk and the intestinal immune response to helminth infections (Kiryluk et al., 2014). Environmental factors interfere with genetic factors to alter disease incidence, but they also affect disease progression, as shown by certain renal diseases linked to the immune response in infection, for example, APOL1-associated kidney disease and trypanosomiasis (Thomson et al., 2014), IgA nephropathy risk and the intestinal immune response to helminth infections (Kiryluk et al., 2014).

18.11 Conclusion

We uncovered in this study that a gene–environment association between xenobiotic-metabolizing genes and OCP levels can increase the risk of CKDu. Although these findings do not prove a cause-and-effect relationship, the high prevalence of the disease calls for further research into the connection and the underlying pathogenetic mechanisms.

18.12 Future Prospects and Research Needs

The phenomenon of repeated chemical exposures is not special to India; it affects people worldwide. Furthermore, the prevalence of CKDu remains high worldwide, but many of the causes are still unclear. It would be essential to see whether genetic vulnerability to pesticides raises the likelihood of CKDu. Such gene–environment association research may shed light on the pathophysiology of CKDu, eventually contributing to improved detection, diagnosis and treatment strategies. Our research focuses on creating a correlation between the prevalence of CKDu and pesticide exposure, emphasizing organochlorine pesticides. This chapter will lead to discovering new genes that will aid in the understanding of pesticide-induced toxicity in CKDu. This will allow us to screen people who are exposed to pesticides at work and in the community. One of the most challenging facets of researching the causes and therapies for complex diseases is that neither the environment nor solely genetic factors can adequately describe the reported disease occurrence and development predictions. Therefore, we must test genetics and environment together in the same studies to estimate risk accurately.

ACKNOWLEDGMENT

Manushi Siddarth, one of the cowriters, would like to express her gratitude to the University Grant Commission of India for offering the Post Doctoral Fellowship 2015–2017.

REFERENCES

Abhilash, P.C., Singh, N., 2009. Pesticide use and application: An Indian scenario. *J. Hazard. Mater.* 165, 1–12. https://doi.org/10.1016/j.jhazmat.2008.10.061.

Achour, A., Zaag, I., Gueddah, L., Trimeche, B., Slama, F.B.H., Zemni, R., 2011. Role of CYP1A1 (T6235C) polymorphism and cigarette smoking in the development of coronary heart disease in Tunisian population. *J. Genet.* 90, 303–307. https://doi.org/10.1007/s12041-011-0058-x.

Agrawal, S., Tripathi, G., Khan, F., Sharma, R., Baburaj, V.P., 2007. Relationship between GSTs gene polymorphism and susceptibility to end stage renal disease among North Indians. *Ren. Fail.* 29, 947–953. https://doi.org/10.1080/08860220701641314.

Ahmad, S.T., Arjumand, W., Seth, A., et al., 2013. Risk of renal cell carcinoma and polymorphism in phase I xenobiotic metabolizing CYP1A1 and CYP2D6 enzymes. *Urol. Oncol.* 31, 1350–1357. https://doi.org/10.1016/j.urolonc.2011.12.009.

Alshagga, M.A., Mohamed, N., Nazrun Suhid, A., Abdel Aziz Ibrahim, I., Zulkifli Syed Zakaria, S., 2011. Frequencies of glutathione s-transferase (GSTM1, GSTM3 AND GSTT1) polymorphisms in a Malaysian population. *Arch. Med. Sci. AMS.* 7, 572–578. https://doi.org/10.5114/aoms.2011.24123.

Bailey, L.R., Roodi, N., Verrier, C.S., Yee, C.J., Dupont, W.D., Parl, F.F., 1998. Breast cancer and CYPIA1, GSTM1, and GSTT1 polymorphisms: Evidence of a lack of association in Caucasians and African Americans. *Cancer Res.* 58, 65–70.

Barsoum, R.S., 2013. Burden of chronic kidney disease: North Africa. *Kidney Int. Suppl.* 3, 164–166. https://doi.org/10.1038/kisup.2013.5.

Bhatnagar, V.K., Kashyap, R., Zaidi, S.S.A., Kulkarni, P.K., Saiyed, H.N., 2004. Levels of DDT, HCH, and HCB residues in human blood in Ahmedabad, India. *Bull. Environ. Contam. Toxicol.* 72, 261–265. https://doi.org/10.1007/s00128-003-9049-9.

Board, P.G., 1981. Biochemical genetics of glutathione-S-transferase in man. *Am. J. Hum. Genet.* 33, 36–43.

Booth, J., Boyland, E., Sims, P., 1961. An enzyme from rat liver catalysing conjugations with glutathione. *Biochem. J.* 79, 516–524. https://doi.org/10.1042/bj0790516.

Botto, L.D., Khoury, M.J., 2001. Commentary: facing the challenge of gene-environment interaction: the two-by-four table and beyond. *Am. J. Epidemiol.* 153, 1016–1020. https://doi.org/10.1093/aje/153.10.1016.

Carreño, J., Rivas, A., Granada, A., Jose Lopez-Espinosa, M., Mariscal, M., Olea, N., Olea-Serrano, F., 2007. Exposure of young men to organochlorine pesticides in Southern Spain. *Environ. Res.* 103, 55–61. https://doi.org/10.1016/j.envres.2006.06.007.

Chandrajith, R., Nanayakkara, S., Itai, K., etal., 2011. Chronic kidney diseases of uncertain etiology (CKDue) in Sri Lanka: Geographic distribution and environmental implications. *Environ. Geochem. Health.* 33, 267–278. https://doi.org/10.1007/s10653-010-9339-1.

Chen, J., Cheng, M., Yi, L., Jiang, C., 2011. Relationship between CYP1A1 genetic polymorphisms and renal cancer in China. *Asian Pac. J. Cancer Prev. APJCP.* 12, 2163–2166.

Codru, N., Schymura, M.J., Negoita, S., Akwesasne Task Force on Environment, Rej, R., Carpenter, D.O., 2007. Diabetes in relation to serum levels of polychlorinated biphenyls and chlorinated pesticides in adult Native Americans. *Environ. Health Perspect.* 115, 1442–1447. https://doi.org/10.1289/ehp.10315.

Crofts, F., Cosma, G.N., Currie, D., Taioli, E., Toniolo, P., Garte, S.J., 1993. A novel CYP1A1 gene polymorphism in African-Americans. *Carcinogenesis.* 14, 1729–1731. https://doi.org/10.1093/carcin/14.9.1729.

El Minshawy, O., Ghabrah, T., El Bassuoni, E., 2014. End-stage renal disease in Tabuk Area, Saudi Arabia: An epidemiological study. *Saudi J. Kidney Dis. Transplant. Off. Publ. Saudi Cent. Organ Transplant. Saudi Arab.* 25, 192–195. https://doi.org/10.4103/1319-2442.124574.

Ercal, N., Gurer-Orhan, H., Aykin-Burns, N., 2001. Toxic metals and oxidative stress part I: mechanisms involved in metal-induced oxidative damage. *Curr. Top. Med. Chem.* 1, 529–539. https://doi.org/10.2174/1568026013394831.

Uboh, F.E., 2011. Endosulfan-induces renal toxicity independent of the route of exposure in rats [WWW Document]. https://doi.org/10.3923/ajbmb.2011.359.367

Ganguli, A., 2016. Uddanam nephropathy/regional nephropathy in India: Preliminary findings and a plea for further research. *Am. J. Kidney Dis. Off. J. Natl. Kidney Found.* 68, 344–348. https://doi.org/10.1053/j.ajkd.2016.04.012.

Garte, S., Gaspari, L., Alexandrie, A.K., et al., 2001. Metabolic gene polymorphism frequencies in control populations. *Cancer Epidemiol. Biomark. Prev. Publ. Am. Assoc. Cancer Res. Cosponsored Am. Soc. Prev. Oncol.* 10, 1239–1248.

Goncharov, A., Haase, R.F., Santiago-Rivera, A., Morse, G., Akwesasne Task Force on the Environment, McCaffrey, R.J., Rej, R., Carpenter, D.O., 2008. High serum PCBs are associated with elevation of serum lipids and cardiovascular disease in a Native American population. *Environ. Res.* 106, 226–239. https://doi.org/10.1016/j.envres.2007.10.006.

Gupta, P.K., 2004. Pesticide exposure – Indian scene. *Toxicology.* 198, 83–90. https://doi.org/10.1016/j.tox.2004.01.021.

Ha, M.-H., Lee, D.-H., Jacobs, D.R., 2007. Association between serum concentrations of persistent organic pollutants and self-reported cardiovascular disease prevalence: Results from the national health and nutrition examination survey, 1999–2002. *Environ. Health Perspect.* 115, 1204–1209. https://doi.org/10.1289/ehp.10184.

Han, E.H., Kim, J.Y., Kim, H.-K., Hwang, Y.P., Jeong, H.G., 2008. o,p'-DDT induces cyclooxygenase-2 gene expression in murine macrophages: Role of AP-1 and CRE promoter elements and PI3-kinase/Akt/MAPK signaling pathways. *Toxicol. Appl. Pharmacol.* 233, 333–342. https://doi.org/10.1016/j.taap.2008.09.003.

Hu, S.I., Klein, M., Carozza, M., Rediske, J., Peppard, J., Qi, J.S., 1999. Identification of a splice variant of neutrophil collagenase (MMP-8). *FEBS Lett.* 443, 8–10. https://doi.org/10.1016/s0014-5793(98)01654-8.

Hunter, D.J., 2005. Gene-environment interactions in human diseases. *Nat. Rev. Genet.* 6, 287–298. https://doi.org/10.1038/nrg1578.

Jain, M., Kumar, S., Rastogi, N., et al., 2006. GSTT1, GSTM1 and GSTP1 genetic polymorphisms and interaction with tobacco, alcohol and occupational exposure in esophageal cancer patients from North India. *Cancer Lett.* 242, 60–67. https://doi.org/10.1016/j.canlet.2005.10.034.

Karami, S., Boffetta, P., Rothman, N., et al., 2008. Renal cell carcinoma, occupational pesticide exposure and modification by glutathione S-transferase polymorphisms. *Carcinogenesis.* 29, 1567–1571. https://doi.org/10.1093/carcin/bgn153.

Khan, D.S., Kumari, D., 2012. Analytical study of histopathological changes induced by of endosulfan in kidney of albino rats. *Asian J. Pharm. Clin. Res.* 5, 2.

Kiryluk, K., Li, Y., Scolari, F., Sanna-Cherchi, et al., 2014. Discovery of new risk loci for IgA nephropathy implicates genes involved in immunity against intestinal pathogens. *Nat. Genet.* 46, 1187–1196. https://doi.org/10.1038/ng.3118.

Kiyohara, C., Hirohata, T., Inutsuka, S., 1996. The relationship between aryl hydrocarbon hydroxylase and polymorphisms of the CYP1A1 gene. Jpn. *J. Cancer Res. Gann.* 87, 18–24. https://doi.org/10.1111/j.1349-7006.1996.tb00194.

Kumar, V., Yadav, C.S., Singh, S., et al., 2010. CYP 1A1 polymorphism and organochlorine pesticides levels in the etiology of prostate cancer. *Chemosphere.* 81, 464–468. https://doi.org/10.1016/j.chemosphere.2010.07.067.

Lallas, P.L., 2001. The Stockholm convention on persistent organic pollutants. *Am. J. Int. Law.* 95, 692–708. https://doi.org/10.2307/2668517.

Li, Y.F., 1999. Global technical hexachlorocyclohexane usage and its contamination consequences in the environment: From 1948 to 1997. *Sci. Total Environ.* 232, 121–158. https://doi.org/10.1016/S0048-9697(99)00114-X.

Li, Y.F., Millikan, R.C., Bell, D.A., Cui, L., Tse, C.-K.J., Newman, B., Conway, K., 2004. Cigarette smoking, cytochrome P4501A1 polymorphisms, and breast cancer among African-American and white women. *Breast Cancer Res. BCR.* 6, R460–R473. https://doi.org/10.1186/bcr814.

LonguemauxLonguemaux, S., Deloménie, C., Gallou, C., et al., 1999. Candidate genetic modifiers of individual susceptibility to renal cell carcinoma: A study of polymorphic human xenobiotic-metabolizing enzymes. *Cancer Res.* 59, 2903–2908.

Makalowska, I., Lin, C.-F., Makalowski, W., 2005. Overlapping genes in vertebrate genomes. *Comput. Biol. Chem.* 29, 1–12. https://doi.org/10.1016/j.compbiolchem.2004.12.006.

Mannervik, B., Alin, P., Guthenberg, C., et al., 1985. Identification of three classes of cytosolic glutathione transferase common to several mammalian species: Correlation between structural data and enzymatic properties. *Proc. Natl. Acad. Sci. U. S. A.* 82, 7202–7206. https://doi.org/10.1073/pnas.82.21.7202.

Masson, L.F., Sharp, L., Cotton, S.C., Little, J., 2005. Cytochrome P-450 1A1 gene polymorphisms and risk of breast cancer: A HuGE review. *Am. J. Epidemiol.* 161, 901–915. https://doi.org/10.1093/aje/kwi121.

Mittal, R.D., Srivastava, D.L., 2007. Cytochrome P4501A1 and microsomal epoxide hydrolase gene polymorphisms: Gene-environment interaction and risk of prostate cancer. *DNA Cell Biol.* 26, 791–798. https://doi.org/10.1089/dna.2007.0630.

Mustafa, M., Pathak, R., Tripathi, A.K., Ahmed, R.S., Guleria, K., Banerjee, B.D., 2010. Maternal and cord blood levels of aldrin and dieldrin in Delhi population. *Environ. Monit. Assess.* 171, 633–638. https://doi.org/10.1007/s10661-010-1307-9.

Nanayakkara, S., Senevirathna, S.T.M.L.D., Abeysekera, T., et al., 2014. An integrative study of the genetic, social and environmental determinants of chronic kidney disease characterized by tubulointerstitial damages in the North Central Region of Sri Lanka. *J. Occup. Health.* 56, 28–38. https://doi.org/10.1539/joh.13-0172-oa.

Ngowi, A.V.F., Mbise, T.J., Ijani, A.S.M., London, L., Ajayi, O.C., 2007. Pesticides use by smallholder farmers in vegetable production in Northern Tanzania. *Crop Prot. Guildf. Surrey.* 26, 1617–1624. https://doi.org/10.1016/j.cropro.2007.01.008.

Ottman, R., 1990. An epidemiologic approach to gene-environment interaction. *Genet. Epidemiol.* 7, 177–185. https://doi.org/10.1002/gepi.1370070302.

Papadopoulou-Mourkidou, E., Milothridou, A., 1990. Residues and persistence of endosulfan in dry tobacco leaves and cigarettes. *Bull. Environ. Contam. Toxicol.* 44, 394–400. https://doi.org/10.1007/BF01701221.

Pathak, R., Suke, S.G., Ahmed, R.S., et al., 2008. Endosulfan and other organochlorine pesticide residues in maternal and cord blood in North Indian population. *Bull. Environ. Contam. Toxicol.* 81, 216–219. https://doi.org/10.1007/s00128-008-9459-9.

Peter, H., Deutschmann, S., Reichel, C., Hallier, E., 1989. Metabolism of methyl chloride by human erythrocytes. *Arch. Toxicol.* 63, 351–355. https://doi.org/10.1007/BF00303122.

Raimondi, S., Paracchini, V., Autrup, H., et al., 2006. Meta- and pooled analysis of GSTT1 and lung cancer: a HuGE-GSEC review. *Am. J. Epidemiol.* 164, 1027–1042. https://doi.org/10.1093/aje/kwj321.

Rajapurkar, M.M., John, G.T., Kirpalani, A.L., et al., 2012. What do we know about chronic kidney disease in India: First report of the Indian CKD registry. *BMC Nephrol.* 13, 10. https://doi.org/10.1186/1471-2369-13-10.

Ramirez-Rubio, O., McClean, M.D., Amador, J.J., Brooks, D.R., 2013. An epidemic of chronic kidney disease in Central America: an overview. *Postgrad. Med. J.* 89, 123–125. https://doi.org/10.1136/postgradmedj-2012-201141rep.

Rutten, G.A., Schoots, A.C., Vanholder, R., DeSmet, R., Ringoir, S.M., Cramers, C.A., 1988. Hexachlorobenzene and 1,1-di(4-chlorophenyl)-2,2-dichloroethene in serum of uremic patients and healthy persons: determination by capillary gas chromatography and electron capture detection. *Nephron.* 48, 217–221. https://doi.org/10.1159/000184915.

Ryan, M.J., Johnson, G., Kirk, J., Fuerstenberg, S.M., Zager, R.A., Torok-Storb, B., 1994. HK-2: an immortalized proximal tubule epithelial cell line from normal adult human kidney. *Kidney Int.* 45, 48–57. https://doi.org/10.1038/ki.1994.6.

Seidegård, J., Pero, R.W., Miller, D.G., Beattie, E.J., 1986. A glutathione transferase in human leukocytes as a marker for the susceptibility to lung cancer. *Carcinogenesis.* 7, 751–753. https://doi.org/10.1093/carcin/7.5.751.

Sergentanis, T.N., Economopoulos, K.P., 2010. Four polymorphisms in cytochrome P450 1A1 (CYP1A1) gene and breast cancer risk: A meta-analysis. *Breast Cancer Res. Treat.* 122, 459–469. https://doi.org/10.1007/s10549-009-0694-5.

Siddarth, M., Datta, S.K., Ahmed, R.S., Banerjee, B.D., Kalra, O.P., Tripathi, A.K., 2013. Association of CYP1A1 gene polymorphism with chronic kidney disease: a case control study. *Environ. Toxicol. Pharmacol.* 36, 164–170. https://doi.org/10.1016/j.etap.2013.03.008.

Siddharth, M., Datta, S.K., Bansal, S., Mustafa, M., Banerjee, B.D., Kalra, O.P., Tripathi, A.K., 2012. Study on organochlorine pesticide levels in chronic kidney disease patients: Association with estimated glomerular filtration rate and oxidative stress. *J. Biochem. Mol. Toxicol.* 26, 241–247. https://doi.org/10.1002/jbt.21416.

Siddarth, M., Datta, S.K., Mustafa, M., Ahmed, R.S., Banerjee, B.D., Kalra, O.P., Tripathi, A.K., 2014. Increased level of organochlorine pesticides in chronic kidney disease patients of unknown etiology: Role of GSTM1/GSTT1 polymorphism. *Chemosphere.* 96, 174–179. https://doi.org/10.1016/j.chemosphere.2013.10.029.

Silva, M.H., Gammon, D., 2009. An assessment of the developmental, reproductive, and neurotoxicity of endosulfan. *Birth Defects Res. B. Dev. Reprod. Toxicol.* 86, 1–28. https://doi.org/10.1002/bdrb.20183.

Singh, A.K., Farag, Y.M.K., Mittal, B.V., et al., 2013. Epidemiology and risk factors of chronic kidney disease in India – results from the SEEK (Screening and Early Evaluation of Kidney Disease) study. *BMC Nephrol.* 14, 114. https://doi.org/10.1186/1471-2369-14-114.

Sobti, R.C., Sharma, S., Joshi, A., Jindal, S.K., Janmeja, A., 2004. Genetic polymorphism of the CYP1A1, CYP2E1, GSTM1 and GSTT1 genes and lung cancer susceptibility in a north Indian population. *Mol. Cell. Biochem.* 266, 1–9. https://doi.org/10.1023/b:mcbi.0000049127.33458.87.

Sprenger, R., Schlagenhaufer, R., Kerb, R., Bruhn, C., Brockmöller, J., Roots, I., Brinkmann, U., 2000. Characterization of the glutathione S-transferase GSTT1 deletion: Discrimination of all genotypes by polymerase chain reaction indicates a trimodular genotype-phenotype correlation. *Pharmacogenetics.* 10, 557–565. https://doi.org/10.1097/00008571-200008000-00009.

Srivastava, D.S., Mandhani, A., Mittal, R.D., 2008. Genetic polymorphisms of cytochrome P450 CYP1A1 (*2A) and microsomal epoxide hydrolase gene, interactions with tobacco-users, and susceptibility to bladder cancer: A study from North India. *Arch. Toxicol.* 82, 633–639. https://doi.org/10.1007/s00204-007-0276-4.

Stanifer, J.W., Muiru, A., Jafar, T.H., Patel, U.D., 2016. Chronic kidney disease in low- and middle-income countries. *Nephrol. Dial. Transplant. Off. Publ. Eur. Dial. Transpl. Assoc. – Eur. Ren. Assoc.* 31, 868–874. https://doi.org/10.1093/ndt/gfv466.

Subramaniam, K., Solomon, J., 2006. Organochlorine pesticides BHC and DDE in human blood in and around Madurai, India. *Indian J. Clin. Biochem. IJCB.* 21, 169–172. https://doi.org/10.1007/BF02912936.

Suryanarayana, V., Deenadayal, M., Singh, L., 2004. Association of CYP1A1 gene polymorphism with recurrent pregnancy loss in the South Indian population. *Hum. Reprod. Oxf. Engl.* 19, 2648–2652. https://doi.org/10.1093/humrep/deh463.

Taioli, E., Ford, J., Trachman, J., Li, Y., Demopoulos, R., Garte, S., 1998. Lung cancer risk and CYP1A1 genotype in African Americans. *Carcinogenesis*. 19, 813–817. https://doi.org/10.1093/carcin/19.5.813.

Thomson, R., Genovese, G., Canon, C., et al., 2014. Evolution of the primate trypan lytic factor APOL1. *Proc. Natl. Acad. Sci. U. S. A.* 111, E2130–E2139. https://doi.org/10.1073/pnas.1400699111.

Waliszewski, S.M., Carvajal, O., Infanzon, R.M., Trujillo, P., Hart, M.M., 2004. Copartition ratios of persistent organochlorine pesticides between human adipose tissue and blood serum lipids. *Bull. Environ. Contam. Toxicol.* 73, 732–738. https://doi.org/10.1007/s00128-004-0487-9.

Wang, X.L., Greco, M., Sim, A.S., Duarte, N., Wang, J., Wilcken, D.E.L., 2002. Effect of CYP1A1 MspI polymorphism on cigarette smoking related coronary artery disease and diabetes. *Atherosclerosis*. 162, 391–397. https://doi.org/10.1016/s0021-9150(01)00723-7.

Wesseling, C., Crowe, J., Hogstedt, C., Jakobsson, K., Lucas, R., Wegman, D.H., 2013. The epidemic of chronic kidney disease of unknown etiology in Mesoamerica: A call for interdisciplinary research and action. *Am. J. Public Health*. 103, 1927–1930. https://doi.org/10.2105/AJPH.2013.301594.

Wijetunge, S., Ratnatunga, N.V.I., Abeysekera, T.D.J., Wazil, A.W.M., Selvarajah, M., 2015. Endemic chronic kidney disease of unknown etiology in Sri Lanka: Correlation of pathology with clinical stages. *Indian J. Nephrol.* 25, 274–280. https://doi.org/10.4103/0971-4065.145095.

Wormhoudt, L.W., Commandeur, J.N., Vermeulen, N.P., 1999. Genetic polymorphisms of human N-acetyltransferase, cytochrome P450, glutathione-S-transferase, and epoxide hydrolase enzymes: relevance to xenobiotic metabolism and toxicity. *Crit. Rev. Toxicol.* 29, 59–124. https://doi.org/10.1080/10408449991349186.

Wright, C.M., Larsen, J.E., Colosimo, M.L., et al., 2010. Genetic association study of CYP1A1 polymorphisms identifies risk haplotypes in nonsmall cell lung cancer. *Eur. Respir. J.* 35, 152–159. https://doi.org/10.1183/09031936.00120808.

Xi, Q., Wang, Z., Zaromytidou, A.-I., et al., 2011. A poised chromatin platform for TGF-β access to master regulators. *Cell*. 147, 1511–1524. https://doi.org/10.1016/j.cell.2011.11.032.

19

Environmental Biotechnology: Public Perception

Harit Jha, Vivekananda Mandal, Kamal Kumar Sen, Roshni Tandey, Kavi Bhushan Singh Chouhan and Rajendra Mehta

CONTENTS

19.1 Introduction

Biotechnology provides tools and has the potential to achieve sustainable changes in society. However, it also presents many socioeconomic, ethical, environmental issues that may impact human society. The common people receive biotechnology products with varying degrees of caution. The therapeutic components and drugs are better accepted compared to any agricultural products. The promise of benefits is not sufficient to convince people to accept new processes and products. The direction and funding of new projects are dependent on the acceptability, opinion and perception of the public. It is thus pertinent to disseminate the information and initiate a building exercise before arriving at a decision. The information on the benefits and risks of a product or process should be made accessible.

To achieve this goal, several mechanisms exist, namely, a biosafety regulatory system, an appellate authority and a judicial system, for redressal of grievances, scientists for proper dissemination of accurate information to the public and trained personnel with scientific temperament in health, education, animal husbandry and agriculture departments, along with media organizations.

However, several challenges need to be identified and overcome to achieve success in improving public perception; the major challenges include various levels of understanding owing to a divergence in literacy, a lack of awareness about biotechnology, the limitation of communication material in local languages, improper science reporting, inefficient coordination between agencies and insufficient interaction among different stakeholders, regulators and policymakers.

There is a need to create awareness in public about opportunities for biotechnology development, the benefits that can be derived from the processes and products. The safety concerns should be also addressed by improving transparency and proper sharing of information and data on quality and toxicity testing.

There are implications of public understanding of biotechnology for policy in technology assessment, education

DOI: 10.1201/9781003186304-19

and information campaigns and where and what decisions are taken. Scientists, industry and government have special responsibilities in the ethical application of biotechnology, consistent with the goals of society and together with the media in providing understandable information to the public consistent with the goals of society.

19.1.1 Biotechnology in India

India has a large pool of engineers and scientists, with a huge network of institutions with economic manufacturing solutions. Publicly funded National Research Laboratories with thousands of scientists, a large number of private companies and thousands of institutes across the country offer diplomas and degrees in biological sciences, biotechnology and bioinformatics.

In 1982, National Biotechnology Board (NBTB) was constituted by the Indian government to identify priority areas and evolve a long-term perspective for biotechnology in India. It was also responsible for creating infrastructure and supporting indigenous development of manpower.

A separate Department of Biotechnology (DBT) was set up in February 1986. The DBT was a Scientific Advisory Committee (SAC). The members of the SAC were from different scientific agencies to improve coordination.

The DBT started many institutes for research in prominent areas: the Biotechnology National Institute of Immunology in 1981; National Facility for Animal Tissue and Cell Culture, Pune in 1986, renamed as National Centre for Cell Science; the Centre for DNA Fingerprinting & Diagnostics; the National Brain Research Centre (NBRC); and the National Institute for Plant Genome Research (NIPGR). Some of the other institutes created and supported by the DBT include the Institute of Bioresources and Sustainable Development and the Institute for Stem Cell Biology and Regenerative Medicine (INstem), the Institute of Life Sciences, the Translational Health Science and Technology Institute (THISTI), the National Agri-Food Biotechnology Institute (NABI) at Mohali and National Institute of Biomedical Genomics (NIBMG) at Kalyani in West Bengal.

It also started Biotech Consortium India Limited (BCIL) in collaboration with various Industries to create a link among government, industries and the end-user public. The DBT, through its different organizations and institutes and the BCIL, is involved in the propagation of information on biotechnology through its annual reports, workshops, seminars and more. The DBT also funds courses on biotechnology, including an MSc in environmental biotechnology at more than 35 institutes across India to train people in frontier areas of biotechnology.

The National Science and Technology Policy of the Government and the Vision Statement on Biotechnology issued by the DBT have provided directions to various agencies and institutions to help development in the areas of biotechnology and life sciences. This created notable changes in terms of human resource generation, support for research and development (R&D) and infrastructure development.

This will require larger investments and effective functioning of the innovation pathway. Capturing new opportunities and the potential economic, environmental, health and social benefits will challenge government policy, public awareness and educational, scientific, technological, legal and institutional framework.

The National Biotechnology Development Strategy takes stock of what has been accomplished and provides a framework for the future, within which strategies and specific actions to promote biotechnology can be taken. The policy framework is a result of wide consultation with stakeholders—scientists, educationists, regulators, representatives of society and others and reflects their consensus. It focuses on crosscutting issues, such as human resource development, academic and industry interface, infrastructure development, lab and manufacturing, promotion of industry and trade, biotechnology parks and incubators, regulatory mechanisms, public education and awareness building. This policy also aims to chalk out the path of progress in sectors, such as agriculture and food biotechnology, industrial biotechnology, therapeutic and medical biotechnology, regenerative and genomic medicine, diagnostic biotechnology, bioengineering, nano-biotechnology, bioinformatics and information technology (IT)–enabled biotechnology, clinical biotechnology, manufacturing and bioprocessing, research services, bioresources, environment and intellectual property rights (IPR) and patent law.

19.2 Environmental Biotechnology

Environmental issues, including afforestation and reforestation, water pollution, soil pollution, air pollution, restoration of contaminated sites and treatment of industrial effluents, are important and affect the public directly. Environmental biotechnology can provide acceptable solutions for monitoring, assessing and restoring polluted water and land bodies. It can also help in reducing, recycling and degrading pollution. It can be a major thrust area to provide green and clean technology for the future. Strict laws and enforcement agencies have to play an important role in creating awareness and adaption of cleaner technologies profitable for industries.

19.2.1 Objectives of Environmental Biotechnology

The objectives of environmental biotechnology are to provide clean alternatives in a cost-effective manner, risk assessment and quality monitoring, bioremediation of wastes, conversion of harmful and toxic recalcitrant chemicals into nontoxic by-products, eco-restoration of degraded habitats, value-added products from biomass, control of biological invasion through biotechnological interventions, greener process technologies and effective *ex situ* conservation strategies.

Environmental biotechnology is a potent process for treating xenobiotics, pollutants and hazardous wastes created due to anthropogenic activities. The methods are economical and adaptable; however, they are slow and susceptible to environmental factors. The technologies invented and standardized in the controlled environment of laboratories need to be vigorously tested and modified for scale-up and *in situ* operations. The successful model may create a positive perception and with broader acceptability in the society. The public and private institutions and companies may accept and use technologies for the remediation of pollutants and may create a demand-based market for such processes and products. R&D is thus essential with continuous improvement and upgradation. Environmental biotechnology is green technology and may contribute significantly to restoring the environment for humans and the flora and fauna of a habitat.

19.2.2 Applications of Environmental Biotechnology

The applications of environmental biotechnology cover a wide range, all of which lead to the protection and remediation of environmental components. Some of the applications are shown in Figure 19.1.

Environmental biotechnology is now emerging as a composite but increasingly integrated discipline, which relies on a detailed understanding of microbial biodiversity in both its natural and its anthropogenic context to help us engineer large-scale systems for environmental quality and, more generally, introduce innovation and prosperity in human society in all the areas affected by microbial life, from the global elemental cycles influencing our climate to the quality and abundance of our food to the health of our gut.

Although the uptake of these technologies is routine in industrialized countries, many developing countries, however, are lagging in adopting some of these modern biotechnologies for the sustainable growth of industry. Furthermore, not all countries are faced with identical challenges and priorities, and each country is at a distinct stage of development, with a wide difference in technological capabilities. Some of the most debated issues concerning biotechnology and biotech products include the issue of ethics, IPR and the implications of patents on the cost of technology in developing countries.

India is reorganized as a mega-biodiversity country, and biotechnology offers opportunities to convert our biological resources into economic wealth and employment opportunities. Innovative products and services that draw on renewable resources bring greater efficiency into industrial processes, check environmental degradation and deliver a more biobased economy.

19.2.2.1 Remediation

Environmental biotechnology is no longer a new field; composting and wastewater treatments are familiar examples of old environmental biotechnologies. However, recent studies in molecular biology and ecology offer opportunities for more efficient biological processes. Notable accomplishments of these studies include the cleanup of polluted water and land areas.

Khanna (1996) proposed that the scope of environmental biotechnology be broadened to address issues such as environmental monitoring, restoration of environmental quality, resource/residue/waste recovery/utilization/treatment, the substitution of the nonrenewable resource base with renewable resources and the prospective applications in these subareas of environmental biotechnology and to document case studies on environmental monitoring (enteric viruses), restoration of environmental quality (oil spill remediation), resource recovery (hydrocarbon recovery from oily sludges, biosurfactants from distillery spent wash, desulfurization of coal and sour gases), and substitution of non-renewable resources with renewables (conversion of lignocellulosics into value-added chemicals).

Some of the major environmental problems and the consequent solutions provided by the biotechnological interventions are discussed later in the chapter.

Soil is a major resource required for agriculture, food, water, economy and environment. It provides support for the growth of flora and fauna. The microbial flora is an indispensable source of humus and provides texture and quality to soil. The indiscriminate exploitation of soil, for the extraction of minerals, ores, groundwater and agriculture, has adversely affected its quality. Industrialization and the dumping of solid waste have wasted land and soil resources, thereby causing a threat to the soil microflora, composition and consequently impacting agriculture and groundwater resources.

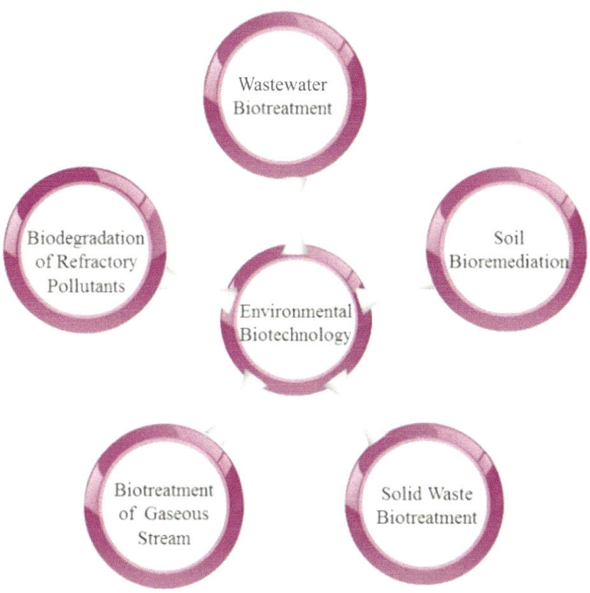

FIGURE 19.1 Applications of environmental biotechnology.

FIGURE 19.2 Soil pollution.

FIGURE 19.3 Water pollution.

It is pertinent to find solutions to soil and land pollution. Microbial bioremediation and phytoremediation (Kensa 2011) may provide technologies for *in situ* and long-term remediation of polluted soils. Phytoremediation is important technology; it involves applying indigenous or exotic plant species to remediate polluted soil. The resistant or tolerant plants during their growth and metabolism tend to metabolize and detoxify the pollutants. The biodegradation renders toxic components to simple metabolites, which can be further reused by soil microflora. Phytoremediation is an economical and easy method that does not require continuous monitoring and interventions.

19.2.2.2 Phytoremediation

Phytoremediation is a collection of methods that utilize plant and their metabolism for cleaning of contaminated sites (Peuke & Rennenberg 2005). The term *phytoremediation* was introduced in 1991, although the practice is much older. Phytoremediation applications may involve bioaccumulation, bioleaching, biovolatilization, biodegradation by an enzymatic system of plant and segregation and sequestration of contaminants by root hair and root network (Ghosh & Singh 2005; Hooda 2007). Phytoremediation also helps reduce soil erosion and improve groundwater levels and quality (Kokyo et al. 2014). The mechanism varies from plant to plant; however, indigenous plants or genetically modified plants may be used as needed. Phytoremediation is economic and is useful for the sites with relatively high level of the pollutants; it also has the advantage of low initial cost and maintenance.

Phytoremediation includes six major plant-oriented techniques to reduce the hazardous waste from the sites, namely, phytoextraction, phytostabilization, phytohydraulics, rhizofiltration, phytovolatilization and phytodegradation.

1. Phytoextraction: The absorption or uptake and translocation of the heavy metals by the plant's root into the subsurface of the soil is called phytoextraction. It can be harvested after maturity and the heavy metal recycled as an ash from the burning of woods.

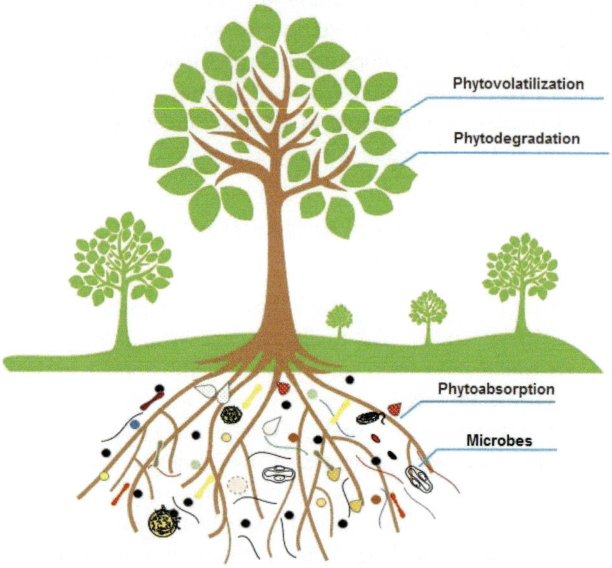

FIGURE 19.4 Phytoremediation.

2. Phytostabilization: *Phytostabilization* is defined as the use of certain plants to immobilize the heavy metals in soil and water through the absorption and accumulation in plant tissues.
3. Phytohydraulics: The trees whereas present deep roots come into contact with groundwater and absorb the contamination, which is dissolved in water through the root tissue.
4. Rhizofiltration: Using plants to clean up different aquatic environments comes under the rhizofiltration.
5. Photovolatilization: Photovolatilization is the transpiration process of the contaminant. The plant uptake the contaminant elements and release to the atmosphere after modification of their form. It occurs mostly on growing trees and shrubs.
6. Phytodegradation: The plant tissues biotransform the contaminants with different metabolic activities called phytodegradation. It depends on the plant type and its occurrence, namely, leaves, stems or roots.

Some of the plants used for phytoremediation include the following:

1. Indian mustard (*Brassicajuncea*) Family: Brassicaeae

   Phytoremediation: Pb, Cd, As, etc.
2. Hybrid poplar (Paras pipal) *Populus deltoids* Family: Salicaceae

   Phytoremediation: As and Cd contamination on the cultivated lands.
3. Sunflower (*Helianthusannuus*) Family: Asteraceae

   Phytoremediation: Polyaromatic hydrocarbons (PAHs) level for root zone of soils.
4. Water milfoil (*Myriophyllum spicatum)* Family: Haloragaeae

Phytoremediation: Contaminant from the water, namely, Pb, Cu, Ni, Cd, Zn, etc.

Mulberry (*Morus alba*) Family: Moraceae

Phytoremediation: Organic and inorganic contaminant such as total petroleum hydrocarbon (TPH), PAHs, pesticides, chlorinated solvents, polychlorinated biphenyls (PCBs), etc.

5. Spearmint (*Menthaspicata*) Family: Lamiaceae

   Phytoremediation: As, Cd, Cr, Cu, and Hs due to their phytostabilization mechanism.

6. Indian grass (*Sorghastrumnutans*) Family: Poaceae

   Phytoremediation: It is able to detoxify some agrochemical contaminants, such as pesticides and herbicides mostly related to metolachlor and atrazine.

7. Alfalfa (*Medicago sativa*) Family: Fabaceae

   Phytoremediation: Se, Hg and As contaminants from the soil and water. It is used in the phytovolatilization process to reduce the heavy metals.

8. White Willow (*Salix* spp.) Family: Salicaceae

   Phytoremediation: They are used for treatment of soil contaminated with Cd, Ni and Pb and even the mixed heavy metals released on industrial sites.

9. Pigeon pea (*Cajanus cajan*) Family: Fabaceae

   Phytoremediation: *C. Cajan* has the potential for the absorption of Ar from soil as well as water.

10. Alpine pennygrass (*Thlaspi caerulescens*) Family: Brassicaceae

    Phytoremediation: It holds good phytoremediation properties to absorb Cd and Zn metal in soil.

Long-term phytoremediation-based treatment can be designed into future site landscaping plans, for example, tree borders used for shading and visual screening can also provide ongoing groundwater remediation and more. Such vegetation can often create valuable ecological niches, particularly in urban industrial areas. The plant species currently being developed for phytoremediation seem capable of effective bioaccumulation of the targeted contaminant, but efficiency might be improved through the use of transgenic (genetically engineered) plants.

19.3 Technologies Developed by Indian Institutes

The National Environmental Engineering Institute (NEERI) and its various centers funded by the Council of Scientific and Industrial Research has succeeded in monitoring, assessing and restoring environmental quality. They routinely monitor various environmental parameters and provide solutions through physico-chemical or environmental Biotechnology intervention.

- Electrolytic defluoridation (EDF) technique
- Hand pump–attachable iron removal (IR) plant
- High-rate transpiration system (HRTS)
- HYDROPLUME—a high-rate secondary clarifier for wastewater treatment
- Microbial culture—for improved efficiency of existing wastewater treatment systems
- NEERDHUR
- NEERI–ZAR—portable instant water filter
- Passive air-rejuvenating system (PARS)
- Phytorid wastewater treatment technology
- Automatic mechanical urinal-toilet flusher "NEERFLUSH" for Swachh Bharat Mission (Table 19.1)

19.3.1 Oilzapper: Microbes to Treat Oil-Contaminated Soil and Sludge

The Oilzapper is a consortium of immobilized bacterial strains along with a carrier material. The Oilzapper is useful in the bioremediation of oil-contaminated soil and sludge and converting them to carbon dioxide and water

19.3.2 TEAM—for Turning Organic Waste into Energy

TERI's Enhanced Acidification and Methanation (TEAM) system is used for formation of biogas by biodegradation of organic waste.

19.3.3 Ceramic Membranes for Treating Industrial Waste

Fly ash–based ceramic membrane filters in submerged membrane bioreactors (MBRs) have been used for sewage treatment. These filters are relatively economical and reusable.

19.3.4 Mycorrhiza-Based Biofertilizer: A Natural Alternative to Chemical Fertilizers

Mycorrhiza-based fertilizer developed by TERI without contamination provides several benefits to plants. It is available in formulations of tablets, granules and powder forms.

19.3.5 Bollcure: A Biopesticide Based on Plant Extract

TERI has developed a unique plant-extract-based biopesticide 'Bollcure'. It is very useful against a variety of pests cotton bollworm (*Helicoverpa armigera*) thrips, jassids and others.

19.3.6 Smokeless, Clean-Combustion Cookstoves

A clean-combustion cookstove with reduced emissions and a better thermal capacity has been designed and subsequently distributed for community cooking purposes (http://www.teriin.org/products).

19.3.7 Private Companies Involved in Environmental Biotechnology Processes and Products

Besides government-funded institutions, a large number of private/public-funded companies are also involved in applications of biotechnology tools in environmental management. Turnkey solutions for pollution and effluent treatment, wastewater treatment plants, wildlife conservation, bioenergy crop research and development, pollution control equipment design, environmental assessment, consultancy and testing services and analysis are now available commercially from companies such as the following:

- Analytik Jena AG
- BÜCHI Labortechnik AG
- Innovative Environmental Technologies Pvt. Ltd. (IETL)
- Eco Services India Pvt. Ltd.
- Xcelrislabs Ltd.
- Five Elements Environment Ventures Pvt. Ltd.
- Colifast AS
- Advanced Biofuel Center
- Mars Bioanalytical Pvt. LTd
- Organica Biotech Pvt. Ltd.
- Kanomax USA, Inc.
- Spectra Agritec
- ENVIROCENTRE
- Flottweg SE
- Shivam Water Treaters Pvt. Ltd.
- M/s Green Solutions (India) Pvt Limited
- Meco Technologies Private Limited
- Sharp Engineering

Private companies play an important role in creating demand and marketing an idea. The competition between various companies and public-sector organizations is helpful in creating newer technologies.

The technologies in environment biotechnology are thus available and being upgraded, improvised and innovated by the trained personnel in the infrastructure available at central- and state-funded universities, engineering colleges, research institutes and others. However, the acceptance and upscaling of the technology depend on perception (Figure 19.5).

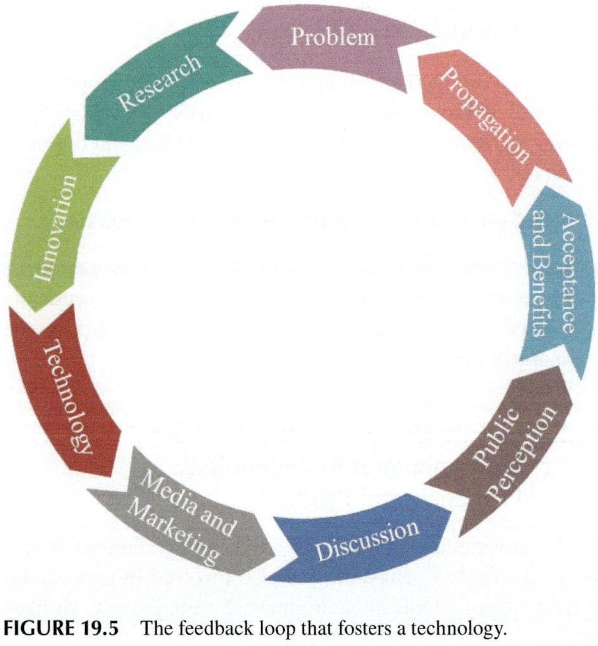

FIGURE 19.5 The feedback loop that fosters a technology.

19.4 Public Perception of Environmental Biotechnology

Public acceptance of biotechnology is also crucial for the advancement of its application in industry. The development of biotechnology into products and resources for human use needs to be researched alongside the ethical, environmental, economic, legal and social implications in order to avoid, for example, the negative publicity that characterized the genetically modified (GM) crops. Discussions on the potential risk and benefits of these technologies should include plans to support research in this area.

19.4.1 Surveys for Understanding Public Perception

The perception of the public on biotechnology can be understood through surveys. Some surveys include online surveys conducted by the Institute of Biological Science University of Tsukuba (Japan, 2000), Australia, Hong Kong, India, Israel, Japan, New Zealand, the Philippines, Russia, Singapore and Thailand. Another survey was conducted by the Asian Food Information Center (AFIC) in urban areas (capital cities) in China, India, the Philippines, Japan and South Korea (2008). The surveys were on attitudes acceptance of genetic engineering and ethics of biotechnology and perceptions of GM food biotechnology.

In the last two decades, public understanding of biotechnology has improved due to an increase in literacy, scientific temperament and input by print and other media (Bauer 2005). The perception of environmental biotechnology and biotechnology is relatively balanced. The benefits and their ethical issues occupy the attention of media consequently of people in equal measure. Public acceptance of biotechnology and environmental biotechnology thus depends on the collective perception and wisdom of people (Aerni 1999). An individual's perspective of a technology depends on their social, religious, ethical beliefs and the perception of the dominant narrative of the country and area. The credibility and the reputation of the institutions introducing technologies also decide the perception. Kelley (1995) proposed that attitudes toward biotechnology and genetic engineering are determined by the benefits offered by the technology, the information available on the technology and the logical risks and concerns on possible threats. Safety and ethical concerns also play a major role in determining the perception of a technology (Einsiedel 1997). Hoban (1997) propounded that acceptance of a technology is influenced by knowledge level, awareness of benefits, confidence and trust.

The environmental problems are multidimensional and affect the ecology, environment, biodiversity, human society. Until now, environmental technology, and, with it, environmental biotechnology, has by no means been designed in a holistic way. Strangely enough, farming, in general, and farmland, in particular, is the crucial endpoint of many waste treatment processes. At the downstream side of the wastewater treatment plant, 'organic fertilizer' or even 'animal feed' appears too often as a miraculous solution.

The scenario of biotechnology consists of three major parts: (1) people who derive financial and commercial benefits from the biotechnology and its applications (industrialists, scientists, agriculturalists and most of those working in R&D); (2) public authorities and public interest groups, nongovernmental organizations, involved with biotechnology sociopolitical reasons; and (3) the common people as recipients of the benefits of biotechnology goods and services as well as the risks and consequences involved. The complexity of the biotech scene is based on the complex interplay among these groups. One possible interplay is the exchange of scientific, technological, economic and other relevant factual data, information and arguments, with a view toward eliminating misunderstandings and the possible rationalization of conflicts.

Most people now get their information from the internet, with zero quality control, and plenty of organizations have an agenda to sell. Television and radio remain major sources, but few programs can provide in-depth analysis of technical issues. There are some excellent science programs available, but they tend to be underrepresented. Popular media remains not only a major source of information but also a major source of misinformation and, therefore, confusion (Macer et al. 1997). Accessible technical journals do exist and help inform a small cadre of nonscientists interested enough to actually subscribe and read the articles. But this group is a very small segment of society and not the group needing help (Davison et al. 1997; Fischhoff & Fischhoff 2001).

The workshop on "Knowledge, Attitudes and Behavior towards Biotechnology" (Brussels, Sept. 10, 1990), with experts from the various fields involved, came to the following conclusions (Diettrich 1992):

1. Scientific or technological knowledge is not a hindrance; rather, mistrust in science and applications creates confusion among people. Scientists tend to offer benefits with sometimes hyped expectations and claims without proper communication on possible threats and limitations. The astute defense of the claims also creates doubts among the public. People have to be convinced that their concerns about and experiences with a product will be also addressed. The information disseminated on a product or process should involve economic, legal and regulatory concerns along with science and technology.
2. The flow of information and dissemination of knowledge be through the trusted institutions and unbiased agencies. If required, organizations with proper technical knowledge and mass interaction (scientific institutions, educational institutions, museums). The selection and funding of such organizations should be also kept transparent.
3. Scientists and the scientific community should be also trained in areas of communication. Communication with the public should be clear, and any message should not be lost in scientific jargon.

   Scientific communication activities aiming at the public have to be professionalized in the sense of public relations in cooperation with social scientists and psychologists (anecdotal approach via actual problems to be solved by biotechnology). A particular point is targeting and segmenting: Different strategies have to be developed according to the various target groups as well as to the various subjects in question.
4. Concerning the high importance of cultural differences, the state of the art in public knowledge and attitudes in the various countries have to be analyzed. It was stated that the role of consensus and the awareness of the need to find it are rather different in various communities and countries. Every country and area demands different approaches for discussing and propagating information. The social, economic and cultural aspects play a major role in any discussion and, thus, perception.

In a study carried out on U.S. data obtained via a national telephone survey of U.S. consumers. A survey instrument was developed at the Food Policy Institute at Rutgers University to gather information on public attitudes toward food biotechnology (Rowe 2004). The survey was designed to gather information on the public awareness of various issues pertaining to the use of biotechnology in food production, public approval of the use of biotechnology in plants and animals and public views about various private and public institutions associated with biotechnology research and product development. The survey obtained information related to the participant's socioeconomic conditions, values, age ideologies, cultural background, view on scientific developments and the scientists and the like. This study examined the relationship between consumer reception of the use of biotechnology in food production and their economic, demographic, and value attributes. The results indicate that a majority of U.S. consumers do not have fixed ideas on the requirement of this technology. Despite having reservations, particularly about its use in animals, Americans are not prepared to reject this technology altogether. Most consumers remained undecided, with a need of further evidence on the safety of GM foods (Hossain et al. 2003).

There are limited data on studies of public perceptions of biotechnology from developing countries; most of the data available are from small and limited sample sizes. They do not reflect the diversity of the opinions and social, economic, literacy and cultural backgrounds (Bauer 1995). The international comparisons, thus, can only be made with developed country studies. The study carried out in South Africa revealed that the public was mostly positive about biotechnology, specifically GM) food, although they had limited information. Europeans tend to have negative perception of GM food compared to South Africans. The knowledge and information on biotechnology and GM foods are less among South Africans compared to people in developed countries, and thus, the perception of the likely environmental impact of GM foods is relatively positive. The level of knowledge required to answer questions pertaining to safety, economic impact and environmental impact of GM food is lower in South Africa than in developed countries (Gastrow et al. 2016)

A survey was carried out on 440 respondents from various expert groups in the Klang Valley region of Malaysia. In the study, respondents were asked about their attitudes on GM soybean that is already available in the Malaysian market. A study confirmed that the perception on issues like GM food should be dealt in totality, including multiple dimensions of economic, social and knowledge base (Latifah et al. 2005)

So we know that knowledge and attitudes are not necessarily positively correlated. The effect of information mainly depends on interpreting preexisting attitudes (Gaskell et al. 2000). It is thus imperative that an environment is created for the proper dissemination of information on environmental biotechnology (Latifah et al. 2007). If possible, small multi-location pilot projects should be initiated with transparent feedback from end users. The participants should be actively involved, and any scientific data should be collected by taking users in confidence. The attitude of the public can be improved only by trust-building exercises and transparency in field experiments, data collection, surveys and dissemination of complete information on products and processes.

19.4.2 Measures of Improving Public Perception of Environmental Biotechnology

Thus, for the diffusion of biotechnologies to be successful and for improving the perception of environmental biotechnology tools and products, the following measures should be put in place:

- **Ensuring linkages between research workers and the user corporate groups**

  This can be achieved by creating an environment for better interaction between researchers and industries. The objective of research and the problems encountered by fieldworkers and industries may be shared with researchers. The approach of investigation should be solution-oriented with the creation of a new process or product.
- **Public and private partnership in research and application of environmental technologies**

  Public–private partnerships are essential to give an impetus to the research and applications. Partnerships will also ensure proper funding and the dissemination of clean technologies through creating a demand and supply chain.
- **Strict enforcement of the 'polluter pays' principle. This would require interaction with law enforcement agencies.**

  Enforcement agencies need to be supported by strong laws, monitoring systems, real-time technological interventions and trained staff. Environmental protection agencies may also coordinate with the industries, law enforcement agencies (police, law officers, civic administration). Real-time monitoring of pollution at the major sites with online detectors, closed-circuit television and the like will also help provide sufficient information to environmental protection agencies.
- **Capacity building and training through workshops, of law enforcement officials, municipal workers, state government functionaries and corporate groups on role and relevance of biotechnology in waste treatment**

  Workshops will help law enforcement officials, municipal workers, state government functionaries and corporate groups understand the importance and applicability of biotechnological tools in the waste treatment. They will also provide feedback on the limitations and feasibility of such technologies.
- **Steps to encourage small and medium-sized businesses and companies in producing eco-friendly products, microbial consortia and the like for wider usage**

  Various programs are supported by the Indian governments, including Small Business Innovation Research Initiative (SBIRI) and Biotechnology Industry Research Assistance Council (BIRAC), for supporting the small and medium-sized industries and start-ups for producing environmentally friendly products and the process for different applications. The process of getting funds is still tedious and requires a long duration. Thus, the process of evaluation and subsequent sanction of funds may be eased. Special tax and other benefits may also be enhanced and provided to the companies involved in eco-friendly processes and products.
- **Building greater awareness for the protection of proprietary rights of microbial consortia and other biological products through appropriate methods (e.g., process patents, trademarks, etc.)**

  One major problem in India was the lack of awareness on IPR laws; this has improved due to intervention of DBT and other government agencies. The patent process has also been eased. Many amendments in the patent laws as per obligations of international treaties have also made them useful and at par with laws of other countries. The IPR is now also part of most of the course curriculum in Biological and Basic sciences along with law schools. Awareness workshops and online technical, as well as financial, support for the patent process have contributed to the recent increase in the patents from India.
- **Coordination among the Department of Biotechnology (DBT), the Ministry of Environment and Forest (MoEF), the Indian Council of Agricultural Research (ICAR), the Council of Scientific and Industrial Research (CSIR), the Central Pollution Control Board, user agencies and industry through a coordinating body**

  The coordination among these agencies can be improved by a task force or a coordinating body, which may decide on the short- and long-term objectives and the work plan after elaborate deliberation with industries, end users and the public. The plan may also incorporate measure for proper dissemination of the information and advantages of environmental biotechnology. An improved perception and awareness by the public will foster demand for newer technologies and reduce any anxiety about the adoption of new products and processes.

19.5 Conclusion

The public perception of environmental biotechnology interventions, processes and products is determined by surveys. Perception also depends on literacy and scientific temperament. Thus, the first step should be a thrust on literacy and scientific awareness. Subsequently, public perception can be improved to enhance acceptability by increasing coordination between different agencies, cooperation between public and private players, encouragement of small entrepreneurs, capacity building and improving awareness. All these measures are inherently linked and contribute synergistically.

REFERENCES

Aerni, P. 1999. Public acceptance of genetically engineered food in developing countries: the case of transgenic rice in the Philippines. *AgBiotechNet1*, November. ABN 031.

Bauer, M. W. 1995. Towards a functional analysis of resistance. In M. Bauer (Ed.), *Resistance to new technology – nuclear power, information technology, biotechnology*, 393–418. Cambridge: Cambridge University Press.

Bauer, M. W. 2005. Public perception and mass media in the biotechnology controversy. *International Journal of Public Opinion Research* 17(1):5–22.

Davison, A., Barns, I. and Schibeci, R. 1997. Problematic publics: A critical review of surveys of public attitude to biotechnology. *Science Technology and Human Values* 22(3):317–348.

Diettrich, O. 1992. Biotechnology and social perception: Particular aspects of the socio-economic integration of biotechnology in the context of activities of the Commission of the European Communities. In R. von Schomberg (Ed.), *Science, Politics and Morality*. Series A: Philosophy and Methodology of the Social Sciences. Kluwer Academic Publishers.

Einsiedel, E. F. 1997. *Biotechnology and the Canadian public: Report on a 1997 national survey and some international comparisons*. University of Calgary, Alberta.

Fischhoff, B. and Fischhoff, I. 2001. Publics' opinion about biotechnologies. *Agbioforum* 4:155–162.

Gaskell, G., Allum, N., Bauer, M., Durant, J., Allansdottir, A., Bonfadelli, H., Boy, D., de Cheveigné, S., Fjaestad, B., Gutteling, J. M., Hampel, J., Jelsøe, E., Jesuino, J. C., Kohring, M., Kronberger, N., Midden, C., Nielsen, T. H., Przestalski, A., Rusanen, T., Sakellaris, G., Torgersen, H., Twardowski, T., and Wagner, W. 2000. Biotechnology and the European public. *Nature Biotechnology* 18(9):935–938.

Gastrow, M., Roberts, B., Reddy, V., and Ismail, S. 2016. Public perceptions of biotechnology in South Africa conducted for the public understanding of biotechnology programme of the South African agency for science and technology advancement education and skills development research programme human sciences research council.

Ghosh, M., and Singh, P. S. 2005. A review on phytoremediation of heavy metals and utilization of its byproducts. *Applied Ecology and Environmental Research* 3(1):1–18.

Hoban, T. 1997. Consumer acceptance of biotechnology: An international perspective. *Nature Biotechnology* 15(3):233.

Hooda, V. 2007. Phytoremediation of toxic metals from soil and waste water. *Journal of Environmental Biology* 28(2):367–376.

Hossain, F., Onyango, B., Schilling, B., and Hallman, W. 2003. Public perceptions of biotechnology and acceptance of genetically modified food. *Journal of Food Distribution Research* 34(3).

Kelley, J. 1995. Public perceptions of genetic engineering: Australia, 1994. Final report to the Department of Industry, Science and Technology.

Kensa, V. M. 2011. Bioremediation – An overview *J. of Industrial Pollution Control* 27(2):161–168.

Khanna, P. 1996. Environmental biotechnology in India: prospects and case studies. *World J Microbiol Biotechnol* 12(5):457–461.

Kokyo, O., Tiehua, C., Tao, L., and Hongyan, C. 2014. Study on application of phytoremediation technology in management and remediation of contaminated soils. *Journal of Clean Energy Technologies* 2:3.

Latifah, A., Abd, R. M. N., Jamaluddin, M. J., Mohamad, O., and Nor, M. 2005. Factors for a socially acceptable gene technology. *Malaysian Journal of Environmental Management* 6:37–146.

Latifah, A., Abd, R. M. N., Jamaluddin, M. J., Mohamad, O., and Nor, M. 2007. Public acceptance of modern biotechnology. *Asia Pacific Journal of Molecular Biology and Biotechnology* 15(2):39–51.

Macer, D. J. R., Bezar, N., Kamada, H., and Macer, N. Y. 1997. Attitudes to biotechnology in Japan and New Zealand in 1997, with international comparisons. *Journal of Asian and International Bioethics* 7:137–143.

Peuke, D. A., and Rennenberg, H. 2005. Phytoremediation. *EMBO Reports* 6(6).

Rowe. 2004. How can genetically modified foods be made publicly acceptable? *Trends in Biotechnology* 22(3):107–109.

20

Biodegradation of Hazardous Contaminants and Pesticides

Marion McClary

CONTENTS

20.1 Overview

As a result of agricultural, commercial and industrial activities conducted in the absence of environmental regulations and enforcement in the past, sediments that are contaminated by organic compounds, heavy metals and other potentially toxic chemicals have accumulated in many of the world's environments (Tomassoni, 2000). These sediment-borne contaminants have eventually become incorporated into aquatic food webs and adversely affect ecological organisms, such as benthic organisms and fish, and ultimately pose a risk to human health (Rice, 1999; Lopes and Furlong, 2001; Slack et al., 2005). In recent years, the need for remediation or management of contaminated sediments has become increasingly evident.

Contaminated sediments can be managed in various ways. They can be removed by dredging (Palermo, 1998; Azcue et al., 1998; USEPA, 2002; USEPA and WDNR, 2002) and treated *ex situ*, as appropriate, prior to disposal (Palermo et al., 1990; Reible et al., 2003), which is challenging because metals and organic contaminants have different fate and transport mechanisms (Sparks, 1995). The process of removal and treatment to remove contaminants is also very costly. Sediments could also be managed *in situ*. *In situ* approaches can include monitored natural recovery, capping or treatment by biological or chemical means. The biological means are the focus of this chapter and are discussed later. For now, the discussion is about capping. *In situ* capping generally involves placing clean sediments or similar materials over contaminated sediments for the purpose of physically and chemically isolating sediment-borne contamination (Thoma et al., 1993). If biodegradation is desired, organisms that do it could be placed in the sediments of the cap as well so that there is a dual purpose. One purpose is isolation by the cap, and the other is biodegradation by organisms in it (Figure 20.1).

In situ capping is attracting growing interest as a sediment management alternative (NRC, 2001) and is widely used (Ling et al., 1996; Azcue et al., 1998; Mohan et al., 2000; Fredette and French, 2004). Advantages of *in situ* capping relative to sediment removal include minimal environmental and habitat disturbance, minimal sediment exposure and handling, minimal release of volatile organics and the elimination of the need for disposal facilities, resulting in lower remediation costs. In addition, *in situ* sediment caps are relatively easy to construct, repair or replace as part of ongoing operation and maintenance.

Granular materials (e.g., sand) can serve as appropriate and adequate capping materials at many sediment remediation sites (Palermo et al., 1989). Granular materials are typically inexpensive, often locally available, relatively easy to place and can provide an adequate substrate for benthic macroinvertebrate organisms (Reible et al., 2003). However, typical granular materials are also relatively easily eroded and permeable to the diffusive and advective movement of dissolved, sediment-borne contaminants (Dortch et al., 1990; Bokuniewicz, 1992; Lohse et al., 1996), although erosion is reduced as the water becomes deeper (Huttel and Webster, 2001). Additionally, sand-based remedial caps are often relatively thick (i.e., on the order of feet rather than inches) in order to adequately overcome these performance limitations. Relatively thick caps may adversely impact waterway navigation at some deepwater sites or adversely affect the hydrology and vegetative communities in wetland ecosystems. Finer grained materials (e.g., clays) are typically more cohesive, less permeable and more reactive than sands. A relatively thin, clay-based cap (i.e., on the order

DOI: 10.1201/9781003186304-20

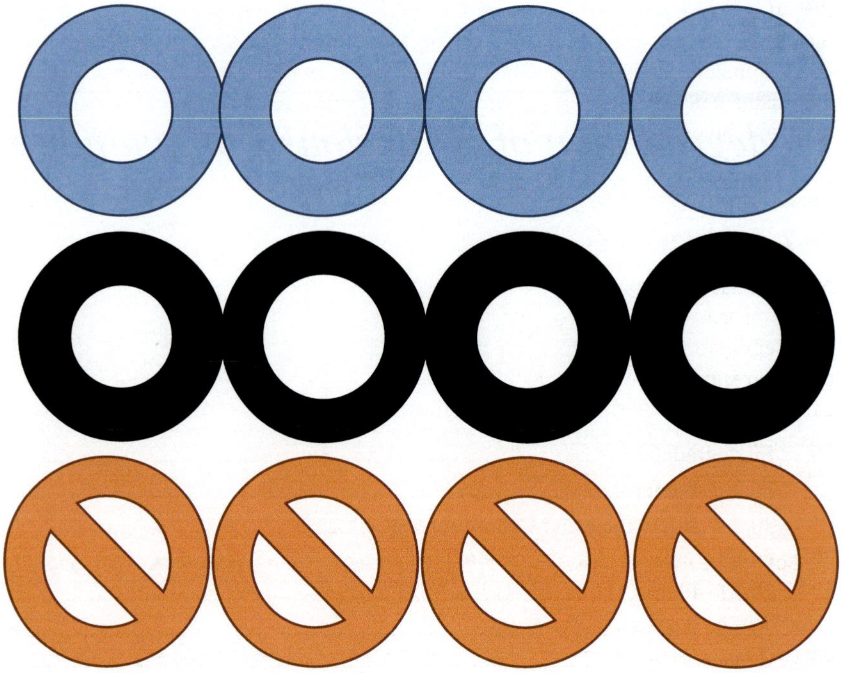

FIGURE 20.1 Diagram of an *in situ* capping treatment.

of inches rather than feet) can provide a better capping remedy less disruptive of wetland quality and function. Additionally, clay has a higher cation exchange capacity than sand and may more effectively trap contaminants, such as heavy metals (Jacobs and Förstner, 1999; Simpson et al., 2002; Jacobs and Waite, 2004; Zimmerman et al., 2004; Murphy et al., 2006; Weber et al., 2006; Reible et al., 2006; McDonough et al., 2007; McDonough and Lowry, 2008). Until recently, it has been logistically difficult to get fine-grained capping material effectively in place across inundated sediment surfaces.

Water quality due to sediment capping shows some differences in water quality. For example, pH was statistically lower (Liu et al., 2005). Dissolved oxygen was found to increase most likely due to lower levels of microorganism activity. Huttunen et al. (2001) and Palermo et al. (2002) found a decrease in aerobic microbial activity after capping but found anaerobic biogas production in surface waters above contaminated sediments in the months immediately after cap placement. Palermo et al. (2002), Richardson (1998) and Rudd and Taylor (1980) found that biogas production results in bubbles to damage caps. This could be because the active microbes in the sediment determine the composition of the biogas (Wang, Kassem et al., 2009) or the lack thereof. Wang, Kassem et al. (2009) found an increase in sediment microbial activity in sand caps compared to uncapped sediments.

Organic contamination such as polychlorinated biphenyls (PCBs) have low solubility in water (Luthy et al., 1997; Jonker and Koelmans, 2002; Murphy et al., 2006) but have an extremely high affinity for organic substances in sediment (Luthy et al., 1997; Jonker and Koelmans, 2002). Luthy et al. (1997), Ponder et al. (2001), Schrick et al. (2004) and Saleh et al. (2007) found that PCBs absorbed to a sediment matrix were hard to desorb to the liquid phase. This is contrary to Eek et al. (2008) who found transport of PCB 28, PCB 52 acenapthylene, fluorine, phenanthrene, anthracene, fluoranthene and pyrene from sediment capped with crushed limestone consisting of >99% $CaCO_3$ or crushed gneiss consisting of quartz, amphibole, albite, chlorite, biotite and orthoclase. This could be due to the difference in cap thickness. Cap thickness does make a difference because the diffusive flux of dissolved contaminants from contaminated sediments can be reduced to less than 1% by a 10-cm cap layer of inert material that is undisturbed by bioturbation (Eek et al., 2008). This could also be due to the lack of organics in the cap (Eek et al., 2008) or residual contaminants being present in the new sediment layer that formed on top of the cap (Eek et al., 2008). Perhaps the use of reactive metals such as Fe and Mg in a zerovalent state could dechlorinate the PCBs and other chlorinated compounds (Morales et al., 2002; Liu et al., 2005; Agarwal et al., 2007a; Agarwal et al., 2007b; Fang and Al-Abed, 2007). Overall, capping can provide a less contaminated substrate in a marsh. Zimmerman et al. (2004) tested 2-mm layers of sand and activated carbon as a cap for PCBs and found 89% retention compared to 23% retention by sand alone, the difference being due to the activated carbon.

Because we do not live in an isolated environment, the use of a sediment cap placed over contaminated sediment does not remove the organisms that live within the contaminated sediment or the organisms that live in the surrounding areas that can migrate to the contaminated sediment. As mentioned earlier, Huttunen et al. (2001) and Palermo et al. (2002) found a decrease in aerobic microbial activity after capping but found anaerobic biogas production in surface waters above contaminated sediments in the months immediately after cap placement and could be due to active microbes in the sediment (Wang, Kassem, et al., 2009). In some ways, that cap attracted the microbes because Wang, Kassem, et al. (2009) found an increase in sediment microbial activity in sand caps compared to uncapped sediments. It is for this reason that the role of organisms cannot be ignored. One role of these microorganisms, and other organisms for that matter, is biodegradation.

20.2 Biodegradation

Biodegradation is the degradation or "breakdown" of a substance or substances by a biological organism or organisms. The degradation is the transformation of the substance from one chemical or physical form into another chemical or physical form that can be less or more hazardous than the previous form. Contaminants are compounds that have found their way into the environment and contaminate the environment. Pesticides are compounds used to control pests, whether they be plants, which are controlled by herbicides, or insects, which are controlled by insecticides. The problem is that the pesticide can also affect organisms other than the pest that is the target.

Biodegradation of hazardous contaminants and pesticides can be done by microorganisms, such as bacteria, but other microorganisms, such as microalgae and fungi, have this capability. Plants are known to perform biodegradation, and certain animals have been used to assess biodegradation. This assessment involves exposing the living animal to the compound or compounds that have undergone biodegradation to see if the organism dies. If the animal does die, it has been exposed to a lethal concentration. If the animal does not die, it may have been exposed to a sublethal concentration. To determine if the concentration is sublethal, its behavior and/or physiology, at the organ, tissue or cell level has to be assessed. Behaviors that can change due to toxic compounds that can be evidence of sublethal effects include feeding behavior, antipredator behavior, habitat selection and/or building of a shelter, navigation and/or migration and mating behavior. If an animal is unable to eat, protect itself or mate, it will not survive as a species.

Other behaviors that can be affected are endogenous biological rhythms, which can be related to feeding behavior, antipredator behavior, selection of habitat and/or building of a shelter, navigation and/or migration and mating behavior. Because these rhythms, usually based on activities or the lack thereof, are endogenous, they can function without the existence of external cues that tell the animal what time of day or night that it is. These external cues can be removed experimentally by placing the animal in conditions of constant light or constant darkness to see if the animal maintains a rhythm of activity and rest approximate to the natural rhythm of activity and rest. If an animal has an endogenous rhythm, it is natural that they will lose it over time and will have a rhythm unlike the natural rhythm. Studies with antidepressants did not kill mosquitofish, but one, venlafaxine, disrupted their circadian rhythm and affected behavior (Melvin, 2017). This is an example of sublethal changes that chemicals can cause.

Physiology that can change due to toxic compounds that can be evident of sublethal effects includes respiration, osmoregulation and nerve conduction, to name a few. If an animal is unable to respire, regulate its body fluids or respond to stimuli, it cannot survive. In terms of respiration, changes in the rate of oxygen uptake can indicate effects. If an animal is placed in a closed system saturated with oxygen, its respiration rate could be independent or dependent on the oxygen available. If the rate is independent, it will proceed at a constant rate, although the partial pressure of oxygen is declining as it is being used up in this closed system. At some point, the animal will realize that the partial pressure of oxygen is too low, known as the critical oxygen tension, and the rate of oxygen uptake will decrease as the partial pressure of oxygen decreases. At this point, the respiration rate is dependent. Large animals tend to have a slow respiration rate, so their respiration rate is mostly independent until reaching the critical oxygen tension. From that point, the respiration rate is dependent on the partial pressure of oxygen. Small animals tend to have a fast respiration rate, so their oxygen uptake is most dependent and has no critical oxygen tension. The effect of a compound can be determined if the respiration rate is abnormal and the critical oxygen tension is different than what is expected. As for osmoregulation, aquatic animals that are osmoregulators have to maintain being either hypoosmotic or hyperosmotic to the environment. A compound that affects an animal's ability to do this could cause the animal and its cells to again water and lyse or lose water and dehydrate.

Changes in physiology and or morphology may also be evident at the organ, tissue and/or cellular level, and traces of the compound may be found at these levels. An example of changes in morphology at the organ level can be found as a result of exposure to tri-n-butyl-tin. Tri-n-butyl-tin can act as a hormone and exposure to it can cause male mud snails to grow a larger than normal penis and can cause female snails to grow a penis. In this case, the animals do not die, but evidence of their exposure to a contaminant can be seen at the organ level. At the tissue and/or cellular level, the production of certain heat-shock proteins are evidence of the contamination due to tri-n-butyl-tin. Figure 20.2 is a flow chart of how animals can be used to assess biodegraded compounds for toxicity. The chart does not include animals that die after an initial exposure to the compound or compounds. If death does occur, the compound should be retested, and if death still results, the concentration of the compound or compounds should be reduced to concentrations that are representative of environmental exposure and below them. Some assessment is based on death alone. Figure 20.2 provides information about the sublethal effects that could result in death in the long run after chronic exposure. It could also be a method of early detection. Early detection could result in remediation before death occurs depending on the method of remediation. Removing contamination could also result in removing the animals, which could also result in their death depending on the method of cleaning the sediments. If the remediation is to be done by biodegradation, the survival of the animals would depend on how fast the biodegradation organisms work. If the death of the animal is slow and the biodegradation process is fast, the animal has a chance of survival. If death of the animal is fast and the biodegradation process is slow, the animal has less of a chance of survival. Enhancing biodegradation with heat may accelerate the degradation process, but this could be harmful to animals that are not removed from the sediment and are exposed to the heat.

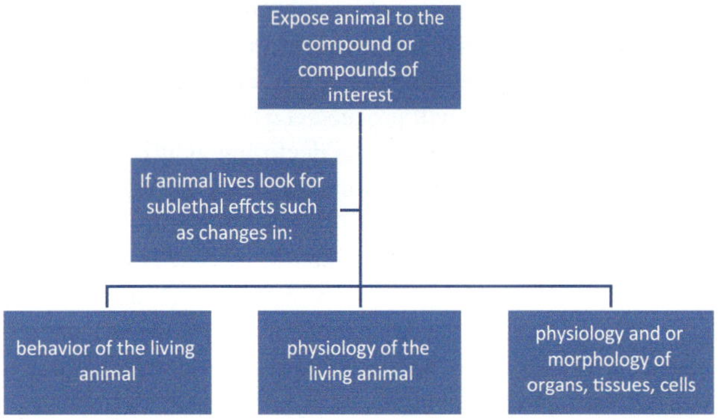

FIGURE 20.2 Flow chart of the use of animals for the assessment of biodegraded compounds.

At the ecological level, biodegraded compounds could have an effect on a segment of the population and not others. For example, only very young animals could be impacted and die at higher numbers than middle-aged or older animals, or the middle-aged or older animals could be affected more so than the very young animals. Based on the age group affected, the population could increase, decrease or become more patchy in its distribution. Some of this could be due to its ability to no longer compete with its competitors or outcompete its competitors, which may or may not be affected by the compound or compounds in the same way. In this way the compound, which is an abiotic factor, may have biotic effects when it comes to its competitors and/or predators.

Contamination can affect diversity or species richness and abundance which is an indication of the environment as a whole. Although some environments are naturally harsh and have lower diversity or species richness and abundance naturally, contaminated environments can also be harsh because of the contamination depending on the tolerance of the species in the environment. The best way to know that the low diversity and richness and abundance are due to the contamination is to know what the diversity and species richness and abundance were before, during and after the contamination was present so that comparisons can be made. Unfortunately, we do not always know which environments will become contaminated beforehand to collect data on species richness and diversity and abundance, so often, these data are lacking for comparison. Without such comparison, we run the risk of underestimating the effect of the contamination. Species diversity can be determined using the Shannon-Weiner Index:

$$H' = -\Sigma N_i/N \ln N_i/N,$$

where N_i, the number of individuals in the ith species, is divided by N, the total number of individuals of all species collected, and multiplied by the logarithm of itself to get a result which is added to the result obtained for each of the ith species.

When it comes to biodegradation by plants, in most cases bacteria and or fungi are also involved which means that plants may not do the biodegradation alone (see Figure 20.4). Although this is the case, this chapter discusses the biodegradation of hazardous contaminants and pesticides by the group of organisms that perform it. These organisms are (1) bacteria, (2) microalgae and cyanobacteria, (3) fungi, (4) plants and (5) animals, which are mostly used to assess the biodegradation. Their placement in a sediment cap as a means biodegradation is also discussed. In this case, there is a dual purpose. One is isolation, which is done by the sediment cap, and the other is biodegradation, which is done by the organism or organisms.

20.2.1 Bacteria

Biodegradation is one way to remove pesticides from the environment (Nawaz et al., 2011). The advantages of it are that it is cost-efficient, accepted by the public and done on-site without the need to transfer material (Niti et al., 2013; Strong and Burgess, 2008) compared to conventional methods of chemical treatment, recycling, burning or placement in landfills, which are costly and toxic (Kumar et al., 2011; Dua et al., 2002; Rockne and Reddy, 2003). Certain strains of bacteria have been used to degrade pesticides (Rockne and Reddy, 2003; Zhongli et al.,2001; Chang et al., 2005; Horne et al., 2002; Chen-Goodspeed et al., 2001). Some strains were able to do so because the pesticide was used as a source that was needed for their metabolic pathway or gene function (Hussain et al., 2015; Kumar et al., 2016; Lal et al., 2010; Nagata et al., 2015). Evidence that a catabolic pathway was involved in the degradation was found by polymerase chain reaction (PCR) (Baelum et al., 2012; Bers et al., 2013). It is widely accepted that indigenous microorganisms are best to use at the site that needs biodegradation (Abdelbagi et al., 2003; Abdelbagi et al., 2000; Elmahi, 1996) and that degradation rates can be enhanced by activators (Elsaid and Abdelbagi, 2010). If a sediment cap is to be used, as mentioned previously, microbes that are known to degrade the contaminants present can be placed in the sediment cap. Degradation can be by mineralization, which is what *Pseudomonas aeruginosa* did to pendimethalin (Shaer et al., 2013), or just degradation, which is what *Bacillus safensis*, *B. subtilis* and *B. cereus* did to chlorpyrifos, in order of efficiency, and what *B. cereus*, *B. subtilis* and *B. safensis* did to malathion and dimethoate, in order of efficiency (Ishag et al., 2016).

Biodegrading contaminated soils by microorganisms can also occur by the use of compost in the process of composting. In this case the microbial community changes over time due to temperature changes. These changes include influences on the gram-positive–to–gram-negative bacterial ratios, which resulted in losses of polyaromatic hydrocarbons (PAHs) in PAH-contaminated soils associated with the gram-positive–to–gram-negative bacterial ratios (MacNaughton et al., 1999; Zeng et al., 2011). Changes could also consist of a change from alpha- and gamma proteobacteria to bacterioidetes and firmicutes (Gandolfi et al., 2010) or an increase in diversity that still results in the removal of PAH (Di Gennaro et al., 2009). A rise in temperature makes organic compounds more bioavailable for microbes to degrade (Kumar et al., 2008). Other factors such as pH, soil moisture, soil texture, organic amendments and the effect of surfactants are reviewed in Odukkathil and Vansudevan (2013).

With the growing human population, farms are required to produce more food. Of this food, 45% of annual food production is lost due to pests (Abhilash and Nandita, 2009). One way to counter this is to use chemical pesticides. The complete breakdown of pesticides into organic components is known as biomineralization, the partial breakdown is known as partial biodegradation and the resultant compound is less toxic than the original compound (Odukkathil and Vansudevan, 2013). An example of partial degradation is dichlorodiphenyltrichloroethane (DDT) being transformed to dichlorodiphenyldichloroethane (DDD) and dichlorodiphenyldichloroethylene (DEE), but in this case, DDD and DEE are still toxic (Foght et al., 2001). In either case, the breakdown occurs because the microorganism uses the pesticide as a source of carbon, nitrogen, or some other element such as sulfur which the *Achromobacter xylosoxidans* strain CS5 uses the pesticide endosulfan and endosulfan sulfate for (Wen et al., 2009). The breakdown happens because bacteria have hydrolytic enzymes to breakdown herbicides such as atrazine. Other enzymes like oxygenases, hydroxylases, hydrolases and isomerases in *Pseudodomonas* and *Alcaligenes sp.* allow them to degrade the herbicide 2,4-D (Mulbry and Kearney, 1991).

Atrazine, also known as 2-chloro-4-ethylamino-6-isoprop ylamino-s-triazine and 6-chloro-N-ethyl-N'-(1-methylethyl) -1,3,5-triazine-2,4-diamine, is an organic compound having the molecular formula $C_8H_{14}ClN_5$. Atrazine is primarily produced by Syngenta Crop Protection, a Swiss-based company. It is a colorless or white, odorless, crystalline powder commonly used as an herbicide all over the world (NIOSH). Like all herbicides, atrazine is used to control the growth of weeds in crops.

In the United States, atrazine is considered the most commonly used herbicide. An estimated 76.4 million pounds of atrazine are applied annually in the United States alone. Approximately 75% of the field corn acreage utilizes atrazine. It is primarily used in Iowa, Illinois, Indiana, Ohio, Nebraska and Delaware. It is also used on lawns in Florida and the Southeast. Atrazine is used on many crops, such as corn, sorghum, sugarcane, wheat, guava, macadamia nuts, hay, pasture and Christmas trees. It is also used residentially and on recreational turf in parks and golf courses (USEPA).

Herbicides are grouped according to their site of action, target site, similarity of induced symptoms or chemical classes. According to the Herbicide Resistance Action Committee (HRAC), atrazine is classified under the chemical family triazine, HRAC group C_1, and WSSA (Weed Science Society of America) Group 5. Atrazine's site of action is the inhibition of photosynthesis at photosystem II. In other words, it binds to the plastoquinone-binding protein in photosystem II. This leads to plant death due to starvation and oxidative damage, which is caused by vanquishing the electron transport process. Atrazine is applied on the affected area via methods of ground spray rig or agricultural aircraft. It is absorbed through the roots of the weeds and then displaced to the actively growing tips and leaves. Killing the weeds via photosynthesis inhibition takes approximately 14 to 21 days.

The use of atrazine is controversial due to evidence that supports that it has endocrine-disruptor activity. An endocrine disruptor is a synthetic chemical that mimics and/or blocks hormones to alter the body's functions when absorbed by the body. It "fools" the body into responding to a stimulus, sometimes in inapt times (USEPA). This phenomenon, in turn, leads to interference with reproduction and development. Additionally, some findings have linked atrazine exposure to cancer.

In a study conducted at North Carolina State University, it was concluded that atrazine stimulates hemoglobin synthesis via the dhb2 gene (hemoglobin gene). Atrazine's ability to activate the hypoxia-inducible factor (HIF) pathway is an explanation for the stimulation of atrazine production. HIFs are oxygen sensors with α and β subunits that react to changes in the level of oxygen, especially the decrease in oxygen, in the cellular environment. In other words, atrazine lowers the oxygen levels available to organisms through the HIF pathway (Rider and LeBlanc, 2006).

When pollution occurs, most likely more than one kind of pollutant is involved at a site. More than one third of contaminated sites have more than one type of pollutant (Tang et al., 2010; Mansour, 2002). Actinobacteria are found in soils and were tested for their ability to degrade soil that was contaminated with Cr (VI) and lindane. All species of actinobacteria tested by Polti et al. (2014) were able to grow and remove both contaminants. *Amycolatopsis tucumanensis* removed the highest amount of Cr (VI) and *Streptomyces* sp. M7 removed the highest amount of lindane (Polti et al., 2014).

Studies on heavy metals in soil have revealed that low concentrations aid microbial growth but become toxic at higher concentrations (Giller et al., 1998). The presence of heavy metals can also increase the tolerance and resistance of the microbes to them based on the ability of microbes to adapt quickly (Hassen et al., 1998; Obbard, 2001) and this resistance increases over time (Ellis et al., 2003). Although Cr, Pb, Cd, Cu and Zn removal can be increased after 50 days of composting, the density of microbes could be significantly reduced in the process (Haroun et al., 2007). If organic amendments are added, microbial activity has been found to increase and can cause a decrease in Cd toxicity (Moreno et al., 2002). This could be due to variations in the population of microbes that respond to changes in pH as heavy metal solubility does (Alburquerque et al., 2011).

20.2.2 Microalgae and Cyanobacteria

Pesticides, including fungicides, herbicides, and insecticides, have been used since the 1940s to eliminate or control pests both within and outside of farms (Schwarenbach et al., 2010). Some insecticides contain organophosphates such as phorate, malathion, parathion, monocrotophos and quinalphos. As one of the most used herbicides on the market, Rodeo is suspected to be one of the causing agents for aquatic mortality. This herbicide is sold by Monsanto in the form of Roundup. A major component of Roundup is the chemical glyphosate. Glyphosate is also sometimes used for the "control of emergent aquatic weeds in ditches, wetlands, and margins of water bodies" and marsh grasses (Solomon and Thompson, 2011). Roundup kills by being absorbed into the soft stalk tissue and leaves of the plant (NPIC, 2010). As it is transported through the plant, it inhibits amino acid metabolism via the shikimic pathway. Finally, it interferes with plant development and growth and causes deterioration, yellowing, wilting, browning and eventually plant death. Because Roundup is applied by spraying, the wind can carry it onto animals that live in near areas or into the surrounding water of those areas. It is possible that the function and life of these animals are affected by the chemical. If it is used directly to kill aquatic weeds, any aquatic life is exposed to its effects.

Relyea (2003) tested the effects of multiple pesticides and insecticides on an aquatic community containing algae and 25 species of animals. It was reported that not only was species fertility reduced by 22% with Roundup in that aquatic community, but it also it eliminated two species of tadpoles and nearly eradicated a third species, which resulted in a 70% decline in the species richness of tadpoles (Relyea, 2003). The effects of Roundup were later tested on more tadpole species, and it was found that it killed 96% to 100% larval species and 68% to 86% of juvenile amphibians (Relyea, (2005). These results, and more, suggest that Roundup can cause high mortality rates among aquatic and terrestrial animals, which can lead to a decline in diversity and population.

Roundup is effective at killing plants of all types but is frequently used to kill weeds and marsh grasses, such as *Spartina alterniflora* (Nieto, 2002). *Spartina alterniflora* threatens to degrade intertidal zones of estuaries because it is an invasive species (Kilbride and Paveglio, 2001). It is also known that this grass can modify water flow and sediment deposition (Neira, 2006). Mussels have a "symbiotic" relationship with *Spartina alterniflora*. The grass accommodates living for the mussels, while growth is stimulated and soil nutrients are increased for *Spartina alterniflora* (Bertness, 1984). Mussels stimulate aboveground and belowground grass production. Growth of the grass emerges more at the edges of a marsh because the mussels filter more food, reproduce sooner and attract more larval settlers. This is beneficial to the mussels because at the marsh edges are favorable growth conditions and early sexual maturation (Franz, 2001). The mussel feed by sucking in and filtering water through its siphon. The effect of Roundup on ribbed mussels has shown a 100% death rate at percentages of 10, 5, 2.5 and 1.25 while living under lab conditions (Nieto, 2002). Although these mussels were exposed to Roundup under laboratory conditions, the effect of glyphosate is still relevant. Depending on the dosage applied, the chemical can possibly kill an entire species.

Microalgae and cyanobacteria are able to metabolize these organophosphates (Ahmed and Casida, 1958; Christie, 1969; Zuckerman et al., 1970; Megharaj et al., 1987). When the insecticide methyl parathion was added to soil samples as enrichment, which can be done if a sediment cap is to be used, a total of three unicellular green algae species and seven cyanobacteria species were found after a number of days (Megharaj et al., 1994). When two microalgae species and four cyanobacteria species were isolated and tested for their ability to use methyl parathion as a source of phosphorus and nitrogen, all four cyanobacteria species grew very well, but the two green algae species did not (Megharaj et al., 1994). Although this was the case, all six species, two microalgae (*Chlorella vulgaris* and *Scendesmus bijugatus*) and four cyanobacteria (*Nostoc linckia*, *N. muscorum*, *Oscillatoria animalis*, and *Phormidium foveolarum*), were able to hydrolyze the insecticide methyl parathion as evidenced by the presence of purine nucleoside phosphorylase (PNP) (Megharaj et al., 1994).

20.3 Fungi

Farms use agrochemicals, such as pesticides, in the environment that contribute to the pollution of the soil and water. Conventional treatments, such as flocculation, filtration, precipitation, absorption and activated sludge, have drawbacks because the pollution has to be transferred from one location to another (Rigot and Matsumura, 2002; Badawy et al., 2006; Assassi et al., 2011) so more effective and cost-efficient methods are being sought. Fungi are able to degrade these contaminants because they have enzymes that can break down pesticide residues, such as chlordane, lindane, DDT (Ezzi and Lynch, 2005; Zhou et al., 2007), dieldrin, endosulfan, pentachloronitrobenzene and pentachlorophenol (Katayama and Matsumura, 1993).

Unlike pesticides, heavy metals cannot be destroyed. Heavy metals, such as cadmium, copper mercury, manganese, zinc and arsenic, find their way into the environment via wastewater, both geochemical and industrial, and as ingredients in pesticides, fertilizers, wood preservatives and activities, such as tanning (Errasquin and Vazquez, 2003; Tripathi et al., 2007). Fungi are able to detoxify these substances because they can tolerate these substances (Doblera et al., 2000). For example, *Trichoderma*, a genus of soil fungi in the order Hypocreales and division Ascomycota, can tolerate heavy metals such as Ni, As, and Zn (Tripathi et al., 2013) and more than one kind as well (Kredics et al., 2001; Errasquin and Vazquez, 2003). Some microorganisms are able to do this because they may place the metals into vacuoles (Volesky et al., 1993) or release them from their membranes (Kamizomo et al., 1989). Fungi such as *Trichoderma* deals with heavy metals by (1) biosorption, (2) bioaccumulation, (3) biovolatilization or (4) phytobial remediation (Tripathi et al., 2013). In biosorption, heavy metals are taken from wastewater and accumulate by binding to negatively charged free groups in biopolymers in the cell wall (Volesky et al., 1993; Gomes et al., 1998; Errasquin and Vazquez, 2003; Cernansky et al., 2007; Ting and Choong, 2009; see Figure 20.3a). In bioaccumulation, metals are moved into cells that are able to tolerate them (Alguacil and Merino, 1998; Gomes et al., 1998; Errasquin and Vazquez, 2003; Ting and Choong, 2009; (Figure 20.3b). Some species of fungi in the genus *Trichoderma* can do this with copper, zinc, cadmium and arsenic (Errasquin and Vazquez, 2003; Harman et al., 2004; Zeng et al., 2010). In biovolatilization, organic and inorganic compounds of the metal are converted into their volatile derivatives by enzymes in a biochemical reaction (Mukhopadhyay et al., 2002; see Figure 20.3c). Fungi, such as *T. asprellum* and *T. viridae*, do this to arsenic (Urik et al., 2007; Zeng et al., 2010; Srivastava et al., 2011). Phytobial remediation includes the involvement of plants (Harman et al., 2004), where the organic contaminant is degraded and detoxified because of the fungal–plant association (Cao et al., 2008; Figure 20.3d). The fungus, *Trichoderma* spp., increases the uptake of the metals in the plant roots with extracts them from

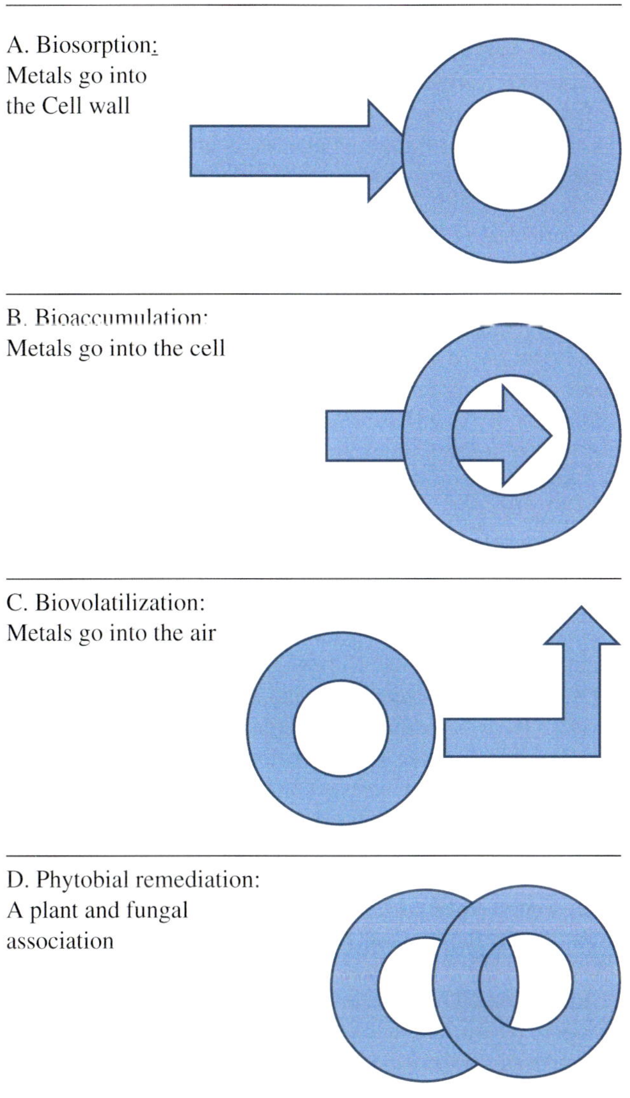

FIGURE 20.3 Diagrams of biosorption, bioaccumulation, biovolatilization and phytobial remediation.

the environment (Harman et al., 2004; Cao et al., 2008). This is done by the fungus *Trichoderma atroviridae* to metals, such as Ni, Zn and Cd, in the plant *Brassica juncea* (Adams et al., 2007).

As with bacteria, the appropriate fungi could be added to the sediment of a cap if the contaminant in the sediment to be capped is known. For example, if the sediment to be capped contains arsenic, fungi such as *T. asprellum* and *T. viridae* could be added to the sediment of the cap. If the sediment to be capped contains Ni, Zn and Cd, the fungus *Trichoderma atroviridae* and the plant *Brassica juncea* could be added to the sediment of the cap since this is phytobial remediation which involves both fungi and plants (see Figures 20.3d and 20.4). Because the growth of plants can be slow, transplantation of the appropriate plant would be quicker than planting seeds of the appropriate plant. The sediment of choice for the fungus and the plant would also have to be suitable to both organisms. This would have to be pre-determined before placement of the sediment cap. If the cap is to cover contaminated sediment that is in the water, the fungi and plants would have to be able to tolerate standing water. The plant might also have to be an aquatic plant depending on the depth of the water. These requirements could limit the options that are available. Experiments would have to determine the best organisms to use in this case.

One type of organic contaminant is PAHs. These compounds find their way into the environment by falling out of the air, being released as sewage from current or previous industries or by accident (Pashin and Bakhitova, 1979; Cerniglia, 1992). Because of their fused benzene rings, they are difficult to degrade by microorganisms (Matsubara et al., 2006). Bacteria (Jin et al., 2011) and many filamentous fungi can degrade PAHs (Das and Chandran, 2011) and fungi of the genus *Trichoderma*are tolerant to crude oil, naphthalene, phenanthrene and benzol[a] pyrene (Argumedo-Delira et al., 2012). Another organic pollutant is pentachlorophenol (PCP), which also has an aromatic ring structure, which makes it remain stable in the soil (Scelza et al., 2008). The fungus *Phanerochaete chrysosporium* can mediate the removal of PCP (Logan et al., 1994; Shim and Kawamoto, 2002; Yu et al., 2011) as can *Trametes versicolor* (Walter et al., 2005), and *T. versicolor, Bjerkanders adusta, Bjerkandera fumosa* and/or *Lopharia spadicea* removed 56% of pyrene from soil (Anastasi et al., 2009).

For an organism to degrade a contaminant, it must first be able to tolerate that contaminant and not die in its presence. Some species have that natural ability, but others do not. Genetic modification can be done to improve this situation. Fungi of the genus *Trichoderma* was genetically modified to improve their tolerance to fungicides, such as benomyl and carbendazim, (Papvizas and Lewis, 1983), heavy metals (Kredics et al., 2001) and the degradation of cyanide (Zhou et al., 2007) and cadmium by phytoremediation with the plant *Brassics napus* (Wang, Liu, et al., 2009). If the fungus is to be placed in a sediment cap, the fungus must also be able to tolerate the sediment of the cap and where the cap is placed. This means if the cap is to be placed over sediment that is underwater, the fungus must be able to tolerate this condition.

20.3.1 Plants

Biodegradation by plants is known as phytoremediation which is the use of green plants and associated microorganisms, such as bacteria and fungi, to remove environmental contaminants (Nawaz et al., 2011). If the degradation is done by plants alone, it is known as phytodegradation, but if it is done together with bacteria or fungi, which were stimulated to act by the plant, it is known as phytostimulation (Nawaz et al., 2011).

Plants in buffer zones or filter strips can be used to recover contaminants from the soil or water (Borner, 1994). A marsh plant known to do this is the reed *Phragmites australis.* It takes in contaminants, such as metals, and holds onto them like a filter. If they were to be removed and replanted, they could be a more effective filter, although such a practice would not be acceptable as *P. australis* is invasive species in many areas. Even after removal, some plants can accumulate and transform toxicants (Dobson et al., 1997).

For an example of phytodegradation, pesticides that enter plants may be converted into less toxic forms for further

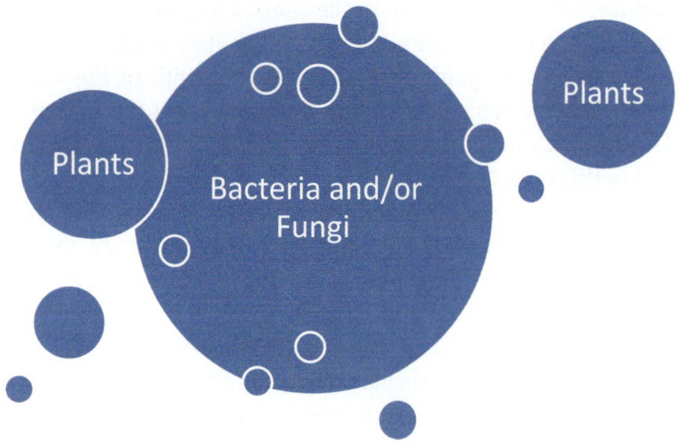

FIGURE 20.4 Phytodegradation (plants alone) versus phytostimulation (plants and bacteria and/or fungi). Circles are conceptual and do not indicate the sizes of the organisms.

degradation, or they may be placed into the waterproofing material called lignin (Castro-Diaz et al., 2001; Conger and Portier, 1997; Davis et al., 2003; Hall et al., 2001; McCutcheon and Schnoor, 2003). A salt marsh plant, the cordgrass *Spartina alterniflora* treats metals like it does salt and moves them to salt glands on their leaves. When it rains, the salt and the metals are washed back into the marsh. In this way, *S. alterniflora* is not the filter that its space competitor, *P. australis*, is but its presence is tolerated where it is not invasive. Volatile compounds in the soil may also be transported into the atmosphere by the plant (Davis et al., 2002, 2003). For an example of phytostimulation, endophytic bacteria that live in the favorable environment of the root zone of plants, the rhizosphere, enter in the plant, promote its growth and aid in the degradation of soil pollutants (Bacon and White, 2000; Sessitsch et al., 2002; Anderson et al., 1993; Cunningham et al., 1996; Curl and Truelove, 1986; Davis et al., 1998, 2003). For bacteria in the soil that do not enter the plant, they transform the pesticides that sorb onto the organic matter created by dead roots (Karthikeyan et al., 2004). Aromatic compounds such as PCBs, PAHs, and pesticides like DDT are degraded by microbes stimulated by plants (Leigh et al., 2002).

Woody trees, such as black walnut, white oak and tulip poplar trees, can move PCBs into their bark after exposure to a landfill contaminated with PCBs that is 14 km away (Meredith and Hites, 1987). Other contaminants found in bark include hexachlorocyclohexanes (HCHs), hexachlorobenzene (HCB), DDT, dieldrin, Aldrin, chlordanes, endrins and endosulfans (Simonich and Hites, 1997). An organic chemical could be held in the plant for long periods, during which transformation by metabolism would take place (Trapp et al., 2001). The metabolism could take place in living parenchyma cells. Chemicals such as dieldrin, chlorobenzene and DDT can be found 1 week after entering the plant (Trapp et al., 1990). The compounds that are volatile could diffuse through the plant and be removed into the atmosphere (Zhang et al., 2001). Mango leaves and pine needles have been found to have HCHs and HCB (Calamari et al., 1995).

Woody trees may be tolerant to pesticides because the compounds are either diluted during transpiration (Sundaram, 1965), lost by diffusion through the leaves and stems or are bound by the plant and metabolized (Akinyemiju et al., 1983; Burken and Schnoor, 1996, 1997, 1998; Field and Peel, 1971a, 1971b, 1972; Lund-Hoie, 1969; Schnabel and White, 2001; Wichman and Byrnes, 1975). The pesticide could also be incorporated into the phloem or bound to roots or other organs, where they could reach toxic levels (Akinyemiju et al., 1983; Dhillion et al., 1968; Hammer and Tukey, 1946; Leonard et al., 1966; Norris and Freed, 1966a, 1966b, 1966c; Pallas, 1963; Sundaram, 1965; Wichman and Byrnes, 1975). In trees that are more tolerant, the chemical moves most in the xylem, but in trees that are less tolerant, there is less movement of the chemical in the phloem, even though it has more phloem movement (Sundaram, 1965). The movement of chemicals in the xylem resulted in herbicides, amitrole and 2,4,5-trichlorophenoxyacetic acid (2,4,5-T) in the leaves but did not make it to the phloem and the roots, possibly due to injury in the phloem (Leonard et al., 1966). In tolerant trees, such as willows, herbicides, 2, 4-dichlorophenoxyacetic acid (2,4-D) and 2,4,5-T were converted to other compounds within 24 hours (Field and Peel, 1971a, 1971b; 1972).

Some plants, such as crops and grasses, can metabolize herbicides (Hamilton, 1964; Harris and Sana, 1967; Lichtenstein, 1959, 1960; Lichtenstein and Schulz, 1960, 1965; Lichtenstein et al., 1965). In general, they bind the pesticide and form residues that are less toxic (Karthikeyan et al., 2004). This amounts to grasses in riparian buffer strips being are able to hold 70% of (2,4-D) (Asmussen et al., 1977), 86% to 90% of trifluralin (Rohde et al., 1980) and 11% to 100% of atrazine in runoff (Arora et al., 1996; Misra et al., 1996). Grasses have been used to make buffer strips along polluted waters (Angier et al., 2002; Barling and Moore, 1994; Dillaha et al., 1988, 1989; Dosskey, 2002; Patterson and Schnoor, 1992; Rankins et al., 1999, 2001). The degradation of herbicides, such as atrazine, by prairie grasses can take place without the help of soil microorganisms and, in fact, becomes ineffective when soil microorganisms are present although the concentration of metolachlor occurs with and without the presence of soil microorganisms (Zhao et al., 2003). The reduction of the amount of metolachlor, rather than atrazine, was because of metolachlor's water solubility and lipophilicity (Zhao et al., 2003). This reduction was mostly due to infiltration into the root zone rather than by uptake (Dosskey, 2002).

Tolerance of enzymes in plants to degrade toxic organic compounds (Kawahigashi, 2009), explosives (Van Aken, 2009), tricholoroethylene (TCE) (James and Strand, 2009), and PCBs (Sylvestre et al., 2009) can be improved by recombinant DNA technology to create transgenic plants. Such technology can make the plant resistant to methyl parathion and could provide a way to degrade organo-phosphorous pollutants (Wang et al., 2008).

20.3.2 Animals

Studies with animals such as earthworms have focused on their use in assessing biodegradation rather than their ability to perform biodegradation. A method for measuring soil toxicity with earthworms is survival assays, which were discussed earlier (Green et al., 2001). Because earthworms are sensitive to soil that has been contaminated with hydrocarbons, such

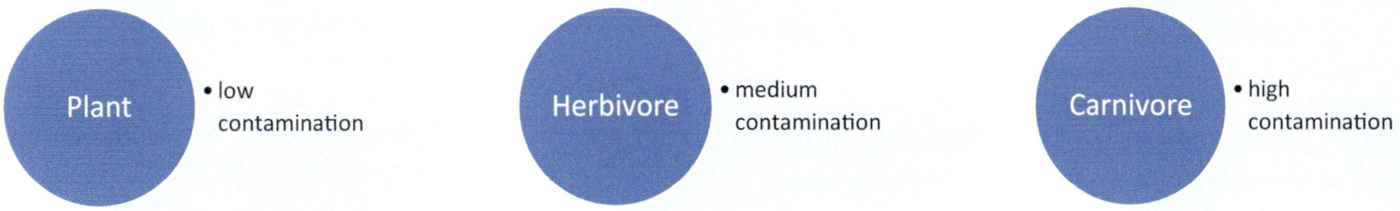

FIGURE 20.5 Bioaccumulation of contaminants due to ingestions up the food chain.

as oil (Shakir et al., 2002), they are able to be used to assess changes in soil toxicity during bioremediation of soil contaminated with PAHs or petroleum hydrocarbons (Athey et al., 1989; Dorn et al., 1998; Marwood et al., 1998; Knoke et al., 1999). Even when total petroleum hydrocarbons (TPHs) have been reduced after treatment, earthworms still had an increase in toxicity, which indicates that the metabolites of the TPHs are toxic (Phillips et al., 2000). Earthworms are 1.4 to 14 times more sensitive microbes to soil contaminated with oil (Dorn et al., 1998).

Although they may not perform degradation themselves, the burrows that earthworms, as well as other animals, make could aid microbes in the degradation process. This is because their burrows could provide oxygen to the soil and speed up the rate of degradation, since oxygen is required (Scholz et al., 1998) and since anaerobic conditions slow down degradation rates (Odukkathil and Vansudevan, 2013). Earthworm burrows could allow more water penetration which would affect the moisture of the soil. Water would make water soluble pesticides available to the microbes (Odukkathil and Vansudevan, 2013) and increases the degradation rate of some contaminants like chlorpyrifos and fenamphifos (Daniel and Timothy, 1991) and flouroxypar (Ling and Hong, 2011) but not others, such as HCH (Mathava and Ligy, 2006) and endosulfan (Arshad et al., 2008; Mathava and Ligy, 2006). Earthworms and other burrowing animals also perform bioturbation, which mixes the sediment, and if it contains contaminants, the contaminants can go into the soil in such a way that it is transported farther than it would have been without the aid of the animal which can spread the contaminants and increase the need for bioremediation.

Another disadvantage that animals and other organisms can cause for biodegradation is bioaccumulation. This is due to the ingestion of contaminants that become concentrated in the digestive system of one animal, which is concentrated again and again as organisms are ingested by others higher up the food chain. This can also include plants and animals. Plants are autotrophs/primary producers. If they take up contaminants and fail to degrade them or degrade them to a more toxic form and are eaten by an animal/herbivore, which is a primary consumer, the contaminants will be concentrated in the digestive system of that animal. When that animal is consumed by another animal/predator/carnivore, that animal, a secondary consumer, will further concentrate the contaminant in its digestive system (Figure 20.5). Figure 20.5 is a very short food chain. In reality, food chains are much longer and are not linear, which results in a food web. This is because animals may feed at more than one trophic level. For example, and animal could herbivorous and feed at the first trophic level, but it may also be a carnivore and feed at the second trophic level or the third trophic level or higher. Doing so would make it an omnivore. In addition to this, the plant that takes in the contaminant does not have to be alive when it is consumed by the animal. In this case, the animal is a scavenger and is consuming detritus.

The longer the food chain or food web means that the contamination could become even higher for those at the top of the food chain or web which are where we are as humans. For example, plastic pollution has caused the breakdown of plastics into micro-plastic particles which are ingested by marine and aquatic animals. Animals such as mussels are filter feeders and consume these microplastic particles. According to Burrows (2018), humans who eat mussels that have ingested microplastics particles consume about 70 micro-particles of plastic for every 3 ounces of mussels eaten. In this way, one can see the impact of being at the top of the food chain/web.

20.4 Conclusion

Biodegradation is the least costly means of remediating a contaminated environment. This is because the biodegrading organisms have biochemical needs and if these needs can be met by the contaminant or pesticide, the contaminant or pesticide can be degraded to a less toxic form, in most cases but not in all cases. Biodegradation is the least costly if the biodegrading organism such as bacteria, microalgae, fungi, and or plant is naturally found in the environment that needs remediation. If the organism or organisms are not found in the environment of interest, the cost will increase because they have to be transported there. Their placement in such an environment can also be within a sediment cap with sediments that are both appropriate for the sediment type of the environment and suitable for the biodegrading organism or organisms. The use of a cap will increase the cost even more because the sediments of the cap would have to be transported to the site and the biodegrading organism or organisms would have to be incorporated into the cap. Although this is the case, the cost will still be below the cost of removing the contaminated sediment, transporting it to an appropriate location of treatment and treating it at that location.

REFERENCES

Abdelbagi, A., Elmahi, M., Osman, D. 2000. Chlorinated hydrocarbon insecticides residues in the Sudanese Soils of limited or no pesticide use. *Arab J. Plant Prot.* 18(1):35–39.

Abdelbagi, A., Elmahi, M., Osman, D. 2003. Organochlorine insecticides residues in Sudanese soils of intense pesticide use and in the surface soil of a Qurashi pesticide store, University of Khartou. *J. Agric. Sci.* 11(1):59–68.

Abhilash, P. C., Nandita, S. 2009. Pesticide use and application: An Indian scenario. *J. Hazard. Mater.* 156(1–3):1–12.

Adams, P., De-Leij, F. A. A. M., Lynch, J. M. 2007. *Trichoderma harzianum* Rifai 1295–22 mediates growth promotion of crack willow (*Salix fragilis*) saplings in both clean and metal-contaminated soil. *Microbial Ecol.* 54:306–313.

Agarwal, S., Al-Abed, S. R., Dionysiou, D. D. 2007a. In situ technologies for reclamation of PCB-contaminated sediments: Current challenges and research thrust areas. *J. Environ. Eng.* 133:1075–1078.

Agarwal, S., Al-Abed, S. R., Dionysiou, D. D. 2007b. Enhanced corrosion-based Pd/Mg bimetallic systems for dechlorination of PCBs. *Environ. Sci. Technol.* 41:3722–3727.

Ahmed, M. K., Casida, J. E. 1958. Metabolism of some organophosphorus insecticides by microorganisms. *J. Econ. Entomol.* 51:59–63.

Akinyemiju, O. A., Dickmann, D., Leavitt, R. 1983. Distribution and metabolism of simazine-tolerant and – intolerant poplar (*Populus* sp.) clones. *Weed Res.* 31:775–778.

Alburquerque, J. A., de la Fuente, C., Bernal, M. P. 2011. Improvement of soil quality after "alperujo" compost application to two contaminated soils characterized by differing heavy metal solubility. *J. Environ. Manag.* 92:733–741.

Alguacil, F. J., Merino, Y. 1998. Biotratamiento de contaminantes de origen inorgáníco. *Rev. Metal Madris.* 34:428–436.

Anastasi, A., Coppola, T., Prigione, V., Varese, G. C. 2009. Pyrene degradation and detoxification in soil by a consortium of basidiomycetes isolated from compost: Role of laccases and peroxidases. *J. Hazard. Mater.* 156:1229–1233.

Anderson, T. A., Guthrie, E. A., Walton, B. T. 1993. Bioremediation. *Environ. Sci. Technol.* 27:2630–2636.

Angier, J. T., McCarty, G. W., Rice, C. P., Bialek, K. 2002. Influence of a riparian wetland on nitrate and herbicides exported from an agricultural field. *J. Agric. Food Chem.* 50:4424–4429.

Argumedo-Delira, R., Alarcon, A., Ferrera-Cerrato, R., Almaraz, J. J., Pena-Cabriales, J. J. 2012. Tolerance and growth of 11 *Trichoderma* strains to crude oil, naphthalene, phenanthrene and benzo(*a*)pyrene. *J. Environ. Manage.* 95:291–299.

Arora, K., Mickelson, S. K., Baker, J. L., Tierney, D. P. 1996. Herbicide retention by vegetative buffer strips from runoff under natural rainfall. *Trans. ASAE.* 39(6):2155–2162.

Arshad, M., Hussain, S., Saleem, M. 2008. Optimization of environmental parameters for biodegradation of alpha and beta endosulfan in soil slurry by *Pseudomonas aeruginosa*. *J. Appl. Microbiol.* 104:1364–5072.

Asmussen, L. E., White, A. W., Jr., Hauser, E. W., Sheridan, J. M. 1977. Reduction of 2,4-D load in surface runoff down a grassed waterway. *J. Environ. Qual.* 6(2):159–162.

Assassi, M., Fourcade, F., Geneste, F., Floner, D. Maachi, R., Amrane, A. 2011. Combined electrochemical and biological treatment for pesticide degradation: Application to phosmet. *Int. J. Environ. Res.* 5:41–46.

Athey, L. A., Thomas, J. M., Miller, W. E., Word, J. Q. 1989. Evaluation of bioassays for designing sediment cleanup strategies at a wood treatment site. *Environ. Toxicol. Chem.* 8:223–230.

Azcue, J. M., Zeman, A. J., Mudroch, A., Rosa, F., Patterson, T. 1998. Concept of subaqueous capping of contaminated sediment with active barrier system (ABS) using natural and modified zeolites. *Water Sci. Technol.* 37:323.

Bacon, C., White, J. 2000. *Microbial Endophytes*. Marcel Dekker, Inc., New York, NY.

Badawy, M. I., Ghaly, M. Y., Gad-Allah, T. A. 2006. Advanced oxidation processes for the removal of organo phosphorus pesticides from wastewater. *Desalination.* 194:166–175.

Baelum, J., Prestat, E., David, M. M., Strobel, B. W., Jacobsen, C. S. 2012. Modeling of phenoxy acid herbicide mineralization and growth of microbial degrades in 15 soils monitored by quantitative real-time PCR of the functional *rfd*A gene. *Appl. Environ. Microbiol.* 78:5305–5312.

Barling, R. D., Moore, I. D. 1994. Role of buffer strips in management of waterway pollution: A review. *Environ. Manage.* 18:543–558.

Bers, K., Mot, R. D., Springael, D. 2013. *In situ* response of the linuron degradation potential to linuron application in an agricultural field. *FEMS Microbiol Ecol.* 85:403–416.

Bertness, M. D. 1984. Ribbed mussels and *Spartina alterniflora* production in a New England Salt Marsh. *Ecology.* 65(6):1794–1807.

Bokuniewicz, H. J. 1992. Analytical description of subaqueous groundwater seepage. *Estuaries.* 15:458–464.

Borner, H. 1994. *Pesticides in Ground and Surface Water*. Chemistry of Plant Protection and Springer-Verlag, New York.

Burken, J. G., Schnoor, J. L. 1996. Phytoremediation: Plant uptake of atrazine and role of root exudates. *J. Environ. Eng.* 122:958–963.

Burken, J. G., Schnoor, J. L. 1997. Uptake and metabolism of atrazine by poplar trees. *Environ. Sci. Technol.* 31:1399–1406.

Burken, J. G., Schnoor, J. L. 1998. Predictive relationships for uptake or organic contaminants by hybrid poplar trees. *Environ. Sci. Technol.* 32(2):3379–3385.

Burrows, S. 2018. Microplastics in mussels: The ocean is feeding our garbage back to us. https://returntonow.net/2018/07/06/microplastics-in-mussels-the-ocean-is-feeding-our-garbage-back-to-us/.

Calamari, D., Tremolada, P., Notarianni, V. 1995. Relationship between chlorinated hydrocarbons in vegetation and socioeconomic indices on a global scale. *Environ. Sci. Tecnol.* 29:2267–2272.

Cao, L., Jiang, M., Zeng, Z., Du, A., Tan, H., Liu, Y. 2008. *Trichoderma atroviride* F6 improves phytoextraction efficiency of mustard [*Brassica juncea* (L.) Coss. Var. foliosa Bailey] in Cd, Ni contaminated soils. *Chemosphere.* 71:1769–1773.

Castro-Diaz, S., Davis, L., C., Erickson, L. E. 2001. Plant enhanced remediated of glycol-based aircraft deicing fluids. *Pract. Periodical Hazard. Toxic. Radioact. Waste Manage.* 5:141–152.

Cernansky, S., Urik, M., Sevc, J., Khun, M. 2007. Biosorption and biovolatilization of arsenic by heat-resistant fungi. *Env. Sci. Pollut. Res.* 14:31–35.

Cerniglia, C. E. 1992. Biodegradation of polycyclic aromatic hydrocarbons. *Biodegradation.* 3:351–368.

Chang, C., Chang, C., Yu, W., Lenie, D., Mario, V., Chung, T. 2005. Species-level identification of isolates of the *Acinetobacter calcoaceticus Acinetobacter baumanni* complex by sequence analysis of the 165–23S rRNA gene spacer region. *J. Clin. Microbiol.* 43(4):1632–1639.

Chen-Goodspeed, M., Sogorb., A., Wu, F., Hong, B., Raushel, M. 2001. Structural determinants of the substrate and stereochemical specificity of phosphotriesterase. *Biochemistry.* 40(5):1325–1331.

Christie, A. E. 1969. Effects of insecticides on algae. *Water Sewage Wks.* 166:172–176.

Conger, R. M., Portier, R. 1997. Phytoremediation experimentation with herbicide bentazon. *Remediation.* 7(2):19–37.

Cunningham, S. D., Anderson, T. A. Schwab, A. P., Hsu, F.C. 1996. Phytoremediation of soils contaminated with organic pollutants. *Adv. Argon.* 56:55–114.

Curl, E., A. Truelove, B. 1986. *The Rhizosphere.* Springer-Verlag, Berlin.

Daniel, R. S., Timothy, B. P. 1991. Effect of moisture on sorption and biodegradation of carbofuran in soil. *J. Agric. Food Chem.* 39:2060–2063.

Das, N., Chandran, P. 2011. Microbial degradation of petroleum hydrocarbon contaminants: an overview. *Biotechnol. Res. Int.* doi:10.4061/2011/941810.

Davis, L. C., Banks, M. K., Schwab, A. P., Narayanan, M., Erickson, L. E. 1998. Plant-based bioremediation. In Sikdar, S. K., Irvine, R. L. (Eds.), *Bioremediation: Principles and Practice. Biodegradation Technology Development.* Technomic Publishing Co., Lancaster, Vol. 2, pp. 183–219.

Davis, L. C., Castro-Diaz, S., Zhang, Q., Erickson, L. E. 2002. Benefits of vegetation for soils with organic contaminants. *Crit. Rev. Plant Sci.* 21(5):457–491.

Davis, L. C., Erickson, L. E., Narayanan, M., Zhang, Q. 2003. Modeling and design of phytoremediation. In McCutcheon, S. C., Schnoor, J. L. (Eds.), *Pytoremediation: Transformation and Control of Contaminants.* Wiley, New York, pp. 663–694.

Dhillion, P. S., Byrnes, W. R., Merritt, C. 1968. Simazine distribution and degradation in red pine seedlings. *Weed Sci.* 16:374–376.

DiGennaro, P., Moreno, B., Annoni, E., Garcia-Rodriguez, S., Bestetti, G., Benitez, E. 2009. Dynamic changes in bacterial community structure and in naphihalene dioxygenase expression in vermicompost-amended PAH-contaminated soils. *J. Hazard. Mater.* 172:1464–1469.

Dillaha, T. A., Reneau, R. B., Mostaghimi, S., Lee, D. 1989. Vegetative filter strips for agricultural nonpoint source pollutant control. *Trans. ASAE.* 32:513–519.

Dillaha, T. A., Sherrard, R., Lee, D., Mostaghimi, S., Shanholtz, V. O. 1988. Evaluation of vegetative filter strips as a best management practice for feed lots. *J. Water Pollut. Con. Fed.* 60:1231–1238.

Doblera, R., Sanera, M., Bachofena, R. 2000. Population changes of soil microbial communities induced by hydrocarbon and heavy metal contamination. *Bioremediation J.* 4:41–56.

Dobson, A. P., Bradshaw, A. D., Baker, A. J. M. 1997. Hopes for the future: Restoration ecology and conservation biology. *Science.* 277:515–522.

Dortch, M. S., Hales, L. Z., Letter, J. V., McAnally, W. H. 1990. Methods of determining the long term fate of dredged material for aquatic disposal sites, Technical Report, D-90–1, U.S. Army Engineer Waterways Experiment Station, Vicksburg, MS.

Dorn, P. B., Vipond, T. E., Salanitro, J. P., Wisniewskie, H. L. 1998. Assessment of the acute toxicity of crude oils in soils using earthworms, microtox, and plants. *Chemosphere.* 37:845–860.

Dosskey, M. G. 2002. Setting priorities for research on pollution reduction functions of agricultural buffers. *Environ. Manage.* 30(5):641–650.

Dua, M., Singh, A., Sethunathan, N. Johri, A. K. 2002. Biotechnology and bioremediation successes and limitations. *Appl. Microbiol. Biotechnol.* 59:143–152.

Eek, E., Cornelissen, G., Kibsgaard, A., Breedveld, G. D. 2008. Diffusion of PAH and PCB from contaminated sediments with and without mineral capping; Measurement and modeling. *Chemosphere.* 71:1629–1638.

Ellis, R. J., Morgan, P., Weightman, A. J., Fry, J. C. 2003. Cultivation-dependent and independent approaches for determining bacterial diversity in heavy-metal-contaminated soil. *Appl. Enviorn. Microbiol.* 69:3223–3230.

Elmahi, A. 1996. Distribution of chlorinated hydrocarbon pesticides residues in Sudan Soil. M.Sc. thesis, University of Khartourn, Sudan.

Elsaid, G., Abdelbagi, A. 2010. Comparative biodegradation of endosulfan by mutants and their native microorganisms. *Res. J. Agric. Biol. Sci.* 6(6):953–961.

Errasquin, E. L., Vazquez, C. 2003. Tolerance and uptake of heavy metals by *Trichoderma atroviride* isolated from sludge. *Chemosphere.* 50:137–143.

Ezzi, M. I., Lynch, J. M. 2005. Biodegradation of cyanide by *Trichoderma* spp. and *Fusarium* spp. *Enxyme Microb. Technol.* 36:849–854.

Fang, Y., Al-Abed, S. R. 2007. Partitioning, desorption, and dechlorination of a PCB congener in sediment slurry supernatants. *Environ. Sci. Technol.* 41:6253–6258.

Field, R. J., Peel, A. J. 1971a. The metabolism and radial movement of growth regulators and herbicides in willow stems. *New Phytol.* 70:743–749.

Field, R. J., Peel, A. J. 1971b. The movement of growth regulators and herbicides into the sieve tubes of willow. *New Phytol.* 71:249–254.

Field, R. J., Peel, A. J. 1972. The longitudinal mobility of growth regulators and herbicides in sieve tubes of willow. *New Phytol.* 71:249–254.

Foght, J., April, T., Biggar, K., Aislabie, J. 2001. Bioremediation of DDT contaminated soils: a review. *Bioremediat. J.* 255:1058–1137.

Franz, D. R. 2001. Recruitment, survivorship, and age structure of a New York ribbed mussel population (*Geukensia Demissa*) in relation to shore level- a nine year study. *Estuaries.* 24(3):319–327.

Fredette, T. J., French, G. T. 2004. Understanding the physical and environmental consequences of dredged material deposal: History of New England and current perspectives. *Mar. Pollut. Bull.* 49:93–102.

Gandolfi, I., Sicolo, M., Franzetti, A., Fontanarosa, E., Santagostino, A., Bestetti, G. 2010. Influence of compost amendment on microbial community and ecotoxicity of hydrocarbon-contaminated soils. *Bioresour. Technol.* 101:568–575.

Giller, K. E., Witter, E., McGrath, S. P. 1998. Toxicity of heavy metals to microorganisms and microbial processes in agricultural soils: A review. *Soil Biol. Biochem.* 30:1389–1414.

Gomes, N. C. M., Mendonca-Hagler, L., Savvaidis, I. 1998. Metal bioremediation by microorganisms. *Rev. Microbiol.* 29:85–92.

Green, J. C., Bartels, C. L., Warren-Hicks, W. J., Parkhurst, B. R., Linder, G. L., Peterson, S. A., Hall, J. C., Hoagland, R. E., Zablotowicz, R. M. 2001. *Pesticide Biotransformation in Plants and Microorganisms. Similarities and Divergences.* ACS Symposium Series 777. ASC, Washington, DC.

Hamilton, R. H. 1964. Tolerance of several grass species to 2-chloro-s-triazine herbicides in relation to degradation and content of benzoxazinone derivatives. *J. Agric. Food Chem.* 12:14–17.

Hammer, C. L., Tukey, H. B. 1946. Herbicidal action of 2–4-dichlorophenoxyacetic acid on several shrubs, vines, and trees. *Bot. Gaz.* 108:379–385.

Harman, G. E., Lorito, M., Lynch, J. M. 2004b. Uses of *Trichoderma* spp. to remediate soil and water pollution. *Adv. Appl. Microbiol.* 56:313–330.

Haroun, M., Idris, A., Omar, S. R. S. 2007. A study of heavy metals and their fate in the composting of tannery sludge. *Waste Manag.* 27:1541–1550.

Harris, C. R., Sana, W. W. 1967. Absorption of organochlorine insecticide residues from agricultural soils by root crops. *J. Agric. Food Chem.* 15:861–863.

Hassen, A., Saidi, N., Cherif, M., Boudabous, A. 1998. Resistance of environmental bacteria to heavy metals. *Bioresour. Technol.* 64:7–15.

Herbicide Resistance Action Committee (HRAC). www.hracglobal.com.

Horne, L., Sutherland, D., Harcourt, L., Russel, J., Oakeshott, G., 2002. Identification of an opd (organophosphate degradation) gene in an Agrobacteriumhc isolate. *Appl. Environ. Microbiol.* 68(7):3371–3376.

Hussain, S., Arshad, M., Springael, D., SøRensen, S. R., Bending, G. D., DeversLamrani, M., Maqbool, Z., Martin-Laurent, F. 2015. Abiotic and biotic processes governing the fate of phenylurea herbicides in soils: A review. *Crit. Rev. Environ. Sci. Technol.* 45:1947–1998.

Huttel, M., Webster, I. T. 2001. Pore water flow in permeable sediments. In Boudreau, B. P., Jørgensen, B. B. (Eds.), *The Benthic Boundary Layer*. Oxford University Press, New York, pp. 144–179.

Huttunen, J. T., Lappalainen, K. M., Saarijarvi, E., Vaisanen, T., Marikainen, P. J. 2001. A novel sediment gas sampler and a subsurface gas collector used for measurement of the ebullition of methane and carbon dioxide from a eutrophied lake. *Sci. Total Environ.* 266:153–158.

Ishag, A. E. S. A., Abdelbagi, A. O., Hammad, A. M. A., Elsheikh, E. A. E., Elsaid, O. E., Hur, J.-H., Laing, M. D. 2016. Biodegradtion of chlorpyrifos, malathion, and dimethoate by three strains of bacteria isolated from pesticide-polluted soils in Sudan. *J. Agri. Food Chem.* 8491–8498.

Jacobs, P. H., Förstner, U. 1999. Concept of subaqueous capping of contaminated sediments with active barrier systems (ABS) using natural and modified zeolites. *Water Res.* 33:2083–2087.

Jacobs, P. H., Waite, T. D. 2004. The role of aqueous iron(II) and manganese(II) in sub-aqueous active barrier systems containing natural clinoptilolite. *Chemosphere*. 54:313–324.

James, C. A., Strand, S. E. 2009. Pytoremediation of small organic contaminants using transgenic plants. *Curr. Opin. Biotechnol.* 20:237–241.

Jin, H. M., Jeong, H., Moon, E. J., Math, R. K. Lee, K., Kim, H. J., Jeon, C. O., Oh, T. K., Kim, J. F. 2011. Complete genome sequence of the polycyclic aromatic hydrocarbon-degrading bacterium *Alteromonas* sp. strain SN2 J. *Bacteriol.* 193:4292–4293.

Jonker, M. T. O., Koelmans, A. A. 2002. Sorption of polycyclic aromatic hydrocarbons and polychlorinated biphenyls to soot and soot-like materials in the aqueous environment: Mechanistic considerations. *Environ. Sci. Technol.* 36:3725–3734.

Kamizomo, A., Nishizawa, M., Teranishi, Y., Murata, K., Kimura, A. 1989. Identification of a gene conferring resistance to zinc and cadmium in the yeast *Saccharomyces cerevisiae. Mol. Gen. Genet.* 219:161–167.

Karthikeyan, R., Lawrence, C. D., Erickson, L. E., Al-Khatib, K., Kulakow, P. A., Barnes, P. L., Hutchinson, S. L., Nuzhanova, A. A. 2004. Potential for plant-based remediation of pesticides-contaminated soil and water using nontarget plants such as trees, shrubs, and grasses. *Critical Rev. Plant Sci.* 23:91–101.

Katayama, A., Matsumura, F. 1993. Degradation of organochlorine pesticides, particularly endosulfan, by *Trichoderma harzianum. Environ. Toxicol. Chem.* 12:1059–1065.

Kawahigashi, H. 2009. Transgenic plants for pytoremediation of herbicides. *Curr. Opinion Biotechnol.* 20:225–230.

Kilbride, K. M., Paveglio, F. L. 2001. Long-term fate of glyphoasate associated with repeated rodeo application to control smooth cordgrass (*Spartina alterniflora*) in Willapa Bay, Washington. *Arch. Environ. Contam. Tox.* 40(2):179–183.

Knoke, K. L., Marwood, T. M., Cassidy, M. B., Liu, D., Seech, A. G., Trevors, J. T. 1999. A comparison of five bioassays to monitor toxicity during bioremediation of pentachlorophenol-contaminated soil. *Water, Air and Soil Pollution*. 110:157–159.

Kredics, L., Antal, L., Manczinger, L., Nagy, E. 2001. Breeding of mycoparasitic *Trichoderma* strains for heavy metal resistance. *Lett. Appl. Microbiol.* 33:112–116.

Kumar, A., Bisht, B. S., Joshi, V. D., Dhewa, T. 2011. Review on bioremediation of polluted environment: a management tool. *Int. J. Environ. Sci.* 1(6):1079–1093.

Kumar, A., Trefult, N., Olaniran, A. O. 2016. Microbial degradation of 2,4-dichlorophenoxyacetic acid: insight into the enzymes and catabolic genes involved, their regulation and biotechnological implications. *Crit. Rev. Microbiol.* 42:194–208.

Kumar, M., Lakshmi, C. V., Khanna, S. 2008. Biodegradation and bioremediation of endosulfan contaminated soil. *Bioresour. Technol.* 9(8):3116–3122.

Lal, R., Pandey, G., Sharma, P., Kumari, K., Malhotra, S., Pandey, R., Raina, V., Kohler, H.-P. E., Holliger, C., Jackson, C., Oakeshott, J. G. 2010. Biochemistry of microbial degradation of hexachlorocyclohexane and prospects for bioremediation. *Microbiol. Mol. Rev.* 74:58–80.

Leigh, M. B., Fletcher, J. S., Fu, X., Schmitz, F. J. 2002. Root turnover: An important source of microbial substrate in rhizosphere remediation of recalcitrant contaminants. *Environ. Sci. Technol.* 36:1579–1583.

Leonard, O. A., Bayer, D. E., Glenn, R. K. 1966. Translocation of herbicides and assimilates in red maple and white ash. *Bot. Gaz.* 127(4):193–201.

Lichtenstein, E. P. 1959. Absorption of some chlorinated hydrocarbon insecticides from soils into various crops. *J. Agric. Food Chem.* 7:430–433.

Lichtenstein, E. P. 1960. Insecticidal residues in various crops grown in soils treated with abnormal rates of Aldrin and heptachlor. *J. Agric. Food Chem.* 8:448–451.

Lichtenstein, E. P., Myrdal, G. R., Schulz, K. R. 1965. Absorption of insecticidal residues from contaminated soils into five carrot varieties. *J. Agric. Food Chem.* 13:126–131.

Lichtenstein, E. P., Schulz, K. R. 1960. Translocation of some chlorinated hydrocarbon insecticides into aerial parts of pea plants. *J. Agric. Food Chem.* 8:452–456.

Lichtenstein, E. P., Schulz, K. R. 1965. Residues of Aldrin and heptachlor in soils and their translocation into various crops. *J. Agric. Food Chem.* 13:57–63.

Ling, H. I., Leshchinsky, D., Gilbert, P. A., Palermo, M. R. 1996. *In-situ* capping of contaminated submarine sediments: Geotechnical considerations. In Kamon, M. (Ed.), *Environmental Geotechnics*. Balkema, Rotterdam, pp. 575–580.

Ling, T., Hong, Y. 2011. Fluroxypyr biodegradation in soils by multiple factors. *Environ. Monit. Asses.* 175(1–4):227–238.

Liu, Y., Choi, H., Dionysiou, D. D., Lowry, G. V. 2005. Trichloroethene hydrodechlorination in water by highly disordered monometallic nanoiron. *Chem Mater.* 17:5315–5322.

Logan, B. E., Allemen, B. C., Amy, G. L., Gilbertson, R. L. 1994. Adsorption and removal of pentaclorophenol by white rot fungi in batch culture. *Water Res.* 28:1533–1538.

Lohse, L., Epping, E. H. G., Helder, W., van Raaphoest, W. 1996. Oxygen pore water profiles in contaminated shelf sediments of the North Sea: Turbulent versus molecular diffusion. *Mar. Ecol. Prog. Ser.* 145:63–75.

Lopes, T. J., Furlong, E. T. 2001. Occurrence and potential adverse effects of semivolatile organic compounds in streambed sediment, United States, 1992–1995. *Environ. Toxico. Chem.* 20(4):727–737.

Lund-Hoie, K. 1969. Uptake, translocation, and metabolism of simazine in Norway spruce (*Picea abies*). *Weed Res.* 9:142–147.

Luthy, R. G., Aiken, G. R., Brusseau, M. L., Cunningham, S. D., Gschwend, P. M., Pignatello, J. J., Reinhard, M., Traina, S. J., Weber, W. J., Westall, J. C. 1997. Sequestration of hydrophobic organic contaminants by geosorbents. *Environ. Sci. Technol.* 31:3341–3347.

MacNaughton, S. J., Stephen, J. R., Venosa, A. D., Davis, G. A., Chang, Y.-J., White, D. C. 1999. Microbial population changes during bioremediation of an experimental oil spill. *Appl. Environ. Microbiol.* 65:3566–3574.

Mansour, S. 2002. Evaluation of residual pesticides and heavy metals levels in conventionally and organically farmed potato tubers in Egypt. In He, Z., Larkin, R., Honeycutt, W. (Eds.), *Sustainable Potato Production: Global Case Studies*. Springer, Netherlands, pp. 493–506.

Marwood, T. M., Knoke, K., Yau, K., Suchorski-Tremblay, A., Fleming, C. A., Hodge, V., Liu, D., Seech, A. G., Lee, H., Trevors, J. T. 1998. Comparison of toxicity detected by five bioassays during bioremediation of diesel-spiked soil. *Environ. Toxicol. Water Quality.* 13:117–126.

Mathava, K., Ligy, P. 2006. Bioremediation of endosulfan contaminated soil and water optimization of operating conditions in laboratory scale reactors. *J. Hazard. Mat.* 354–364.

Matsubara, M., Lynch, J. M., De Leij, F. A. A. M. 2006. A simple screening procedure for selecting fungi with potential for use in the bioremediation of contaminated land. *Enz. Microbial Tech.* 39:1365–1372.

McCutcheon, S. C., Schnoor, J. L. (Eds.). 2003. *Phytoremediation: Transformation and Control of Contaminants*. Wiley, New York.

McDonough, K. M. F. J., Lowry, G. 2008. Adsorption of polychlorinated biphenyls to activated carbon: Equilibrium isotherms and a preliminary assessment of the effect of dissolved organic matter and biofilm loadings. *Water Res.* 42(3):575–584.

McDonough, K. M. F. J., Murphy, P., Olsta, J., Zhu, Y. W., Reible, D. D., Lowry, G. V. 2007. Development and placement of a sorbent-amended thin layer sediment cap in the Anacostia River. *Soil Sed. Contam.* 16(3):313–322.

Megharaj, M., Madhavi, D. R., Sreenivasulu, C., Umamaheswari, A., Venkateswarlu, K. 1994. Biodegradation of methyl parathion by soil isolates of microalgae and cyanobacteria. *Bull. Environ. Contam. Toxicol.* 53:292–297.

Megharaj, M., Venkateswarlu, K., Rao, A. S. 1987. Metabolism of monocrotophos and quinalphos by algae isolated from soil. *Bull. Environ. Contam. Toxicol.* 39:251–256.

Melvin, S. D. 2017. Effect of antidepressants on circadian rhythms in fish: Insights and implications regarding the design of behavioural toxicity tests. *Aquat. Tox.* 182:20–30.

Meredith, M. L., Hites, R. A. 1987. Polychlorinated biphenyl accumulation in tree bark and wood growth rings. *Environ. Sci. Technol.* 21:709–712.

Misra, A. K., Baker, J. L., Mickelson, S. K., Shang, H. 1996. Contributing area and concentration effects on herbicide removal by vegetative buffer strips. *Trans. ASAE.* 39(6):2105–2111.

Mohan, R. K., Brown, M. P., Barnes, C. R. 2000. Design criteria and theoretical basis for capping contaminated marine sediments. *Appl. Ocean Res.* 22:85–93.

Morales, J., Hutcheson, R., Cheng, F. 2002. Dechlorination of chlorinated phenols by catalyzed and uncatalyzed Fe(0) and Mg(0) particles. *J. Hazard. Mater.* 90:97–108.

Moreno, J. L., Hernandez, T., Perez, A., Garcia, C. 2002. Toxicity of cadmium to soil microbial activity: Effect of sewage sludge addition to soil on the ecological dose. *Appl. Soil Ecol.* 21:149–158.

Mukhopadhyay, R., Rosen, B. P., Phung, L. T., Silver, S. 2002. Microbiol. Arsenic: From geocycles to genes and enzymes. *FEMS Microbiol. Rev.* 26:311–325.

Mulbry, W., Kearney, P. C. 1991. Degradation of pesticides by micro-organisms and the potential for genetic manipulation. *Crop. Prot.* 10(5):334–346.

Murphy, P., Marquette, A., Reible, D. D., Lowry, V. 2006. Predicting the performance of activated carbon-, coke-, and soil-amended thin layer sediment caps. *J. Environ. Eng.* 132:787–794.

Nagata, Y., Tabata, M., Ohtsubo, Y., Tsuda, M. 2015. Biodegradation of organochlorine pesticides. In Yates, M., Nakatsu, C., Miller, R., Pillai, S. (Eds.), *Manual of Environmental Microbiology*, 4th ed. ASM Press, Washington, DC, pp. 5.12–5.12–30.

The National Institute for Occupational Safety and Health (NIOSH). http://www.cdc.gov/niosh/npg/npgd0043.html.

National Research Council (NRC). 2001. *A Risk-Management Strategy for PCB-Contaminated Sediments*. National Academies Press, Washington, DC.

Nawaz, K., Hussain, K., Choudary, N., Majeed, A., Ilyas, U., Ghani, A., Lin, F., Ali, K., Afghan, S., Raza, G., Lashari, M. I. 2011. Eco-friendly role of biodegradation against agricultural pesticides hazards. *African J. Microbiol. Res.* 5(3):177–183.

Neira, C. 2006. Mechanisms generating modification of benthos following tidal flat invasion by a *Spartina* Hybrid. *Ecological Applications.* 16(4):1391–404.

Nieto, V. 2002. Toxicity tests of roundup on the fiddler crab *Uca Pugnax* (Smith) and the Ribbed Mussel *Geukensia Demissa* (Dillwyn). Fairleigh Dickinson University- University Honors Thesis.

Niti, C., Sunita S., Kamlesh, K., Rakesh, K. 2013. Bioremediation: an emerging technology for remediation of pesticides. *Res. J. Chem. Environ.* 17:88–105.

Norris, L. A., Freed, V. H. 1966a. The absorption and translocation characteristics of several phenoxyalkyl acid herbicides in bigleaf maple. *Weed Res.* 6(3):203–211.

Norris, L. A., Freed, V. H. 1966b. The metabolism of a series of chlorophenoxyalkyl acid herbicides in bigleaf maple. *Weed Res.* 6(3):212–220.

Norris, L. A., Freed, V. H. 1966c. The absorption, translocation, and metabolism characteristics of 4-(2,4-dichlorophenoxy) butyric acid in bigleaf maple. *Weed Res.* 6(4):283–291.

NPIC. 2010. *Glyphosate General Fact Sheet. National Pesticide Information Center*. Oregon State University, September 2010. http://npic.orst.edu/factsheets/glyphogen.pdf.

Obbard, J. 2001. Ecotoxicological assessment of heavy metals in sewage sludge amended soils. *Appl. Geochem.* 16:1405–1411.

Odukkathil, G., Vansudevan, N. 2013. Toxicity and bioremediation of pesticides in agricultural soil. *Rev. Enviorn. Sci. Biotechnol.* 12:421–444.

Palermo, M. R. 1998. Design consideration for *in* situ capping of contaminated sediment. *Water Sci. Technol.* 37:315.

Palermo, M. R., Homziak, J., Teeter, A. M. 1990. Evaluation of clamshell dredging and barge overflow, military ocean terminal, Sunny Point, NC. Tech. Rep. No. D-90–6, U.S. Army Engineer Waterways Experiment Station, Vicksburgh, MI.

Palermo, M. R., Shafer, R. A., Brannon, J. M., Truitt, C. L., Zappi, M. E., Skogerboe, J. G., Adamec, S. A., Sturgis, T. C., Wade, R., Gunnison, D., Myers, T. E. 1989. Evaluation of dredged material disposal alternatives for US navy homeport at Everett, Washington," Technical Report, EL-89–1, US Army Engineer Waterways Experiment Station, Vicksburg, MS.

Palermo, M. R., Thompson, T. A., Swed, F. 2002. *In-Situ* capping as a remedy, component for the lower fox river. Response to a document by the Johnson Company: Ecosystem rehabilitation plan-An integrated plan for habitat enhancement and expedited exposure reduction in the lower Fox River and Green Bay. White Paper No. 6B, Wisconsin site record of decision, operable units 1 and 2; Wisconsin Department of Natural Resources: Madison, Wisconsin; U.S. Environmental Protection Agency Region 5: Chicago, IL.

Pallas, J. E. Jr. 1963. Absorption and translocation of the trimethylamine salt of 2,4-D and 2,4,5-T in four woody species. *For. Sci.* 9:485–491.

Papvizas, G. C., Lewis, J. A. 1983. Physiological and biocontrol characteristics of stable mutants of *Trichoderma viride* resistant to MBC fungicides. *Phytopathology*. 73:407–411.

Pashin, Y. V., Bakhitova, L. M. 1979. Mutagenic and carcinogenic properties of polycyclic aromatic hydrocarbons. *Environ. Health Pers.* 30:185–189.

Patterson, K. G., Schnoor, J. L. 1992. Fate of alachlor and atrazine in a riparian zone field site. *Water Environ. Res.* 64(3):274–283.

Phillips, T. M., Liu, D., Seech, A. G., Lee, H., Trevors, J. T. 2000. Monitoring bioremediation in creosote-contaminated soils using chemical analysis and toxicity tests. *J. Ind. Microbiol. Biotechnol.* 24:132–139.

Polti, M. A., Aparicio, J. D., Benimeli, C. S., Amoroso, M. J. 2014. Simultaneous bioremediation of Cr(VI) and lindane in soil by actinobacteria. *International Biodeterioration & Biodegradation* 88:48–55.

Ponder, S. M., Darab, J. G., Bucher, J., Caulder, D., Craig, I., Davis, L., Edelstein, N., Lukens, W., Nitsche, H., Rao, L., Shuh, D. K., Mallouk, T. E. 2001. Surface chemistry and electrochemistry of supported zerovalent iron nanoparticles in the remediation of aqueous metal contaminants. *Chem. Mater.* 13:479–486.

Rankins, A., Shaw, D. R., Boyette, M. 2001. Perennial grass filter strips for reducing herbicide losses in runoff. *Weed Sci.* 49(5):647–651.

Rankins, A., Shaw, D. R., Shankle, M. W. 1999. Phytotoxic effects of selected herbicides on perennial grasses used as filter strip. *Weed Sci. Soc. Am.* 39:171

Reible, D. D., Hayes, D., Lue-Hing, C., Patterson, J., Bhowmik, N., Johnson, M., Teal, J. 2003. Comparison of the long-term risks of removal and in-situ management of contaminated sediments in the Fox River. *Soil Sediment Contam.: An International Journal*. 12(3):325–344.

Reible, D. D., Lambert, D., Constant, D. W., Mutch, R. D., Zhu, Y. 2006. Active capping demonstration in the Anacostia River, Washington, DC. *Remediation*. 17(1):39–53.

Relyea, R. A. 2003. Predator cues and pesticides: A double dose of danger for amphibians. *Ecol. Appl.* 13(6):1515–1521.

Relyea, R. A. 2005. The impact of insecticides and herbicides on the biodiversity and productivity of aquatic communities. *Ecol. Appl.* 15(2):618–627.

Rice, K. C. 1999. Trace-element concentrations in streambed sediment across the conterminous United States. *Environ. Sci. Technol.* 33(15):2499–2504.

Richardson, M. D. 1998. *Coastal Benthic Boundary Layer: A Final Review of the Program*. NRL Memorandum Report. Marine Geosciences Division-Naval Research Laboratory, Washington, DC.

Rider, C. V., LeBlanc, G. A. 2006. Atrazine stimulates hemoglobin accumulation in *Daphnia magna*: Is it hormonal or hypoxic? *Tox. Sci.* 93(2):443–449.

Rigot, J., Matsumura, F. 2002. Assessment of the rhizosphere competency and Pentachlorophenol-metabolizing activity of a pesticide degrading strain of *Trichoderma harzianum* introduced into the root zone of corn seedlings. *J. Environ. Sci. Health.* 37:201–210.

Rockne, J., Reddy, R. 2003. Invited theme paper, bioremediation of contaminated sites. International e-Conference on Modern Trends in Foundation Engineering: Geotechnical Challenges and solution, 22 pp.

Rohde, W. A., Asmussen, L. E., Hauser, E. W., Wauchope, R. D., Allison, H. D. 1980. Trifluralin movement in runoff from a small agricultural watershed. *J. Environ. Qual.* 9(10):37–42.

Rudd, J. W. M., Taylor, C. D. 1980. Methane cycling in aquatic environments. *Adv. Aquat. Microbiol.* 2:77–150.

Saleh, N., Sirk, K., Liu, Y., Phenrat, T., Dufour, B., Matyjaszewski, K., Tilton, R. T., Lowry, G. V. 2007. Surface modifications enhance nanoiron transport and NAPL targeting in saturated porous media. *Environ. Eng. Sci.* 24:45–57.

Scelza, R., Rao, M. A., Gianfreda, L. 2008. Response of an agricultural soil to pentachlorophenol (PCP) contamination and the addition of compost or dissolved organic matter. *Soil. Biol. Biochem.* 40:2162–2169.

Schnabel, W. E., White, D. M. 2001. The effect of mycorrhizal fungi on the fate of Aldrin: phytoremediation potential. *Int. J. Phytoremediat.* 3(2):221–241.

Scholz, J. M., Schwedes, J., Deckwer, W. D. 1998. Use of a rotary drum reactor with grinding beads for microbial soil remediation. *Chem. Eng. Technol.* 479–483.

Schrick, B., Hydutsky, B. W., Blough, J. L., Mallouk, T. E. 2004. Delivery vehicles for zerovalent metal nanoparticles in soil and groundwater. *Chem. Mater.* 13:479–486.

Schwarenbach, R. P., Egli, T., Hofstetter, T. B., von Gunten, U., Wehrli, B. 2010. Global water pollution and human health. *Annu. Rev. Environ. Resour.* 35:109–136.

Sessitsch, A., Reiter, B., Pfeifer, U., Wihelm, E. 2002. Cultivation-independent population analysis of bacterial endophytes in three potato varieties based on eubacterial and *Actinomycetes*-specific PCR of 16S rRNA genes. *FEMS Microbiol. Ecol.* 39:3–32.

Sharef, I., Abdelbagi, A., Elsheikh, E., Ahmed, A., Osama, G. 2013. Biodegradation of pendimethalin by three strains of bacteria isolated from pesticides polluted soils. *U. of K. J. Agris. Sci.* 21(2):233–252.

Shakir, H., Safwat, H., Weaver, R. W. 2002. Earthworm survival in oil contaminated soil. *Plant and Soil.* 240:127–132.

Shim, S. S., Kawamoto, K. 2002. Enzyme production activity of *Phanerochaete chrysosporium* and pentachlorophenol in a bioreactor. *Water Res.* 36:4445–4454.

Simonich, S. L., Hites, R. A. 1997. Relationships between socioeconomic indicators and concentration of organochlorine pesticides in tree bark. *Environ. Sci. Technol.* 31(4):999–1003.

Simpson, S. L., Pryor, I. D., Mewburn, B. R., Batley, G. E., Jolley, D. 2002. Considerations for capping metal-contaminated sediments in dynamic estuarine environments. *Environ. Sci. Technol.* 36:3772–3778.

Slack, R. J., Gronow, J. R., Vouvoulis, N. 2005. Household hazardous waste in municipal landfills contaminant in leachate. *Sci. Total Enviorn.* 337:119.

Solomon, K., Thompson, D. 2011. Ecological risk assessment for aquatic organisms from over-water uses of glyphosate. *J. Tox. Environ. Health Part B.* 6(3):289–324.

Sparks, D. 1995. *Environmental Soil Chemistry*. Academic Press, San Diego.

Srivastava, P. K., Vaish, A., Dwivedi, S., Chakrabarty, D., Singh, N., Tripathi, R. D. 2011. Biological removal of arsenic pollution by soil fungi. *Sci. Tot. Environ.* 409:2430–2442.

Strong, J., Burgess, E. 2008. Treatment methods for wine-related and distillery wastewaters: A review. *Biorem. J.* 12(2):70–87.

Sundaram, A. 1965. A preliminary investigation of the penetration and translocation of 2,4,5-T in some tropicaltrees. *Weed Res.* 5:213.

Sylvestre, M., Macek, T., Mackova, M. 2009. Transgenic plants to improve rhizoremediation of polychlorinated biphenyls (PCBs). *Curr. Opin. Biotechnol.* 20:242–247.

Tang, X., Shen, C., Shi, D., Cheema, S. A., Khan, M. I., Zhang, C., Chen, Y. 2010. Heavy metal and persistent organic compound contamination in soil from Wenling: An emerging e-waste recycling city in Taizhou area. *China. J. Haz. Mater.* 173:653–660.

Thoma, G. J., Reible, D. D., Valsaraj, K. T., Thibodeaux, L. J. 1993. Efficiency of capping contaminated sediments in situ. 2. Mathematics of diffusion adsorption in the capping layer. *Environ. Sci. Technol.* 27(12):2412–2419.

Ting, A. S. Y., Choong, C. C. 2009. Bioaccumulation and biosorption efficacy of *Trichoderma* isolates SP2F1 in removing copper (Cu II) from aqueous solutions. *World J. Microbiol. Biotechnol.* 25:1431–1437.

Tomassoni, G. A. 2000. A federal statutory/regulatory/policy perspective on remedial decision-making with respect to ground-water/surface-water interaction. In *Ground-Water/Surface-Water Interactions Workshop*. EPA, Washington, DC.

Trapp, S., Matthies, M., Scheunert, I., Topp, E. M. 1990. Modeling the bioconcentration of organic chemicals in plants. *Enviorn. Sci. Technol.* 24:1246–1252.

Trapp, S., Miglioranza, K. S. B., Mosbek, H. 2001. Sprtion of lipophlic organic compounds to wood and implications for their environmental fate. *Environ. Sci. Technol.* 35(8):1561–1566.

Tripathi, P., Singh, P. C., Mishra, A., Chauhan, P., S., Dwivedi, S., Bais, R. T., Tripathi, R. D. 2013. *Trichoderma:* A potential biomediator for environmental clean up. *Clean Techn. Environ. Policy.* 15(4):541–550.

Tripathi, R. D., Srivastava, S., Mishra, S., Singh, N., Tuli, R., Gupta, D. K., Maathuis, F. J. M. 2007. Arsenic hazards: Strategies for tolerance and remediation by plants. *Trends Biotechnol.* 25:158–165.

Urik, M., Cernansky, S., Sevc, J., Simonovicova, A., Littera, P. 2007. Biovolatilization of arsenic by different fungal strains. *Water Air Soil Poll.* 186:337–342.

USEPA (United States Environmental Protection Agency). http://www.epa.gov/opp00001/factsheets/atrazine_background.htm.

USEPA and WDNR (U.S. Environmental Protection Agency and Wisconsin Department of Natural Resources). 2002. *Record of Decision Operable Units 3, 4, and 5 Lower Fox River and Green Bay*. Wisconsin. CERCLIS ID: WID000195481. http://www.dnr.state.wi.us/org/water/wm/lowerfox/rifs/whitepapers/ROD%20for%20OU%203-5%20final.pdf.

Van Aken, B. 2009. Transgenic plants for enhanced phytoremediation of toxic explosives. *Curr. Opin. Biotech.* 20:231–236.

Volesky, B., May, H., Holan, Z. R. 1993. Cadmium biosorption by *Saccharomyces cerevisiae*. *Biotechnol. Bioeng.* 41:826–829.

Wang, B., Liu, L., Gao, Y., Chen, J. 2009. Improved phytoremediation of oilseed rape (*Brassica napus*) by *Trichoderma* mutant constructed by restriction enzyme-mediated integration (REMI) in cadmium polluted soil. *Chemosphere.* 74(10):1400–1403. https://doi.org/10.1016/j.chemosphere.2008.11.07

Wang, Q., Kassem, I. I., Sigler, V., Gruden, C. 2009. Short-term effect of capping on microbial communities in freshwater sediments. *Water Env. Res.* 81(4):441–449. https:// doi.org/10.2175/106143008X370458

Wang, X., Wu, N., Guo, J., Chu, X., Tian, J., Yao, B., Fan, Y. 2008. Pytodegradation of organophosphorus compounds by transgenic plants expressing a bacterial organophosphorus hydrolase. *Biochem. Biophysical Res. Comm.* 365(3):453–458.

Walter, M., Boyd-Wilson, K., Boul, L., Ford, C., McFadden, D., Chong, B., et al. 2005. Field-scale bioremediation of pentachlorophenol by *Trametes versicolor* Int. *Biodeterior Biodegrad.* 56:51–57.

Weber, W. J., Tang, J. X., Huang, Q. G. 2006. Development of engineered natural organic sorbents for environmental applications. 1. Materials, approaches, and characterizations. *Environ. Sci. Technol.* 40(5):1650–1656.

Weed Science Society of America (WSSA). www.wssa.net.

Wen, L., Yun. D., Beibei, X., Yingying, L., Xiang, P., Zhangg, J., Yanchun, Y. 2009. Bioremediation and detoxification of endosulfan in aqueous medium and soil by *Achromobacter xylosoxidans* strain CS5. *J. Hazard. Mater.* 167:209–216.

Wichman, J., Byrnes, W. 1975. Uptake, distribution, and degradation of simazine by black walnut and yellow poplar seedlings. *Weed Sci.* 23:448–453.

Yu, Z., Zeng, G. M., Chen, Y. N., Zhang, J. C., Yu, Y., Li, et al. 2011. Effects of inoculation with *Phanerochaete chrysosporium* on remedial of pentachlorophenol-contaminated soil waste by composting. *Process Biochem.* 46:1285–1291.

Zeng, G. M., Yu, Z., Chen, Y. N., Zhang, J. C., Li, H., Yu, M., et al. 2011. Response of compost maturity and microbial community composition to pentachlorophenol (PCP)-contaminated soil during composting. *Bioresour. Technol.* 102:5905–5911.

Zeng, X., Su, S., Jiang, X., Li, Bai, L., Zhang, Y. 2010. Capability of pentavalent arsenic bioaccumulation and biovolatilization of three fungal strains under laboratory conditions. *Clean: Soil, Air, Water.* 38:238–241.

Zhang, Q., Davis, L. C., Erickson, L. E. 2001. Transport of methyl tert-butyl ether (MTBE) through alfafa plants. *Environ. Sci. Technol.* 35:725–731.

Zhao, S., Arthur, E. L., Coats, J. R. 2003. The use of native prairie grasses to degrade atrazine and metolachlor in soil. In Coats, J. R., Yamamoto, H. (Eds.), *Environmental Fate and Effects of Pesticides*. ACS Symposium Series 853. ACS, Washinton, DC, pp. 157–167.

Zhongli, C., Shunpeng, L., Gouping, F. 2001. Isolation of methyl parathion-degrading strain M6 and cloning of the methyl parathion hydrolase gene. *Appl. Environ. Microbiol.* 67(20):4922–4925.

Zhou, X., Xu, S., Liu, L., Chen, J. 2007. Degradation of cyanide by *Trichoderma* mutants constructed by restriction enzyme mediated integration (REMI). *Biores. Technol.* 98:2958–2962.

Zimmerman, J. R., Ghosh, U., Millward, R. N., Bridges, T. S., Luthy, R. G. 2004. Addition of carbon sorbents to reduce PCB and PAH bioavailability in marine sediments: physicochemical tests. *Environ. Sci. Technol.* 38:5458–5464.

Zuckerman, B. M., Deubert, K., Mackiewieg, M., Gunner, H. B. 1970. Studies on the biodegradation of parathion. *Plant Soil.* 33:273–281.

Index

E

F

G